ENCYCLOPEDIA OF WARSHIPS

ENCYCLOPEDIA OF WARSHIPS

CHRIS MCNAB

amber
BOOKS

First published in 2026

Published by
Amber Books Ltd
United House
North Road
London N7 9DP
United Kingdom

www.amberbooks.co.uk
Facebook: amberbooks
YouTube: amberbooksltd
Instagram: amberbooksltd
X(Twitter): @amberbooks

ISBN: 978-1-83886-489-7

Project Editor: Michael Spilling
Designers: Lewis Hughes-Batley and Fredy Alexandrakis
Picture Research: Terry Forshaw

Printed in Malaysia

Contents

Introduction

The history of warships is one characterized by millennia of marginal gains in performance and capabilities, followed by two extraordinary centuries of technological progress that utterly transformed naval warfare on, above and below the waves.

ABOVE:
The Battle of Trafalgar, fought on 21 October 1805, ranks among the most renowned naval clashes in history. In this scene, the Spanish warship *Santísima Trinidad* rolls heavily in the water, dismasted by British firepower.

In the 2nd millennium BCE, warring empires, or city-states, began converting small trade and fishing vessels into fighting platforms. At first, this process amounted to little more than packing civilian ships with infantry and missile troops. Over the following 1000 years, however, purpose-built warships emerged – primarily oared galleys, sometimes sail-assisted – designed for ramming and boarding. These vessels grew in size and strength; a Greek trireme around 480 BCE, for example, carried roughly 170 oarsmen. By the 5th century BCE, the ancient world had developed its first navies, and shipbuilding had become an established part of military spending.

The next major advance in naval technology unfolded in the Middle Ages (c. 500–c. 1500) as reliance on oars gave way to sail power. This granted warships oceanic range and the size and stability needed to carry the first gunpowder artillery. From the 16th to the 19th centuries – the so-called 'Age of Sail' – these developments reached their peak in mighty, multi-decked, ships-of-the-line, plus a host of smaller specialist combat vessels. Naval battles in this era were slow-moving and thunderous affairs, epics of endurance, seamanship and gunnery.

AGE OF TRANSFORMATION

The 19th century brought the technologies of the Industrial Revolution into ship design and naval warfare. Wooden hulls were replaced with steel hulls; sail power with steam engines; muzzle-loading guns with breech-loading guns mounted on rotating turrets; and forecastles with armoured bridges. These innovations laid the foundations for the age of the great battleships, beginning in 1906 with the launch of HMS *Dreadnought* and lasting roughly until the end of World War II.

Fearsome World War I naval dramas, such as the Battle of Jutland in 1916, seemed to confirm the battleship as the pinnacle of naval technology. But World War II (1939–45) shattered that conclusion. Battleships drifted into creaking obsolescence, while the decisive instruments of naval power became aircraft carriers and submarines, supported by new generations of amphibious landing craft and transport ships.

ABOVE:
A side view of the Russian pre-dreadnought battleship *Tsesarevich*, displaying two of its 152mm (6in) gun turrets.

From 1945 onwards, computerization, nuclear propulsion and guided missile systems reshaped the design and capabilities of modern navies. A single US *Nimitz*-class aircraft carrier could alter the strategic balance of an entire region. Today, unmanned surface vessels (USVs) signal the approach of the next revolution in naval warfare.

The vessels featured in this book stand as the wood and steel benchmarks of naval evolution. We should, however, remember that they were also the homes, and sadly at times the resting places, of hundreds of thousands of sailors, aviators and other personnel. For all its sails, armour and guns, naval warfare remains at its core a human story.

ABOVE:
Navies are adopting missile-armed fast attack craft, more heavily armed than earlier patrol boats. The Chinese Houbei class, also known as Type 022, is catamaran-hulled, with more than 60 examples currently in service.

BELOW:
A 2004 aerial view of the USS *Ronald Reagan* (CVN-76), the ninth ship of the post-World War II *Nimitz*-class carriers.

Age of Sail

The Age of Sail ran roughly from the mid-15th to the mid-19th century. As its name suggests, it was a time in which the mastery of sail and rigging defined the performance of both warships and commercial vessels. Although the fleets of the day depended entirely on wind and waves to carry them across the oceans, the refinement of sailing warship design ushered in the first true blue-water navies. In particular, the Portuguese, Spanish, French and British navies traversed the world's oceans, forging empires even as they fought one another in the process. Multi-decked 'ships-of-the-line', rated by the number of muzzle-loading guns they carried, were formidable instruments of firepower and diplomacy. Yet their success depended heavily on the skill, discipline and professionalism of officers and crew who lived and fought within the confined world of the ship.

NAVAL ACTION
The diminutive 38-gun British frigate HMS *Terpsichore* is pictured here attacking the mighty Spanish first-rate, the 136-gun *Santísima Trinidad* on 1 March 1797 in the Mediterranean Sea.

Ancient war galley (1st century BCE)

TYPE • *Galley* **COUNTRY** • *Greece*

SPECIFICATIONS

Dimensions:	Length: 35m (115ft); Beam: 4.6m (15ft); Draught: 1m (3ft)
Displacement:	36 tonnes (40 tons); 54 tonnes (60 tons) full load
Rig:	Single mast, square sail
Propulsion:	Sails
Speed:	6 knots (11km/h; 6.9mph)
Armament:	Bow ram; ballista; personal crew weapons
Crew:	190 oarsmen plus officers and fighting men

Up to the Middle Ages (and even later in the Mediterranean and Baltic Seas), the oared warship was the primary fighting vessel. Hull design went ahead faster than that of sails and steering gear, and the heavy galley was employed for close-quarters encounters.

Right: Greek penteconter, 750 BCE. This light and manoeuvrable galley type had 25 oars per side, arranged in a single bank.

Rowers
The individual rowers sat on benches atop sliding leather cushions that were strapped to their legs and buttocks, to enable them to use thigh muscles to drive each oar stroke.

Roman Trireme
The single sail enabled the trireme to utilize wind power to take the strain off the rowers, although these craft were mostly short-distance operators.

Prow ram
The ram at the bow of the boat was levelled to strike and hole the enemy ship just below the waterline. Ramming speed would be kept below maximum speed, to prevent the ship getting embedded in its opponent.

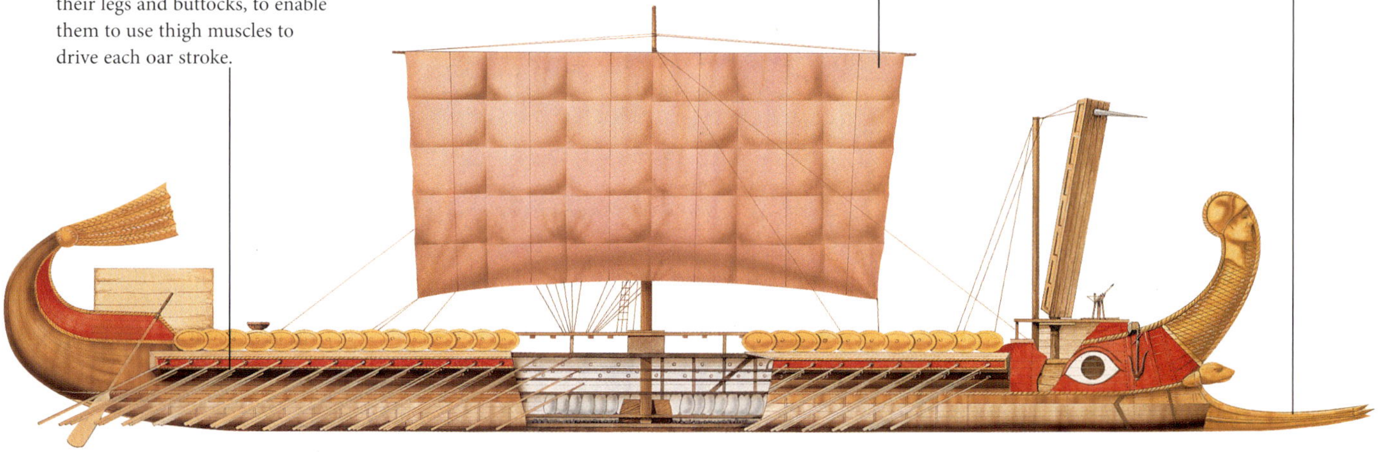

ROMAN WARSHIP
In their long war with Carthage, the Romans copied Carthaginian designs in the form of the heavy trireme (three banks of oars), which could have a crew of about 200 men. In addition to the pro ram and the onboard ballista's stone missiles, they could project fire-pots into enemy ships and were also equipped with a heavy boarding platform, whose spike could be lodged in an enemy hull.

TRIREME ON THE NILE
The famous Roman Nile Mosaic of Palestrina is a magisterial floor mosaic from the 1st century BCE. This image of a trireme features in the bottom right corner. The lookout stationed at the bow was called the *proreta*.

Viking longship (9th–10th century CE)

TYPE • *Galley* COUNTRY • *Scandinavia*

SPECIFICATIONS

Dimensions:	Length: 23.8m (78ft 1in); Beam: 5.2m (17ft 1in); Draught: 1.2m (3ft 11in)
Displacement:	25 tonnes (27.6 tons); 30 tonnes (33.1 tons) full load
Rig:	Single mast, square sail
Propulsion:	Sails and oars
Speed:	10 knots (19km/h; 12mph)
Armament:	Individual crew weapons
Crew:	44–50 oarsmen

The longship was the defining instrument of Viking power from the 8th to 11th centuries CE. It was a perfect raiding vessel, capable of long-distance journeys that could end in littoral landings. Its shallow draught allowed it to navigate rivers and penetrate deep inland.

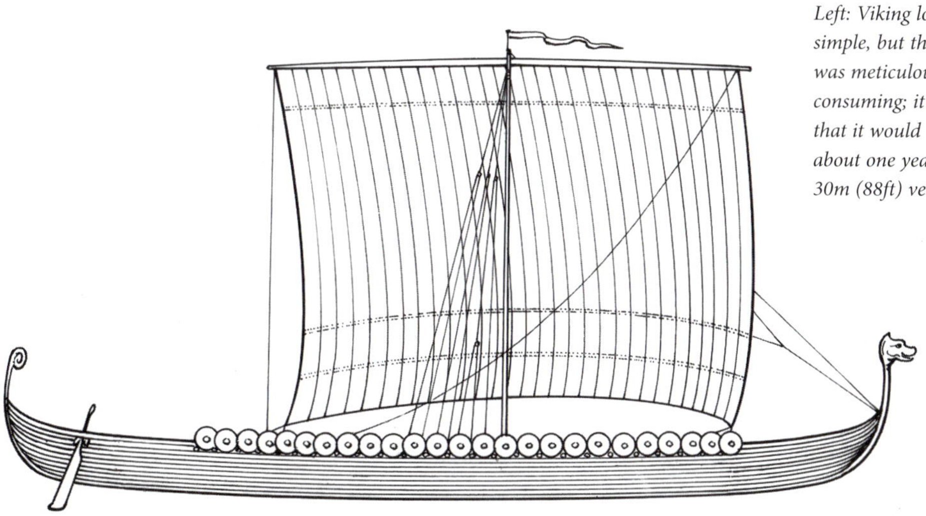

Left: Viking longships appear simple, but their construction was meticulous and time-consuming; it is estimated that it would take 100 men about one year to build a 30m (88ft) vessel.

Steering oar
Up until the 1100s, most Viking longships were steered via a side-mounted steering oar attached to the hull and controlled by a tiller. From the 12th century, hinged stern rudders gradually became more common.

Sail
The sail of the longship was made from rough wool cloth. It was commonly formed in a striped pattern, as seen here, or checks.

9th-century longship
The shallow-draught hull of the Viking longship meant that it could operate in waters that were only 1m (3ft) deep.

LONGSHIP DESIGN
The Vikings were raiders or traders, and the longship was a fine vessel for either purpose. From its early development, it showed the northern practice of clinker-building (overlapping planks), with the mast set in a strong oak keel. It could carry men or animals, but the open hull was not well-suited for dry cargoes; storage was the main limitation on distance of operations.

VIKING SHIP MUSEUM, OSLO
The Oseberg ship is a complete Viking longship dating back to the 9th century CE. It was excavated from a Viking burial mound in Norway. Built of oak, under full sail the ship could have reached speeds of about 10 knots (19km/h; 12 mph).

Medieval cog (13th century)

TYPE • *Cog* **COUNTRY** • *North Sea*

SPECIFICATIONS

DIMENSIONS:	Length: 24m (78ft 9in); Beam: 7.6m (24ft 11in); Draught: 2m (6ft 7in)
DISPLACEMENT:	110 tonnes (121 tons) normal load; 200 tonnes (220 tons) full load
RIG:	One mast, single large square sail
PROPULSION:	Sails
SPEED:	7 knots (13km/h; 8mph)
ARMAMENT:	Personal crew weapons
CREW:	6–8

Growing populations, technological innovations and expanding international trade brought new developments in ship design in the medieval age. Sails took over from oars. At this time, a merchant vessel could be quickly converted for military purposes.

Right: The cog had a large stern-mounted central rudder, hanging from a long stern post. The hull was clinker-built.

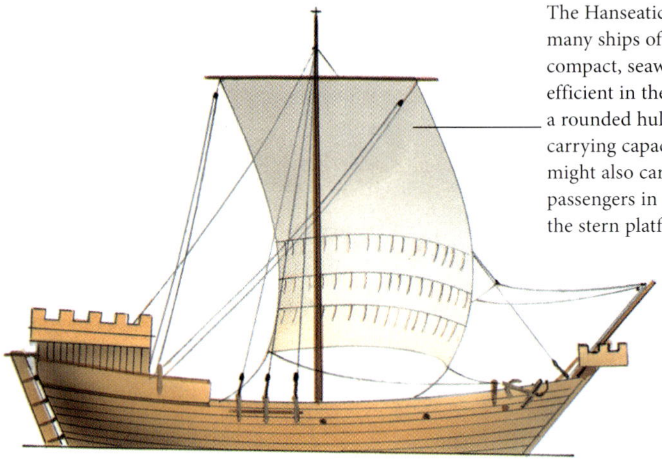

HANSEATIC COG

The Hanseatic League had many ships of this type: compact, seaworthy and efficient in their time, with a rounded hull to maximize carrying capacity. They might also carry a few passengers in a cabin below the stern platform.

SWEDISH CARGO SHIP

Found in mud at Kalmar Bay in the 1930s, this is one of the oldest rudder-fitted ships known. The former steering oar is now a solid fixture. The ship is clinker-built and part-decked. The short bowsprit suggests a staysail was carried. The ship also has a windlass fitted, to pull in a net as well as to raise the anchor.

ITALIAN CARGO SHIP

By 1200, this was becoming an old-world design, although some ships now had two masts (as illustrated) and so harnessed greater wind power. Mediterranean ships were carvel-built (planks meeting edge to edge without overlap).

DEVELOPMENT

The cog emerged in northern Europe in the 12th century. It became the standard cargo ship of the Hanseatic League, a powerful commercial and political network of merchant cities. The main priority was a spacious cargo hold, but the vessel also had a shallow draught to enable it to anchor in all depths of coastal waters. It could be converted to a warship by carrying troops and mounting weapons on the deck and the fort at the stern.

WARSHIPS UNDER SAIL

This 14th-century artwork depicts French and Genoese soldiers crowded about the decks of medieval cogs and other vessels, embarked on an expedition to Barbary (North Africa). Note the flat-bottomed troop transport shown at the bottom of the illustration.

Henry Grâce à Dieu (1514)

TYPE • *Carrack* **COUNTRY** • *England*

SPECIFICATIONS

DIMENSIONS:	Length: 57.9m (190ft); Beam: 15.2m (50ft); Draught: 6.1m (20ft)
DISPLACEMENT:	1000 tonnes (1102 tons); 1500 tonnes (1524 tons) full load
RIG:	Four masts, square rig fore and main, lateen mizzen
PROPULSION:	Sails
SPEED:	8 knots (15km/h; 9mph)
ARMAMENT:	21 bronze cannons; 130 iron guns; 100 hand guns (1545)
CREW:	c. 700

Named for its royal owner, King Henry VIII of England, this was a large ship for its time, intended both as a man of war and as a transport vessel for the king and his entourage. As such, it had to convey the status and splendour of monarchy.

Below: Topgallant sails, providing a third tier of wind drive above the main yards, were still a recent innovation, dating from around 1500.

SAILS
The yellow sails (for a gold effect) were probably a special suit made for Henry's visit to King Francis I of France in 1520.

GRAPNEL
A grapnel on the bowsprit could be hooked on to an enemy ship, to draw it close for boarding actions by sailors.

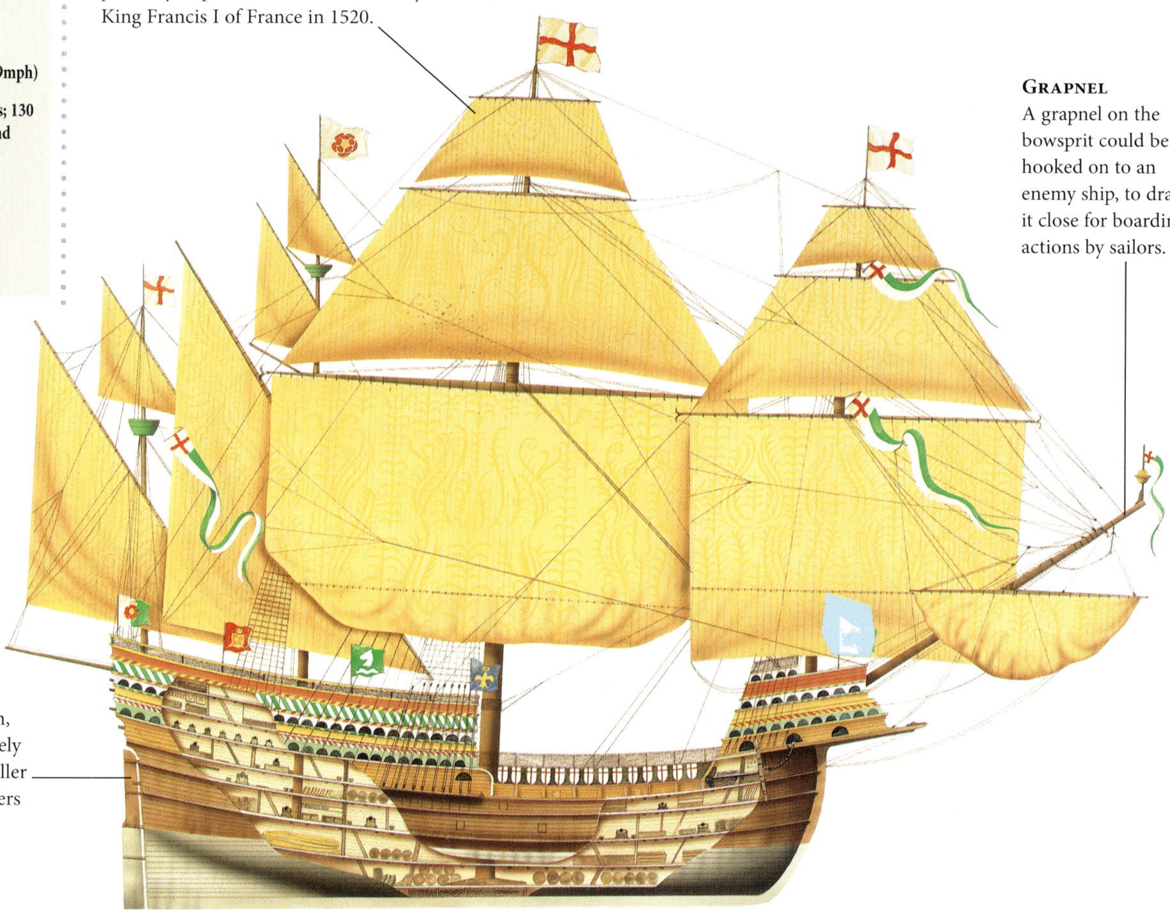

DECKS
The forecastle, four decks high, would have appeared immensely imposing to observers on smaller ships. But it also enabled archers to fire downwards.

CONSTRUCTION
Constructed at the Woolwich Dockyard on the River Thames, where King Henry could watch its progress, this ship was of the carrack type, with the typical rounded hull shape and with a carvel-built hull, the planks set edge to edge rather than overlapping. The massive forecastle was also a typical feature, as was the squared-off stern. Soldiers and sailors lived at stern end.

IN BATTLE
The ship was not engaged in military action until the Battle of the Solent in 1545, when the French attacked Portsmouth with a fleet consisting largely of galleys. The unwieldy 'Great Harry' found it difficult to counter their ability to execute swift turns and sudden stops.

San Martín (1580)

TYPE • *Galleon* **COUNTRY** • *Portugal*

SPECIFICATIONS

Dimensions:	Length: 55m (180ft); Beam: 12m (40ft); Draught: est. 4m (13ft)
Displacement:	c. 1000 tonnes (1102 tons) normal load
Rig:	Four masts, square rig fore and main, lateen mizzens
Propulsion:	Sails
Speed:	c. 8 knots (15km/h; 9mph)
Armament:	48 × heavy guns, including 17 × cannons and culverins on lower deck and 17 × lighter guns on upper deck
Crew:	350 seamen and gunners; 302 soldiers

San Martín's career as a flagship began successfully, when it led the Spanish fleet at the Battle of Terceira, off the Azores, on 25 July 1582. In this engagement, the Spanish and Portuguese defeated a French fleet of 75 ships, sinking 10 and losing none of their own ships.

Below: Two gun decks were incorporated into the design from the beginning. This was a ship purposefully built for fighting.

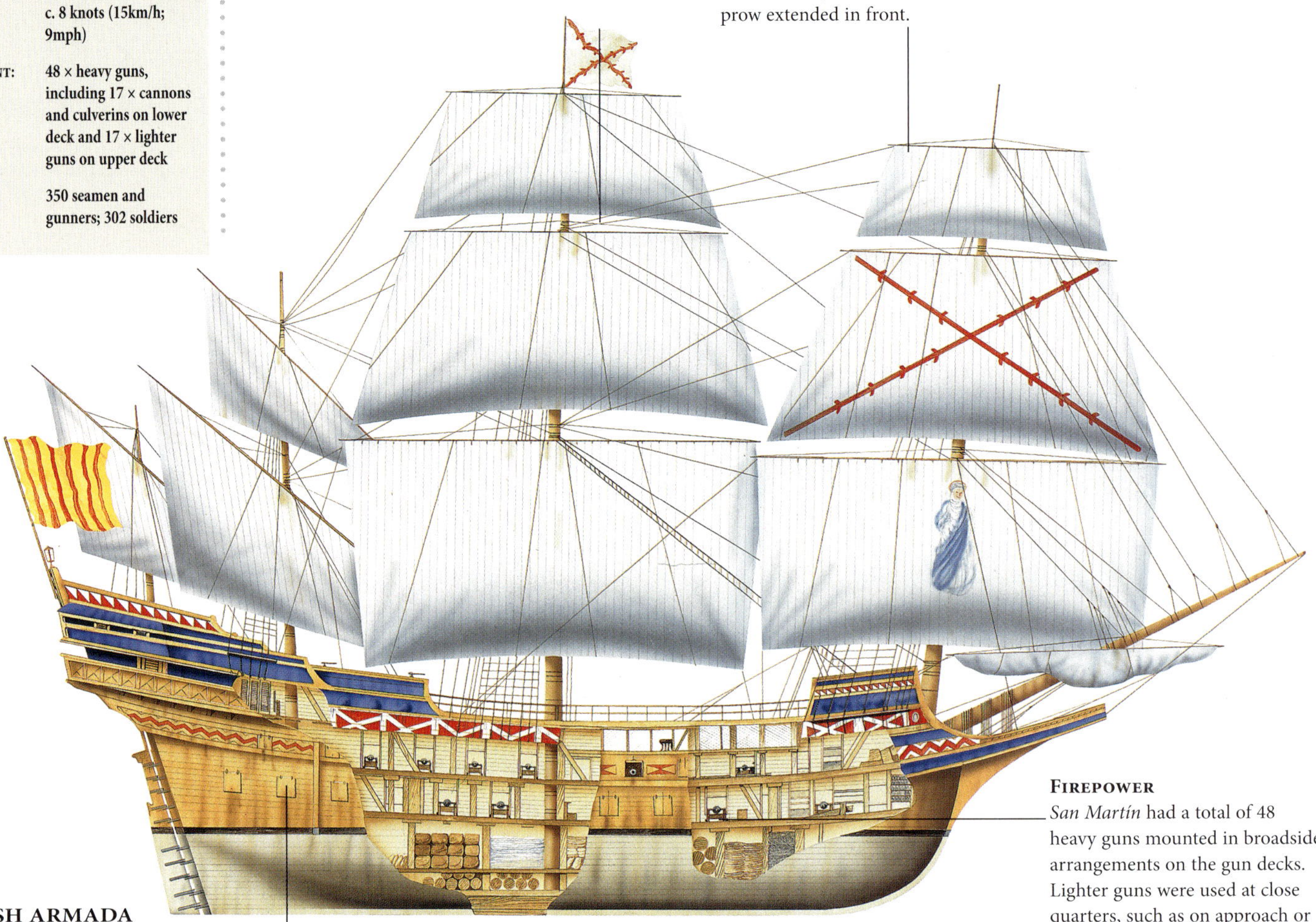

Galleon
San Martín was a galleon, a term that covers a range of ship sizes. Typical features included a narrow ratio of length to beam and a low forecastle compared to earlier ships, with a beaklike prow extended in front.

Firepower
San Martín had a total of 48 heavy guns mounted in broadside arrangements on the gun decks. Lighter guns were used at close quarters, such as on approach or during boarding actions.

Gunports
Gunports (here shown in the closed position, protecting the guns inside), were first used in France, around 1501, but were quickly copied by other nations.

SPANISH ARMADA
This celebrated ship was actually built in Portugal and brought into the Spanish Navy in 1580 when Portugal was annexed by Spain. Appreciating its size and qualities, the Spanish admiralty made it a *capitana*, or flagship, and it led the vast fleet known as the Spanish Armada against England in May 1588. This ill-fated expedition ended in disaster, with only half the ships returning. *San Martín* was one of them, reaching Santander on 23 September 1588.

'THE INVINCIBLE ARMADA'
This 1601 painting by Hendrick Cornelisz Vroom offers a panoramic view of the battle of the Spanish Armada. *San Martín* was in the thick of the fighting, notably in an hour-long duel with the English flagship *Ark Royal* on 3 August 1588, and in a rearguard action off Gravelines on the French side of the English Channel on 8 August.

Vasa (1627)

TYPE • *Galleon* **COUNTRY** • *Sweden*

SPECIFICATIONS	
DIMENSIONS:	Length: 69m (226ft); Beam: 11.7m (38ft); Draught: 4.8m (16ft)
DISPLACEMENT:	1210 tonnes (1334 tons) normal load
RIG:	Three masts, square rig
PROPULSION:	Sails, 1,275 sq m (13,720 sq ft)
SPEED:	Est. 8 knots (15km/h; 9mph)
ARMAMENT:	48 × 24-pounders; 8 × 3-pounders; 2 × 1-pounders; 1 × 16-pounder; 2 62-pounders; 3 × 35-pounder guns
CREW:	145 sailors; 300 soldiers

The pride of the Swedish royal fleet, *Vasa* was planned as part of a large-scale strategy to ensure Swedish control of the Baltic Sea at a time when the country, under King Gustavus II Adolphus, was heavily engaged in warfare in Germany and Poland.

Left: On its maiden voyage on 10 April 1628, Vasa *heeled over and sank only a few minutes after setting sail, and less than a mile from the shore.*

WEIGHTING
The weight distribution of the *Vasa* was top-heavy, contributing to its sinking on its maiden voyage. It took some decades before world navies understood fully how to spread weight across gun decks.

ARMAMENT
Vasa could fire 267kg (588lb) of shot with every broadside. The 24-pounder guns were the principal armament along the two gun decks.

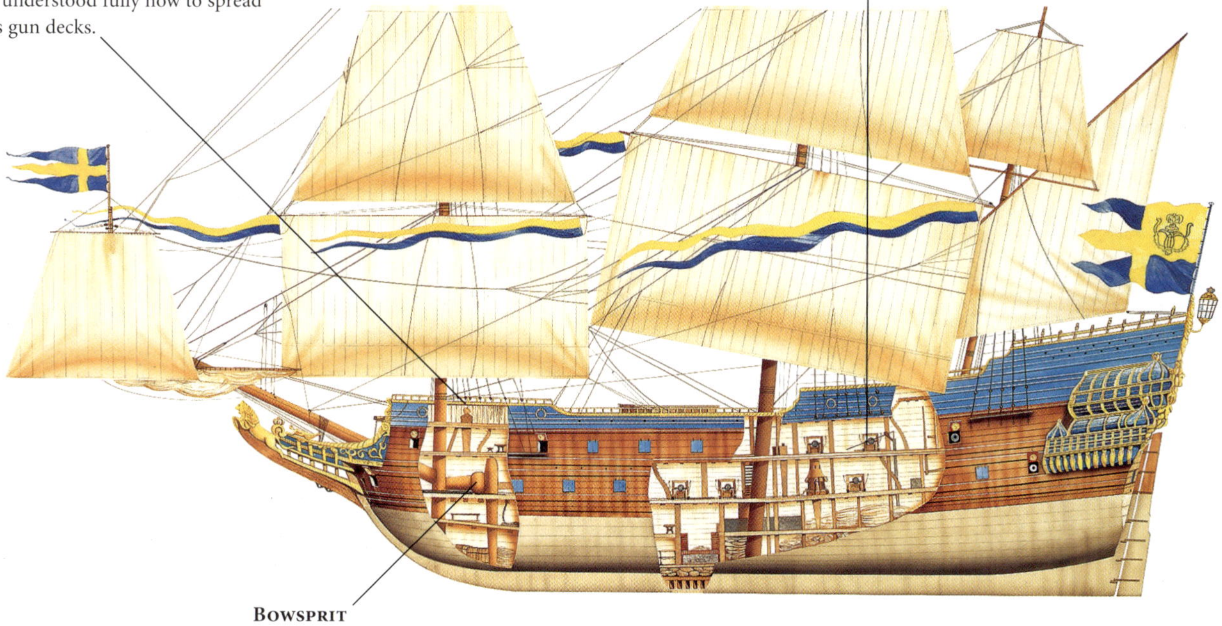

BOWSPRIT
The solid seating needed for the massive bowsprit, with its spritsail and topsail, is clearly shown.

ORIGINS

In the early 17th century, the Dutch were considered the best shipbuilders in Europe, and a Dutch architect, Henrik Hybertsson de Groot, was employed to design and supervise construction of *Vasa*. Typically for the time, the ship was lavishly ornamented to emphasize the wealth and power of the Vasa, Sweden's ruling family, but it was intended for hard tasks in an era of international warfare.

MUSEUM SHIP

The wreck of the *Vasa* lay forgotten until it was located in 1956, when it was found to be largely intact and a recovery operation was launched. On 24 April 1960, *Vasa* appeared again above the surface and since then has been preserved in a specially constructed museum building.

Sovereign of the Seas (1637)

TYPE • *Ship of the line* **COUNTRY** • *England*

SPECIFICATIONS

DIMENSIONS:	Length: 38.7m (127ft); Beam: 14.2m (46ft 6in); Draught: c. 6.5m (21ft 4in)
DISPLACEMENT:	c. 1520 tonnes (1676 tons
RIG:	Three masts, square rig
PROPULSION:	Sails
SPEED:	c. 6 knots (11km/h; 7mph)
ARMAMENT:	102 guns, including 60-pounders, 30-pounders (lower deck), 20-pounders, and 10-pounders (middle deck)
CREW:	Not recorded

Masterwork of the famous English shipwright Phineas Pett, this 'great new ship' of King Charles I was launched from Woolwich Dockyard, on the River Thames, in 1637. In many ways, the design broke with convention: it had three decks of guns and new-type sails, and was exceptional in sheer size.

Below: In a long active career, Sovereign of the Seas *saw considerable action in the Anglo-Dutch wars of the 17th century before being accidentally destroyed by fire in 1703.*

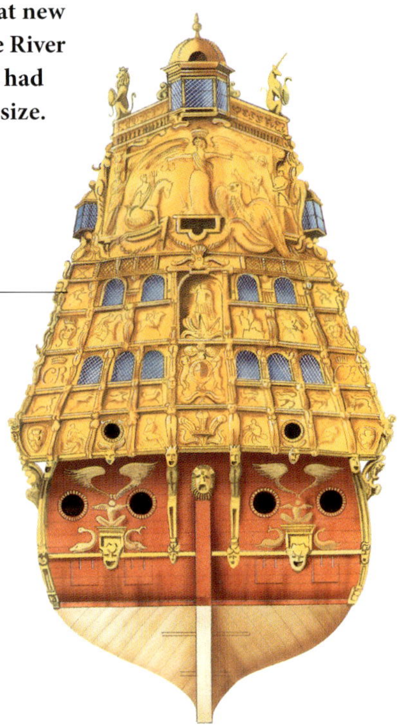

ORNAMENTATION
Like other capital ships of the time, the vessel was elaboratel decorated with carvings, paint in gold and other colours, nor of which had the slightest value in a fight but which looked magnificent.

SAILS
This was the first, or one of the first, vessels to carry royals above the topgallants. Heavier and higher masts were needed, with a corresponding need for more substantial staying and rigging.

ARMAMENT
In total 102 guns were carried, apart from 'murdering pieces' intended to scatter fire on an enemy deck, and there were many musketry positions behind loopholes.

NAMING
The design of *Sovereign of the Seas* had no shortage of critics when it was ordered in 1634. But Charles I was determined to have a ship that expressed England's oceanic claim to sea power. During the Commonwealth period, after Charles I's execution in 1649, it was renamed *Sovereign*, and following the restoration of the monarchy in 1660 became *Royal Sovereign*, following a rebuild.

ACTIVE DUTY
Sovereign of the Seas was comparatively slow and unwieldy in its performance on the waves, but its bulk and firepower made it effective in battle, and it featured in numerous conflicts in four decades of active duty between 1652 and 1692.

Le Soleil Royal (1669)

TYPE • *Ship of the line* **COUNTRY** • *France*

SPECIFICATIONS

Dimensions:	Length: 61m (200ft); Beam: 15.6m (15ft 2in); Draught: c. 7.6m (24ft 9in)
Displacement:	2500 tonnes (2755 tons); normal load
Rig:	Three masts, square rig
Propulsion:	Sails
Speed:	c. 8.5 knots (16km/h; 10mph)
Armament:	104 bronze cannons (from 1689)
Crew:	1200

Le Soleil Royal **was the product of a modernization and revitalization of France's navy by the King's minister Jean-Baptiste Colbert. Only a wealthy and ostentatious monarchy could have afforded to construct warships on such a sumptuous scale.**

Below: Built at Brest by the King's master builder Laurent Hubac, the ship was launched on 13 December 1669 and commissioned in August 1670.

Tops
With the advent of musketry, fighting tops became important locations for sharpshooters, who would engage enemy crew at close quarters.

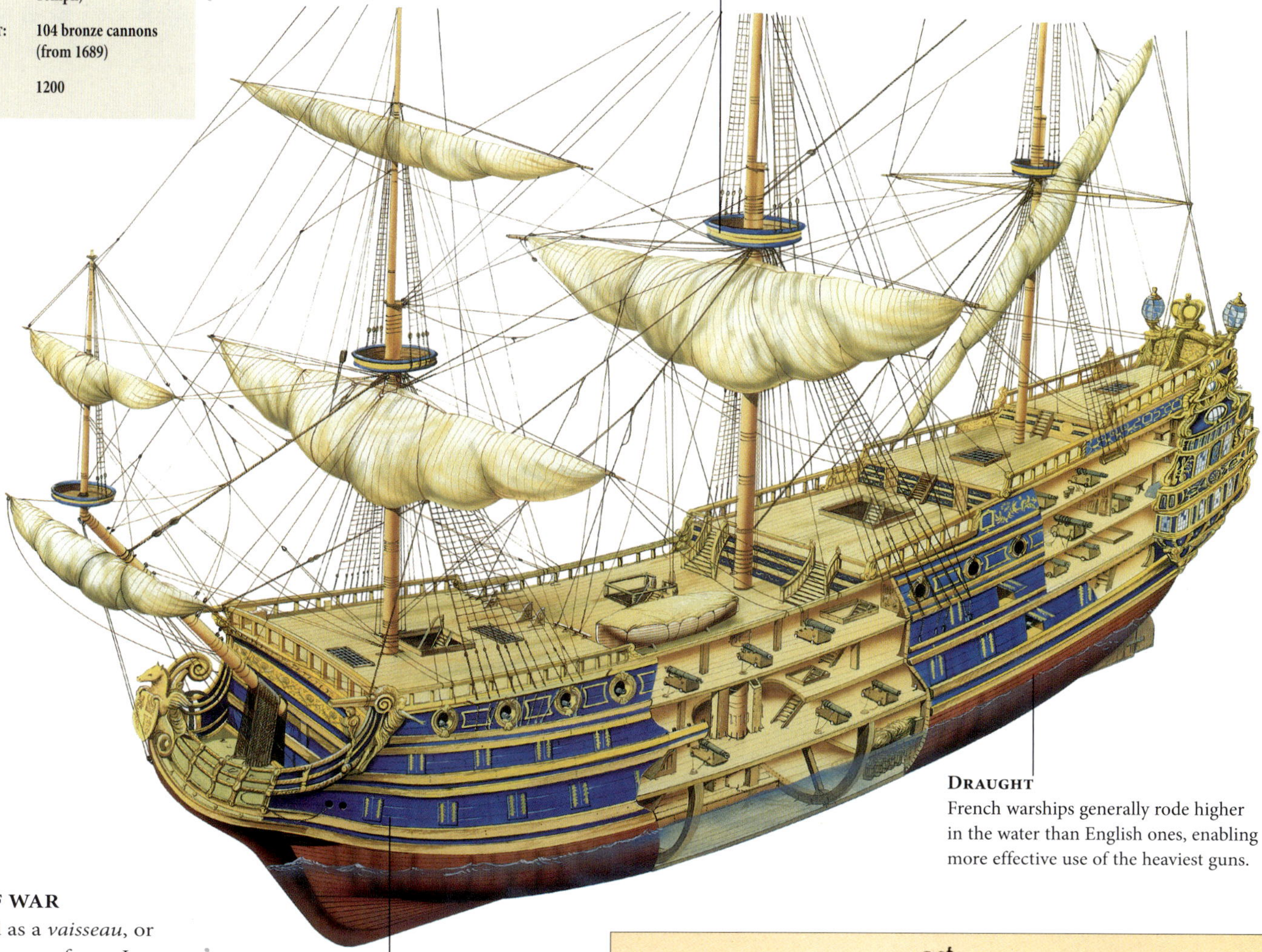

Draught
French warships generally rode higher in the water than English ones, enabling more effective use of the heaviest guns.

Hull colour
The hull was painted in its distinctive royal blue in honour of King Louis XIV.

MAN OF WAR
Classified as a *vaisseau*, or first-class man of war, *Le Soleil Royal* (the 'Royal Sun') was named for Louis XIV of France, the 'Sun King,' and like other iconic ships of the era was decorated as a floating palace, although crew conditions were far more spartan than those endured by the King's own servants. As flagship of France's navy, it participated in the victory at Bévéziers (1690), but two years later the Battle of Barfleur (1692) resulted in its grounding and total destruction.

FLAGSHIP
Le Soleil Royal underwent a refit in 1689, with the number of guns increased to 104, and on completion it was made flagship of France's northern fleet, based at Brest in Brittany under the Comte de Tourville as Admiral.

Venus (1755)

TYPE • *Ship of the line* COUNTRY • *Spain*

SPECIFICATIONS

DIMENSIONS:	Length: 40.9m (147ft 1in); Beam: 10.5m (48ft 9in); Draught: 4.5m (14ft 7in)
DISPLACEMENT:	Not available
RIG:	Three masts, square rig
PROPULSION:	Sails
SPEED:	11 knots (20km/h; 12.4mph)
ARMAMENT:	22 × 18-pounders; 8 × 6-pounder guns (final arrangement)
CREW:	c. 230

Like other Spanish warships not named after saints, *Venus* had a second 'protective' name, *Santa Brígida*. Built at La Carraca, Cádiz, its long career shows how active a medium-sized Spanish warship could be without ever taking part in a major battle.

Below: In 1805 Venus *made its last transatlantic voyage, back to Cádiz, and was held there in ordinary until it was finally broken up in 1809.*

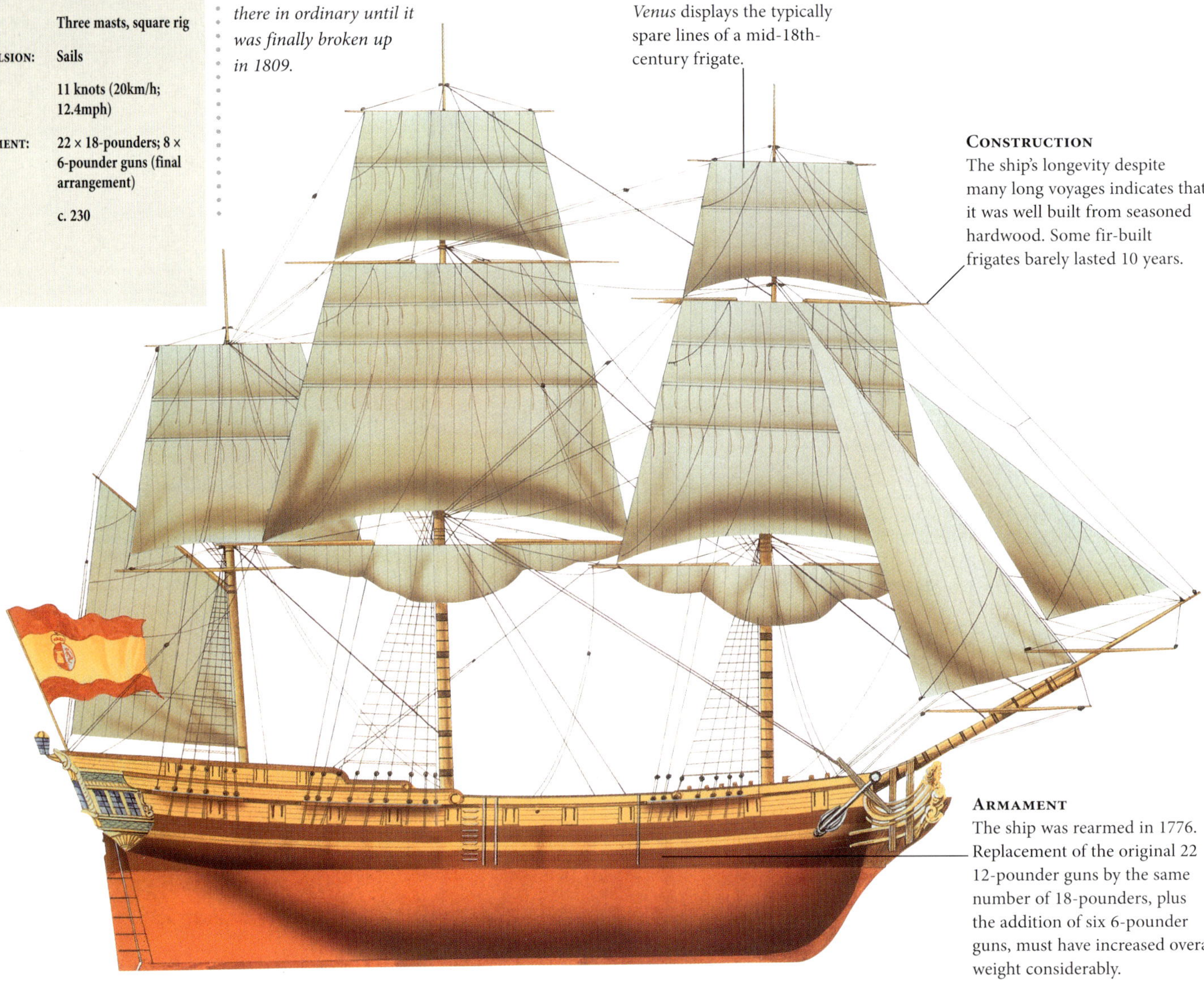

DESIGN
Venus displays the typically spare lines of a mid-18th-century frigate.

CONSTRUCTION
The ship's longevity despite many long voyages indicates that it was well built from seasoned hardwood. Some fir-built frigates barely lasted 10 years.

ARMAMENT
The ship was rearmed in 1776. Replacement of the original 22 12-pounder guns by the same number of 18-pounders, plus the addition of six 6-pounder guns, must have increased overall weight considerably.

COMMISSIONING

Launched on 9 September 1755, *Venus* was one of four almost identical frigates of 26 12-pounder guns, the class leader being *Industria*, and all built at La Carraca. Its early years are not recorded and it may have lain in ordinary for some time. In August 1759, with Don José de Somaglia as captain, it formed part of the fleet sent to Naples to bring the new king Carlos III to Barcelona.

ON DEPLOYMENT

Here, *Venus* is shown as part of the fleet sent to Naples to bring king Carlos III to Barcelona in 1759. From 1788, *Venus* was based at Havana and Veracruz, and sailed on 26 August 1796 from Havana to Cádiz with sugar and tobacco. The employment of warships for transporting high-value cargoes may reflect that they were for government use, or merely that no space was wasted on expensive voyages.

HMS *Southampton* (1757)

TYPE • *Ship of the line* **COUNTRY** • *Great Britain*

SPECIFICATIONS	
Dimensions:	Length: 37.9m (124ft 4in); Beam: 10.7m (35ft; Draught: c. 3.6m (11ft 10in)
Displacement:	1000 tonnes (1102 tons) full load
Rig:	Three masts, full square rig
Propulsion:	Sails
Speed:	c. 13 knots (24km/h; 15mph)
Armament:	26 × 12-pounders; 6 × 6-pounder guns
Crew:	210

With the four ships of the *Southampton* class, the Royal Navy for the first time had a consistent frigate type – well armed on a single gun deck. Classified as fifth-rates, they were not intended to form part of the line of battle but instead to act as scouting and raiding ships.

Masts
The three-part construction of the fore- and mainmasts is clearly visible here. Frigates were designed for speed and manoeuvrability.

Below: Southampton*'s last prize was USS* Vixen, *taken in the War of 1812, but both ships were wrecked on an uncharted reef off Conception Island, Bahamas, on 27 November 1812.*

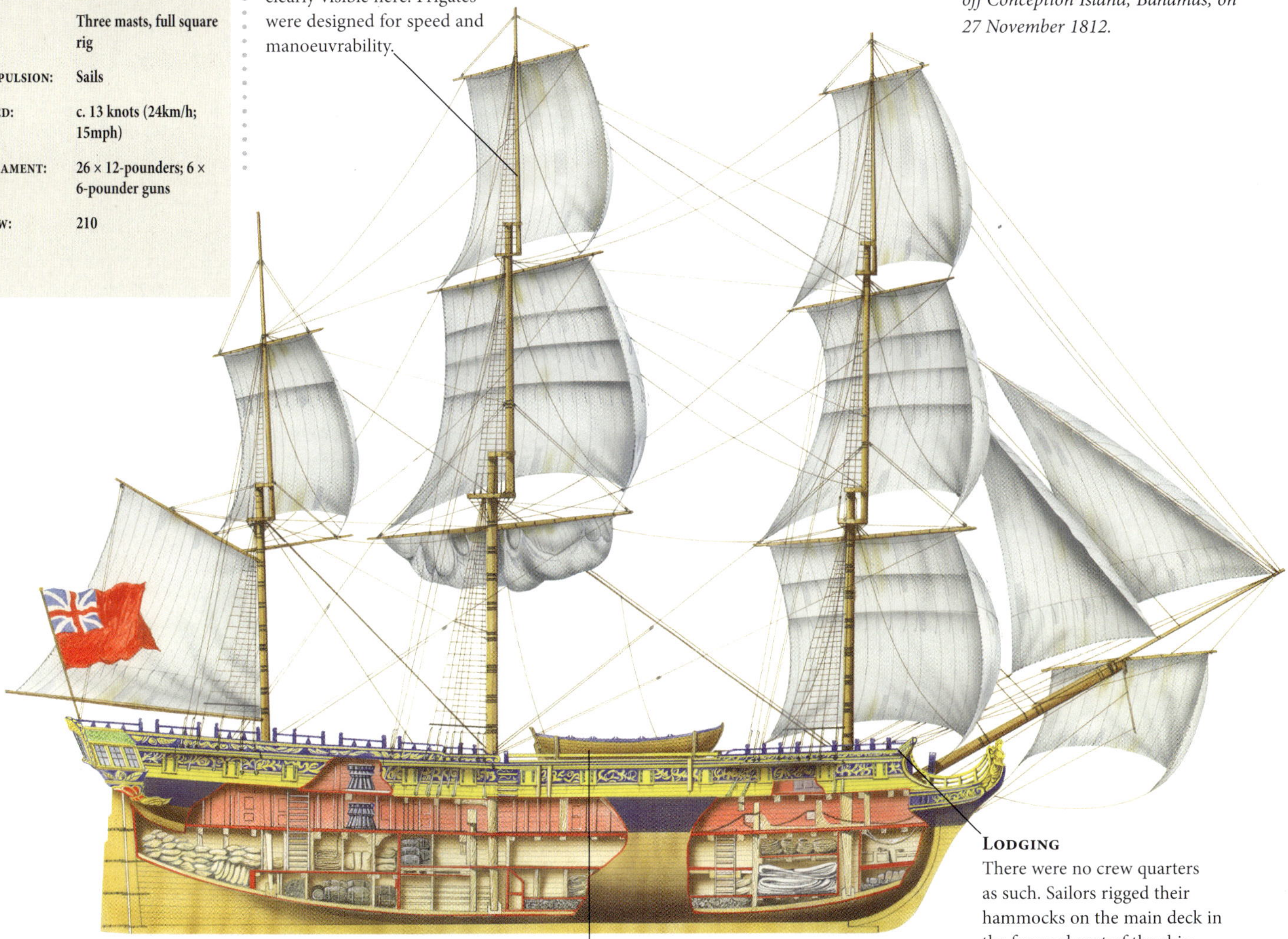

Lodging
There were no crew quarters as such. Sailors rigged their hammocks on the main deck in the forward part of the ship.

Longboat
Every British warship of this time carried a longboat. This small rowed craft was for harbour use, towing and passing supplies and personnel between ships. It was not seen as a lifeboat.

FIRST IN CLASS

HMS *Southampton*, the second Royal Navy ship to bear the name, was the first in the class to be ordered on 12 March 1756. The keel was laid in April at Robert Inwood's yard, Rotherhithe, on the Thames. It was launched on 5 May 1757, and completion followed on 19 June 1757 at Deptford. Three others followed: *Minerva*, *Vestal* and *Diana*. *Southampton* was to be the longest-serving of the quartet.

ROYAL VISIT
In August 1791, King George III made a royal visit to Plymouth to review the fleet. The King did so from aboard *Southampton*, here pictured on the right with HMS *Magnificent* in attendance.

Mortar ketch (1760)

TYPE • *Ketch* **COUNTRY** • *Great Britain*

SPECIFICATIONS

Dimensions:	Length: 35m (115ft); Beam: 10.9m (32ft); Draught: c. 4.1m (13ft 5in)
Displacement:	101 tonnes (111 tons) normal load
Rig:	Two masts, ketch rig
Propulsion:	Sails
Speed:	c. 7 knots (13km/h; 8mph)
Armament:	1 × 330mm (13in) mortar; 1 × 254mm (10in) mortar
Crew:	c. 36

Known as bomb ships (or 'bombs'), mortar ketches were intended for firing explosive shells at fixed targets like shore stations. Originally mounting a single gun, 18th- and 19th-century versions carried two, which could be trained and angled.

Left: Their specialized construction made the bomb ketches slow sailers that tended to wallow in anything other than a calm sea.

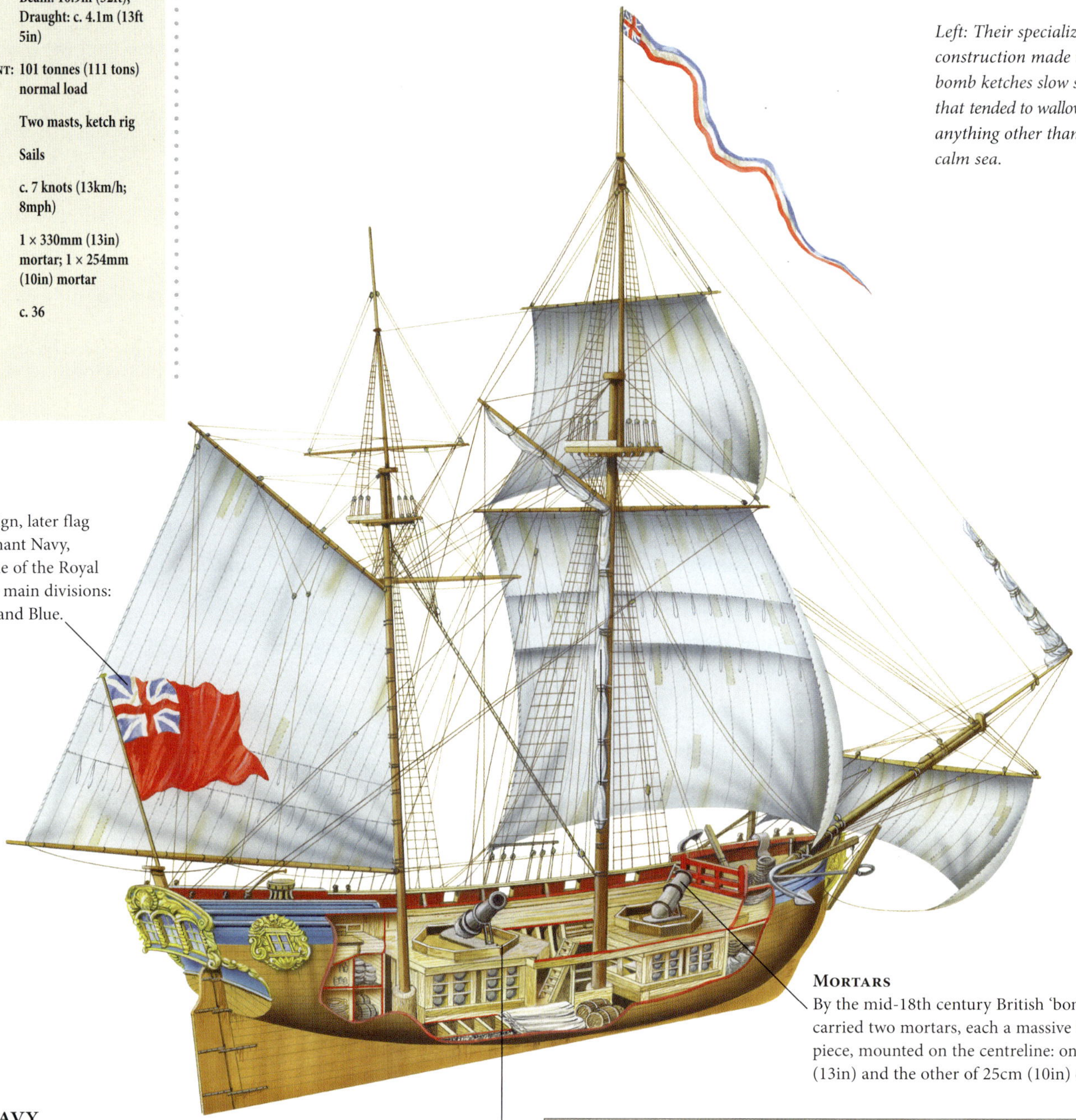

FLAG
The red ensign, later flag of the Merchant Navy, indicated one of the Royal Navy's three main divisions: White, Red and Blue.

MORTARS
By the mid-18th century British 'bombs' carried two mortars, each a massive firing piece, mounted on the centreline: one of 33cm (13in) and the other of 25cm (10in) calibre.

FRAMES
Since the mortars had no recoil system, the ships had to be built with exceptionally strong frames to absorb the shocks of discharge. They also had to be as stable as possible: two anchors were provided to enable them to hold position in order to fire as precisely as the gunners could achieve.

ROYAL NAVY
The Royal Navy had few mortar ketches until the 1790s. The ketches normally only sailed under the protection of other vessels. Shore fortresses were the main target, but the British used mortars to attack concentrations of stationary ships, as in 1799 when five Spanish warships trapped in the Basque Roads were bombarded by HMS *Explosion*, *Sulphur* and *Volcano*.

TRAFALGAR VICTORY
Dwarfed by the great ships of the line around them, bomb ketches salute HMS *Victory* after the Battle of Trafalgar on 21 October 1805. Mortars were not used in ship-to-ship combat.

HMS *Victory* (1765)

TYPE • *Ship of the line* **COUNTRY** • *Great Britain*

SPECIFICATIONS

DIMENSIONS:	Length: 57m (186ft); Beam: 15.8m (51ft 10in); Draught: 8.76m (28ft 9in)
DISPLACEMENT:	3500 tonnes (3858 tons) full load
RIG:	Three masts, full square rig
PROPULSION:	Sails, 5440 sq m (58,556 sq ft)
SPEED:	11 knots (20km/h; 13mph)
ARMAMENT:	30 × 32-pounders; 28 × 24-pounders; 44 × 12-pounder cannons; 2 × 68-pounder carronades (1805)
CREW:	c. 850 sailors and marines

A first-rate, carrying up to 102 guns, HMS *Victory* was intended as an admiral's ship; as Admiral Nelson's flagship at the Battle of Trafalgar in 1805, it became the most iconic of all British warships.

Left: From 1824, Victory *was designated as flagship of the commander in chief at Portsmouth, and although now a museum ship, it remains in commission.*

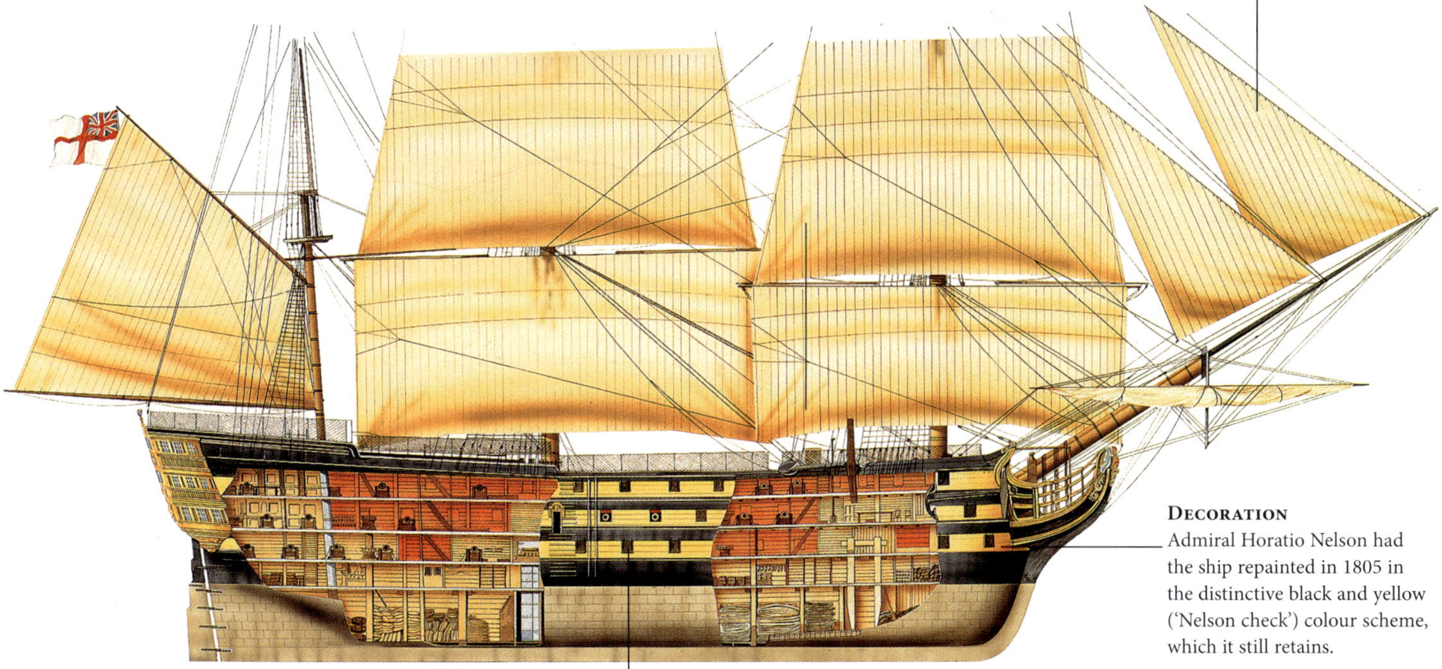

SAILS
Full sail extent was 5440 sq m (58,556 sq ft), making ships of this type both faster and more manoeuvrable than older vessels.

DECORATION
Admiral Horatio Nelson had the ship repainted in 1805 in the distinctive black and yellow ('Nelson check') colour scheme, which it still retains.

GUNS
The heavy guns are in staggered positions from the centreline, with alternate placings on each deck, rather than stacked right above one another.

DESIGNED TO SERVE

HMS *Victory* was designed by Sir Thomas Slade, Surveyor of the Navy, at a cost of £63,176, and was of oak construction, 60cm (2ft) thick at the waterline, with a keel of elm. Laid down at Chatham Naval Dockyard at a time of war in 1759, and launched in 1765, *Victory* was not commissioned until 1778, when war with France was renewed and the Royal Navy was brought to full operational strength.

SOMBRE VICTORY
HMS *Victory* is here depicted off the Isle of Wight by John Winston Carmichael, in his painting of 1847. The artwork shows *Victory* having returned from the Battle of Trafalgar with the body of Nelson on board.

Santísima Trinidad (1769)

TYPE • *Ship of the line* **COUNTRY** • *Spain*

SPECIFICATIONS

Dimensions:	Length: 61.3m (201ft); Beam: 16.2m (53ft); Draught: 8m (26ft)
Displacement:	4826 tonnes (4750 tons); 5200 tonnes (5730 tons) full load
Rig:	Three masts, square rig
Propulsion:	Sails
Speed:	8 knots (15km/h; 9mph)
Armament:	30 × 36-pounders; 32 × 24-pounders; 32 × 12-pounders; 18 × 8-pounder guns (1768)
Crew:	950

***Santísima Trinidad*, the seventh Spanish warship of the name, was confirmed by a royal order on 12 March 1768. Sometimes claimed as the largest warship of the 18th century, with four decks of guns, it was known as the '*Escorial* [royal palace] of the Seas.'**

Below: Santísima Trinidad *was built at the Royal Shipyards at Havana, Cuba, and fought in two of history's great naval battles, at Cape St Vincent and Trafalgar.*

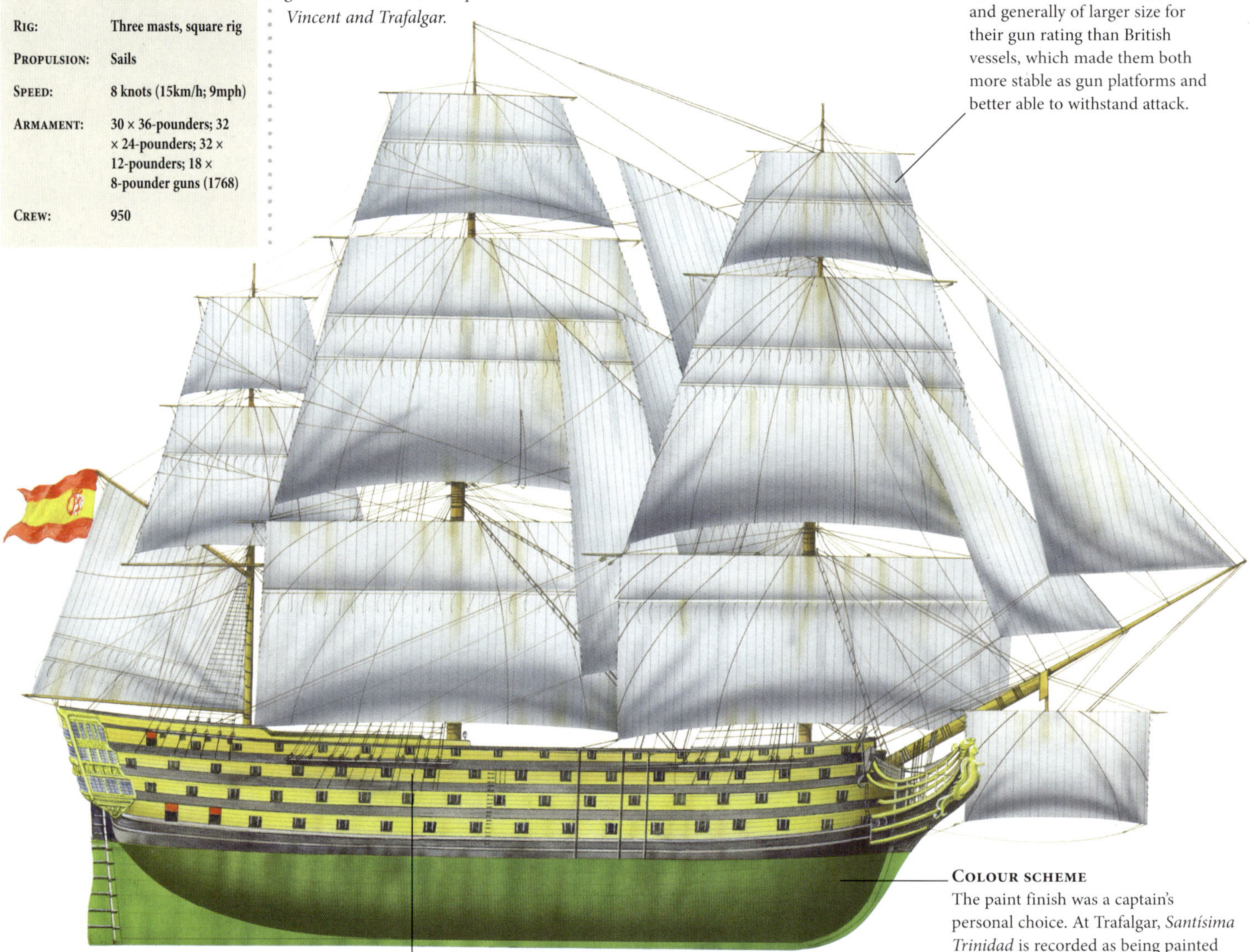

Design
Spanish ships were strongly built and generally of larger size for their gun rating than British vessels, which made them both more stable as gun platforms and better able to withstand attack.

Colour scheme
The paint finish was a captain's personal choice. At Trafalgar, *Santísima Trinidad* is recorded as being painted dark red, with white stripes.

Fourth deck
In 1795 a fourth deck was installed, joining the forecastle to the quarterdeck and raising the number of cannons from 112 to 136. This made *Santísima Trinidad* the most heavily armed ship of its time.

EARLY ACTIONS

In its first years, *Santísima Trinidad* was probably not in commission. With the declaration of war by Spain on Britain in July 1779, it entered service as flagship of the Spanish fleet, operating with allied French ships in the English Channel and the Western Approaches. In August 1780, it led an action that resulted in the capture of 55 British merchant vessels from a convoy. In 1782, it fought in the second siege of Gibraltar.

REPLICA VESSEL

A replica of the *Santísima Trinidad* can be visited in the harbour of Alicante in Spain. This recreation is of the ship before 1795: built by the Irish shipwright Matthew (or Mateo) Mullan, *Santísima Trinidad* was launched as a three-decker.

Turtle (1775)

TYPE • *Submersible* **COUNTRY** • *United States*

SPECIFICATIONS	
Dimensions:	Length: 2.29m (7ft 6in); Beam: 0.9m (35in)
Displacement:	Surfaced 0.55 tonnes (0.6 tons); Submerged 1 tonne (1.1 tons)
Rig:	N/A
Propulsion:	Two-blade foot-worked paddle-propeller
Speed:	Walking pace
Armament:	68kg (150lb) gunpowder keg with clockwork fuse
Crew:	1

The world's first submarine craft created for military action was deployed off New York Harbour on 6 September 1776. Its mission was to sink a British man-of-war with a displacement a thousand times greater than its own.

Right: The large number of levers and handles shows the problem of one-man operation: to maintain depth and trim while also working the awl would have been extremely difficult.

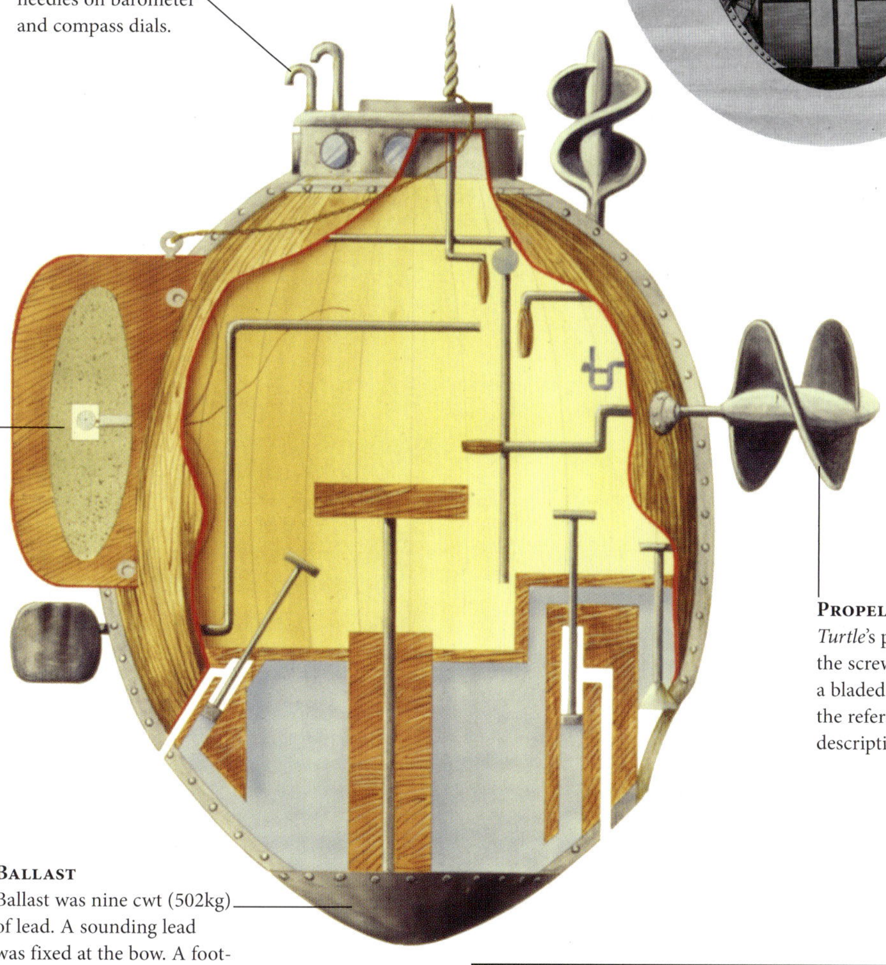

Air tubes
The air tubes are among *Turtle* firsts; also the use of water as dischargeable ballast (at the base of the craft); and the use of 'glow in the dark' needles on barometer and compass dials.

Bomb and detonator
Two clockmakers worked with inventor David Bushnell on the firing mechanism. The operator would set this device when releasing the mine from the submarine, allowing himself enough time to get well away.

Propeller
Turtle's propeller was not of the screw type as shown, but a bladed paddle type, hence the reference to 'oars' in descriptions from the period.

Ballast
Ballast was nine cwt (502kg) of lead. A sounding lead was fixed at the bow. A foot-operated valve admitted water to a ballast tank from which it could be pumped out again.

DAVID BUSHNELL
Turtle was the brainchild of David Bushnell, who was determined to create the most effective submersible vessel yet built. The son of a Connecticut farmer, he became a mature student at Yale where his interest in scientific matters, including explosives, was well known. Work at a Saybrook shipyard to pay for his college course may have sparked his interest in submersibles. The concept of a submarine was unfamiliar, and one contemporary reference refers to the craft as a 'water machine'.

TORTOISE BODY
A detailed description of this remarkable craft survives in letters from Dr Benjamin Gale dated November 1775 to February 1776. He noted that: 'The Body, when standing upright in the position in which it is navigated, has the nearest resemblance to the two upper shells of a Tortoise joined together.'

USS *Hancock* (1776)

TYPE • *Frigate* **COUNTRY** • *United States*

SPECIFICATIONS

DIMENSIONS:	Length: 41.6m (136ft 7in); Beam: 10.8m (35ft 6in); Draught: c. 4.5m (14ft 9in)
DISPLACEMENT:	775 tonnes (854 tons) normal load
RIG:	Four masts, square rig
PROPULSION:	Sails
SPEED:	c. 12 knots (22km/h; 14mph)
ARMAMENT:	24 × 12-pounders; 8 × 6-pounders
CREW:	290

The second ship to be named for the first signatory of the Declaration of Independence, the USS *Hancock* was one of the 13 'Continental frigates' ordered by US Congress on 13 December 1775. It would serve under American, British and French flags.

Below: Ultimately, Hancock *was hulked at the Toulon naval base and used as a gunpowder store. The British and Spanish occupied the port in the second half of 1793, and as they withdrew in December they blew up* Iris *(the renamed* Hancock*).*

CONSTRUCTION
Hancock's career was dramatic but it was also relatively short, which may have been due to the use of less-durable unseasoned wood in the construction.

LATEEN SAIL
A short sternpost mast with lateen-type sail was fitted. The lateen sail enabled a ship to sail at a wider range of angles into the wind.

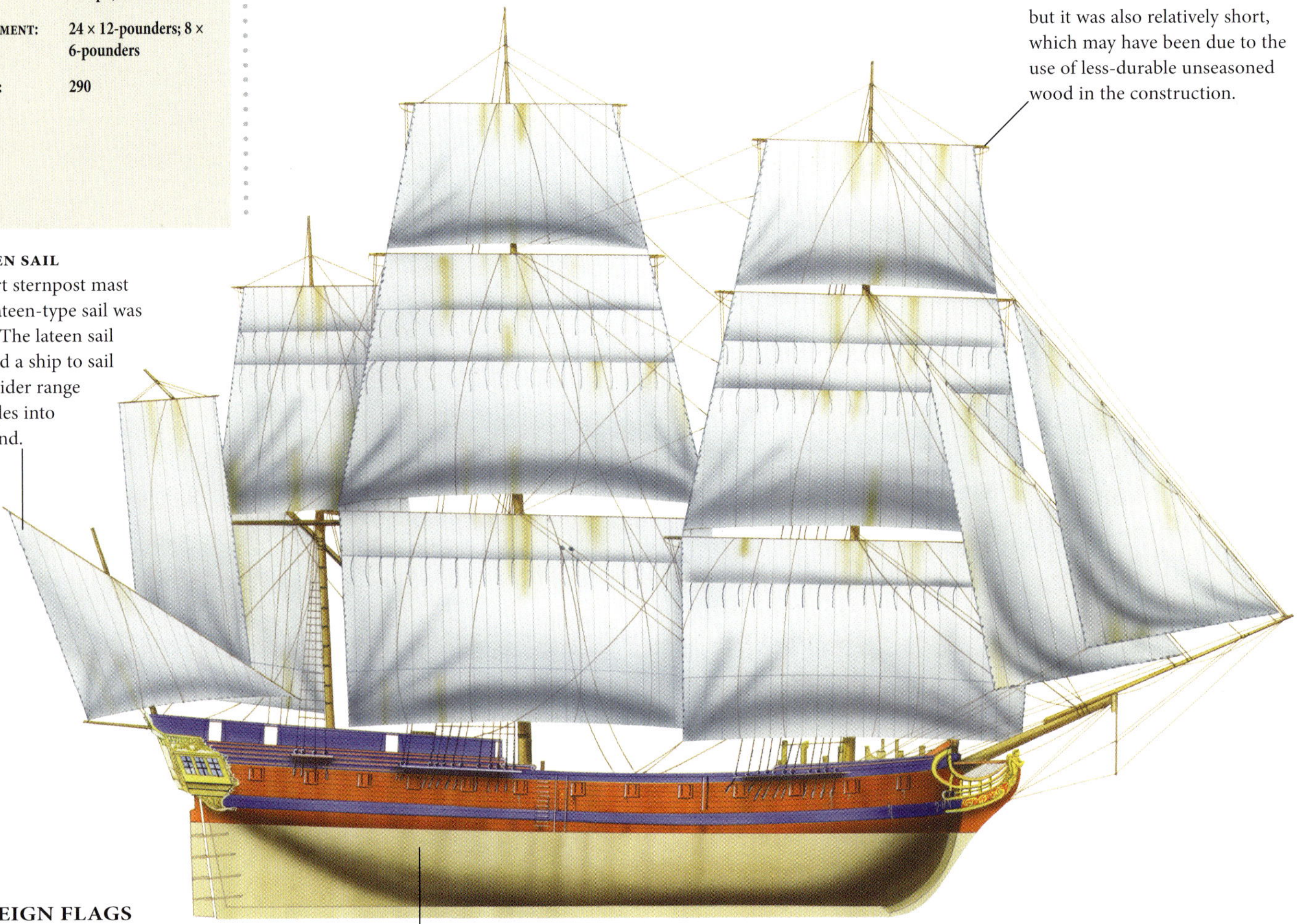

FOREIGN FLAGS
On 7 July 1777, *Hancock*, *Boston* and *Fox* were intercepted by three British frigates. *Hancock* was forced to surrender. Renamed HMS *Iris*, the ship was taken into the Royal Navy, where it was used to capture numerous American vessels. On 11 September of that year, however, *Iris* was captured by the French after the Battle of the Virginia Capes. Now under a third flag, it retained the name of *Iris* and served in the French Navy.

HULL PROFILE
The hull of *Hancock* was slender and designed to reduce draught and increase speed. The colour scheme shown here is speculative, as blue paint was expensive at the time.

***HANCOCK* UNDER SAIL**
The American frigates of the later 1770s were celebrated as fine sailers and excellent warships despite their relatively small size. *Hancock* was considered the most handsome, as well as the most effective, of the 1775 ships.

Santa Ana (1784)

TYPE • *Ship of the line* **COUNTRY** • *Spain*

SPECIFICATIONS

Dimensions:	Length: 56.1m (184ft 2in); Beam: 15.5m (50ft); Draught: c. 7.8m (25ft 7in)
Displacement:	2534 tonnes (2793 tons); normal load
Rig:	Full square rig with staysails and studding sails
Propulsion:	Sails
Speed:	c. 8 knots (15km/h; 9mph)
Armament:	30 × 36-pounders; 32 × 24-pounders; 10 × 8-pounder cannons; 10 × 48-pounders; 2 × 32-pounders; 6 × 24-pounder howitzers; 4 × 4-pounder swivel guns (1805)
Crew:	1048 (wartime)

Flagship of the Spanish fleet at Trafalgar, the 112-gun *Santa Ana* was in the middle of the French–Spanish line. Dismasted, it was forced to surrender and was taken as a prize, but it was reclaimed in a bold operation by French ships on 23 October 1805.

Below: Based at Cádiz, like other capital ships Santa Ana *had lengthy spells out of commission – five years out of its 21-year career.*

Lightning rods
Santa Ana was among the first ships fitted with lightning rods, to conduct lightning strikes to the waterline.

Firepower
Santa Ana was used for testing six large-calibre carronade guns from the Carron factory in Scotland, and also for testing explosive shells in 1785. In peacetime its 36-pounder guns were replaced by 24-pounders.

Construction
Above the keel, the ballast was formed from cannon balls and broken pieces of iron, surrounded by stone and cemented into place. The hull was coppered in September 1798 and the ship re-keeled in 1800.

CONSTRUCTION

A first-rate *navío de línea* (ship of the line), *Santa Ana* was constructed by José Romero y Fernández de Landa to the design of Miguel de la Puente, its keel laid in June 1783, and launched at Ferrol Navy Yard on 29 September 1784. Sea tests were found to be highly satisfactory and the ship was commissioned under Captain Felix de Tejada on 28 February 1785.

TRAFALGAR

At Trafalgar, under Captain Gardoqui, *Santa Ana* was the flagship of Admiral Don Ignacio Alava at the head of the Spanish division. Engaged by Admiral Collingwood's 112-gun *Royal Sovereign* of the British lee division, *Santa Ana* suffered a double-shotted broadside through the vulnerable stern, putting seven guns out of action and causing carnage inside.

Montañes (1793)

TYPE • *Ship of the line* COUNTRY • *Spain*

SPECIFICATIONS

DIMENSIONS:	Length: 52.9m (173ft 6in); Beam: 14.4m (37ft 3in); Draught: 7.49m (24ft 7in)
DISPLACEMENT:	2807 tonnes (3094 tons) normal load
RIG:	Three masts, full square rig
PROPULSION:	Sails
SPEED:	12 knots (22km/h; 14mph)
ARMAMENT:	28 × 36-pounders; 30 × 18-pounders; 8 × 8-pounder guns; 10 × 30-pounder obuses (1805)
CREW:	686

Leader of a class of four 74–80 gun ships of the line, all of which fought at Trafalgar, *Montañes* was financed by money raised among hilltowns and villages, hence the name. As usual with Spanish ships, it also had a saint's name: *Santo Toribio de Mogrovejo*.

Below: A diagram showing the profile of 18th-century French frigates. Julián Martín de Retamosa designed and supervised the construction of Montañes *at Ferrol, where the keel was laid in 1793 and launching was on 24 November 1793.*

CABINS
Some illustrations of the ship show a cabin constructed on the poop deck, but this was not an original feature.

CAPSTAN
A capstan (essentially a large vertical winch) extended through two decks, enabling it to be turned by more men: up to 200 might be needed.

BEAM
Maximum beam was at the waterline, with a distinct tumblehome reducing to almost vertical at the level of the bulwarks.

HULL MAINTENANCE
Between 16 March and 24 June 1805, *Montañes* was given what was doubtless a much-needed careen and the hull was coppered.

FIGHTING VESSEL
On 30 March 1795, *Montañes* was one of the ships in action against eight French ships of the line and two frigates off Cape San Sebastian, and in June it was assigned to Admiral de Alava's fleet at Cádiz.

OCEANIC JOURNEYS
On 11 November *Montañes* left Cádiz with a squadron bound for Callao, Peru, by way of Cape Horn: a dangerous voyage but successfully accomplished. In 1796 it crossed the Pacific to Manila. By 1798, it was back at Cádiz, but made a further voyage to the Philippines in 1802–03, returning via the Cape of Good Hope, thereby making it one of the comparatively few vessels of the period to circumnavigate the globe.

USS *Constitution* (1797)

TYPE • *Frigate* **COUNTRY** • *United States*

SPECIFICATIONS

Dimensions:	Length: 53m (175ft) between perpendiculars; Beam: 13.3m (43ft 6in); Draught: 7m (23ft)
Displacement:	1430 tonnes (1576 tons) normal load; 2235 tonnes (2464 tons) full load
Rig:	Three masts, square rig
Propulsion:	Sails, 3995 sq m (43,000 sq ft)
Speed:	13 knots (24km/h;154mph)
Armament:	28 × 24-pounders; 10 × 12-pounder guns
Crew:	450

USS *Constitution* was solidly built, but at a time when a warship's service life rarely exceeded two or three decades, its builders would likely have been amazed to know that the ship would remain a commissioned vessel in the US Navy more than 200 years after its launch.

Below: One of six frigates authorized by Congress on 27 March 1794, Constitution *was designed by Joshua Humphreys and built by the Hartt Shipyard, Boston.*

Sails
The maximum sail area was over 3995 sq m (43,000 sq ft). Spacing of the masts enabled further sails to be hoisted between fore- and mainmasts.

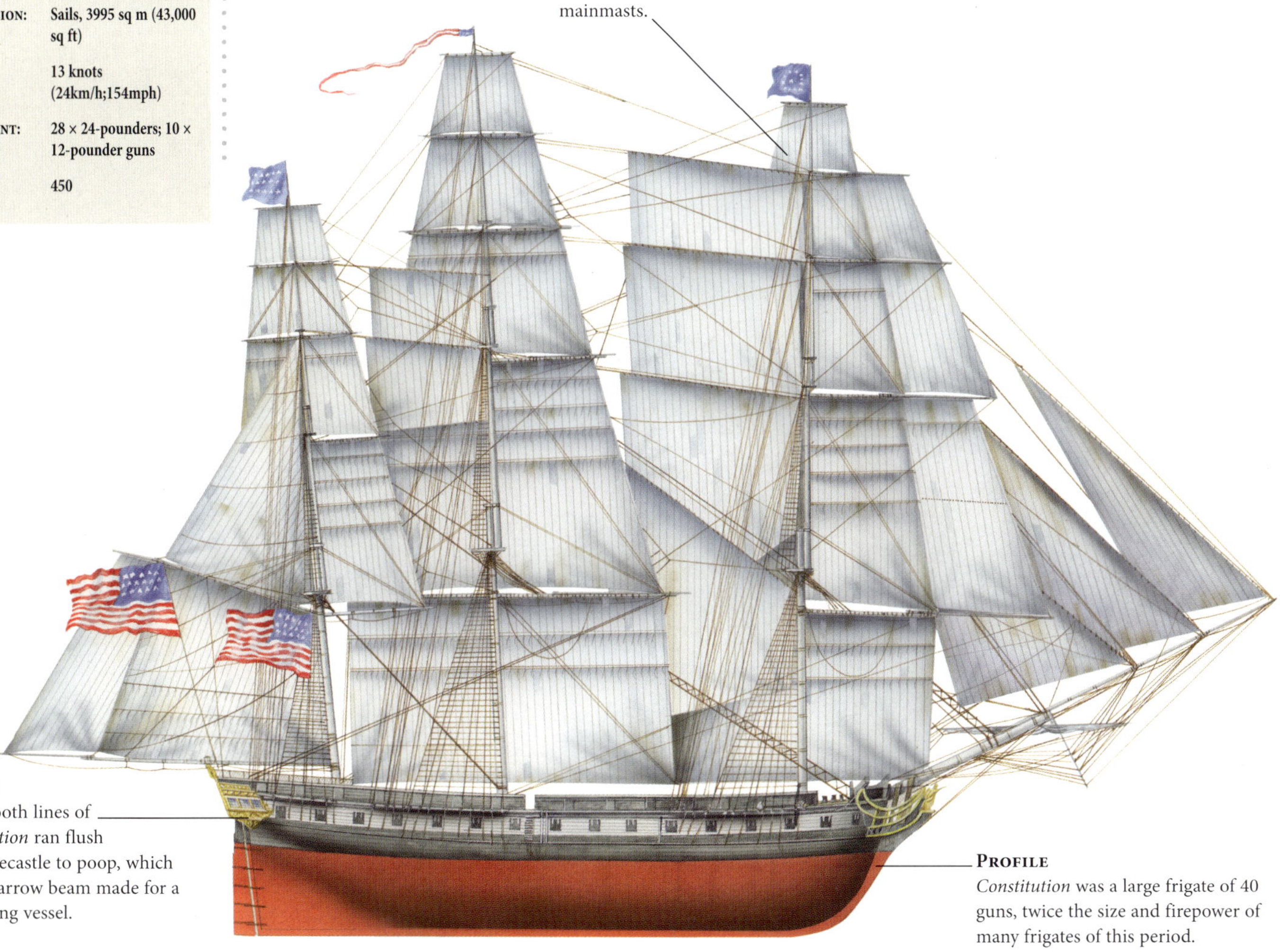

Design
The smooth lines of *Constitution* ran flush from forecastle to poop, which with a narrow beam made for a fast-sailing vessel.

Profile
Constitution was a large frigate of 40 guns, twice the size and firepower of many frigates of this period.

WAR OF 1812
In 1811, *Constitution* carried the US Minister Joel Barlow to France, and during the War of 1812 took five warships, including HMS *Guerrière* (49 guns), *Java* (38 guns) and *Cyane* (22 guns), as well as nine merchantmen either taking them as prizes or destroying them. A long period in ordinary followed until it was recommissioned in May 1821 for two Mediterranean cruises, ending in July 1828.

PRESERVED IN BOSTON
Constitution was a training ship from 1877 to 1882, then made a receivingship at Portsmouth, New Hampshire. Scrapping was narrowly avoided in 1905. On 1 July 1931, it was recommissioned and sailed around both US coasts, returning to Boston on 7 May 1934. It is preserved at the former Charlestown Navy Yard, in Boston.

USS *Essex* (1799)

TYPE • *Frigate* **COUNTRY** • *United States*

SPECIFICATIONS	
Dimensions:	Length: 42.2m (138ft 7in); Beam: 11.4m (37ft 3in); Draught: 3.7m (12ft 3in)
Displacement:	864 tonnes (952 tons); normal load
Rig:	Three masts, full square rig
Propulsion:	Sails
Speed:	11.4 knots (21km/h; 13mph)
Armament:	26 × 12-pounders; 10 × 6-pounder guns (1799)
Crew:	300

The well-travelled USS *Essex* served in the Mediterranean Sea against the Barbary pirates and played a distinguished part in the War of 1812, until it was captured by British frigates at Valparaiso, Chile, and subsequently brought into the Royal Navy.

Below: Essex *was named for Essex County, Massachusetts, which raised the necessary $139,362 to build the ship.*

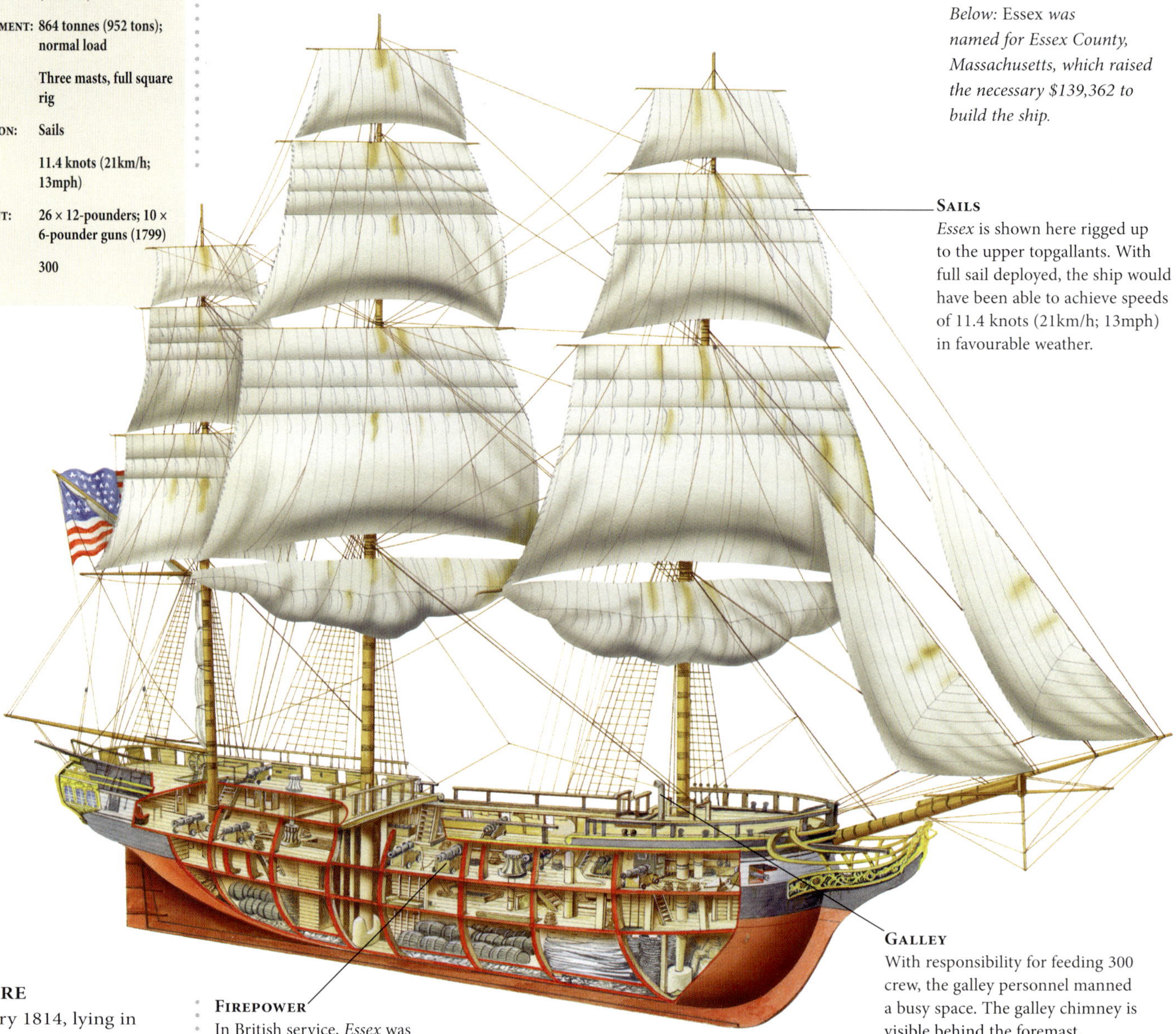

Sails
Essex is shown here rigged up to the upper topgallants. With full sail deployed, the ship would have been able to achieve speeds of 11.4 knots (21km/h; 13mph) in favourable weather.

Galley
With responsibility for feeding 300 crew, the galley personnel manned a busy space. The galley chimney is visible behind the foremast.

Firepower
In British service, *Essex* was re-gunned with 26 18-pounders on the gun deck and 12 32-pounder carronades on the quarterdeck, also two 9-pounder long guns and two 32-pounder carronades on the forecastle.

CAPTURE

In January 1814, lying in Valparaiso Roads, *Essex* was blocked by the British frigate *Phoebe* (36 guns) and sloop *Cherub* (18 guns). Captain Porter attempted to break out, but was forced back by a storm. Following him into Chilean territorial waters, *Phoebe* and *Cherub* opened fire and a stiff battle ensued. After two-and-a half hours, with 155 men lost, Porter struck his flag. Repaired by the British, *Essex* was sailed to England and taken into the Royal Navy as HMS *Essex*.

EQUATORIAL CROSSING

Essex was built at Salem, Massachusetts, by Enos Briggs and launched on 30 September 1799. Presented to the US Navy on 17 December, its inaugural captain was Edward Preble. *Essex* was the first US Navy vessel to cross the equator.

Chasse-marée (1800)

TYPE • *Schooner* **COUNTRY** • *France*

SPECIFICATIONS

DIMENSIONS:	Length: c. 22.8m (75ft); Beam: 6m (20ft); Draught: c. 1.5m (5ft)
DISPLACEMENT:	50 tonnes (55 tons); normal load
RIG:	Three or two masts, lug rig, often with topsails
PROPULSION:	Sails
SPEED:	c. 10 knots (18.5km/h; 11.5mph)
ARMAMENT:	From one to six light cannons, usually 3-pounders
CREW:	c. 50–60

Capable of great speed, which made it extremely useful as a smugglers' boat, the *chasse-marée* ('tide-chaser') could also operate as a privateer to capture merchant vessels, but was fast and agile enough to escape most warships.

Left: The shallow draught of the chasse-marée *and their crews' familiarity with coastlines helped these vessels to evade and survive encounters with more powerful ships.*

SAIL
Carrying a large extent of sail on three or sometimes two masts, the *chasse-marée* was the French equivalent of the American schooner and the British cutter.

LODGING
Virtually no accommodation was provided: these vessels were intended for quick in-and-out actions.

ARMAMENT
Smaller boats had a cannon fitted in a reinforced prow; larger ones might have additional lateral guns to boost the firepower.

MILITARY USE

The *chasse-marée* had developed as a fast vessel to carry fresh-caught fish to port, but its value as a swift, elusive craft led to its conversion to military use, with bracing in the prow to support a cannon. The largest *chasse-marées* carried a lug topsail on the mainmast. A relatively large crew was needed to work the ship and to provide a prize crew on a captured vessel.

SEA HANDLING

Sailing into the wind, the weatherly qualities of a *chasse-marée* made it hard to beat, but with a following wind it could be run down by a frigate or sloop of war.

Dhow (1800)

TYPE • *Dhow* **COUNTRY** • *Tunis*

SPECIFICATIONS	
DIMENSIONS:	Length: c. 21.3m (70ft); Beam: c. 4.3m (14ft); Draught: c. 1.7m (5ft 7in)
DISPLACEMENT:	Up to 150 tonnes (165 tons)
RIG:	One to three masts, lateen or settee rig
PROPULSION:	Sails
SPEED:	9 knots (17km/h; 10mph)
ARMAMENT:	Not usually armed, but could mount one to six light cannons
CREW:	10–50

Not known in Arabic, the term 'dhow' first appeared in English around 1802, referring to a small sailing vessel, usually single-masted and with a lateen or settee rig, operating in the Persian Gulf, Red Sea and Indian Ocean.

Right: Dhows were often built on open beaches rather than in yards, using basic tools. Larger dhows could be used for military purposes, carrying troops and light guns, but essentially they were merchant vessels.

MASTS
Masts were relatively short, raked forwards, stepped into the keel timber, and fitted with a very wide yard normally built up from four or five separate pieces.

LATEEN SAIL
The lateen sail was of quadrilateral, not triangular, form and is more correctly termed a settee sail in the context of a dhow.

CONSTRUCTION
Iron nail fastenings began to supplant sewn planks from the mid-16th century, although sewn hulls were still found in the 19th century. The dhow hull had no tumblehome: if anything it was angled outwards from waterline to bulwarks.

ORIGINS

Indigenous to the coasts of the Arabian Peninsula, India and East Africa, the earliest dhows were simple dugouts with teak planks sewn to their sides to form a hull. Gradually, larger vessels evolved, based on a keel to which additional planking was sewn, and originally steered by a steering oar until the rudder came into use. Dhow types are generally defined by their hull shape. They include the *ghanjah*, a large vessel with a curved stem and a sloping, ornately carved transom.

SLAVE DHOW
A British man of war firing on a slave dhow, depicted in *The History of the Universities' Mission to Central Africa, 1859–1898* (1899). While dhows could still be seen at this time, they were fewer than the hundreds visible around 1800.

Pirate junk (1800)

TYPE • *Junk* **COUNTRY** • *Malay archipelago*

SPECIFICATIONS	
Dimensions:	Length: c. 27.4m (90ft); Beam: c. 7.3m (24ft); Draught: c. 1.2m (4ft)
Displacement:	20.3 tonnes (22 tons) normal load
Rig:	Two or three masts, junk rig
Propulsion:	Sails
Speed:	8 knots (15km/h; 9mph)
Armament:	8–10 small cannons
Crew:	20–60

The junk has a long history. The name provides only a general indication of the vessel type: traditionally built, featuring a built-out platform bow, a central well and a high poop, and carrying one to three masts with sails divided horizontally.

Below: The high poop deck of a typical dhow provided a better forward view and made it easier to board merchant ships.

Sails and masts
The junk sail is divided by rigid members or battens, which extend it forward of its mast. Size varied from small single-masted vessels to ocean-going ships with up to eight masts.

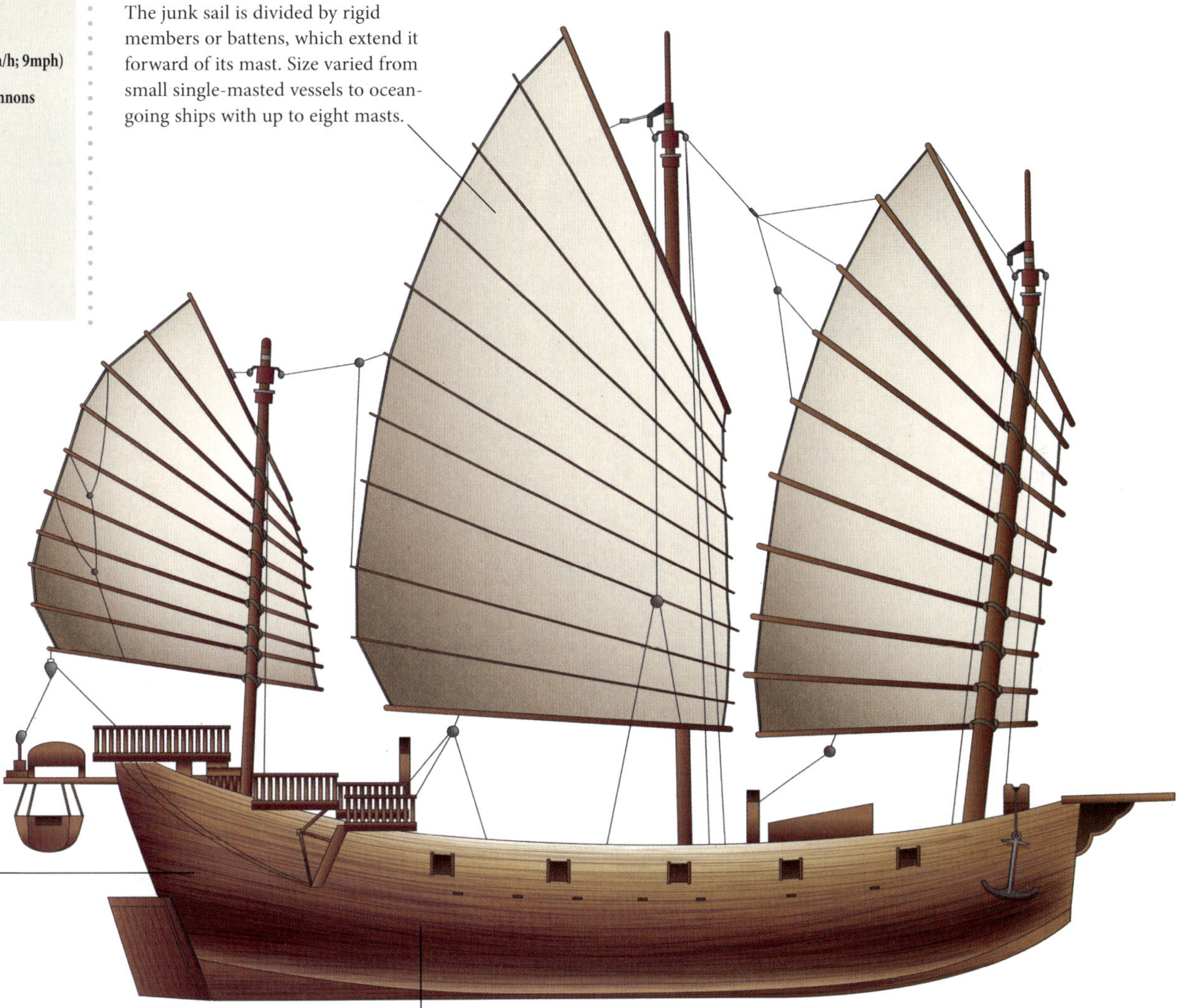

Design
The longevity of the junk is explained by its effectiveness: although the upper hull may look bulky, its underwater lines are sleek, and its sail arrangement makes it fast, easily controlled and capable of sailing close to the wind.

Hull
The traditional hull is built of softwood planks with several separate internal compartments that improve buoyancy and prevent flooding.

PIRACY RIFE

Piracy in the seas off China was endemic, particularly at times of weak government. A large three-masted junk made a highly effective pirate ship. Pirates often combined in fleets to attack merchant vessels sailing in convoy or to extort payment for safe passage. These junks could carry a large number of armed pirates, and were armed with eight to 10 small cannons.

UNDER ATTACK
This 1860 illustration depicts a Chinese junk under attack by Malay *prahus* (another type of small merchant craft often used in piracy). The establishment of British naval stations and the use of naval escorts led to a rapid decline in international pirate attacks by around 1810.

Xebec (1800)

TYPE • *Xebec* **COUNTRY** • *Algiers*

SPECIFICATIONS

Dimensions:	Length: 31m (103ft 9in); Beam: 6.7m (22ft); Draught: 2.5m (8ft)
Displacement:	c. 193 tonnes (213 tons) normal load
Rig:	Three masts, lateen rig
Propulsion:	Sails
Speed:	10 knots (19km/h; 11.5mph)
Armament:	15–18 × 12-pounder guns
Crew:	24

Associated with the Barbary pirates, the xebec was developed from the late 17th century into the 19th century. Powered both by wind and oars, it was a versatile craft in terms of handling and utility.

Below The xebec had a shallow draught, enabling it to run close inshore if pursued by a larger ship, and its fine underwater lines gave it a good turn of speed.

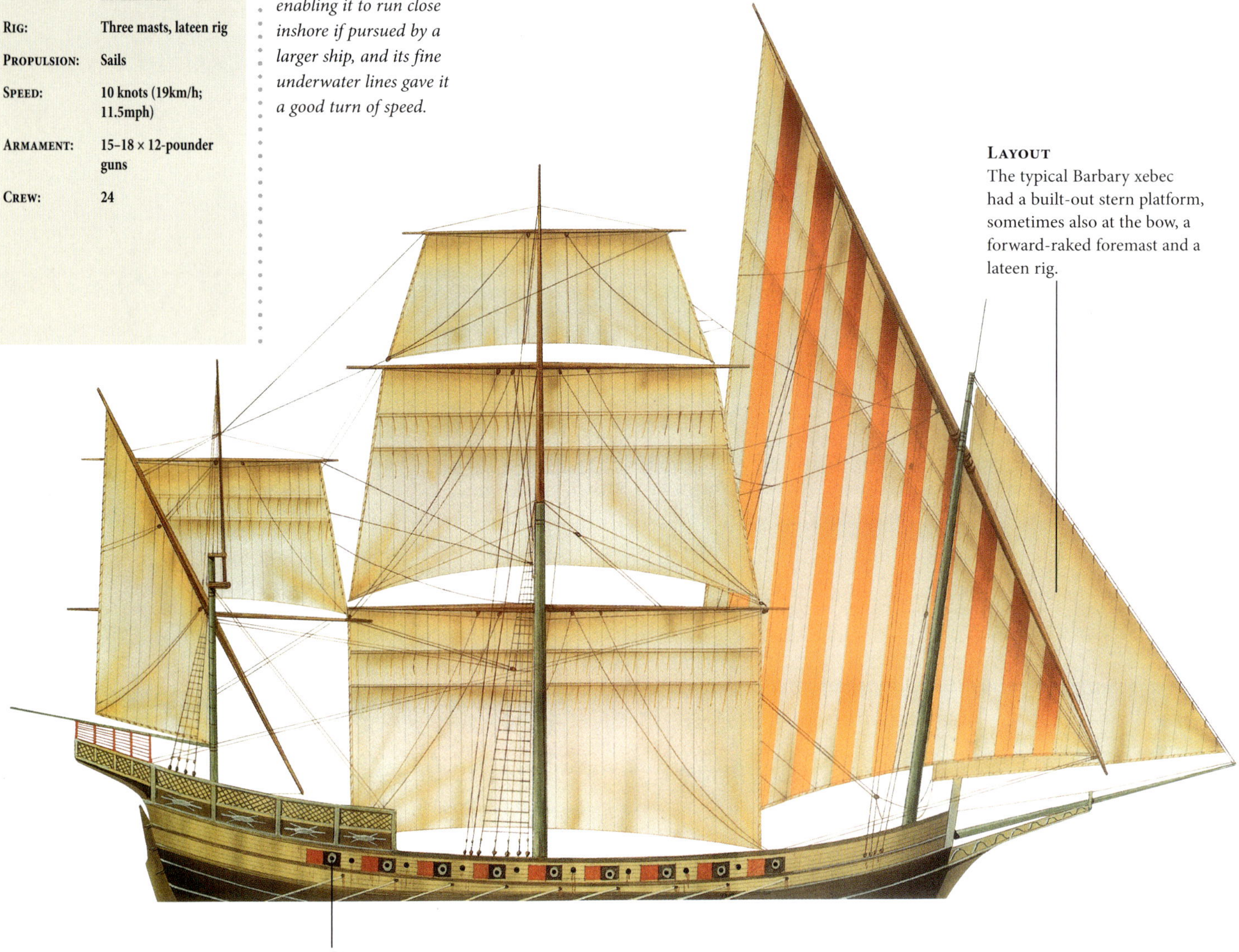

Layout
The typical Barbary xebec had a built-out stern platform, sometimes also at the bow, a forward-raked foremast and a lateen rig.

Oars
An important feature were the oars, nine on each side, which enabled the xebec to pull towards a victim that was becalmed or moving slowly in a light wind.

HISTORY
The xebec began as a merchant vessel, but in the 18th century was already being used as a naval craft by both Spain and France, and as a pirate ship by the corsairs of the North African coast. In these forms they carried guns and were comparable in force to a sloop or a light pre-1750 frigate. By 1800, the Spanish Navy had large xebec-frigates of 610 tonnes (600 tons).

PIRATE SHIP
This image illustrates an action between massed ships of the Dutch fleet and Barbary pirates. The xebec was the standard vessel used by the lawless pirate groups operating from Algiers, Tunis and Tripoli under the protection of local governors.

Bucentaure (1803)

TYPE • *Ship of the line* **COUNTRY** • *France*

SPECIFICATIONS	
DIMENSIONS:	Length: 59.3m (194ft 5in); Beam: 14m (45ft 11in); Draught: 6m (19ft 8in)
DISPLACEMENT:	3868 tonnes (4264 tons) full load
RIG:	Three masts, full square rig
PROPULSION:	Sails
SPEED:	10 knots (18.5km/h; 11.5mph)
ARMAMENT:	30 × 36-pounders; 32 × 24-pounders; 18 × 12-pounder guns; 6 × 36-pounder howitzers
CREW:	840

The French Navy made far more use of the 80-gun two-decker than the British, and *Bucentaure* was the first of a class of 16 launched between 1803 and 1815, with a further five up to 1824.

Below: After Trafalgar, the demasted ship was blown onto a reef in Cádiz Bay and sank, with only a handful of survivors.

STERN
It was well known that the stern was a weak area, but naval officers of all nations clung to their windows and galleries.

FIREPOWER
The ship was bristling with firepower: 30 36-pounders, 32 24-pounders, 18 12-pounder guns; and six 36-pounder howitzers.

CHARACTERISTICS
The class of ship was longer than 74-gun three-deckers, as well as able to fire a heavier broadside. They were also faster and handled better to windward: always a weak point of the high-sided three-decker.

STATE OF THE ART

Bucentaure was laid down at Toulon in November 1802, launched on 13 July 1803 and commissioned in January 1804 under Captain Jean-Jacques Magendie. Its name commemorates the French conquest of Venice in May 1797 – the state barge of the Doge of Venice was *Bucintoro*. Jacques-Noël Sané designed the class, which some experts consider to have been the finest ships of the line ever built.

EMBATTLED AT TRAFALGAR

At Trafalgar, *Bucentaure* was central in the French-Spanish line. The British flagship HMS *Victory* cut across Villeneuve's line just behind *Bucentaure*, firing a devastating broadside through the transom. Completely demasted, *Bucentaure* ultimately surrendered to HMS *Conqueror*.

HDMS *Christian VII* (1803)

TYPE • *Ship of the line* **COUNTRY** • *Denmark*

Flagship of the Danish Navy, this big 90-gun two-decker ship was the masterpiece of the Danish designer, Frantz Hohlenberg. However, its Danish service was brief and uneventful before it was captured by the British in 1807.

SPECIFICATIONS

DIMENSIONS:	Length: 57m (187ft 2in); Beam: 15m (50ft 10in); Draught: 6m (18ft 6in)
DISPLACEMENT:	2162 tonnes (2383 tons) normal load
RIG:	Three masts, full square rig
PROPULSION:	Sails and screw propeller etc
SPEED:	12 knots (22km/h; 14mph)
ARMAMENT:	30 × 32-pounders; 32 × 18-pounder guns; 12 × 32-pounders; 2 × 18-pounders; 4 × 12-pounder carronades (from 1810)
CREW:	670

Below: Christian VII *was the flagship of three admirals and had six captains from 1808 to 1813.*

MASTS
The ship had three masts in a full square rig arrangement. Under favourable winds, the ship could reach a speed of 12 knots (22km/h; 14mph).

STERN DESIGN
The stern design was intended to accommodate guns that could fire over a wide field against approaching gunboats or similar craft, by narrowing the transom and dispensing with the stern and quarter galleries.

REFIT
In British hands, the ship's broadside discharge was reduced from 560kg (1236lb) to 466kg (1029lb). This reduced strains on the hull that would have been exacerbated by the blockade duty and heavily armed patrolling required of Royal Navy ships.

CAPTURE

Built at the Nyholm Dockyard and launched in 1803, *Christian VII* appears to have gone straight into ordinary, despite a state of war with Great Britain. Under Admiral James Gambier, on board HMS *Prince of Wales*, a large British fleet appeared off Copenhagen on 2 September 1807, and bombarded the city until it surrendered. The British then took over the naval dockyard. Every usable ship and piece of equipment was carried off, including *Christian VII.*

SHIP MODEL

Having completed refitting at Portsmouth, at a cost of £27,230 including new guns, *Christian VII* served in the North Sea where its modest draught (the ship was designed for Baltic waters) was ideal, as well as in the English Channel.

HMS *Shannon* (1806)

TYPE • *Frigate* **COUNTRY** • *United Kingdom*

SPECIFICATIONS

DIMENSIONS:	Length: 45.8m (150ft 2in); Beam: 12.17m (39ft 11in); Draught: 4.5m (14ft 9in)
DISPLACEMENT:	1082 tonnes (1193 tons) normal load
RIG:	Three masts, full square rig
PROPULSION:	Sails
SPEED:	13 knots (24km/h; 15mph)
ARMAMENT:	30 × 36-pounders; 32 × 24-pounders; 18 × 12-pounder guns; 6 × 36-pounder howitzers
CREW:	330

Shannon **was a powerful 19th-century Royal Navy frigate. Launched on 5 May 1806, and completed at Chatham the following August, the 38-gun ship was commanded for its first seven years by Captain Philip Broke.**

Left: Shannon *remained in service until 1830, when it became a receiving hulk at Sheerness. Renamed* St Lawrence *on 1 March 1844, it had been broken up at Chatham by 12 November 1859.*

DESIGN
The overall layout reflected that of the captured French 38-gun frigate *Hébé*, being a full-rigged ship with heavy firepower and a good turn of speed, although it was prone to rolling in poor weather.

FIREPOWER
Total armament amounted to 30 36-pounders, 32 24-pounders, 18 12-pounder guns and six 36-pounder howitzers, giving the ship flexible firepower from long to close ranges.

QUARTERDECK
The quarterdeck of the ship carried substantial firepower: eight 9-pounder long guns and 14 32-pounder carronades.

CONGREVE ROCKETS
On 8 October 1806, *Shannon* joined a squadron under Commodore Edward Owen, firing Congreve rockets (an early form of explosive shell, propelled by a charge from a launching frame) at French ships in Boulogne harbour.

BATTLE OFF BOSTON

On 1 June 1813, *Shannon* fought the American three-masted heavy frigate *Chesapeake* off Boston. The fight was brutal, with the ships only about 33m (110ft) apart. Very soon they ran together, still exchanging broadsides, then the British sailors boarded *Chesapeake* for hand-to-hand fighting. Less than 15 minutes after the start, *Chesapeake* surrendered. In that time 48 Americans and 30 Britons were killed, with 154 men wounded from both sides.

***SHANNON* OFF CALCUTTA, C. 1858**
By the 1800s frigates could be quite large ships, mounting 38 guns or even 44 in the case of the USS *Constitution* and its companion ships. A single gun deck was still a defining factor, and speedand manoeuvrability the most desired characteristics.

USS *Oneida* (1809)

TYPE • *Brig* **COUNTRY** • *United States*

SPECIFICATIONS

Dimensions:	Length: 26m (86ft 6in); Beam: 6.8m (22ft 6in); Draught: c. 4.9m (16ft)
Displacement:	247 tonnes (272 tons); normal load
Rig:	Two masts, square rig
Propulsion:	Sails
Speed:	c. 11 knots (20km/h; 13mph)
Armament:	18 × 24-pounder carronades
Crew:	117

Oneida **(named after a local Native American tribe) was laid down in 1808 at Oswego, on the southeast shore of Lake Ontario, and built by the shipwright Christian Bergh with the assistance of the New York shipbuilder Henry Eckford.**

Below: Oneida *was launched on 31 March 1809. Completion and commissioning were not until late 1810.*

Brig layout
The ship is classed as a brig, meaning that it had two square-rigged masts. *Oneida* had no galley, which limited its capacity for staying out on the lake.

Design
Oneida's original design provided for the fitting of 'sweeps' (long oars) as another source of motive power, but there is no record of their installation.

Pivot gun
A pivot gun in the bows was part of the ship's original design, to bolster the ship's forward defences. If installed, however, it was removed at an early date.

BACKGROUND
Lake Ontario, forming part of the border between the independent United States and the British colony of Canada, was an inevitable battleground in the event of a British–American war. As relations became critical towards 1812, both sides built warships specifically to gain control of the lake. *Oneida* was the first of a squadron that confronted the British more often than any other US ships in the War of 1812.

WAR OF 1812
This artwork depicts the American attack on Kingston Harbour, Lake Ontario, on 8 November 1812, with Commodore Isaac Chauncey aboard USS *Oneida* leading the fight alongside six armed schooners.

Montebello (1812)

TYPE • *Ship of the line* **COUNTRY** • *France*

SPECIFICATIONS

Dimensions:	Length: 63.2m (207ft 4in); Beam: 17.5m (57ft 5in); Draught: 8.14m (26ft 8in)
Displacement:	5095 tonnes (5616 tons) full load
Rig:	Three masts, full square rig
Propulsion:	Sails and Indret 104.4kW (140hp) auxiliary engine and screw propeller
Speed:	10 knots (19km/h; 12mph)
Armament:	32 × 36-pounders; 34 × 24-pounders; 52 × 8-pounder guns; 6 × 26-pounder carronades
Crew:	1079

Launched while Napoleon Bonaparte was still Emperor of France, *Montebello*'s long career of 45 years was not unusual for a French first-rate. It survived long enough, and its hull remained sufficiently robust, to have a steam engine installed in 1852.

Masts
Montebello carried 2265 sq m (24,380 sq ft) of sail. The mainmast, in three sections, measured 71m (233ft) from keel to top.

Below: Many historians consider all 16 ships of the Océan type, including Montebello, *as forming a single class, since they shared the same defining characteristics.*

Design
The ship had clean and simple lines, a minimal forecastle, a very modest raised poop deck and tumblehome sides. It gave stability as a gun platform and offered good sailing qualities.

Guns
Guns were carried on a spar deck connecting the forecastle and quarterdeck, a French feature not found on British first-rates.

Auxiliary engine
In 1852 an Indret 140hp auxiliary engine and screw propeller were fitted, but the addition did nothing to improve speed and made the ship more difficult to handle.

HISTORY

Montebello was laid down in 1810 and launched in December 1812 at Toulon. The same layout was used by *Souverain* (1819) and *Friedland* (1840), a tribute to the lasting impact of Jacques-Noël Sané's design. *Montebello* served until 1821 before a major refit, effectively a rebuild, was considered necessary. In 1860 it was used as a training ship for gunners at Toulon, and later it was converted to a floating barracks and used as such until it was finally broken up in 1889.

MONTEBELLO IN HARBOUR
Montebello was deployed to the Black Sea in the Crimean War of 1855, when France and Great Britain were in alliance against Russia, as flagship of Admiral Bruat. It participated in shore bombardments at Sevastopol and Fort Kinburn, and in the landings at Kerch, but its crew was ravaged by cholera, with 150 dying.

USS *Java* (1815)

TYPE • *Frigate* **COUNTRY** • *United States*

SPECIFICATIONS

Dimensions:	Length: 53.3m (175ft); Beam: 13.5m (44ft 6in); Draught: 4.17m (13ft 8in)
Displacement:	1535 tonnes (1692 tons) normal load
Rig:	Three masts, full square rig
Propulsion:	Sails, 2600 sq m (28,000 sq ft)
Speed:	c. 12 knots (22km/h; 14mph)
Armament:	33 × 32-pounder guns; 20 × 42-pounder carronades
Crew:	400

The USS Java ship was a large 50-gun spar-decked frigate laid down in Baltimore in 1814, launched the following year. Its active service was mostly in the Mediterranean Sea, demonstrating American naval strength to the Barbary pirate fleets of North Africa.

Below: From 25 January to 2 March 1817, Java *crossed from Gibraltar back to Boston, and was then laid up at Newport, Rhode Island. Recommissioned in 1827, it returned to the Mediterranean for two years.*

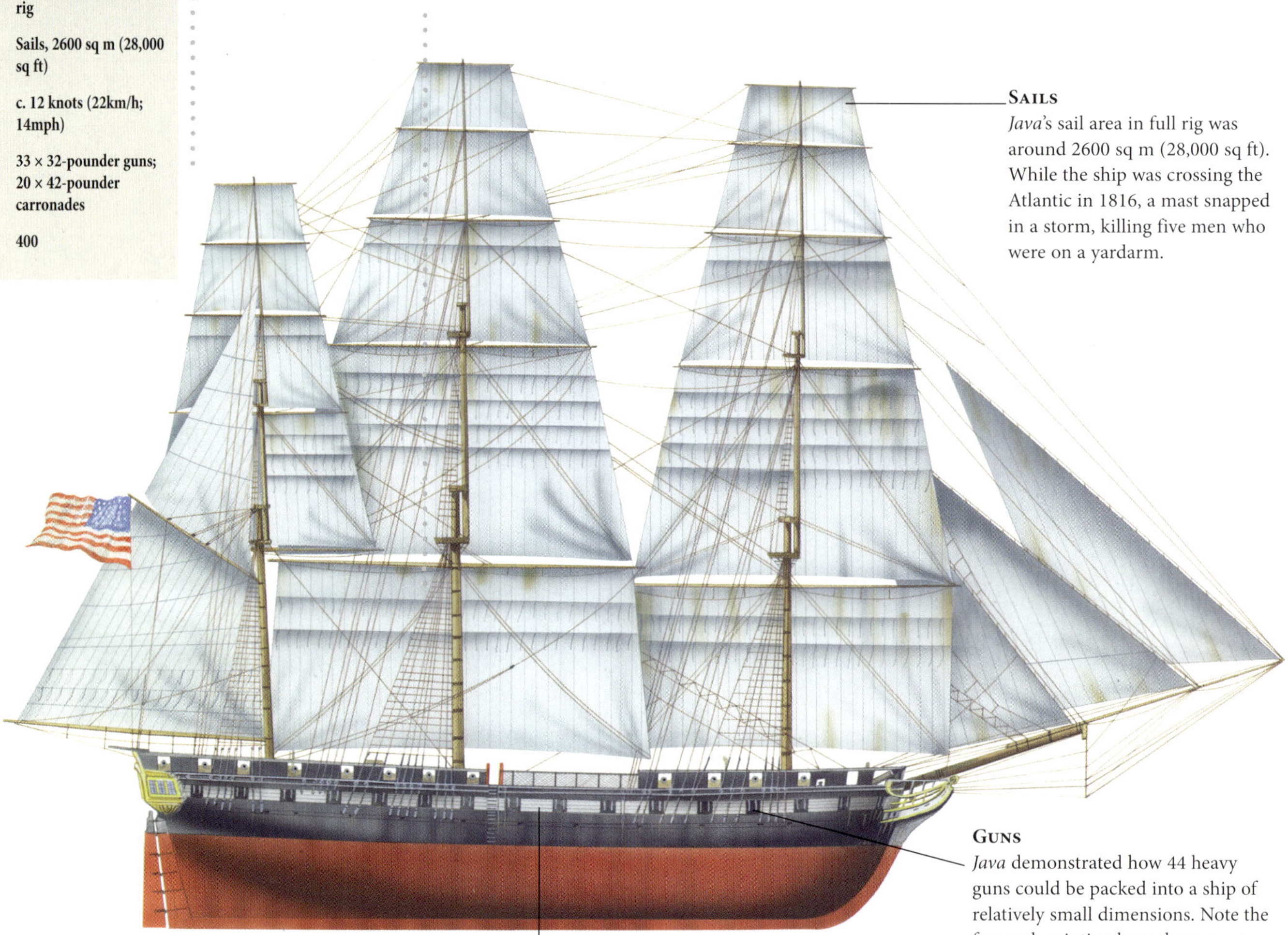

Sails
Java's sail area in full rig was around 2600 sq m (28,000 sq ft). While the ship was crossing the Atlantic in 1816, a mast snapped in a storm, killing five men who were on a yardarm.

Guns
Java demonstrated how 44 heavy guns could be packed into a ship of relatively small dimensions. Note the forward-pointing bow chaser gun.

Design
Java was built with a spar deck over the central well, but no guns were mounted on it, as experience had shown that placing armament there impaired the sailing performance of large frigates.

HERO COMMANDER
Ranked as a first-class frigate, *Java* was launched and ready for sea by 5 August 1815. In July 1814 its command was offered to Commodore Oliver Hazard Perry, a national hero following his naval exploits on Lake Erie, and he supervised the outfitting of the vessel. Java was finished too late for the War of 1812–15, but was deployed to the Mediterranean from 22 January 1816.

CAPTURE OF HMS *JAVA*
Java was a prize-name in the US Navy, commemorating the capture and destruction of HMS *Java* by USS *Constitution* off Brazil in 1812, as depicted in this artwork.

USRC *Morris* (1831)

TYPE • *Cutter* **COUNTRY** • *United States*

SPECIFICATIONS

DIMENSIONS:	Length: 22.4m (73ft 4in); Beam: 6.3m (20ft 6in)
DISPLACEMENT:	113 tonnes (125 tons) normal load
RIG:	Two masts, topsail schooner
PROPULSION:	Sails
SPEED:	12 knots (22km/h; 14mph)
ARMAMENT:	6 x 9-pounder guns (extendable to 14)
CREW:	24

The United States Revenue Cutter *Morris*, named after a signatory of the Declaration of Independence, was one of 13 ships of the *Morris–Taney* class to be launched, all named after Secretaries of the Treasury.

Below: Samuel Humphreys, Chief Constructor for the Navy between 1826 and 1846, designed the cutters for roles that might include fighting pirates and privateers, combatting smugglers and operating with naval forces.

RIGGING
Rigging was kept as simple as possible, but a topsail and mizzen topsail were still carried as speed was essential.

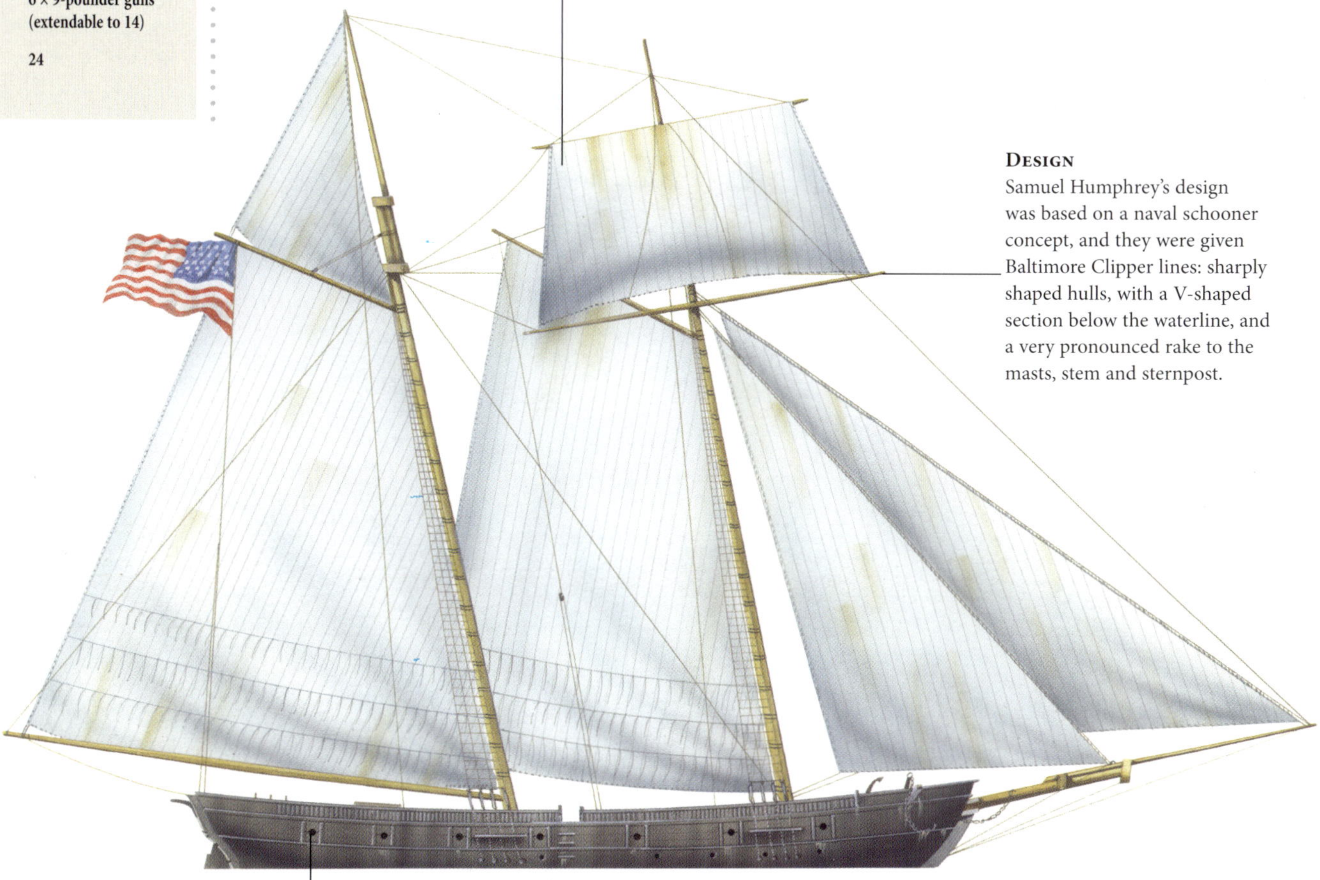

DESIGN
Samuel Humphrey's design was based on a naval schooner concept, and they were given Baltimore Clipper lines: sharply shaped hulls, with a V-shaped section below the waterline, and a very pronounced rake to the masts, stem and sternpost.

FIREPOWER
These cutters were the backbone of the Revenue Service for more than a decade. In normal service they mounted six 9-pounder guns but could carry up to 14 guns.

RATIONALE
With the tradition of small, handy, well-armed vessels firmly established from the time of the War of Independence, during the 19th century the sphere of action in the United States moved from patriotic defence to the suppression of illegal smuggling. In response, the US Revenue Service built fast cutters that could intercept or chase contraband runners. Some, including *Morris*, were provided with additional gunports to serve, if required, with the US Navy.

REVENUE CUTTER
The USRC *Morris* is depicted in this engraving from *Ballou's Pictorial*, 20 January 1855. The ships were known as the *Morris–Taney* class, although the first to be launched were *Dexter*, *Crawford*, *Hamilton* and *Gallatin*, all in 1830.

Age of Steam

Small coastal and river steamboats were already in service by the 1820s, but from the 1860s the first generations of ocean-going, steam-powered 'ironclads' entered commission, their wooden hulls sheathed in iron plating. Steam propulsion revolutionized naval warfare, not least by liberating fleets from the unpredicability of nature. Manoeuvres were no longer reliant on wind and speeds could be maintained as long as their furnaces were fed with coal. Alongside steam, other transformative technologies reshaped the battlefield: breech-loading, rifled and turret-mounted guns; high-explosive shells; all-steel hulls; onboard electrical power and early rangefinders and fire-control systems. Compared to the sail ships of yesteryear, these steam-powered warships were vastly faster and far more destructive instruments of war.

JAURÉGUIBERRY
The French pre-dreadnought battleship *Jauréguiberry* was commissioned in February 1897. Powered by two triple-expansion steam engines, it could reach a speed of 17.5 knots (32.4km/h; 20.1mph).

USS *Mississippi* (1841)

TYPE • *Paddle frigate* **COUNTRY** • *United States*

SPECIFICATIONS

Dimensions:	Length: 70m (229ft); Beam: 12m (40ft); Draught: 5.8m (19ft)
Displacement:	3271 tonnes (3549 tons) normal load
Propulsion:	Sails, three masts, barque rig, plus 7-stroke side-lever engine powering paddle wheels
Speed:	8 knots (15km/h; 9mph)
Range:	Dependent on stores
Armament:	10 x 254mm (10in) guns; 8 x 203mm (8in) guns (1841)
Complement:	257

***Mississippi*, a ship-rigged side-wheel steamer, was laid down at the Philadelphia Navy Yard in 1839, built under the supervision of Commodore Matthew C. Perry, and commissioned on 22 December 1841, with Captain W. D. Salter in command. It was launched in January 1842.**

Right: During the American Civil War (1861–65), Mississippi *ran the Confederate ram CSS* Manassas *ashore, wrecking it with two broadsides.*

U. S. STEAMER "MISSISSIPPI" ATTEMPTING TO RUN DOWN CONFEDERATE RAM "MANASSAS."

Masts
Despite its paddle wheels, *Mississippi* was a ship of elegant lines, enhanced by the slight rake to the masts.

Paddle wheel
The obvious advantage of the paddle wheel was motive power without the wind. The chief disadvantage was poor performance of the wheels in rough sea conditions.

Bulwarks
The forward bulwarks, located between the bowsprit socket and the foremast, could be dropped to allow the guns to bear.

FAR EAST SERVICE

Mississippi cruised the Mediterranean during 1849–51, then returned to the United States to prepare for service as flagship in Perry's momentous first voyage to Japan. The squadron entered Tokyo Bay on 8 July 1853. Perry proceeded, in one of the most significant naval/diplomatic missions ever recorded, to negotiate a trade treaty with the Japanese. After further cruising in the Far East, the squadron returned to Japan on 12 February 1854, and on 31 March the Treaty of Kanagawa was signed.

STEAM AND PADDLE
USS *Mississippi* is shown here under steam power in this 1903 drawing by R.G. Skerritt, with the sails stowed away. The ship was one of the transitional types bridging the ages of sail and steam.

USS *Powhatan* (1850)

TYPE • *Paddle frigate* COUNTRY • *United States*

SPECIFICATIONS

DIMENSIONS:	Length: 77.3m (253ft 8in); Beam: 14m (45ft); Draught: 5.64m (18ft 6in)
DISPLACEMENT:	3825 tonnes (4216 tons) normal load; 15,138 tonnes (16,687 tons) full load
PROPULSION:	Sails, three masts, brigantine rig, plus side paddles, steam engine, 1119kW (1500hp)
SPEED:	11 knots (20.3km/h; 12.6mph)
ARMAMENT:	1 × 280mm (11in); 10 × 230mm (9in) SB guns; 5 × 12-pounder guns
COMPLEMENT:	289

Painted black with a broad white stripe, *Powhatan* was the flagship of Commodore Matthew Perry's squadron of 'black ships' on his second Japanese mission in 1854. It was the last and one of the largest of the US Navy's paddle frigates, rigged as a barque and later as a brigantine.

Below: Powhatan*'s final service was in commerce protection off Cuba in early 1886. On 2 June of that year the ship was decommissioned; it was sold on 30 July for breaking up at Meriden, Connecticut.*

RIG
The ship had three masts, configured in a brigantine rig. The sailing rig did not extend to topsails.

SIDE WHEELS
Paddle wheels, because of their vulnerability to hostile fire, were not popular with naval captains. Despite this, the US Navy built around 47 side-wheel gunboats between 1862 and 1865.

ARMAMENT
Powhatan's armament, which consisted of 16 guns of various calibres, was concentrated on the upper deck of the ship.

EARLY SERVICE
Built at the Gosport Navy Yard, Virginia, launched on 14 February 1850, and commissioned at Norfolk Navy Yard on 2 September 1852, *Powhatan* cost $785,000. It served first as flagship of the Home Squadron under Commodore John T. Newton, and carried the new US Minister to Veracruz, Mexico, in October 1852; then from 15 June 1863, it was assigned to the Far East Squadron under Commodore Perry.

EARLY PHOTOGRAPH
During the Civil War, *Powhatan* helped relieve Fort Pickens, Florida, in April 1861, and later took part in the blockade of Mobile and the Mississippi Delta. From October 1862 to August 1863 it served in the blockade of Charleston, and from November 1863 to September 1864 acted as Rear Admiral James L. Lardner's flagship in the West Indies.

USS *Merrimack* (1855)

TYPE • *Steam frigate* **COUNTRY** • *United States*

SPECIFICATIONS

Dimensions:	Length: 84m (275ft); Beam: 11.7m (38.5ft); Draught: 7.3m (24ft)
Displacement:	2903 tonnes (3200 tons) normal load
Propulsion:	Sail and steam; single screw; horizontal back-acting steam engine
Speed:	12 knots (22.2km/h; 13.8mph)
Range:	N/A
Armament:	14 × 203mm (8in) guns; 2 × 254mm (10in) guns; 24 × 229mm (9in) guns
Armour:	None (wooden-hulled steam frigate)
Complement:	c. 750

The Union screw frigate *Merrimack* is best known for what it later became: during the Civil War, after being burned at Norfolk to prevent capture, it was raised and reconstructed by the Confederates and rebuilt as a casemate ironclad.

Right: The success of the Merrimack/Virginia, *or at least the promise of it, blinded the Confederates to any type of warship other than the casemate ironclad, and many vessels of this type were placed under construction.*

Sail power
With combined steam and sail power, and under favourable winds, the *Merrimack* could make 12 knots (22.2km/h; 13.8mph).

HISTORY
The ship began its life as the US 40-gun screw frigate *Merrimack*, launched at Boston on 15 June 1855. The vessel was burnt to prevent its capture by Confederate forces at Norfolk in 1861, but it was refloated by Confederate Navy and rebuilt at Gosport Navy Yard as an ironclad ram named *Virginia*, completed in February 1862. *Virginia* famously fought *Monitor* at the Battle of Hampton Roads on 9 March 1862, history's first engagement between ironclad warships.

Machinery
The Confederates discovered that *Merrimack*'s single-shaft returning rod connecting engine was still intact and no plans were made to fit a new powerplant. However, Merrimack's engine had not been efficient to begin with, and struggled to cope with the *Virginia*, made much heavier by the addition of tons of armour.

BEFORE THE WAR
A painting of the USS *Merrimack*, showing the ship before it was converted into the Confederate ironclad CSS *Virginia*. The image is a pre-Civil War illustration.

USS *Brooklyn* (1858)

TYPE • *Sloop of war* **COUNTRY** • *United States*

SPECIFICATIONS

DIMENSIONS:	Length: 71m (233ft); Beam: 13m (43ft); Draught: 4.95m (16ft 3in)
DISPLACEMENT:	2572 tonnes (2835 tons) normal load
PROPULSION:	Sails, three masts, square rig; one steam engine driving a single screw
SPEED:	12 knots (22km/h; 14mph)
RANGE:	N/A
ARMAMENT:	1 × 254mm (10in); 20 × 228mm (9in) guns
COMPLEMENT:	335

During a 30-year career, this wooden-hulled screw sloop first served in the Caribbean, saw active service in the American Civil War – most notably in the fierce fighting around Mobile – and afterwards undertook cruising duties in the West Indies, Europe, the East African coast, Patagonia and the Far East.

Right: One of five screw sloops authorized by the US Congress on 3 March 1857, Brooklyn *was laid down at the New York yard of Jacob A. Westervelt and Son that year, launched in 1858, and commissioned on 26 January 1859.*

DAMAGE
During the attack on Forts St. Philip and Jackson on 24 April 1862, the Confederate ram *Manassas* rammed *Brooklyn*, punching a 7.3m (24ft) hole in the starboard side.

FUNNEL
The funnel from the ship's steam engine could be retracted when the ship was under full sail, reducing drag and minimizing the risk of embers catching the sails and rigging.

GUNS
Brooklyn's guns, most of which were 228mm (9in) in calibre, were all of the smoothbore, muzzle-loading type.

CIVIL WAR
Brooklyn served with the West Gulf Blockading Squadron from 1861 to 1864, taking part in the relief of Fort Pickens, Florida, and joining other ships in pursuit of the elusive CSS *Fort Sumter.* Between July and December 1861, it underwent repairs and a refit at Philadelphia Navy Yard and on 19 December it sailed to rejoin Farragut's squadron. It took part in the heavy fighting on the lower Mississippi, assisting in the capture of New Orleans on 25 April 1862 and engaging in battles around Vicksburg.

***BROOKLYN* IN HARBOUR**
Brooklyn's first captain was David C. Farragut, the future admiral, and in February 1859 its first cruise was to Haiti and Panama. From April 1859 to July 1860 it assisted the US mission to President Benito Juarez's government in Mexico.

HMS *Victoria* (1859)

TYPE • *Ship of the line* **COUNTRY** • *United Kingdom*

SPECIFICATIONS

DIMENSIONS:	Length: 79.2m (260ft); Beam: 18.3m (60ft 1in); Draught: 6.5m (21ft 2in)
DISPLACEMENT:	7070 tonnes (7793 tons) normal load
PROPULSION:	Sails, three masts, full square rig; Maudslay 3283kW (4403hp) steam engine, 8203kW (11,000shp)
SPEED:	11.8 knots (22km/h; 13.6mph)
RANGE:	N/A
ARMAMENT:	62 × 200mm (8in); 32 × 30-pounders; 36 × 32-pounders; 1 × 68-pounder gun
COMPLEMENT:	1000

The largest wooden-hulled warship ever to enter service, and the last of the three-deckers, *Victoria* had an impressive appearance despite its outdated design. *Victoria* was laid down at Portsmouth on 1 April 1856 and launched on 12 November 1859.

Below: The last time the great wooden-hulled ships went on parade was at the Spithead Naval Review of July 1867, before Queen Victoria and the Ottoman Emperor. In August 1867, Victoria *was paid off.*

GUNS
The ship was armed with an impressive 127 guns, making it capable of destroying any ship afloat at the time. Fire could be directed on more than one target at a time.

ENGINE
The early engines were bulky. The weight of machinery, low down in the hull, enabled builders to greatly reduce the amount of ballast carried.

HULL
The hull was heavily strapped on the inner side with diagonal iron riders, 127mm (5in) wide and 25mm (1in) thick, to hold the planking together against the vibrations from machinery and screw.

DESIGN RATIONALE

Victoria was planned from the start as a sail-assisted steamship. In the British Admiralty's typical knee-jerk reaction to French developments, it was intended to rival the 130-gun French three-decker *Bretagne*, which had been laid down as a sailing ship but was converted to steam propulsion while being built, and launched in February 1855.

QUEEN OF THE SEAS

Victoria has been described as 'the acme of the three-decker,' at twice the tonnage of Nelson's flagship HMS *Victory* and with a far higher destructive capacity.

Gloire (1859)

TYPE • *Ship of the line* **COUNTRY** • *France*

SPECIFICATIONS

Dimensions:	Length: 77.25m (254ft 5in); Beam: 16.16m (53ft); Draught: 8.48m (27ft 10in)
Displacement:	5766 tonnes (6356 tons); normal load
Propulsion:	Sails, three-masted rig) – light barquentine rig, later full-ship rig; 1 × horizontal return connecting-rod steam engine driving a single screw, 1864kW (2500ihp)
Speed:	13.5 knots (25km/h; 15.5 mph)
Range:	4000km (2500 miles) at 8 knots (15 km/h; 9.2 mph)
Armament:	36 × 164.7mm (6.5in) rifled muzzle-loading guns (as built)
Complement:	550

Classed as a *frégate cuirassée* (armoured frigate), *Gloire* was designed by the celebrated naval architect Dupuy de Lôme, and as the first oceangoing 'ironclad', represented a new milestone in warship development. Despite its large machinery, *Gloire* carried a full set of sails.

Below: As with all technical innovations, the first ironclad initiated an era in which gunmakers and ship designers (and metallurgists) competed with each other to outdo the latest advance made by the other side.

Sails
Although provided with one of the most powerful engines yet fitted to a ship, *Gloire* carried a full suit of sails, being changed from the original barquentine rig to full square rig.

Armour
The armour, in an upper belt of 10cm (4in) and a lower belt of 12cm (4.7in), weighed 823 tonnes (907 tons). To offset this weight, a second gun deck, which would normally be standard on a ship of this size, was omitted.

SINGLE GUN DECK
Gloire ('Glory'), eighth French warship of the name, owed its 'frigate' designation to its single gun deck. In armament, protection, and speed it outclassed any other warship at the time of its commissioning. Laid down at Toulon in April 1858, launched on 24 November 1859 and commissioned in August 1860, its building was rapid for a large ship but the general design was on conventional lines.

Guns
Its new MLR guns of 1866 could pierce 12cm (4.7in) iron plates at 805m (2640ft) but were replaced by BL guns in 1886.

CAREER
Gloire did not have an eventful career, stationed mostly at Cherbourg as if to remind the Royal Navy, just across the Channel, that the *Marine Nationale* was a force to contend with. The ship was decommissioned in 1879 and scrapped in 1883.

HMS *Warrior* (1860)

TYPE • *Ironclad* **COUNTRY** • *United Kingdom*

SPECIFICATIONS

Dimensions:	Length: 128m (420ft); Beam: 17.78m (58ft 4in); Draught: 7.9m (26ft)
Displacement:	9283 tonnes (10,233 tons) normal load
Propulsion:	Sails, three masts, full square rig including stunsails; Penn horizontal trunk steam engine, 3874kW (5267ihp), single screw
Speed:	16.3 knots (30km/h; 19mph)
Armour:	114mm (4.5in) steel plating
Armament:	4 × 203mm (8in); 28 × 178mm (7in) ML guns
Complement:	707

The sail-steam ironclad *Warrior* was planned as a direct response to France's *Gloire*. Its iron hull marked a clear break with the wooden-hulled tradition. However, the design showed compromises between the new and the old.

Left: Warrior's length was notable, as greater length was associated with higher speed, and care had been taken in designing the underwater lines.

Powerplant
The Penn horizontal trunk engine was the most powerful yet installed on a warship. Under combined power on 15 November 1861, the ship made 16.3 knots (30km/h; 19mph).

Figurehead
Warrior was one of the last three Royal Navy ships to be given a carved figurehead, ending centuries of naval tradition and craftsmanship.

Armour
The 114mm (4.5in) steel plating was impenetrable by any naval gun of the time. By 1863, guns would be introduced to pierce such armour.

SERVICE HISTORY

From 1861 to 1864 *Warrior* served with the Channel Fleet, then was out of commission while re-arming (1864–67) before rejoining the Channel Fleet. Another refit in 1872–75 provided it with the poop deck and steam capstan. It was a coastguard ship at Portland, then in 1881 was reclassified as an armoured cruiser, and served until 1884 as a training ship for reservists on the River Clyde.

PRESERVED SHIP

In 1904 *Warrior* was adapted for service with the HMS *Vernon* torpedo school, and the cut-down hull was finally transferred to Pembroke to provide a pier for an oil pipeline. Restored in the 1980s, *Warrior* is shown here preserved at Portsmouth naval base, England.

CSS *Manassas* (1861)

TYPE • *Ironclad* **COUNTRY** • *Confederate States of America*

SPECIFICATIONS

Dimensions:	Length: 44m (143ft); Beam: 10m (33ft); Draught: 5.2m (17ft)
Displacement:	351 tonnes (387 tons) normal load
Propulsion:	Steam engine driving a single screw propeller
Speed:	c. 8 knots (15km/h; 9mph) maximum
Range:	N/A
Armour:	38mm (1.5in) of iron plate
Armament:	1 × 64-pounder gun (later reportedly replaced by a 32-pounder)
Complement:	33

The *Manassas* began its career as a privateer. Businessman John Stevenson raised funds by subscription to buy the *Enoch Train*, a New England icebreaker, with the intention of using it for this purpose. It was originally built in 1855 as a twin-screw towboat.

Left: At the time, the Manassas *was a strange, futuristic-looking vessel. Admiral David Porter called it 'the boat with the iron horn.'*

Ram
The *Manassas* carried a crew of 33, but only one gun, a stationary 32-pounder carronade. In many ways the gun was a nuisance because the main weapon of the *Manassas* was an underwater iron ram designed to tear holes in the sides of the Federal wooden fleet.

Iron deck
The oak deck was sheathed with 38mm (1.5in) of iron plate, which when allied to the slope of the exposed deck meant the ship could deflect away most incoming shot.

Bow
The bow was filled in with timber to give it the requisite weight and density for ramming impact. Including the timber, the ram was 6m (20ft) long.

RENAMING
Boatbuilders in New Orleans expanded the *Enoch Train*'s dimensions significantly. They rebuilt the ship with massive 432mm (17in) beams to make a solid bow of 6m (20ft). Over this lumber, a covering of thin iron plates rendered the exposed part of the vessel shot-proof. The result was a strange-looking turtle – or egg-shaped vessel. Locals informally called it 'the Turtle' until it was officially renamed the *Manassas* in honour of the Confederate victory there.

IN ACTION
An artwork entitled *The Ram* depicts a battered *Manassas* after an action in April 1862. On 12 October the year before, around New Orleans, the *Manassas* drove a 127mm (5in) hole into the hull of the USS *Richmond*, marking the first wartime use of an ironclad ram.

USS *Housatonic* (1861)

TYPE • *Sloop of war* **COUNTRY** • *United States*

SPECIFICATIONS

DIMENSIONS:	Length: 62.48m (205ft); Beam: 11.58m (38ft); Draught: 5.02m (16ft 6in)
DISPLACEMENT:	1754 tonnes (1933 tons) normal load
PROPULSION:	Sails, three-masted barque rig; screw propeller, 2 Martin boilers, 522kW (700ihp) steam engine
SPEED:	10 knots (18.5km/h; 11.5mph)
ARMAMENT:	1 × 163mm (6.4in) Parrott rifle; 3 × 107mm (4.2in) Parrott rifles; 1 × 280mm (11in) Dahlgren smoothbore; 2 × 160mm (6.3in) 32-pdrs; 2 × 148mm (5.82in) 24-pdr howitzers; 1 × 117mm (4.62in) 12-pdr howitzer; 1 × 76mm (3in) 12-pdr rifle
COMPLEMENT:	214

Four wooden-hulled screw sloops of the *Ossipee* class were ordered for the Union Navy in 1861, and *Housatonic* was laid down in that year. Its career was short and the ship holds the grim distinction of being the first ever sunk by a submarine, marking a pivotal point in naval warfare.

Right: This print shows the loss of the Housatonic *off the beach inlet, South Carolina, on 17 February 1864. After the Civil War, several operations were conducted to submerge the wreck further, reducing it as a navigational hazard.*

DESIGNATION
The *Ossipee* class were the first US cruising warships designated as screw sloops, appearing just before the slightly larger *Sacramento* class of 1862–63.

GUNS
Housatonic was armed with Parrott rifles. While these guns were accurate with solid shot, they were notorious for being unsafe.

HULL
The ship was wooden-hulled. Its eventual fate sped up the provision of armoured waterline protection for warships.

SUB ATTACK
In early 1864, *Housatonic* was singled out for attack by the pioneer submarine *H.L. Hunley*, armed with a spar torpedo. On the evening of 17 February 1864, the submarine approached and the torpedo was lodged against the starboard hull and successfully detached from its spar. The explosion blew a hole beneath the waterline and *Housatonic* rapidly filled with water and sank.

BLOCKADE SHIP
Built at the Boston Navy Yard, *Housatonic* was launched on 20 November 1861 and commissioned on 29 August 1862. In September it was deployed to the Charleston Blockading Squadron.

USS *Commodore Perry* (1861)

TYPE • *Side-wheel ferry* **COUNTRY** • *United States*

SPECIFICATIONS

Dimensions:	Length: 44m (143ft); Beam: 10m (33ft); Draught: c. 3m (10ft)
Displacement:	464 tonnes (512 tons) normal load
Propulsion:	Steam engine driving side-wheel paddle propulsion (side-wheel steamer)
Speed:	7 knots (13km/h; 8mph)
Range:	N/A
Armament:	2 × 230mm (9in) guns; 2 × 32-pdr smoothbores; 1 × 12-pdr howitzer
Complement:	125

The *Commodore Perry* was built in 1859 in Williamsburg, New York, and purchased by the Navy on 2 October 1861. It was commissioned later in the month with Acting Master F. J. Thomas in command of the vessel.

Below: Commodore Perry *joined the North Atlantic Blockading Squadron in time to take part in the 7–8 February 1862 attack on Roanoke Island in North Carolina.*

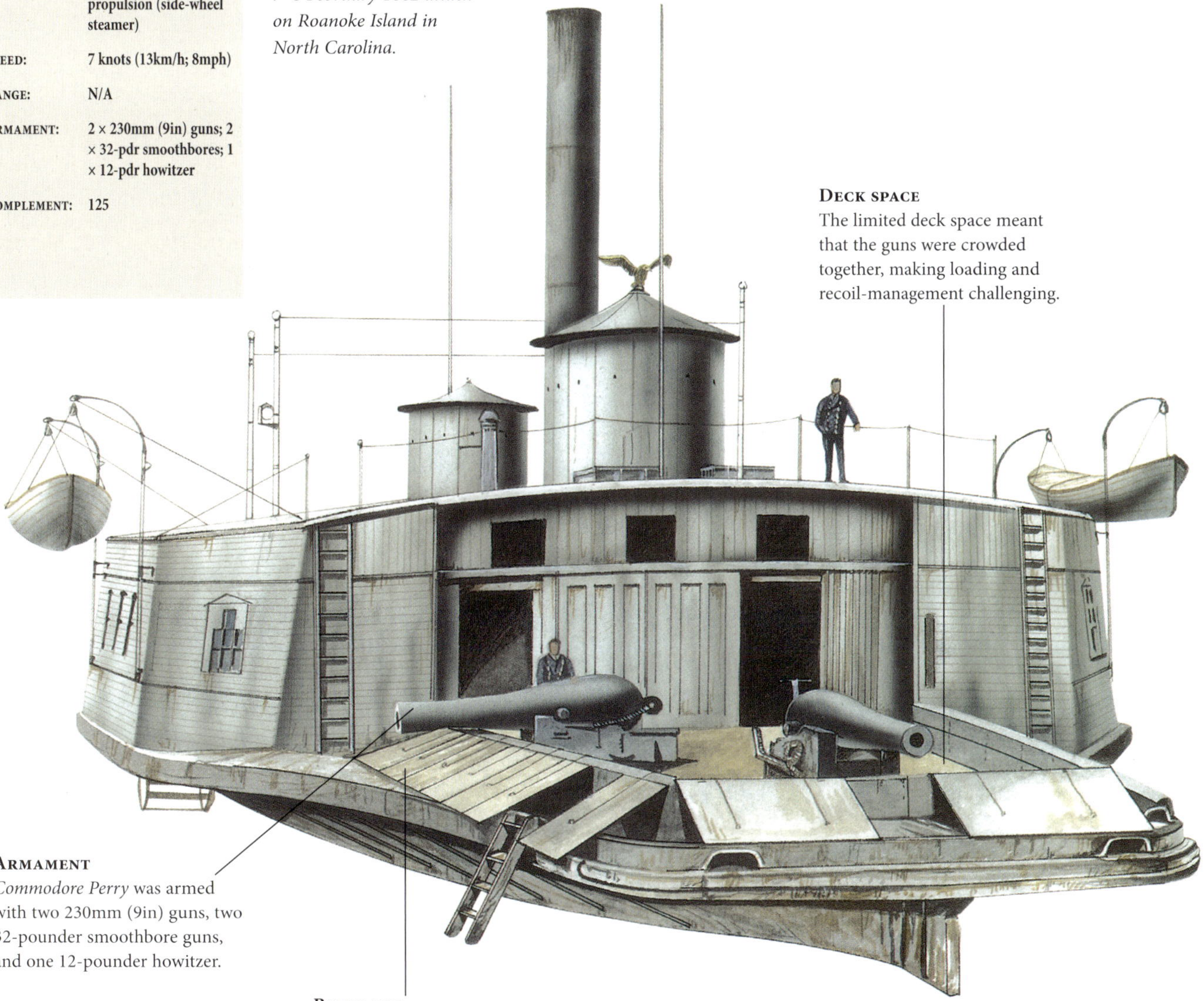

Deck space
The limited deck space meant that the guns were crowded together, making loading and recoil-management challenging.

Armament
Commodore Perry was armed with two 230mm (9in) guns, two 32-pounder smoothbore guns, and one 12-pounder howitzer.

Bulwarks
The hinged bulwarks, erected when the ship was not in action, are shown here in their lowered position, allowing the heavy guns to be run forward.

RUN AGROUND

A converted ferry boat, the *Commodore Perry* was ideal for steaming through shallow waterways. One such mission occurred on 3 October 1862 when it took part in an Army–Navy expedition against Franklin, Virginia. The crew lost control of the ship at a narrow, sharp bend and ran aground under fire. Quick action by Lieutenant William Cushing saved the day.

CIVIL WAR STEAMER
Commodore Perry saw extensive service in the American Civil War. On 10 December 1864, it participated in the attack on Plymouth, North Carolina, and on 30 January 1863 was part of a joint expedition against Hertford.

CSS *Pioneer* (1862)

TYPE • *Submersible* **COUNTRY** • *United States*

SPECIFICATIONS

Dimensions:	Length: 9m (30ft); Beam: 3m (10ft); Draught: unknown (likely c. 1.2–1.5m/4–4ft 11in, based on hull form)
Displacement:	Unknown
Propulsion:	Manually operated crank shaft, screw propeller
Speed:	Unknown
Range:	N/A
Armament:	1 × clockwork torpedo
Complement:	2–3

In March 1862, Robert Barrow, James McClintock and Baxter Watson applied for a letter of marque at New Orleans for a submarine named *Pioneer*. The craft was made from riveted iron plates that had been cut from old boilers and was propelled by a crank shaft.

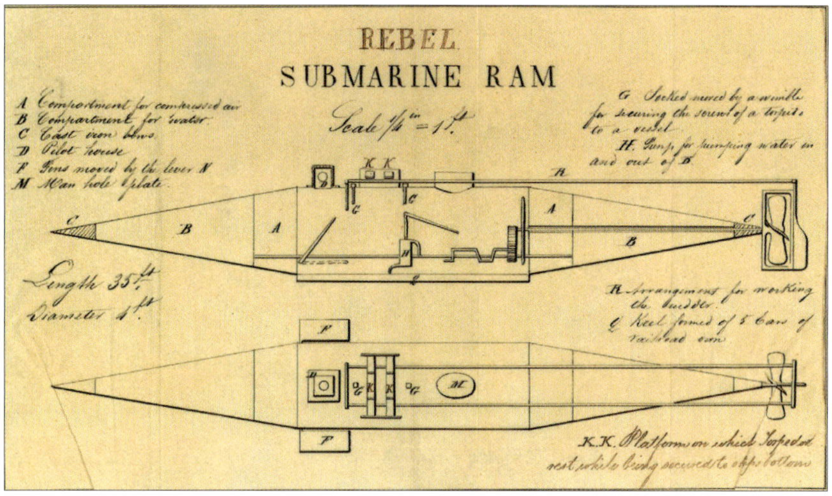

Left: A 19th-century prototype drawing illustrates how basic the craft was internally.

HULL
The *Pioneer* was constructed from quarter-inch (6mm) riveted iron plates that had been cut from old boilers.

PROPULSION
A single crew member propelled the submarine by turning the manual crank of the screw. The other crew member acted as the pilot.

TORPEDO
The vessel was armed with a clockwork torpedo carried on top of the submarine. It was intended to attach the torpedo to the hull of a ship, the clockwork-timed detonator giving the crew time to get away.

DESIGN

In a post-Civil War interview, McClintock described the *Pioneer* as being 9m (30ft) long with a 3m (10ft) cylindrical midship section. Tapered conical ends extended from the cylindrical section, resulting in a cigar shape. It had a conning tower with manholes in the top and small windows of circular glass in the sides. It was armed by a clockwork torpedo, carried on top of the submarine and intended to be screwed into the bottom of the enemy ship by gimlet-pointed screws of tempered steel.

ON DISPLAY
CSS *Pioneer* (1862) at State Home for Confederate Soldiers, New Orleans, 1926. The craft was never tested in combat because it was scuttled to prevent it falling into Federal hands when Admiral David Farragut captured New Orleans in May 1862.

CSS *Virginia* (1862)

TYPE • *Ironclad* **COUNTRY** • *Confederate States of America*

SPECIFICATIONS

Dimensions:	Length: 83.8m (275ft); Beam: 15.6m (51ft 2in); Draught: 6.4m (21ft)
Displacement:	3629 tonnes (4000 US tons)
Propulsion:	Two horizontal back-acting steam engines, one shaft, four boilers
Speed:	10 knots (19km/h; 11.5mph)
Armament:	2 × 178mm (7in) Brooke rifles; 2 × 160mm (6.4in) Brooke rifles; 6 × 228mm (9in) Dahlgren smoothbores; 2 × 12-pounder howitzers
Armour:	Belt 25–76mm (1–3in); deck 25mm (1in); casemate 101mm (4in)
Complement:	150

The CSS *Virginia* began its life as the *Merrimack*, a 317-tonne (350-ton), 40-gun steam frigate that the Federals burned and scuttled when they abandoned Gosport Navy Yard on 20 April 1861.

Below: Workers took the wreck of the Virginia *and cut the hull down to the berth deck and built a casemate with slanting sides and ports for 10 guns.*

Gun ports
There were 14 elliptical gun ports cut into the ship's structure, housing 12 guns in a variety of calibres.

Casemate
The casemate walls contained 609mm (24in) of oak and pine timber encased in 101mm (4in) of armour plating, making them highly resistant to enemy fire.

Rudder
Virginia was a strong vessel, but it was also an ungainly one. It steered badly, needing about 40 minutes to turn 180 degrees.

REBIRTH
Confederate engineers raised *Merrimack*'s hulk and found it to be in good shape, except for the upper works, which had been destroyed by the fire. Confederate Secretary of the Navy Stephen Mallory proposed equipping the Merrimack with armour and using it to break the ever-tightening Federal blockade. Naval Constructor John Luke Porter and Lieutenant John Mercer Brooke then designed an ironclad ram that completely changed naval warfare during the Civil War.

UNDER DEVELOPMENT
Virginia under development at Gosport Navy Yard. On 8 March 1862, the *Virginia*, sailed into Hampton Roads, where Commodore Franklin Buchanan saw five Federal blockade ships lying in anchor. First the *Virginia* went after the *Congress* and the *Cumberland*, engaging them both in battle. The ship survived the day and sank two enemy ships.

CSS *Alabama* (1862)

TYPE • *Sloop of war* **COUNTRY** • *Confederate States of America*

SPECIFICATIONS

Dimensions:	Length: 67.1m (220ft); Beam: 9.7m (31ft 9in); Draught: 5.38m (17ft 8in)
Displacement:	1067 tonnes (1177 tons) normal load
Propulsion:	Sails, three-masted barque rig; screw propeller driven by a direct-acting 447kW (600ihp) engine
Speed:	13 knots (24km/h; 15mph)
Armament:	6 × 32-pounders; 1 × 110-pounder gun; 1 × 68-pounder gun
Complement:	148

***Alabama* was a highly successful commerce raider, with a total of 55 ships captured and destroyed, and a further 10 boarded and seized, with their value estimated at around $6 million. This was accomplished during a career of only two years before it was sunk in a two-ship duel.**

Right: By late 1863, the ship was badly in need of a refit, which Captain Semmes hoped to have done at Cherbourg, but French policy towards the Confederacy meant he could only take on coal and provisions.

Rigging
The rig did not extend beyond the topsails. *Alabama* did much of its voyaging under sail power alone.

Insurance
The success of *Alabama* and other ships such as CSS *Florida* and *Shenandoah* caused insurance rates for US-registered vessels to rise by 900 per cent.

PREDATOR
Built at Birkenhead, England, ostensibly as the merchant vessel *Enrica*, *Alabama* left port on 29 July 1862 and in a rendezvous at the Azores was armed and equipped as a warship. It was commissioned by Captain Raphael Semmes on 24 August as *Alabama*. Over a wide sweep of the Atlantic it sank 27 ships before the end of the year. In the course of 1863 it sailed into the Indian Ocean, reaching Singapore before returning to European waters.

Hull
The *Alabama* was described as an 'auxiliary bark'. Its sleek hull was designed along the lines of a sloop, and it could achieve speeds of 13 knots (24km/h; 15mph).

COMMANDING OFFICER
Captain Raphael Semmes, *Alabama*'s commanding officer, standing by his ship's 110-pounder rifled gun during a visit to Cape Town in August 1863. The ship was later sunk by USS *Kearsage* on 19 June 1864, off Cherbourg.

Magenta (1862)

TYPE • *Ironclad* COUNTRY • *France*

SPECIFICATIONS

DIMENSIONS:	Length: 91.9m (282ft 1in); Beam: 17.3m (56ft 8in); Draught: 8.44m (27ft 8in)
DISPLACEMENT:	6822 tonnes (7414 tons) normal load
PROPULSION:	Sails, three masts (barquentine then barque rig); screw propeller driven by a horizontal return-rod steam engine, 2604kW (3540ihp)
SPEED:	12 knots (22.2km/h; 13.8mph)
ARMAMENT:	10 × 240mm (9.45in); 4 × 190mm (7.48in); 50 × 30-pounder MLR guns
ARMOUR:	Belt 119mm (4.7in); battery: 109–120mm (4.3–4.7 in)
COMPLEMENT:	681

***Magenta* and its companion ship *Solférino* were distinguished as the only broadside ironclads to have two decks of guns, and the first to have spur rams. Despite their heavy armament, as wooden-hulled ships – only partially armoured – they were vulnerable to explosive shells.**

Left: Magenta *(front) and* Solférino *(third back) stand either side of the 90-gun ship of the line* Napoléon III *in Brest, France.*

RIGGING
Originally rigged as a barquentine, *Magenta* carried 1709.4 sq m (18,400 sq ft) of sail, but in 1864 the arrangement was changed to a barque rig and the sail area increased to 1951 sq m (21,000 sq ft).

ARMOUR
A complete belt of wrought-iron armour, 119mm (4.7in) thick, protected the waterline, but the battery deck's armour did not extend to the ends of the ship.

GUNS
The double deck of guns was intended to rival the new British iron ships HMS *Warrior* and *Black Prince*, in a ship of much shorter length.

DEVELOPMENT
Built to a design drawn up by Stanislas Dupuy de Lôme, *Magenta* (named after the French victory over Austria in June 1859) was laid down at Brest Naval Dockyard on 22 June 1859, launched on 22 June 1861, and completed on 2 January 1862. Numerous changes to the ship's armament were made in the course of *Magenta*'s career, particularly in relation to its armament configuration.

ILLUMINATED SHIP
The International Naval Festival at Cherbourg on 15 August 1865 was a spectacular occasion. Here *Magneta* is shown richly illuminated, as fireworks explode in the background.

USS *New Ironsides* (1862)

TYPE • *Ironclad* **COUNTRY** • *United States*

SPECIFICATIONS

Dimensions:	Length: 70.7m (232ft); Beam: 17.53m (57ft 6in); Draught: 4.57m (15ft 8in)
Displacement:	4186 tonnes (4120 tons) full load
Propulsion:	Sails, barque rig, without royals; screw propeller, 4 Martin boilers, 522kW (700ihp)
Speed:	7 knots (12.9km/h; 8mph) under steam-and-sail
Armament:	2 × 203mm (8in) Parrott rifles; 2 × 130mm (5.1in) Dahlgren rifled guns; 1 × 86mm (3.4in) rifled ML gun
Armour:	Belt 114–76mm (4.5–3in); battery sides 114mm (4.5in); deck 25mm (1in) wrought-iron
Complement:	449

Named in honour of USS *Constitution*, this ship was literally iron-sided, with a belt of armour and an armoured deck, although the hull was of wood. Its career was nowhere near as long as its namesake's – *New Ironsides* was laid up in 1865 and destroyed by fire on 16 December 1866.

Right: Laid up at League Island Navy Yard, Philadelphia, after the Civil War, New Ironsides was destroyed in an accidental fire, started by a night watchman's stove, on 16 December 1866.

Handling
Flat-bottomed and of relatively shallow draughtt (although this was not unusual with American warships, which often had to operate in coastal waters), the ship had a tendency to roll, and the masts were reduced in height.

Hull
The side armour withstood a Confederate 70-pound spar torpedo. Some 13.7m (45ft) of the wooden hull was sprung.

TOUGH SHIP
In wartime haste in 1862, *New Ironsides* was assigned to the South Atlantic Blockading Squadron as flagship of Rear Admiral Samuel Du Pont, and stationed off Charleston, South Carolina. In April the shore forts were bombarded and returned fire, with the ship taking 55 noted hits without suffering serious damage. In the bombardment of Morris Island, 214 hits were recorded, again doing little damage.

Tower
An armoured low conning tower was fitted between the main- and mizzen-masts, later replaced by a cylindrical 'pillbox' tower.

SAIL AND STEAM
New Ironsides was contracted with Merrick & Sons of Philadelphia, who built the engine, the hull construction being subcontracted to William Cramp & Son's yard in Philadelphia. The ship was launched on 10 May 1862, and commissioned on 21 August.

USS *Intelligent Whale* (1863)

TYPE • *Submersible* **COUNTRY** • *United States*

SPECIFICATIONS

DIMENSIONS:	Length: 8.74m (28ft 8in); Beam: 2.1m (7ft); Draught / Depth: 2.7m (9ft)
DISPLACEMENT:	1.8 tonnes (2 tons) surfaced
PROPULSION:	Manually hand-cranked screw propeller
SPEED:	4 knots (7.4km/h; 4.6mph)
RANGE:	N/A
ARMAMENT:	Diver-deployed mine
COMPLEMENT:	6–13

Civil War submarining is most closely associated with the Confederacy, which relied on an asymmetric threat to counter the Federal blockade. The Union, however, also explored the development of a submarine capability, making two attempts: *Alligator* and *Intelligent Whale*.

Left: Intelligent Whale *was built at Newark, New Jersey by the American Submarine Company according to a design developed by Scovel Merriam. This photo was taken at the New York Navy Yard on 27 July 1915.*

DIVING
The vessel could be submerged by filling compartments with water, then expelling the water from the tanks by pumps and compressed air to surface again.

PERFORMANCE
Intelligent Whale contained enough air to stay submerged for up to 10 hours and could make a speed of about 4 knots (7.4km/h; 4.6mph.

CREW CAPACITY
The submarine's bulbous shape allowed it to accommodate a crew of 13, but only six personnel were required to operate her.

DEMONSTRATION
Litigation and difficulty in finding crews delayed progress. *Intelligent Whale* was tested, however, in April 1866 by Brigadier General Thomas William Sweeny, who submerged the submarine, exited in a diving suit, planted an explosive charge on a target scow, and then reentered the vessel. After *Intelligent Whale* had moved to a safe distance, Sweeny exploded the charge by a lanyard and destroyed the scow.

CURSED SHIP
The *Intelligent Whale* gained the reputation of being the 'Disastrous Jonah', sinking three times and drowning its crews. The Navy ultimately abandoned the project, and the *Intelligent Whale* is now on display at the National Guard Militia Museum of New Jersey in Sea Girt, New Jersey.

SMS *Kaiser Max* (1863)

TYPE • *Armoured frigate* **COUNTRY** • *Austria*

SPECIFICATIONS

DIMENSIONS:	Length: 70.8m (232ft 2in); Beam: 14m (45ft 11in); Draught: 6.3m (20ft 9in)
DISPLACEMENT:	3605 tonnes (3548 tons) normal load
PROPULSION:	Sails, three-masted barquentine rig; single screw, 2-cylinder horizontal steam engine, 1436kW (1926ihp)
SPEED:	11.4 knots (21.1km/h; 13.1mph)
RANGE:	N/A
ARMAMENT:	15 × 24-pounders; 16 × 48-pounders; 1 × 12-pounder; 1 × 6-pounder guns
ARMOUR:	Belt: 110mm (4.3in)
COMPLEMENT:	386

Lead ship of a class of three broadside ironclads, *Kaiser Max* was an enlarged version of the preceding *Drache* class of 1861. All were engaged in the Battle of Lissa, between Austria–Hungary and Italy, the first naval battle fought between fleets of ironclad warships.

Right: A photograph of SMS Kaiser Max, *likely taken in the late 1890s. The ship was stricken from the Navy List on 30 December 1905, and used as an accommodation ship at Cattaro Bay. In 1920 it was ceded to Yugoslavia and renamed* Tivat.

GUNS
The ship had more than 30 guns installed. A stern casemate was fitted for the installation of a stern-chaser gun.

REBUILD
In 1873 *Kaiser Max* was returned to STT at Trieste for a drastic rebuild, and emerged as a centre-battery ship, with its rig reduced in 1880 from 1633.15 sq m (17,579 sq ft) to 1158.6 sq m (12,471 sq ft). It was eventually removed altogether.

SEAKEEPING
The class was not considered effective, with poor seakeeping ability. A partial rebuild was undertaken in 1867 when the open bow section was plated to prevent seawater washing in.

ITALIAN THREAT

Kaiser Max was laid down at the Stabilimento Tecnico Triestino (STT) Yard in October 1861, launched on 14 March 1862 and completed in early 1863. It was designated as a second-class armoured frigate. Two companion ships, and *Juan de Austria*, were built simultaneously to the same design: the Austro-Hungarian Empire was in a hurry to have a war fleet to match that of newly independent Italy.

IN HARBOUR

By 1866, the Empire was at war with Italy and Prussia, and on 20 July that year the Austrian and Italian fleets met off the island of Lissa (Vis), in the Adriatic Sea, for the first major battle on the high seas since Trafalgar in 1805. Better handled, the Austrian fleet won a convincing victory, although *Kaiser Max* was badly damaged.

USS *Choctaw* (1863)

TYPE • *Ironclad* **COUNTRY** • *United States*

SPECIFICATIONS

DIMENSIONS:	Length: 79m (260ft); Beam: 14m (45ft); Draught: 2.4m (8ft)
DISPLACEMENT:	911 tonnes (1004 tons) normal load
PROPULSION:	Two steam engines with 2.4m (8ft) strokes, each driving a side-wheel paddle)
SPEED:	9 knots (17km/h; 10mph)
ARMAMENT:	1 × 100-pounder rifle; 3 × 228mm (9in) smoothbores; 2 × 30-pounder rifles; fitted with a ram
ARMOUR:	Casemate: 25.4mm (1in)
COMPLEMENT:	106

The USS *Choctaw* was originally built in 1856 in New Albany, Indiana, as a side-wheel merchant steamer. It was acquired by the US Government in September 1862 and began to be converted to a warship using plans developed by Commodore William Porter.

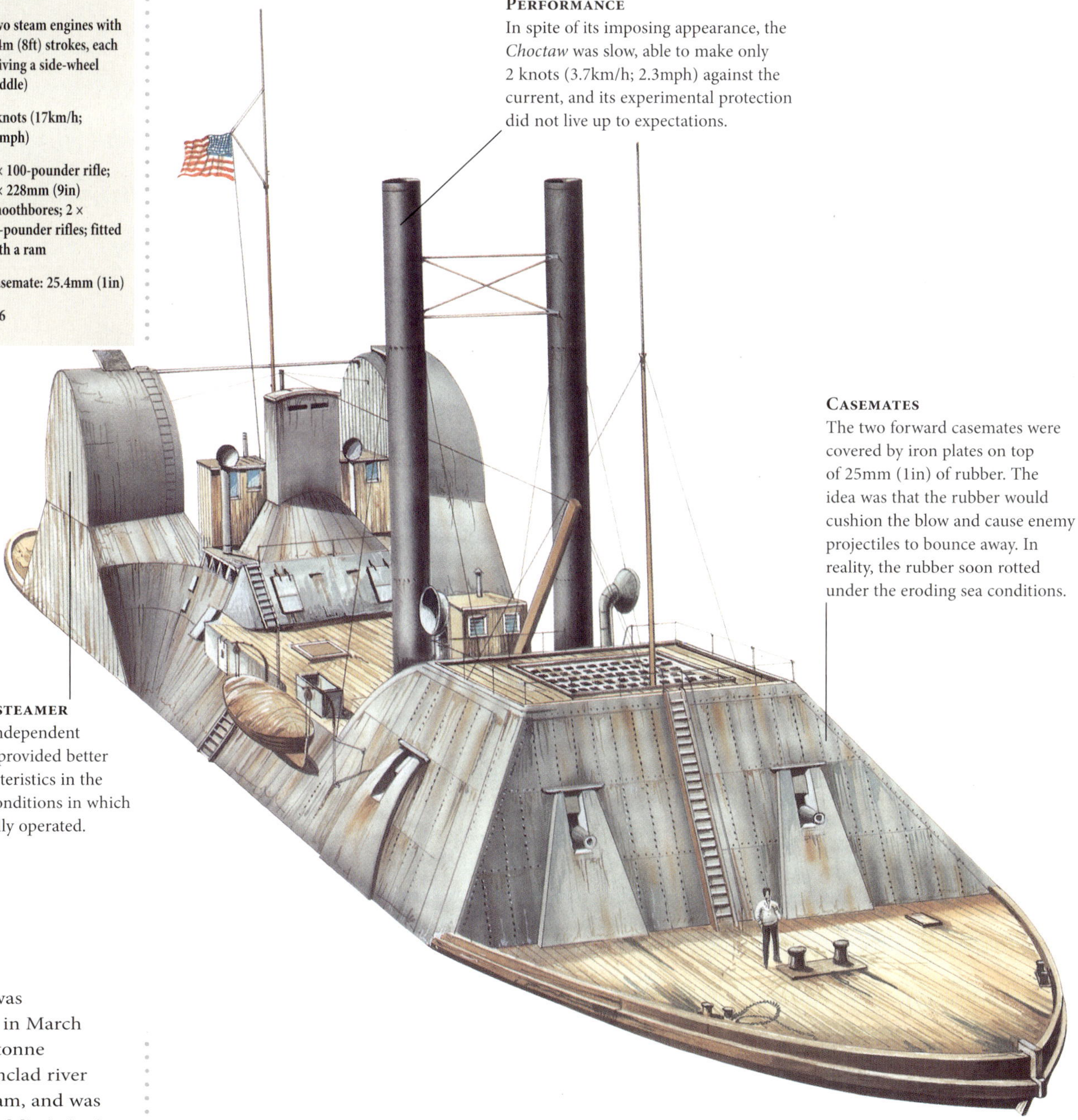

PERFORMANCE
In spite of its imposing appearance, the *Choctaw* was slow, able to make only 2 knots (3.7km/h; 2.3mph) against the current, and its experimental protection did not live up to expectations.

CASEMATES
The two forward casemates were covered by iron plates on top of 25mm (1in) of rubber. The idea was that the rubber would cushion the blow and cause enemy projectiles to bounce away. In reality, the rubber soon rotted under the eroding sea conditions.

SIDE-WHEEL STEAMER
Choctaw had independent wheels, which provided better steering characteristics in the narrow river conditions in which the ship typically operated.

DESIGN

The *Choctaw* was commissioned in March 1863 as a 911-tonne (1004-ton) ironclad river gunboat and ram, and was assigned to the Mississippi River Squadron. Porter's original design was altered during the conversion process, and the *Choctaw*'s transformation was a radical one: its length was extended from 74m to 85m (245ft to 280ft). Instead of building one long casemate, as was common in previous vessels, the *Choctaw*'s constructors erected three.

OFF VICKSBURG, CIVIL WAR
Choctaw fought in several Civil War engagements. On 30 April 1863, during a diversionary attack near Chickasaw Bluff, Tennessee, for example, *Choctaw* was struck 59 times by Confederate shore fire.

CSS *Hunley* (1863)

TYPE • *Submersible* **COUNTRY** • *Confederate States of America*

The *Hunley* was a submarine privately built in the spring of 1863 in the machine shop of Park and Lyons at Mobile, Alabama, under the direction of Lieutenants William Alexander and George Dixon, two Confederate Army engineers.

SPECIFICATIONS

Dimensions:	Length: 12.2m (40ft); Beam: 1.17m (3ft 10in); Draught: 1.22m (4ft)
Displacement:	7.5 tonnes (8.3 tons) submerged
Propulsion:	Hand-cranked propulsion powering a single screw
Speed:	2.5 knots (4.6km/h; 2.9mph)
Range:	N/A
Armament:	1 × spar torpedo (black powder charge on a 5m/16ft spar)
Complement:	1 officer, 7 enlisted

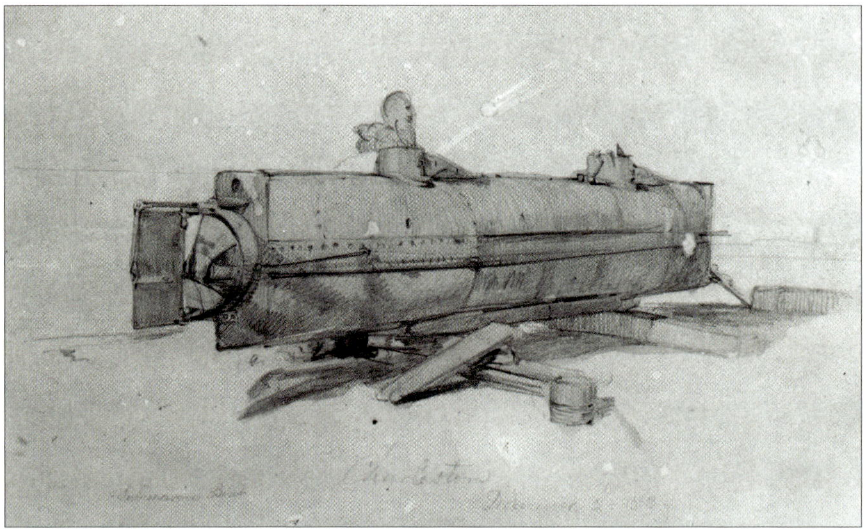

Right: Hunley *was based on plans furnished by Horace Hunley, James McClintock and Baxter Watson, who had earlier worked on the* Pioneer *submarine.*

Spar torpedo
The spar torpedo was only 5m (16ft) in front of the ship's bow, which made the moment of detonation perilous for *Hunley*'s crew as well as that of the enemy ship.

Hull design
The *Hunley* was 12m (40ft) long, 1.1m (3.5ft) wide and 1.2m (4ft) deep. The main hull was made out of a boiler, cut lengthwise, with a 305mm (12in) wide longitudinal strip added to increase the height.

Bow
A pointed bow and stern attached to the boiler gave the *Hunley* a distinctive cigar shape. It was designed for a crew of nine: one man to steer and the other eight to power the vessel by hand-turning a crankshaft that moved the propeller.

NIGHT ACTION

On the night of 17 February 1864, *Hunley* approached the Federal sloop *Housatonic* in Charleston harbour. The *Housatonic* spotted the *Hunley* and engaged it with small arms and tried to escape, but it was too late. The *Hunley* exploded its 130-pounder spar torpedo, and the *Housatonic* became the first ship in the history of naval warfare to be sunk by a submarine. The blast, however, likely damaged the *Hunley* as well, and it sank, with the entire crew lost.

WORKSHOP RESTORATION

The wreckage of the *Hunley* was discovered off of Sullivan's Island in 1995 and recovered in 2000. It is currently on display at the Warren Lasch Conservation Center on the old Navy Base in North Charleston.

CSS *Tennessee* (1864)

TYPE • *Ironclad* **COUNTRY** • *Confederate States of America*

SPECIFICATIONS

Dimensions:	Length: 64m (209ft); Beam: 15m (48ft); Draught: 4.3m (14ft)
Displacement:	1155 tonnes (1273 tons) normal load
Propulsion:	2 geared non-condensing steam engines with 610mm (24in) cylinders and a 2.1m (7ft) stroke, driving a single screw
Speed:	5 knots (9.3km/h; 5.8mph)
Armament:	2 × 203mm (8in) Brooke rifles; 4 × 160mm (6.4in) Brooke rifles?
Armour:	Sides 127mm (5in); forward end 152mm (6in); deck 51mm (2in)
Complement:	133

The CSS *Tennessee* was an ironclad ram built in Selma, Alabama, and launched in February 1863. It sailed to Mobile to complete its outfitting and was commissioned in February 1864, the most costly ship built by the Confederacy and the one that took the longest to complete.

Left: A vintage photograph of USS Tennessee. *The ship was decommissioned on 19 August 1865, sold on 27 November, and then scrapped.*

Shutters
Iron shutters over the gun ports were supposed to close when the gun was run back, but they often jammed in the heat of action.

Arms and armour
Tennessee had 152mm (6in) of armour on its casemate, 127mm (5in) on the sides and 51mm (2in) on the deck, along with firepower provided by six Brooke rifles.

Engines
The *Tennessee*'s engines had been taken from a riverboat steamer. She was inadequately powered for its weight and therefore hard to manoeuvre.

BATTLE
The *Tennessee* posed a serious threat to the Union's plan to attack Mobile. Admiral David Farragut's Federal fleet, reinforced by four monitors, attacked Mobile Bay on 5 August 1864. Although the *Tennessee* managed to inflict some damage on the Federal ships, it was soon surrounded, hit by multiple shells and repeatedly rammed. Facing inevitable destruction, the ship was forced to surrender.

RESTORED VESSEL
After the battle of Mobile Bay, the Federals repaired *Tennessee*, rechristened it the USS *Tennessee*, and used it to help capture Fort Morgan later in August 1864. She was then sent to New Orleans, Louisiana for further repairs and served with the Mississippi Squadron until after the end of the Civil War.

HMS *Research* (1864)

TYPE • *Ironclad* **COUNTRY** • *United Kingdom*

SPECIFICATIONS

DIMENSIONS:	Length: 59.4m (195ft); Beam: 11.76m (38ft 7in); Draught: 4.4m (14ft 6in)
DISPLACEMENT:	1771 tonnes (1743 tons) normal load
PROPULSION:	Sails, three-masted barque rig; screw propeller driven by a 2-cylinder horizontal direct-acting steam engine, 775.5kW (1040ihp))
SPEED:	10.3 knots (19km/h; 11.8mph)
ARMAMENT:	4 x 100-pounder smoothbore guns (later replaced in 1870 by 4 x 180mm/7in muzzle-loading rifles)
ARMOUR:	Belt 114mm (4.5in); armoured box battery 114mm (4.5in)
COMPLEMENT:	150

Despite the name, *Research* was not one of the Royal Navy's survey or exploration vessels, but a small ironclad sloop with a central battery, heavily armed for its size, and the first ship of its kind to be built for the British fleet.

Below: Research*'s design was by Sir Edward Reed, appointed Chief Constructor to the Royal Navy in 1863 and responsible for introducing the concept of the central 'box' battery instead of the traditional broadside arrangement.*

ENGINE
Research was fitted with a two-cylinder Boulton and Watt horizontal single-expansion direct-acting steam engine. Steam was provided by two tubular boilers, and the screw could be hoisted clear of the water for better performance under sail.

RIG
The rig was barque-style with a sail area of 1695.5 sq m (18,250 sq ft). Under sail alone it made around 6 knots (11.1 km/h; 6.9 mph); combined power gave around 10 knots (19km/h; 11mph).

REFIT
In a refit in 1869–70, the 100-pounder SB guns were replaced by four 180mm (7in) MLR guns, a spar deck was installed between forecastle and quarterdeck, and the funnel was moved forward out of the armoured box holding the guns.

CONVERSION
The *Research* was laid down at Pembroke Dockyard on 3 September 1861 and completed in April 1864. Originally intended as a 17-gun wooden screw sloop to be named HMS *Trent*, a change in Admiralty policy in 1862 ordered its conversion to an ironclad, lengthened by 3m (10ft), widened by 1.6m (5ft 6in) and with the standard sloop bow and stern converted into a ram and rounded stern.

IN HARBOUR
HMS *Research*'s seakeeping left a lot to be desired. It had a reputation as a heavy roller and a poor sea boat in rough conditions, and was normally laid up in the winter months.

HMS *Achilles* (1864)

TYPE • *Armoured frigate* **COUNTRY** • *United Kingdom*

SPECIFICATIONS

DIMENSIONS:	Length: 120m (380ft); Beam: 17.75m (58ft 3in); Draught: 8.3m (27ft 3in)
DISPLACEMENT:	9986 tonnes (11,010 tons) normal load
PROPULSION:	Sails, four masts with full square rig; screw propeller driven by a Penn trunk engine, 4265.4kW (5720ihp)
SPEED:	14.3 knots (26.4km/h; 16.4mph)
ARMAMENT:	4 × 110-pounders; 16 × 100-pounder guns
ARMOUR:	110mm (4.5in) waterline belt, 3.9m (13ft) high with 457mm (18in) wood backing; 114mm (4.5in) battery sides; 114mm (4.5in) transverse bulkheads
COMPLEMENT:	709

***Achilles* was an improved version of HMS *Warrior*, laid down before the previous ship was completed, which remedied several of its deficiencies. Both ships were one-offs, built while the British Admiralty was still assessing the qualities of all-iron armoured warships.**

Left: In 1902 Achilles *was stationed at Malta as a base ship, named* Hibernia, *and renamed* Egmont *in 1904. In 1914 it returned to Chatham as a depot ship, before being named* Egremont *in 1918 and* Pembroke *in 1919. It was finally sold for breaking in 1925.*

HULL SHAPE
New design ideas had been applied to both the hull and the armour protection. A simplified blunt bow, capable of ramming, was provided, with a new rounded stern that enabled the steering gear to be protected.

COMPARTMENTS
The hull was divided into 106 watertight compartments and had a double bottom.

RIG CHANGES
Achilles carried four masts, the first British warship to do so since the bonaventure masts of the 16th century, but all fully rigged. There were problems with the rig, and in 1865 the bow mast and bowsprit were removed, with the headsails rigged to the stem. In 1866 the bowsprit was restored and the foremast moved 7.6m (25ft) forward. The 1867 arrangement allowed for a sail area of 2799.4 sq m (30 133 sq ft). In 1877 the rig was reduced to a barque rig.

BELT ARMOUR
A waterline armour belt 3.9m (13ft) high ran along the entire hull, 110mm (4.5in) thick with 457mm (18in) wood backing; the battery was protected by a 114mm (4.5in) lateral belt and by transverse bulkheads of the same thickness at each end.

SEAHANDLING
Great length in relation to beam did not make for easy steering, but *Achilles* was a more handy sea boat than *Warrior*. It was on the active list until 1885 and then laid up.

CSS *Stonewall* (1865)

TYPE • *Ironclad* **COUNTRY** • *Confederate States of America*

SPECIFICATIONS	
Dimensions:	Length: 57m (187ft); Beam: 9.9m (32ft 8in); Draught: 4.36m (14ft 4in)
Displacement:	1422 tonnes (1400 tons) normal load
Propulsion:	Sails, two-masted brig rig; twin screw propellers driven by a return-connecting-rod steam engine, 895kW (1200ihp)
Speed:	10 knots (18.5km/h; 11.5mph)
Armament:	1 × 254mm (10in) RML; 2 × 163mm (6.4in) RML guns
Armour:	Belt 114mm (4.5in); turrets 140mm (5.5in)
Complement:	130

***Stonewall* was classed as an 'ironclad ram'. Ordered by the Confederate States, built in France as a ship of war, it saw virtually no service with the Confederate Navy, but played an important part in Japanese naval history.**

Right: Stonewall *is shown here at Ferrol, Spain, in 1865. On 24 March 1865, it encountered the Union sloops USS* Niagara *and* Sacramento *off Ferrol, but they evaded a battle with the impregnable-looking* Stonewall.

ARMOURED RAM
An armoured belt extended 1.5m (5ft) below the waterline. The hull was copper-sheathed and a massive plough ram was fitted, with detrimental effect on the ship's handling.

RUDDERS
Twin rudders gave the ship a tight turning circle, which was a useful manoeuvre capability for ramming actions.

GUNHOUSE
The aft gunhouse, oval in shape, held two 163mm (6.4in) guns, one on each side; with only a vestigial poop and folding bulwarks, they could be trained to fire over a wide arc.

HISTORY

Contracted on 16 July 1863 with the Bordeaux shipbuilders L. Arman, *Stonewall* was launched on 21 June 1864, and commissioned in January 1865. The French government refused to deliver the ship to the Confederacy, and under the name *Sphinx* it was instead sold to Denmark for use in war against Prussia. Arriving too late to take part, it was finally sold via a middleman to the Confederacy in December 1864.

UNDER SAIL, 1865
At the end of the American Civil War, *Stonewall* was passed to the US government, who sold it to the Imperial Japanese government in 1867. Immediately on arrival in Japan, the ship was renamed *Kotetsu*.

HMS *Bellerophon* (1866)

TYPE • *Ironclad* **COUNTRY** • *United Kingdom*

SPECIFICATIONS

Dimensions:	Length: 91.4m (300ft); Beam: 17m (56ft 1in); Draught: 7.52m (24ft 8in)
Displacement:	7672 tonnes (8457 tons) normal load
Propulsion:	Sails, three masts (square rig to 1884, then barque rig); screw propeller driven by a Penn 2-cylinder horizontal steam engine, 4862.7kW (6521ihp)
Speed:	14.17 knots (26.2km/h; 16.3mph)
Armament:	10 × 229mm (9in) MLR; 5 × 178mm (7in) MLR guns (as built)
Armour:	Belt 152mm (6in)
Complement:	650

An ironclad with centrally mounted battery, this ship had many features that made it among the most advanced of its time in terms of design, power and armament. Out of active service in 1904, it was finally discarded by the Royal Navy in 1922.

Below: After rearmament, Bellerophon *was the sole British ironclad to have only breechloading guns: 10 203mm (8in) BLs in the main battery and four 152mm (6in) as chase guns, the forward ones being set in new embrasures on the forecastle.*

Sails
Bellerophon carried 2200 sq m (23,800 sq ft) of sail, but was not a fast sailer, its best recorded speed under sail alone being 10 knots (18.5km/h; 11.5mph).

Double bottom
Constructed on the 'bracket frame' system, *Bellerophon* was the first Royal Navy ship with a complete double bottom. Wrought-iron armour protection was substantial.

Guns
Apart from the battery-mounted guns, there were two 178mm (7in) muzzle-loader chase guns at both the forward and aft ends of the main deck, and a further 7in MLR at the aft end of the upper deck.

SERVICE

The Royal Navy's third *Bellerophon* was launched on 26 May 1865. Initial service was with the Channel Fleet until 1871, then with the Mediterranean Fleet until 1872. During a refit in 1872 a poop deck was added, and *Bellerophon* took over as flagship on the North America station, remaining until 1881 when a further major refit was carried out, until 1884, with new boilers and a complete rearmament.

COASTAL PATROL
Bellerophon served at the North America station at Halifax, Nova Scotia, until 1892. Paid off that year at Plymouth, it then served as guard ship at Pembroke until 1903, before being converted as a stokers' training ship in 1904, renamed *Indus III*.

HMS *Hercules* (1868)

TYPE • *Armoured frigate* **COUNTRY** • *United Kingdom*

SPECIFICATIONS

DIMENSIONS:	Length: 99m (325ft); Beam: 18m (59ft); Draught: c. 7.3m (24ft)
DISPLACEMENT:	8816 tonnes (9718 tons) normal load
PROPULSION:	Sails, three-masted full square rig; screw propeller driven by a Penn trunk steam engine, 5353kW (7178ihp)
SPEED:	14.69 knots (27.2km/h; 16.9mph)
ARMAMENT:	8 × 254mm (10in) MLR; 2 × 230mm (9in) MLR guns
ARMOUR:	Waterline belt: 229mm (9in); central-battery box: 203–152mm (8–6in)
COMPLEMENT:	638

Comparison with *Bellerophon* shows how battleship design was progressing. *Hercules* carried more armour and heavier guns, with hull embrasures fore and aft of the battery to allow wider arcs of fire. Progressive improvements were made through the ship's active lifetime.

Right: Hercules *was the sixth Royal Navy ship to carry the name, and was commissioned on 21 November 1868.*

RIG
Hercules was ship-rigged, with a total sail area of 4589 sq m (49,400 sq ft), including stunsails, but – like other central-battery ships – did not make great speed under sail, 11 knots (20.3km/h; 12.6mph) being its best performance.

IMPROVEMENTS
In 1874–75 steam-powered steering was installed (prior to that it could take from eight to 12 men to effect a full turn), two 356mm (14in) torpedo tubes were mounted in 1878, and a net defence system fitted in 1886, reflecting the new threat of the submarine.

ARMAMENT
It was the first ship to carry the latest guns of the time, 250mm (10in) muzzle-loaders. Although they would soon be obsolete, they could fire 181kg (400lb) explosive shells.

HISTORY

Hercules served with the Channel Fleet until 1874, and pulled HMS *Agincourt* off the Pearl Rock at Gibraltar in 1871. During a gale in 1872 it sustained damage by colliding with HMS *Northumberland*. In 1875–77 it was flagship of the Mediterranean Fleet. After the 1893 modernization, *Hercules* was designated as 'battleship, third class,' and held with the reserve fleet at Portsmouth until 1904.

PEAK SERVICE
This photograph shows *Hercules* in pristine shape, as for review. In 1904 the ship was renamed *Calcutta*; it went to Gibraltar as a depot ship until 1914. Towed back to Portsmouth in 1914, it was used as an engineers' training depot as *Fisgard II*, by then being merely a mastless hulk.

SMS *König Wilhelm* (1868)

TYPE • *Armoured frigate* **COUNTRY** • *Prussia*

SPECIFICATIONS

DIMENSIONS:	Length: 112m (367ft 5in); Beam: 18.3m (60ft); Draught: 8.56m (28ft 1in)
DISPLACEMENT:	10,761 tonnes (11,862 tons) full load
PROPULSION:	Sails, three-masted square rig; screw propeller, Maudslay horizontal 2-cylinder engine, 6000kW (8000ihp)
SPEED:	14.7 knots (28km/h; 17mph)
ARMAMENT:	33 × 72-pounder smoothbore ML guns
ARMOUR:	Belt 152–305mm (6–12in); battery 150mm (5.9in)
COMPLEMENT:	730

Classified by the Prussian Navy as a *Panzerfregatte* or armoured frigate, *König Wilhelm* was flagship of the Prussian fleet, and after 1871 the largest battleship in the navy of the newly unified German Empire. It retained that status for 20 years.

Below: König Wilhelm *was a symbol of the newfound confidence of the unified German Empire.*

POWERPLANT
Steam power came from a Maudslay horizontal two-cylinder engine. Eight boilers were set in two rooms, each with a funnel.

MODERNIZATION
A complete modernization was undertaken in Hamburg in 1896. The sail rig was taken down and replaced by two new-type naval masts, while new boilers and triple-expansion engines were installed. The ship was completely re-gunned.

NAMING
The battleship was originally ordered as *Fatikh* by the Imperial Ottoman Navy from the Thames Ironworks at Blackwall, London, but when the Turks cancelled the order the Prussians bought it on the stocks in February 1867. Launched on 25 April 1868, it was named for King Wilhelm I. From 1893 to 1897, it was flagship of Division II of the Imperial German Navy, at Wilhelmshaven.

PLATES
Wrought-iron plates were fixed to a teak backing. At the midships waterline, plate thickness was 305mm (12in) on the outer armour, with 178mm (7in) hull plates and 250mm (9.8in) of teak.

UPGRADED SHIP
This photograph shows the ship after completion of its 1895–96 modernization by Blohm Und Voss, Hamburg, but prior to the removal of the mizzen-mast in 1898.

HMS *Cerberus* (1870)

TYPE • *Ironclad* **COUNTRY** • *United Kingdom*

SPECIFICATIONS

Dimensions:	Length: 68.6m (225ft); Beam: 13.7m (45ft); Draught: 4.4m (14ft 4in)
Displacement:	3393.6 tonnes (3741 tons) normal load
Propulsion:	Twin shafts; Maudslay steam engines, 1022kW (1370ihp)
Speed:	29.75 knots (18km/h; 11.2mph)
Armament:	4 x 254mm (10in) MLR guns
Armour:	Sides 203–152mm (8–6in); breastwork 228–203mm (9–8in); turrets 254–228mm (10–9in)
Complement:	155

The first of Britain's 'breastwork monitors' – small but armoured, heavily gunned ships designed for coastal and harbour protection, with gun turrets mounted on a raised breastwork – *Cerberus* was intended as a guard ship for the port of Melbourne, Australia.

Left: In later years Cerberus *was a depot ship at Williamstown naval base in Victoria, Australia. In 1936 it was scuttled and incorporated into a breakwater.*

Twin screws
Twin screws, each powered by separate engines, made up for the lack of sail power, and a balanced rudder was fitted.

Breastwork
The armoured breastwork raised the turrets to a suitable firing level and enclosed the funnel uptakes, ventilation shafts and hatchways that would otherwise have had to have their own protective armour.

Layout
A central superstructure rose above the breastwork, the gun turrets were mounted on the centreline at each end of the breastwork, each with a 180-degree angle of fire, and there was a single pole mast, used only for signalling purposes.

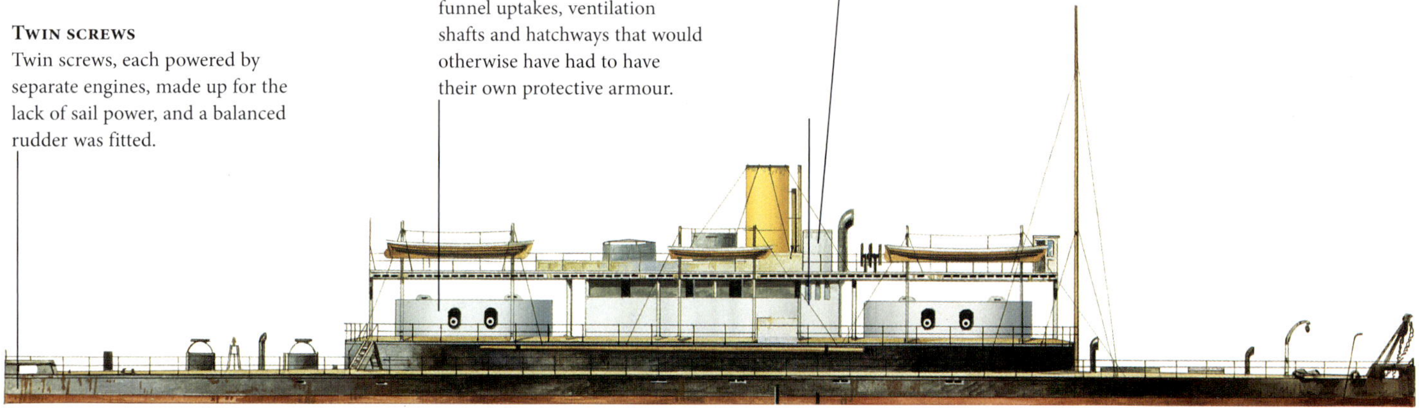

DESIGN
British naval vessels traditionally had a high freeboard to enable them to meet all sea conditions. However, several reasons combined to make *Cerberus* sit low in the water: this made for a more stable gun platform, it saved on armour plating and was adequate for operations in relatively sheltered waters. In addition, the design dispensed with conventional masts and yards, which required a higher freeboard.

IN OPEN SEA
For *Cerebus*'s single long voyage, out to Melbourne, temporary bulwarks extended fore and aft from the breastwork, and a full three-masted rig was fitted. As coal capacity was only 213 tonnes (235 tons) the voyage was made largely under sail.

HMS *Devastation* (1873)

TYPE • *Ironclad* **COUNTRY** • *United Kingdom*

SPECIFICATIONS

Dimensions:	Length 86.9m (285ft), Beam 19m (62ft 4in), Draught 8.4m (27ft 6in)
Displacement:	8464 tonnes (9330 tons) normal load
Propulsion:	Penn direct-acting trunk engines of 4959kW (6650hp), 2 screws
Speed:	13.84 knots (25.6km/h; 15.9mph)
Armament:	(1873) 4 × 31.7-tonne (35 tons) 305mm (12in) MLR guns
Armour:	Belt 305–216mm (12–8.5in), Turrets 356–250mm (14–10in), Deck 76–50mm (3–2in)
Complement:	358

HMS *Devastation*, the first seagoing turret ship, carrying the heaviest guns yet, was a highly controversial design. Yet its big-gun layout, and its reliance on engines alone, without sails, proved successful against the critics' warnings of impending failure.

Below: The low freeboard of a monitor-type ship improved steadiness and required less armour than a higher-sided ship.

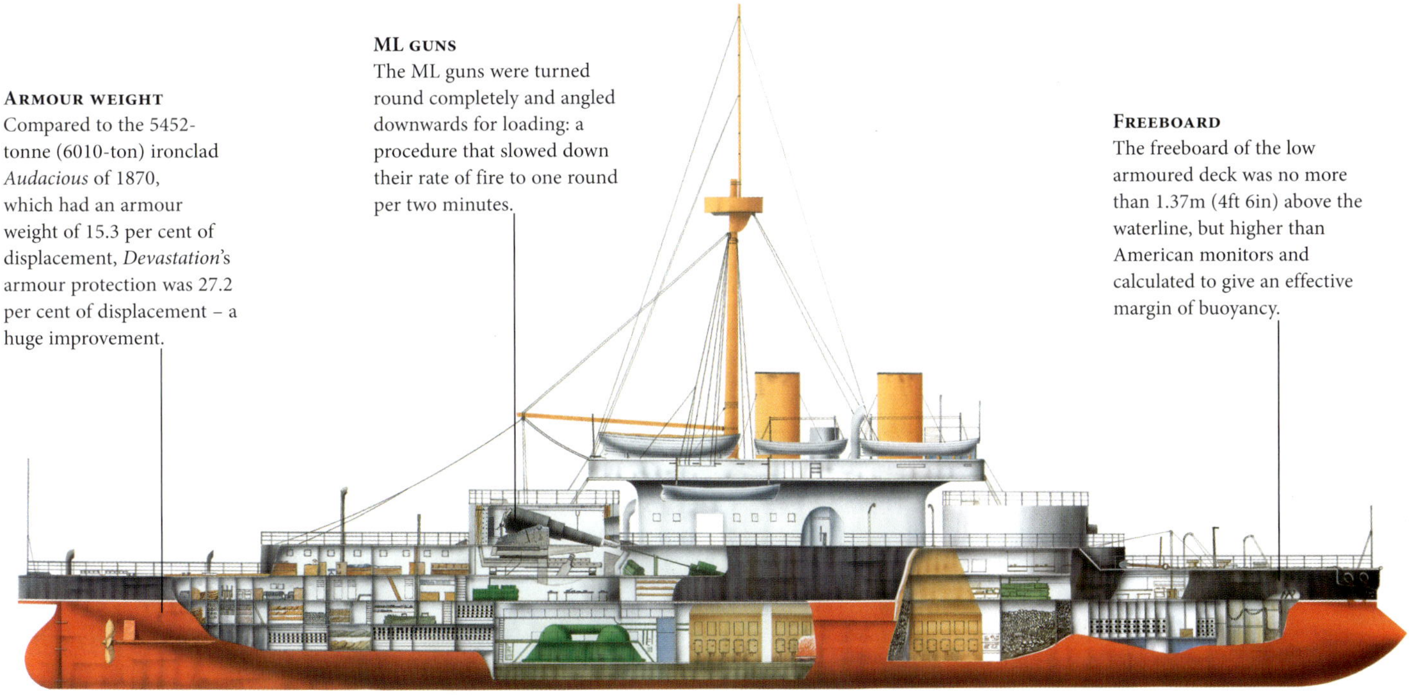

Armour weight
Compared to the 5452-tonne (6010-ton) ironclad *Audacious* of 1870, which had an armour weight of 15.3 per cent of displacement, *Devastation*'s armour protection was 27.2 per cent of displacement – a huge improvement.

ML guns
The ML guns were turned round completely and angled downwards for loading: a procedure that slowed down their rate of fire to one round per two minutes.

Freeboard
The freeboard of the low armoured deck was no more than 1.37m (4ft 6in) above the waterline, but higher than American monitors and calculated to give an effective margin of buoyancy.

DESIGN INPUT

Devastation was the creation of Sir Edward Reed, Chief Constructor at the Admiralty since 1862, who was closely involved in the technical, as well as the financial and political, aspects of warship design in a period of radical change and development. Numerous post-1870 alterations were made to the design by his successor Sir Nathaniel Barnaby.

SILHOUETTE
This photograph of *Devastation* shows the ship's high and strange silhouette, the mast towering above the low main deck. The ship was laid down at Portsmouth on 12 November 1869, launched on 12 July 1871 and completed on 19 April 1873, at a cost of £361,438.

Koning der Nederlanden (1874)

TYPE • *Ironclad* **COUNTRY** • *Netherlands*

The Dutch Navy's largest ship until the 20th century, *Koning der Nederlanden* was built not for service in the British-dominated North Sea, but to patrol and protect the Netherlands' distant colonies in the Dutch East Indies.

SPECIFICATIONS

Dimensions:	Length: 85.03m (279ft); Beam: 15.14m (49ft 8in); Draught: 5.94m (19ft 6in))
Displacement:	5400 tonnes (5951 tons) normal load
Propulsion:	Sails, three-masted barquentine rig; 2 screws, Penn compound reciprocating engines, 3450kW (4630ihp)
Speed:	12 knots (22km/h; 14mph)
Armament:	4 × 280mm (11in) MLR; 4 × 120mm (4.7in) BL guns
Armour:	Belt 200–150mm (8–6in); turrets 305–230mm (12–9in)
Complement:	256

Right: The Koning der Nederlanden *had a service career spanning nearly six decades. In World War II, as the Japanese advanced on Java, the old ship was set on fire and scuttled on 2 March 1942 to prevent its capture.*

Gun turrets
The main guns were 25.4-tonne (30-ton) Armstrong Whitworth rifled muzzle-loaders, mounted in pairs in hydraulically operated turrets.

Deck
The deck was unarmoured, but at this time naval guns were limited in range and could not fire at high trajectories, so 'plunging fire' was not yet a concern.

Hull armour
Incorporated in the design was a 1.22m (4ft) ram, and a full-length waterline wrought-iron armoured belt was fitted.

HISTORY

Laid down at the Rijkswerf shipyard in Amsterdam on 31 December 1871, originally named *Matador*. The ship was launched on 20 October 1874 and commissioned on 16 February 1877, at a cost of 3,220,170 florins. In April 1895, it was decommissioned and afterwards used as a barrack and guard ship at Surabaya, Indonesia, the Royal Netherlands Navy's main base in the East Indies.

IN SINGAPORE HARBOUR

The *Koning der Nederlanden* spent its working life in East Asia. Recommissioned on 1 December 1899, despite its virtually obsolete condition, with secondary armament only, it served as a 'stand-by' ship until 1916, when, again at Surabaya, it was converted to a submarine tender and served as such until 1923.

Redoutable (1876)

TYPE • *Ironclad* COUNTRY • *France*

French naval designers could always be relied upon for their originality. *Redoutable* was a highly distinctive ship of the central battery type in appearance and dimensions, but was also the first warship to be built largely of steel.

SPECIFICATIONS	
Dimensions:	Length: 97.13m (318ft 8in); Beam: 19.6m (64ft 6in); Draught: 7.8m (25ft 7in)
Displacement:	9372 tonnes (9224 tons) normal load
Propulsion:	Sails, three-masted square rig (later barquentine); screw propeller, horizontal compound steam engine, 4623kW (6200ihp)
Speed:	14.7 knots (27.2km/h; 17mph)
Armament:	8 × 274mm (10.8in); 6 × 140mm (5.5in); 12 × 1-pounder guns
Armour:	Belt 350mm (13.8in); battery 300–240mm (11.8–9.5in); deck 60–45mm (2.4–1.8in)
Complement:	705

Right: The Redoutable *photographed after its modernization. In the early 1900s the ship was deployed in the Far East during a period of high international tension, with war between Japan and Russia.*

Layout
Compared to other central battery ships, it was of wider beam and with a shortened battery, holding only four main guns, corner-mounted.

Magazines
The magazines for the battery were considered by some experts to be dangerously close to the boiler rooms.

Armour
Redoutable had a partial armoured deck, 61mm (2.4in) thick on each side of the battery, but the upper part of the battery was not armoured.

IRON AND STEEL
Launched in September 1876, *Redoutable*, a celebrated name in the French Navy, was completed in December 1878. Its designer was the engineer Louis de Bussy. France led the world in steelmaking with the Siemens process at that time, and only the bottom plates of the vessel were of iron, the rest of the hull being steel, with consequent improvement in structural strength and reduction in weight.

IN PORT, BREST, 1882
After this photo, in the mid-1890s, a major modernization and reconstruction was completed, with new guns, new boilers and a revised superstructure. Electric circuits were installed. The rig was dismantled and replaced by two military masts, and torpedo-defence netting was fitted.

Petr Veliki (1876)

TYPE • *Ironclad* COUNTRY • *Russia*

SPECIFICATIONS

DIMENSIONS:	Length: 101.7m (333ft 8in); Beam: 19.2m (63ft); Draught: 7.5m (24ft 9in)
DISPLACEMENT:	10,573 tonnes (11,640 tons) normal load
PROPULSION:	Sails, three masts, partial barquentine rig; 2 shafts, horizontal return-connecting-rod steam engines
SPEED:	13 knots (24km/h; 15mph)
ARMAMENT:	4 x 305mm (12in); 6 x 86mm (3.4in); 6 x 4-pdr guns; 1 x 25mm (1in) machine gun; 4 spar torpedoes
ARMOUR:	Belt 356–203mm (14–8in); turrets 356mm (14in); citadel 356mm (14in)
COMPLEMENT:	441

Named for Tsar Peter the Great, and originally intended as a breastwork monitor for coastal defence, *Petr Veliki* passed through many alterations and enlargements. During its long career, the main events were reconstructions and replacements rather than naval action.

Left: An archive photograph of the Petr Veliki. *The ship's planning went back to 1867 and was characterized by afterthoughts and additions, including the name. It was changed to its final form shortly before the launch in August 1872.*

INTERNAL ARRANGEMENT
Internally there were nine transverse and two wing watertight bulkheads, and a longitudinal centre bulkhead separating the boilers and engine rooms.

ENGINES
The original boilers and engines were defective and replaced by vertical compound engines at Elder's Yard in Glasgow in 1881–82.

DESIGN
In final design it was a turret ship, with a turret at each end of a short but massively protected citadel surrounding the single funnel and the ventilators.

REBUILD
Apart from a Mediterranean cruise in 1882, *Petr Veliki* served entirely with the Baltic Fleet. By the late 1890s, the ship was obsolescent and the turrets were taken out in 1898. In 1903 it was marked for conversion into a gunnery training ship, and by 1908 this was complete. Changes included an entirely new superstructure, new side armour, two funnels, two pole masts and four 203mm (8in) sponson-mounted guns.

RUSSIAN FLEET, 1892
Petr Veliki was assigned to the Gunnery Training Detachment until 1917. After the Russian Revolution it served as a submarine depot ship, first at Kronstadt, then in Helsinki, before returning to Kronstadt in April 1918. Hulked on 21 May 1921, the ship was used as a stores and accommodation vessel until its final disposal in 1959 for scrap.

HMS *Alexandra* (1877)

TYPE • *Ironclad* COUNTRY • *United Kingdom*

SPECIFICATIONS

DIMENSIONS:	Length: 104.8m (344ft); Beam: 19.4m (63ft 8in); Draught: 8.1m (26ft 6in)
DISPLACEMENT:	9642 tonnes (10,628 tons) normal load
PROPULSION:	Sails, three-masted barque rig; twin screws, vertical inverted compound steam engine, 6337kW (8498ihp)
SPEED:	15.09 knots (28km/h; 7.4mph)
ARMAMENT:	2 × 280mm (11in); 10 × 254mm (10in) MLR guns
ARMOUR:	Belt 305–152mm (12–6in); main battery 305mm (12in); upper battery 200mm (8in); deck 38–25mm (1.5–1in)
COMPLEMENT:	674

As a central battery ironclad, *Alexandra* was effectively obsolescent even when commissioned, with turreted warships already afloat. However, it also had new features and served in the Royal Navy, often as a flagship, for 23 years, mostly with the Mediterranean Fleet.

Right: From 2 January 1877, Alexandra *was flagship of the Mediterranean Fleet and kept the role to 1889.*

INNOVATIONS
An innovation was two 450kW (600ihp) auxiliary engines, which kept the screws turning while the ship was under sail. It was the last British battleship to carry its main guns below the top deck, with the typical box-battery arrangement.

ARMOUR
The protective belt was 3.2m (10ft 6in) high and covered the whole waterline length.

HULL
The iron hull was divided into 115 watertight compartments, with a central longitudinal bulkhead 74.7m (245ft) long, and a double bottom.

BATTERY SHIP
By the mid-1870s the efficiency and power of marine steam engines had greatly improved. However, the British Admiralty was not yet prepared to give up the practice of fully rigging the ships intended for blue-water service, and *Alexandra* was rigged as a three-masted barque.
It was launched on 7 April 1875 and completed on 31 January 1877. *Alexandra* was regarded as one of the most successful central-battery ships.

MASTED SHIP
The first Royal Navy ship to have vertical compound engines, making it the fastest large warship of its day, *Alexandra* was also a good sailer, although somewhat slow under sail alone, as all central battery ships seem to have been.

Fuso (1878)

TYPE • *Ironclad* COUNTRY • *Japan*

SPECIFICATIONS

DIMENSIONS:	Length: 67m (220ft); Beam: 14.6m (48ft); Draught: 5.5m (18ft 4in)
DISPLACEMENT:	3777 tonnes (4160 tons) normal load
PROPULSION:	Sails, three-masted barque rig; twin shafts driven by two pairs of Penn reciprocating steam engines, 2932kW (3932ihp)
SPEED:	13 knots (24km/h; 15mph)
ARMAMENT:	4 × 240mm (9.4in); 2 × 170mm (6.7in); 6 × 76mm (3in) guns; 1 × Nordenfelt 4-barrel machine gun
ARMOUR:	Belt 229–163mm (9–6.4in); battery 203mm (8in); bulkheads 178mm (7in)
COMPLEMENT:	295

Japan's most powerful ship of war in its new navy was designed and built in England, and fitted with German guns. At first an imperial flagship, by the time of its discarding in 1908 Japan had a substantial and proven battle fleet.

GUNS
As built, the gun layout was similar to that of *Redoutable*, with the four main guns at the corners of the battery, which protruded from the sides in order to widen the field of fire.

FORWARD CONTROL
Traditionally warships were commanded and navigated from the afterdeck, but *Fuso* had a forward control cabin mounted on a platform. This allowed for voicepipe communication with the engine room.

REFIT
A refit in 1894, just before the Japanese–Chinese war, saw the mainmast removed, and military tops placed on the fore- and mizzen-masts. Armament was also heavily revised.

ORIGINS
The modern Japanese Navy began with the American-built *Adzuma* of 1869, but it was with the programme of 1875 that development got seriously under way, spurred by the prospect of a Japanese–Korean war. Most powerful of a trio of new vessels, *Fuso* was designed in London by Sir Edward Reed and laid down in September 1875 at Samuda Brothers' yard at Poplar on the Thames. Launched on 14 April 1877, it was completed in January 1878.

BRITISH INFLUENCE
A central-battery ironclad typical of the time, *Fuso* was modelled on the Royal Navy's *Audacious* class of 1870–71, although smaller and more lightly armed.

Italia (1880)

TYPE • *Ironclad* **COUNTRY** • *Italy*

SPECIFICATIONS

Dimensions:	Length: 124.7m (409ft); Beam: 22.5m (74ft); Draught: 8.7m (28ft 8in)
Displacement:	15,900 tonnes (17,545 tons) full load
Propulsion:	Twin shafts, 24 boilers, two vertical compound steam engines, 11,780kW (15,797ihp)
Speed:	17.8 knots (33km/h; 20.5mph)
Range:	9260km (5754 miles)
Armament:	4 × 432mm (17in) BL; 7 × 150mm (5.9in); 4 × 119mm (4.7in) guns; 4 × 356mm (14in) torpedo tubes
Armour:	Redoubt 483mm (19in); conning tower 102mm (4in); deck 102–76mm (4–3in)
Complement:	701

On commissioning, *Italia* was the world's most formidable warship, mounting the heaviest guns ever deployed at sea. It was the lead ship in the *Italia* class. Along with its companion ship *Lepanto*, it remained the largest and fastest warship in the world for years after entering service.

Left: Photographed at Genoa on 17 July 1886 during the review of the 'Permanent Squadron' for King Umberto I of Savoia (Savoy), the battleship Principe Amedeo *is just behind* Italia, *followed by the cruiser* Giovanni Bausan.

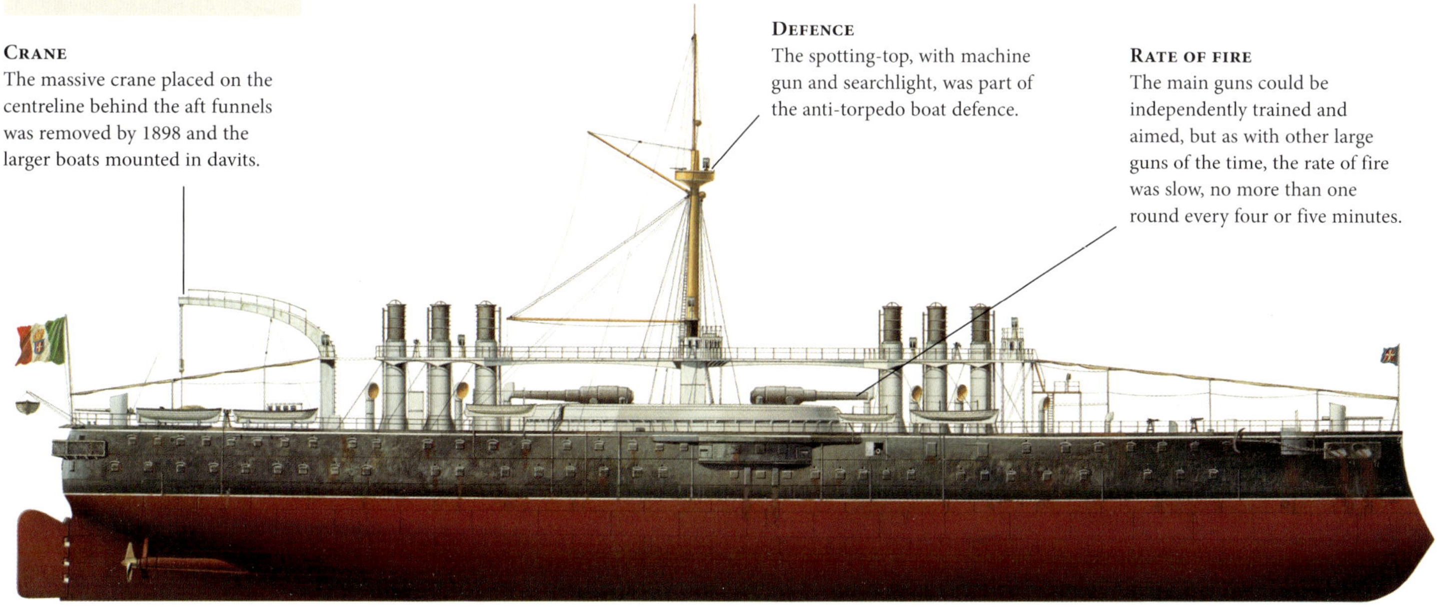

Crane
The massive crane placed on the centreline behind the aft funnels was removed by 1898 and the larger boats mounted in davits.

Defence
The spotting-top, with machine gun and searchlight, was part of the anti-torpedo boat defence.

Rate of fire
The main guns could be independently trained and aimed, but as with other large guns of the time, the rate of fire was slow, no more than one round every four or five minutes.

INFANTRY WARSHIP
Defeat by the Austrians at the Battle of Lissa, 1866, drove the Italians to further expand their fleet. Laid down at Castellammare in 1877 and launched in 1880, *Italia* was designed to be a fast, heavily armed ship capable of transporting an infantry division of 10,000 men along with its equipment. Between 1905 and 1908 *Italia* was rebuilt, losing two funnels and with the tall single mast replaced by two, forward and aft of the funnels.

IN PROFILE
Still in commission during World War I, but renamed as *Stella d'Italia,* the ship was based at Taranto and Brindisi, Italy, for gunnery training until 1917, when it was disarmed and transferred to the mercantile marine as a grain transport.

HMS *Inflexible* (1881)

TYPE • *Ironclad* **COUNTRY** • *United Kingdom*

HMS *Inflexible* was intended to surpass the Italian ships *Caio Duilio* and *Dandolo*, the most heavily armoured and fastest warships of their time. Laid down at Portsmouth on 24 February 1874, *Inflexible* was launched on 27 April 1876 but not completed until 18 October 1881.

Above: A deck plan of Inflexible. *A 76mm (3in) armoured deck ran the length of the ship, beneath the main deck and below the waterline level.*

SPECIFICATIONS

Dimensions:	Length: 97.5m (320ft); Beam: 22.9m (75ft); Draught: 8.08m (26ft 6in)
Displacement:	12,071 tonnes (13,306 tons) normal load
Propulsion:	Twin shafts; 12 boilers; two Elder 3-cylinder compound steam engines, 6269kW (8407hp)
Speed:	14.75 knots (27.3km/h; 17mph)
Range:	10,370km (6440 miles) at 10 knots
Armament:	4 × 406mm (16in) MLR; 6 × 20-pounder guns; 4 × 356mm (14in) torpedo tubes
Armour:	Citadel 610–406mm (24–16in); turrets 432–406mm (17–16in); deck 76mm (3in)
Complement:	470

Torpedo boats
At the stern, two single-tube torpedo boats were carried, launched by a derrick from the aftermast.

Armour
The armour anticipated the later American 'all or nothing' system, with heavy protection round the central citadel (a third of the hull length) protecting magazines and machinery, while the ends of the ship were unarmoured.

Turrets
The turrets were 10.3m (33ft 10in) in diameter and weighed 762 tonnes (840 tons) apiece; it took a full minute for them to make a rotation.

DEPLOYMENT

Perhaps as a gesture to the Italians, *Inflexible* was dispatched to join the Mediterranean Fleet in October 1881. At Alexandria in July 1882, it helped suppress an Egyptian uprising against the British-backed Khedive, and took a few hits from shore forts. It remained in the Mediterranean until 1885, and after refitting and a few years spent largely in reserve, returned there in 1890–93.

ADVANCED VESSEL
Inflexible was the first armoured ship to mount underwater torpedo tubes, with two placed experimentally in the bows. Other technical advances were compound expansion engines, and dynamos for electric lighting (the first warship so fitted).

Ting Yuen (1883)

TYPE • *Ironclad* **COUNTRY** • *China*

SPECIFICATIONS

Dimensions:	Length: 93.88m (308ft); Beam: 17.98m (59ft); Draught: 6.1m (20ft)
Displacement:	7336 tonnes (8087 tons) normal load; 7793 tonnes (8590 tons) full load
Propulsion:	Two masts, square rig on foremast; twin shafts, horizontal compound reciprocating steam engines, eight boilers, 5592.75kW (7500ihp))
Speed:	15.4 knots (28.5 km/h; 17.7 mph))
Range:	4500nmi (8300km; 500 miles) at 10 knots
Armament:	4 × 305mm (12in) guns; 2 × 150mm (5.9in) guns; 3 × 356mm (14in) torpedo tubes
Armour:	Belt 356mm (14in); barbettes 356–305mm (14–12in); casemates and conning tower 203mm (8in)
Complement:	350

With its companion ship *Chen Yuen*, the delivery of this central-citadel turret ship marked a phase of reorganization and modernization of the Chinese Navy in the early 1880s. It was sunk in the Battle of Wei-Hai-Wei in 1894.

Left: The Ting Yuen *is now commonly referred to as the* Dingyuan. *On 4–5 February 1895, a Japanese torpedo from* TB-23 *struck the stern. The ship was run ashore and possibly disabled by its crew to prevent its use by the Japanese. Later the wreck was scrapped.*

Layout
Designed with 'soft ends' on the form derived from the Italian *Duilio*, *Ting Yuen* had a steel hull, with two turrets mounted in echelon forward of the centre, between the foremast and the leading funnel.

Secondary armament
This consisted of two 150mm (5.9in) guns mounted at the bow and stern, and three 356mm (14in) torpedo tubes were fitted, two forward of the citadel, one in the stern.

Coal bunkers
The ship's lateral coal bunkers were also designed as part of its protective system.

INFLUENCES
Laid down at the Vulcan Yard in Stettin (now Szczecin, Poland), in March 1881, launched on 28 December the same year, *Ting Yuen* ('Eternal Peace') was completed by May 1883. Vulcan had just completed the central-citadel ironclad *Württemberg*, of the *Sachsen* class, but in its gun arrangement the Chinese ship more closely resembled the British *Inflexible*.

BATTLE OF YALU RIVER, 1894
At the Battle of the Yalu River, *Ting Yuen* was the flagship of Admiral Ting Ju-Chang, and was hit nearly 200 times, but its unarmoured bow and stern were not destroyed. Fourteen men were killed and a further 25 wounded.

Aquidaban (1885)

TYPE • *Ironclad* COUNTRY • *Brazil*

SPECIFICATIONS

DIMENSIONS:	Length: 85.34m (280ft); Beam: 15.85m (52ft); Draught: 5.59m (18ft 4in)
DISPLACEMENT:	5000 tonnes (5510 tons) normal load
PROPULSION:	Sails, three-masted square rig; twin shafts, inverted compound steam engines, eight cylindrical boilers, 4847kW (6500ihp)
SPEED:	15.8 knots (29km/h; 18mph)
RANGE:	2590km (1610 miles)
ARMAMENT:	4 × 234mm (9.2in); 4 × 140mm (5.5in); 13 × 1-pounder guns; 5 × 356mm (14in) torpedo tubes
ARMOUR:	Belt 279–178mm (11–7in); turrets and conning tower 254mm (10in)
COMPLEMENT:	277

Two battleships built for the Brazilian Navy in the 1880s – the *Aquidaban* and *Riachuelo* – helped establish it as the strongest fleet in the Americas until the US Navy began building its own capital ships. However, a naval revolt turned the fleet against its own country.

Below: Aquidaban's career ended in disaster on 21 January 1906 when the powder magazine exploded at Jacuecanga Bay, near Rio de Janeiro. Within three minutes the ship sank. Of those on board, 212 were killed and 36 injured, with 98 survivors.

FULL SHIP RIG
Aquidaban was a three-masted ship with a full rig, including a bowsprit – an almost obsolete feature in an otherwise technically modern vessel. Propulsion came from two inverted compound engines.

REFIT
In a further refit in 1904, the massive masts were removed and replaced by lighter masts. After this modification the ship was used primarily as a training vessel.

VENTILATORS
Intended for use in equatorial waters, the ship was well provided with ventilators, although boiler-room conditions must have been almost intolerable.

NAVAL REVOLTS
Aquidaban was caught up in the turbulence of internal politics, taking part in two naval revolts in 1891 and 1893–94. In the *Revolta da Armada*, led by Admiral Custodio de Mello against President Floriano Peixoto, it came under attack from torpedo boats on 16 April 1894, was hit twice and sank in shallow water. This incident greatly accelerated the development of anti-torpedo boat defences. The ship was refloated in June 1894.

NAVAL REVIEW, 1890s
The battleship was twice renamed, as *Dezesseis de Abril,* then *Vinte Quatro de Mayo*. Between 1897 and 1898 it was extensively refitted at Stettin (Szczecin), Poland and Newcastle, England, with the sailing rig replaced by two hollow pole masts. The original name was restored in 1900.

HMS *Collingwood* (1887)

TYPE • *Ironclad* **COUNTRY** • *United Kingdom*

Breech-loading guns mounted in barbettes fore and aft, secondary guns grouped in lateral batteries, a steaming speed of 16 knots, and big guns mounted higher than ever before in a British ship, were the prime features of HMS *Collingwood*.

SPECIFICATIONS

DIMENSIONS:	Length: 99m (325ft); Beam: 20.7m (68ft); Draught: 8.03m (26ft 4in)
DISPLACEMENT:	8618 tonnes (9500 US tons) normal load
PROPULSION:	12 boilers; 2 sets of Humphreys inverted-compound 3-cylinder triple-expansion steam engines
SPEED:	16.8 knots (31.1km/h; 19.3mph)
RANGE:	12,964km (8055 miles) at 10 knots
ARMAMENT:	4 x 305mm (12in) BL guns; 6 x 152mm (6in) BL; 12 x 6-pounder; 14 small quick-firing guns; 4 x 356mm (14in) torpedo tubes
ARMOUR:	Belt 457–203mm (18–8in); barbettes 292–250mm (11.5–10in); deck 76–50mm (3–2in)
COMPLEMENT:	498

Left: The ship's career did not allow it the chance to prove its qualities in battle. Briefly commissioned for the Queen's Jubilee Review in July 1887, it was immediately put in reserve. Its longest stint of duty was nine years with the Mediterranean Fleet from November 1889 to March 1897, including a refit at the Malta dockyard in 1896.

TORPEDO NETS
Torpedo nets, hung out at waterline level from lateral booms, were fitted to all major ships and used when lying at anchor.

FORCED DRAUGHT
This was the first RN ship to employ forced draught: injecting air into the 36 furnaces at above atmospheric pressure, to increase combustion and steam production.

BARBETTES
Collingwood was Britain's first 'barbette battleship', with its main guns set in armoured barbettes raised above the open deck fore and aft of the superstructure. A barbette could mount the gun higher; *Collingwood*'s main guns were 8.5m (28ft) above the waterline, keeping drier in a sea, and with a better position for plunging fire. It gave less protection than a turret, but many gun crews disliked the restricted space and foul air of a turret.

BELT
Although it was termed a 'citadel ship', the armoured area was in fact a belt 42.7m (140ft) long and 2.3m (7ft 6in) high protecting the barbettes and machinery.

FREEBOARD
This illustration of *Collingwood* reveals the ship's lack of freeboard. Even in a modest sea, or at speed, the deck forward of the barbette would be awash. Note the capstan and anchor hoist.

Formidable (1887)

TYPE • *Ironclad* **COUNTRY** • *France*

With its companion ship *Amiral Baudin*, this barbette ship initiated a new phase in French naval design, combining a high freeboard and 'tumblehome' sides. *Formidable* was launched on 16 April 1885 and commissioned on 29 December 1888.

SPECIFICATIONS

Dimensions:	Length: 101m (331ft 4in); Beam: 21m (68ft 11in); Draught: 8.5m (27ft 10in)
Displacement:	11,910 tonnes (13,127 tons) normal load
Propulsion:	Twin shafts; 12 boilers; two vertical compound steam engines, 7233kW (9700ihp)
Speed:	16 knots (29.6km/h; 18.4mph)
Range:	N/A
Armament:	3 x 370mm (14.5in); 12 x 138mm (5.4in) guns; 6 x 450mm (17.7in) torpedo tubes
Armour:	Belt 550mm (21.6in); magazine 560mm (22in); deck 100mm (3.9in)
Complement:	650

Below: Viewed head-on, the hull was shaped like a pear with the top sliced off: the beam at the waterline was considerably wider than on the open deck.

Topmost
The topmast on the mainmast was removed for balloon-flying tests: balloon-flying was for signalling purposes.

Funnel caps
Funnel caps, designed to stop flying lumps of clinker and burning coal, were essential, particularly when forced draught was used.

Profile
The profile was dominated by a huge single funnel and two tall masts, with a pronounced ram below. A hexagonal flying deck enclosed both the foremast and the funnel.

SERVICE CAREER

On commissioning, the ship joined the Mediterranean Squadron at Toulon, as flagship. In 1890 it was used to conduct tests of captive balloons, for reconnaissance and signalling purposes. On two occasions it carried President Sadi Carnot, during manoeuvres in 1890 and on a visit to Corsica in April 1891. Later in 1891 it was transferred to the Atlantic Squadron, based at Brest.

FORMIDABLE, 1897
Numerous modifications were made during the later 1890s. A small navigation control house was mounted on the flying deck forward of the foremast, and multiple fighting tops were placed on both masts. The central barbette was removed in 1896–98 and an armoured redoubt was installed with four 162mm (6.4in) guns.

HMS *Victoria* (1890)

TYPE • *Ironclad* **COUNTRY** • *United Kingdom*

SPECIFICATIONS

Dimensions:	Length: 103.6m (340ft); Beam: 21.34m (70ft); Draught: 8.15m (26ft 9in)
Displacement:	11,200 tonnes (12,355 tons) normal load
Propulsion:	Twin shafts; two Humphreys & Tennant triple-expansion steam engines, 6000kW (8000ihp)
Speed:	16 knots (30km/h; 18mph)
Range:	Radius 11,265km (7000 miles) at 10 knots
Armament:	2 × 413mm (16.25in) BL; 1 × 250mm (10in) BL; 12 × 152mm (6in) BL guns; 6 × 360mm (14in) torpedo tubes
Armour:	Belt 457mm (18in); turrets 429mm (17in); deck 76mm (3in)
Complement:	430

With its companion ship HMS *Sans Pareil*, *Victoria* marked the British Admiralty's return to the turret-ship concept. The first battleship to have triple-expansion engines, it had a short career ending in disaster while on manoeuvres.

Left: Comparison of this ship with the French Hoche *and* Marceau *is interesting, as they were designed with the envisaged possibility of fighting each other. Although some features were common to both, the British ships had differently shaped hulls, longer forecastles, and a different gun arrangement.*

Layout
Lateral twin funnels were placed just behind the foremast and the navigation house, and the superstructure, with boats on top, extended to the stern, placing the rear gun higher than the main turret.

Armour
The side armour covered 46.3m (152ft) and was 457mm (18in) thick, with a bulkhead of 406mm (16in) at each end. In all, *Victoria* carried 2997 tonnes (3303 tons) of armour.

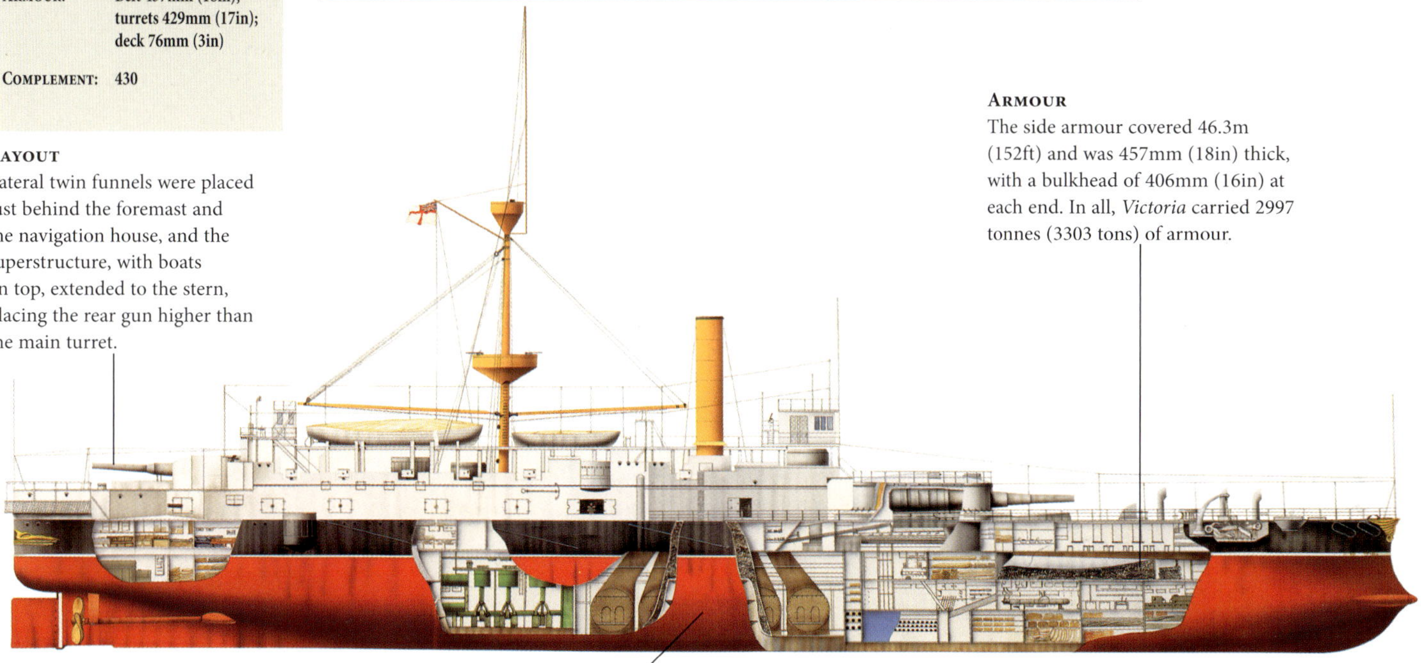

Engines
Triple-expansion engines, fitted for the first time, proved highly effective, but *Victoria*'s funnels had to be heightened by 5.2m (17ft) in August 1890 to improve the draught.

RAMMED
Victoria was flagship of Sir George Tryon, commander-in-chief of the Mediterranean Fleet. Approaching Tripoli on 22 June 1893, Tryon gave the inexplicable order for the first division to alter course 16 points to starboard, and the second 16 points to port. As a result the lead ships converged. *Camperdown* rammed *Victoria* behind the bow, penetrating 2.7m (9ft) at a depth of 3.6m (12ft) below the waterline.

SINKING, 1893
Following the impact with *Camperdown*, watertight doors on *Victoria* could not be secured in time. Despite an attempt to reach shallow water, the ship capsized and sank with the loss of 359 crew, including Tryon.

HMS *Royal Sovereign* (1892)

TYPE • *Ironclad* **COUNTRY** • *United Kingdom*

SPECIFICATIONS

DIMENSIONS:	Length: 115.8m (380ft); Beam: 22.86m (75ft); Draught: 8.5m (28ft)
DISPLACEMENT:	14,138 tonnes (15,585 tons) full load
PROPULSION:	8 single-ended boilers; 2 sets of 3-cylinder vertical triple-expansion steam engines, 6711kW (9000hp); twin screws)
SPEED:	18 knots (33.3km/h; 20.7mph)
RANGE:	4720nmi (8741km; 5429 miles) at 10 knots
ARMAMENT:	4 × 343mm (13.5in) guns; 10 × 152mm (6in) QF guns; 16 × 6-pounder QF guns; 12 × 3-pounder QF guns; 7 × 356mm (14in) torpedo tubes
ARMOUR:	Belt 457–356mm (18–14in); deck 76–63mm (3–2.5in); barbettes 432–279mm (17–11in);
COMPLEMENT:	712

This was the first British battleship to exceed 10,886 tonnes (12,000 tons) displacement, and the first to have steel-plate armour. In the words of an expert commentator, it was the first of a class of seven that 'sat the water with majesty and distinction'.

Right: HMS Royal Sovereign *underway in 1913, the last year of its service. It was sold off in October the same year.*

ANCHORS
Double anchors were on the starboard only, not yet secured in the hawse-holes but with embrasures at deck level to hold them.

GUNS
The 152mm (6in) quick-firing guns were mounted in casemates for the first time in a British capital ship.

BILGE KEELS
Bilge keels had to be fitted to all the *Royal Sovereign* class to cure their tendency to excessive rolling.

NAVAL EXPANSION

Britain's Naval Defense Act of 1889 provided for the construction of eight large battleships. At the time, Britain possessed 22 first-class battleships and 15 second-class; France had 14 and 7 respectively, and Russia had 7 and one. Germany was about to expand its battleship strength. *Royal Sovereign* was launched on 26 February 1891, and completed in March 1892.

IN SERVICE

Initially assigned to the Channel Fleet, *Royal Sovereign* was part of the British squadron at the opening of the Kiel Canal in June 1895. From June 1897 to August 1902 it served in the Mediterranean and was then guard ship at Portsmouth until 1905.

SMS *Wörth* (1893)

TYPE • *Ironclad* **COUNTRY** • *Germany*

SPECIFICATIONS

DIMENSIONS:	Length: 115.7m (379ft 7in); Beam: 19.5m (64ft); Draught: 7.6m (24ft 11in)
DISPLACEMENT:	10,670 tonnes (11,760 tons) normal load
PROPULSION:	Twin shafts; triple-expansion steam engines, 7457kW (10,000ihp)
SPEED:	17 knots (31km/h; 20mph)
RANGE:	N/A
ARMAMENT:	4 × 280mm (11in); 2 × 280mm (11in); 8 × 105mm (4.1in); 8 × 86mm (3.4in) guns; 6 × 450mm (17.7in) torpedo tubes
ARMOUR:	Belt 406–305mm (16–12in); turrets 230mm (9in); deck 76mm (3in)
COMPLEMENT:	568

Imperial Germany's naval expansion began with the four *Brandenburg*-class battleships, all laid down in 1890. *Wörth* was the first of the 'Brandenburgs' to be completed, on 31 October 1893.

Left: A depiction of SMS Wörth *ploughing through stormy waters. By 1914 the ship was obsolete and although assigned to the Fifth Squadron of the High Seas Fleet, it played no part in World War I.*

TURRETS
The turrets, while imposing, were relatively lightly armoured. They could be reloaded only in fore-and-aft (or straight lateral) position.

HULL EXTENSIONS
Lateral hull extensions on each side of the forward turret carried forward-firing 86mm (3.4in) guns, a feature not incorporated in subsequent German ships.

GUNS
Wörth was armed with six 280mm (11in) guns, mounted in three rounded turrets; the fore and aft guns were of 40 calibre, in the central turret they were 35 calibre, the shorter barrels being necessary to let the turret to fully revolve.

PRE-DREADNOUGHTS

Battleships of the 1890s came to be called 'pre-dreadnoughts'. Following a complete reorganization of the German Admiralty in the late 1880s, a new fleet was planned, including ships of long-range capacity able to protect, or preserve, Germany's new colonies. *Wörth* was planned by Alfred Dietrich, chief designer of the Imperial Navy, and his aim was to provide modern battleships on a par with those of other navies, especially Britain's.

WÖRTH, KIEL, 1899
In February 1899, an edict of the German Imperial Cabinet altered the status of the Brandenburgs from *Panzerschiff* ('armoured ship') to *Linienschiff* ('ship of the battle line'). In 1900–01 all four were sent to China in a show of force at the time of the Boxer Rebellion.

Re Umberto (1893)

TYPE • *Ironclad* COUNTRY • *Italy*

SPECIFICATIONS

DIMENSIONS:	Length: 127.6m (418ft 7in); Beam: 23.4m (76ft 10in); Draught: 9.3m (30ft 6in)
DISPLACEMENT:	15,702 tonnes (17,314 tons) full load
PROPULSION:	Twin shafts; vertical compound steam engines; 18 cylindrical boilers, 14,541kW (19,500ihp)
SPEED:	18.5 knots (34.3km/h; 21.3mph)
ARMAMENT:	4 x 343mm (13.5in); 8 x 152mm (6in); 16 x 120mm (4.7in); 16 x 57mm (2.2in); 10 x 37mm (1.5in) guns; 5 x 450mm (17.7in) torpedo tubes
ARMOUR:	Belt and casemate 102mm (4in); barbettes 349mm (13.75in); deck 76mm (3in)
COMPLEMENT:	733

Italy's first all-steel battleship, *Re Umberto* embodied many of the ideas of the naval modernizer Benedetto Brin, representing a significant advance over the earlier *Italia* and *Ruggiero di Lauria* classes.

Right: Re Umberto *was classified as a first-class battleship. Completion of the third in the class,* Sardegna, *in 1895 gave Italy 10 ships designated 'first class', compared with four German, five Russian, 20 French and 19 British.*

GUNS
The 343mm (13.5in)/30-calibre Mk IV guns were manufactured by Armstrongs in England, based on those supplied for the 'Admirals'.

FUNNELS
The ship had triple funnels, the front two abreast of each other, the third placed well to the stern, with a single tall pole mast equidistant between them.

GUN SHIELDS
Gun shields were mounted on the barbettes, giving the appearance of turrets although they were only 102-mm (4-in) thick, so gave limited protection to their crews.

BARBETTE SHIP
Among the largest warships of the time, *Re Umberto* was laid down at Castellammare Navy Yard, Naples, on 10 July 1884, launched on 17 October 1888 and completed on 16 February 1893. It was a barbette ship, resembling the British *Admiral* class, with big guns placed in two barbettes, fore and aft, built up from a low-freeboard hull. The secondary guns were placed in a slope-sided redoubt amidships.

***RE UMBERTO* OFF SICILY**
Re Umberto's service consisted primarily of fleet manoeuvres and diplomatic visits. It was on active service in the Italian–Turkish war of 1911–12, which resulted in Italy's acquisition of Libya, but in October 1912 it was laid up at Genoa and used as a depot ship there and at La Spezia, Italy. Decommissioned in 1914, it was recommissioned on 9 December 1915, and converted to a floating battery.

Gustave Zédé (1893)

TYPE • *Submarine* **COUNTRY** • *France*

SPECIFICATIONS

DIMENSIONS:	Length: 48.5m (159ft); Beam: 3.2m (10ft 6in); Draught: 3.2m (10ft 6in)
DISPLACEMENT:	264 tonnes (291 tons) surfaced; 274 tonnes (302 tons) submerged
PROPULSION:	2 Sautter-Harlé 268kW (360hp) electric motors; battery powered; single screw
SPEED:	Surfaced 12 knots (22km/h; 14mph); submerged 6.5 knots (12km/h; 7mph)
RANGE:	Surfaced 220nmi (407km; 253 miles) at 5.5 knots; submerged 105nmi (194km; 121 miles) at 4.5 knots
ARMAMENT:	1 × 450mm (17.7in) torpedo tube; 4 × Whitehead torpedoes
COMPLEMENT:	19

The French were the most innovative submarine designers of the late 19th century, with marine engineer Gustave Zédé as a leading figure. His electrically propelled *Gymnote* became the French Navy's first submarine in 1888 and went on to complete more than 2000 dives.

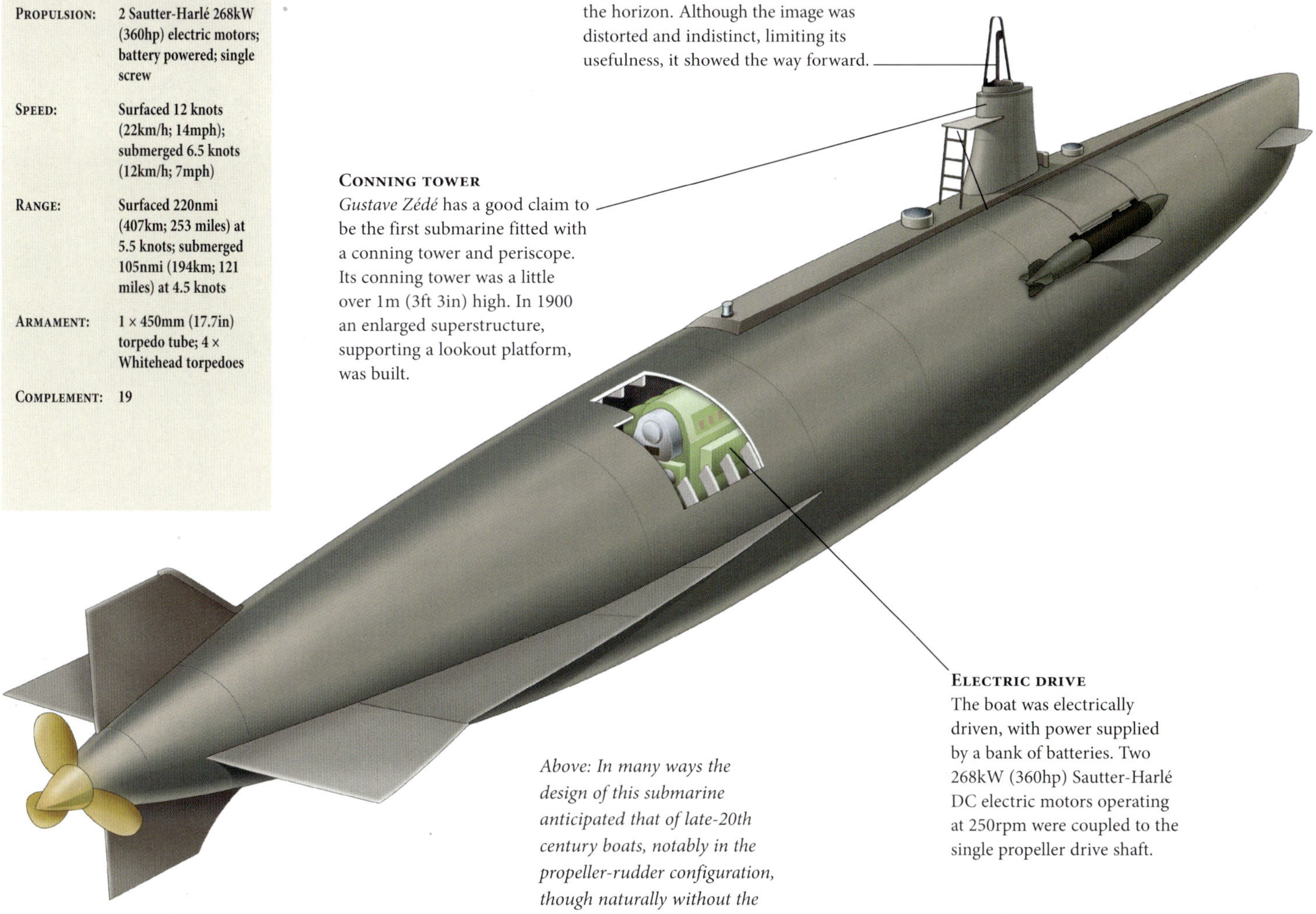

PERISCOPE
The periscope was about 35.5cm (14in) in diameter. The field of view was 20 degrees above and 7 degrees below the horizon. Although the image was distorted and indistinct, limiting its usefulness, it showed the way forward.

CONNING TOWER
Gustave Zédé has a good claim to be the first submarine fitted with a conning tower and periscope. Its conning tower was a little over 1m (3ft 3in) high. In 1900 an enlarged superstructure, supporting a lookout platform, was built.

ELECTRIC DRIVE
The boat was electrically driven, with power supplied by a bank of batteries. Two 268kW (360hp) Sautter-Harlé DC electric motors operating at 250rpm were coupled to the single propeller drive shaft.

Above: In many ways the design of this submarine anticipated that of late-20th century boats, notably in the propeller-rudder configuration, though naturally without the technical sophistication of later submarines.

COMBAT TRIALS

On the evening of 2 July 1901, the submarine left Marseille. A naval tug towed it to a position off Ajaccio harbour in Corsica, where it submerged in the early morning of the 3rd, entering the harbour undetected. A battleship squadron was just weighing anchor to leave on practice manoeuvres, and *Gustave Zédé* fired an unarmed torpedo at the *Charles Martel*, flagship of the Second Division, striking the hull.

LOW PROFILE
Gustave Zédé's hull barely showed above the water when surfaced, with only the walk-platform visible. For 10 years after launching, it was the largest submarine in the world. To contemporaries, it was 'a giant'.

USS *Maine* (1895)

TYPE • *Armoured cruiser* **COUNTRY** • *United States*

SPECIFICATIONS

Dimensions:	Length: 98.9m (324ft 4in); Beam: 17.4m (57ft); Draught: 6.5m (21ft 5in)
Displacement:	6062 tonnes (6682 US tons) normal load
Propulsion:	8 boilers; 2 N.F. Palmer inverted vertical triple-expansion steam engines, 6930kW (9293hp); twin screws
Speed:	16.45 knots (30.5km/h; 19mph)
Range:	3,600nmi (6700km; 4100 miles) at 10 knots
Armament:	4 × 254mm (10in) guns; 6 × 152mm (6in) guns; 7 × 6-pounder; 8 × 1-pounder guns; 4 × 457mm (18in) torpedo tubes
Armour:	Belt 305mm (12in); turrets 203mm (8in); conning tower 254mm (10in); deck 102mm (4in)
Complement:	374

Despatched to Havana, Cuba, amid heightened international tension to 'show the flag' and protect US interests, *Maine* was destroyed by an explosion whose cause remained a controversial for more than three-quarters of a century.

Right: Although the USA had pioneered ironclad warships in the 1860s, these were not large, ocean-going craft capable of extended cruises. As a result, Maine *drew heavily on British and European battleship design.*

Fighting tops
Each fighting top held a 1-pounder gun, part of the anti-torpedo boat defences (note also the searchlights). These guns fired 0.5kg (1.1lb) shells to a range of approximately 3200m (10,500ft).

Armament
152mm (6in) guns were fitted in casemates at the bow and stern, to give supplementary fore and aft fire.

Coal bunkers
Coal bunkers were placed between the forward 254mm (10in) magazine and the outer hull. Spontaneous combustion here may have caused the fatal explosion.

EXPLOSION

In 1897 Cuba was still a Spanish colony, but an independence struggle was underway. *Maine* was despatched to Havana as a deliberate show of force, arriving on 25 January 1898, anchoring in the harbour. On 15 February at 21:40, there was a massive explosion in the forward part of the hull and the ship was left a half-submerged wreck, with 252 crew dead and missing. Given the tensions of the time, an attack could not be ruled out.

CAPITAL SHIP

Maine was authorized by Congress on 3 August 1886 and laid down in the New York Navy Yard as an armoured cruiser on 17 October 1888, but was redesignated as a battleship of the second class and launched on 18 November 1890. It was not commissioned until 17 September 1895.

USS *Indiana* (1895)

TYPE • *Armoured cruiser* **COUNTRY** • *United States*

SPECIFICATIONS

DIMENSIONS:	Length: 106.96m (350ft 11in); Beam: 21.11m (69ft 3in); Draught: 8.2m (27ft)
DISPLACEMENT:	9333 tonnes (10,288 tons) normal load
PROPULSION:	4 double-ended boilers; 2 inverted vertical triple-expansion engines developing 7295kW (9738hp); 2 screws
SPEED:	15.6 knots (28.9km/h; 18mph)
RANGE:	9100km (4900nmi)
ARMAMENT:	4 × 330mm (13in); 8 × 203mm (8in); 4 × 152mm (6in) guns; 20 × 6-pdr; 6 × 1-pdr guns; 4 × 457mm (18in) torpedo tubes
ARMOR:	Belt 460–220mm (18–8.5in); main turrets 380mm (15in); turrets 152mm (6in); deck 76mm (3in)
COMPLEMENT:	473

A first-class battleship described by a British expert as 'distinctly superior to any European vessels of the same displacement', Indiana and its two companion ships were classified as 'coastal battleships' partly for domestic political reasons.

Right: Heavily armed for their size, Indiana *and its classmates* Oregon *and* Massachusetts *carried 330mm (13in), 203mm (8in) and 152mm (6in) guns. The main guns were mounted in pill-box turrets that were not counterbalanced, making the ship tilt when the guns were trained abeam and making the turrets themselves unstable.*

LIVERY
Standard paint finish for the USN battleships at this time was a white hull and superstructure, with buff-coloured upperworks, including funnels and masts.

MAST
The mast was a massive pole construction with the navigation bridge built around it. A basketstyle aftermast was added in 1910.

TURRETS
The 330mm (13in) and 203mm (8in) turrets had all-round loading, an improvement on European practice.

SPANISH-AMERICAN WAR
Indiana was deployed to the North Atlantic Squadron based at Key West and saw action in the Spanish–American War of 1898, shelling shore targets and working on interception missions against Spanish movements. On 3 July *Indiana* was part of the US force that sank the Spanish destroyers *Pluton* and *Furor*, although engine problems prevented it from chasing two Spanish cruisers, which escaped.

USS *INDIANA* (BB-1)
Indiana was classified as Battleship No. 1, and twin ships *Massachusetts* and *Oregon* were Battleships 2 and 3. From 1907 the designation was also expressed as B-1, etc. The initials BB were used from 17 July 1920.

USS *Oregon* (1896)

TYPE • *Armoured cruiser* **COUNTRY** • *United States*

The US Navy's Battleship No. 3 made an historic voyage to engage Spanish ships in the Spanish–American War of 1898, and had a long and eventful career. *Oregon* was launched on 26 October 1893, commissioned on 15 July 1896 and then fitted out for duty on the Pacific Station.

SPECIFICATIONS

Dimensions:	Length: 107m (351ft 2in); Beam: 21.1m (69ft 3in); Draught: 7.3m (24ft)
Displacement:	11,875 tonnes (13,086 tons) normal load
Propulsion:	Twin shafts; vertical inverted triple-expansion engines; four double-ended Scotch boilers, 8285kW (11,111ihp)
Speed:	16.8 knots (31.1km/h; 19.3mph)
Range:	4900nmi (9100km; 5600 miles)
Armament:	4 × 330mm (13in)/35-calibre; 8 × 203mm (8in)/35-calibre; 4 × 152mm (6in)/40-calibre; 20 × 6-pounder; 6 × 1-pounder guns; 6 × 457mm (18in) torpedo tubes
Armor:	Belt 460–203mm (18–8in); main turrets 380mm (15in); conning tower 254mm (10in); deck 110mm (4.5in)
Complement:	473

Left: USS Oregon *(BB-3) returns to San Francisco, California, 16 May 1896, with flags flying, upon the completion of its trials.*

Freeboard
Oregon's freeboard was one of its design limitations: it was too low to allow for the traditional sternwalk often fitted to capital ships.

Design
The back slope design of the ship's keel behind the stem was an distinctive feature, fitted to improve the ship's manoeuvrability in tight turns and to reduce draught without reducing displacement.

Torpedoes
Torpedo doors were fitted at each side of the ship's stern. In total the ship had six torpedo tubes fitted.

SERVICE CAREER

From 1898 *Oregon* served across the Pacific Ocean and the US West Coast, until decommissioning on 27 April 1906. Recommissioned on 29 August 1911, the ship was in and out of reserve over the next nine years, and on 17 July 1920, was redesignated BB–3. Struck from the Navy List on 2 November 1942, sold on 7 December, it was returned to the Navy to be used as a storage hulk.

IN HEAVY SEAS
During a typhoon on 14–15 November 1948, *Oregon* (by this time serving as a storage hulk) broke loose and drifted to sea. It was located on 8 December some 805km (500 miles) southeast of Guam and was towed back.

Jauréguiberry (1896)

TYPE • *Pre-dreadnought* **COUNTRY** • *France*

SPECIFICATIONS

DIMENSIONS:	Length: 112.6m (377ft 4in); Beam: 22.15m (72ft 6in); Draught: 8.45m (27ft 9in)
DISPLACEMENT:	10,919 tonnes (12,036 tons) full load
PROPULSION:	24 Lagrafel d'Allest water-tube boilers; 2 vertical inverted triple-expansion engines, 10 769kW (14,441hp); 2 screws
SPEED:	17.07 knots (31.6km/h; 19.6mph)
RANGE:	7260km (3920nmi)
ARMAMENT:	2 x 305mm (12in) guns; 2 x 274mm (10.8in) guns; 8 x 138.6mm (5.4in) guns; 8 x 100mm (3.9in) guns; 16 x 3-pdr guns; 4 x 457mm (18in) torpedo tubes
ARMOUR:	Waterline belt 400–160mm (15.7–6.3in); upper belt 170–120mm (6.7–4.7in); armoured deck 90mm (3.5in); main turrets 370–280mm (15–11in);
COMPLEMENT:	597

Battleships were controversial in the French Navy, with some strategists and naval architects arguing that they were unnecessary and even outmoded as warships. The *Jauréguiberry* class were the first French battleships to have guns mounted in turrets rather than in barbettes.

Below: Deck plan: the Jauréguiberry *carried eight gun turrets on a cramped deck space that included more than 30 guns of various sizes in total.*

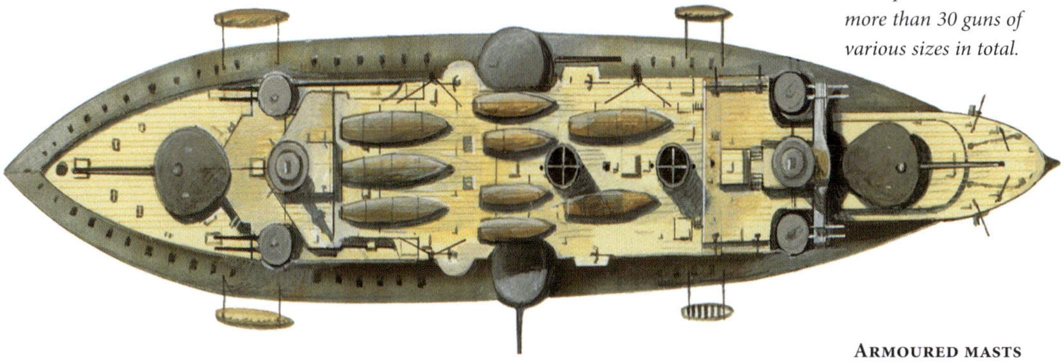

ARMOURED MASTS
The armoured 'military' masts had internal elevators to the lookout, control and gun positions. The relatively massive constructions on the masts would have made vulnerable targets in a gunnery action.

BOAT-LAUNCHING GEAR
One of the many variations between it and other ships in its class was the ingenious but complicated boat-launching gear, needed to clear the tumblehome sides.

TWIN SCREWS
Jauréguiberry was the last twin-screw battleship in the French Navy, which then adopted triple-propulsion.

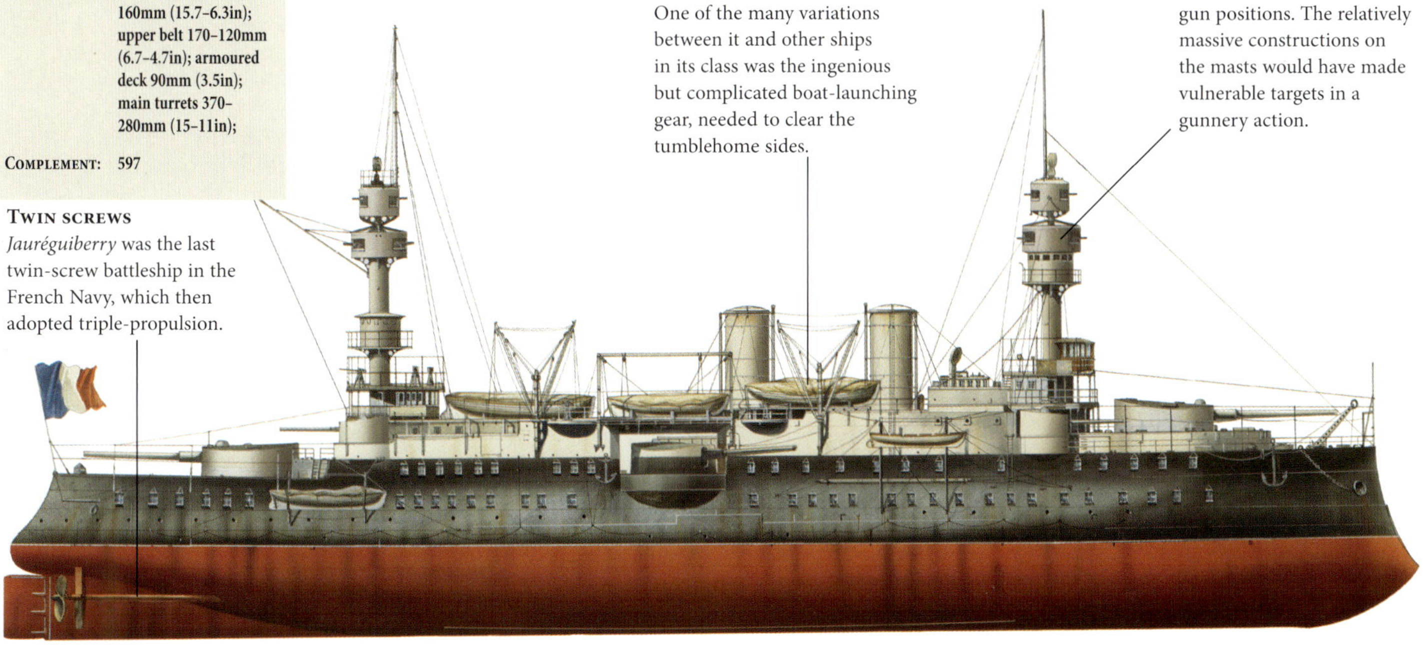

FIREPOWER
The positioning of the big guns gave the ship a field of fire in all directions, with up to three able to fire a broadside. The firing arc of the 305mm (12in) guns was 250 degrees, and at maximum elevation of 15 degrees could send a 340kg (750lb) shell 12,000m (3937ft), which was rather more than the maximum range anticipated for ship-to-ship fighting in the years before 1910.

FULL STEAM
A side view of the French battleship *Jauréguiberry* taken early in its career. *Jauréguiberry* and its companion ships were deliberately designed to look like floating steel fortresses. A consequence of this was greatly restricted deck space compared to other battleships.

HMS *Renown* (1897)

TYPE • *Pre-dreadnought* **COUNTRY** • *United Kingdom*

SPECIFICATIONS

DIMENSIONS:	Length: 125.7m (412ft 3in); Beam: 22m (72ft 4in); Draught: 8.3m (27ft 3in)
DISPLACEMENT:	13,071 tonnes (14,378 tons) full load
PROPULSION:	Twin shafts; vertical triple-expansion engines; eight boilers; 7457kW (10,000ihp)
SPEED:	19 knots (35km/h; 23mph)
RANGE:	8500nmi (15,700km; 9800 miles)
ARMAMENT:	4 × 254mm (10in) BL; 10 × 152mm (6in) QF; 12 × 12-pdr QF; 12 × 3-pdr QF guns; 5 × 457mm (18in) torpedo tubes
ARMOUR:	Belt 203–152mm (8–6in); bulkheads 254–152mm (10–6in); barbettes 254mm (10in); turrets 152–76mm (6–3in); conning tower 229–76mm (9–3in); deck 76–50mm (3–2in)
COMPLEMENT:	674

Although a handsome ship, HMS *Renown* had main guns that were of modest calibre compared with its rivals. In terms of secondary armament and armour protection, however, the ship was well equipped.

Right: Renown *was laid down at Pembroke Dock in February 1893, launched on 8 May 1895 and completed in January 1897, at a cost, including guns, of £751,206.*

GUNS
Renown was originally intended to be given 305mm (12in) guns as the first of a new class, but suitable guns had not been developed by the time of fitting out, and it received 254mm (10in) guns, the same as its immediate predecessors.

PERFORMANCE
Renown was a fast ship, capable of 19 knots (35km/h; 23mph) with forced draught applied to its triple expansion engines, as well as a good sea boat.

ARMOUR
In terms of armour, *Renown* was an innovative ship. Previous warships had a level protective deck inside the hull, but here the protective deck sloped downwards at the sides to meet the lower edge of the armour belt.

REVIEW FLAGSHIP

Renown was flagship at the 1897 Jubilee Review, and on the West Indies and North America station from 1897 to 1899, and of the Mediterranean Fleet from 1899 to 1902. In 1902 the main deck guns were removed for a ceremonial visit by the Duke and Duchess of Connaught to India and those on the upper deck taken out during a refit in 1904 preparatory to another royal visit to India.

STEADY SAILER

Renown's steadiness and good steaming qualities made it suitable for non-military purposes. In reserve between 1902–04 and 1906–07, when it was assigned to the 4th Division of the Home Fleet, the ship's final role was as a stokers' training vessel at Portsmouth, 1909 to 1913.

Charles Martel (1897)

TYPE • *Pre-dreadnought* **COUNTRY** • *France*

The first of a series of five ships of broadly similar displacement and form, all armed with two 305mm (12in) main guns, differing greatly in details, the *Charles Martel* was unmistakably of French design.

SPECIFICATIONS

Dimensions:	Length: 115.49m (378ft 11in); Beam: 21.64m (71ft); Draught: 8.38m (27ft 6in)
Displacement:	11,880 tonnes (13,100 tons) full load
Propulsion:	2 shafts; vertical triple-expansion engines; 24 Lagrafel d'Allest boilers; 11,100kW (14,900ihp)
Speed:	18 knots (33km/h; 21mph)
Range:	2218nmi (4108km; 2552 miles)
Armament:	2 × 305mm (12in) guns; 2 × 274mm (10.8in) guns; 8 × 137mm (5.5in) QF guns; 4 × 9-pdrs; 12 × 3-pdrs; 8 × 1-pdr revolvers; 2 × 457mm (18in) torpedo tubes
Armor:	Belt 457–254mm (18–10in); upper belt 102mm (4in); main turrets 375mm (15in); conning tower 228mm (9in)
Complement: 644	

Left: In British opinion, although the ship was impressive in appearance, the stability of the Charles Martel *in a seaway was dubious.*

Design
Midships, the hull sides were drawn up in a high tumblehome formation covering the central battery and topped by a flying deck through which the two funnels and tall ventilators projected, and which stretched back to form a platform around the mainmast.

Masts
Successive captains complained, without result, about the ship's instability caused by the sheer weight of the towering armoured masts.

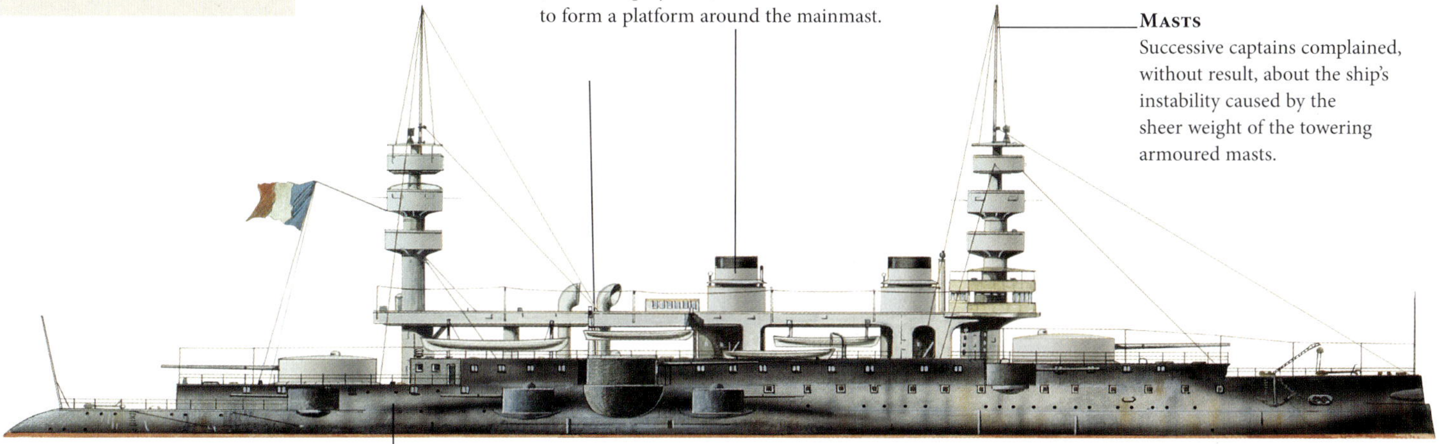

Armour
Protection was of nickel steel, concentrated at the waterline belt and on the gun turrets. The internal armoured deck was of 68mm (2.7in) steel on two layers of 10mm (0.4in).

MEDITERRANEAN

The ship's service was almost entirely in the Mediterranean Squadron, either active or in reserve, apart from one joint exercise in the English Channel with the Northern Squadron in 1900. In 1900 it was flagship of the 2nd Division. In fleet manoeuvres of 1901 it was struck by a dummy torpedo fired from the submarine *Gymnote*. By August 1914 *Charles Martel* had been laid up at the Brest naval base for some years.

FLOATING FORTRESS
The *Charles Martel* pictured in its heyday. During 1914–18 it was used as a floating barracks, and was formally taken off the Navy list in 1922. It was sold that year and towed to Holland for breaking up.

USS *Iowa* (1897)

TYPE • *Pre-dreadnought* **COUNTRY** • *United States*

SPECIFICATIONS

Dimensions:	Length: 110m (360ft); Beam: 22m (72ft 1in); Draught: 7.3m (24ft)
Displacement:	11,528 tonnes (12,703 tons) normal load
Propulsion:	2 shafts; Cramp vertical triple-expansion engines; 12 boilers; 8203kW (11,000ihp)
Speed:	17.9 knots (31.6km/h; 19.7mph))
Range:	5140nmi (9520km; 5920 miles)
Armament:	4 × 305mm (12in) guns; 8 × 203mm (8in) guns; 6 × 102mm (4in) guns; 20 × 6-pdr; 4 × 1-pdr guns; 4 × 360mm (14in) torpedo tubes
Armour:	Belt 356–102mm (14–4in); barbettes 381–318mm (15–12.5in); main turrets 432–381mm (17–15in);
Complement:	727

The United States' fourth 'modern' battleship, the USS *Iowa* was a one-off, its designers building on the experience the US Navy had gained with the USS *Oregon* and two ships of the pioneering *Indiana* class.

Right: Iowa *seen in 1897 or early 1898, with the cruiser USS* Columbia *in the right background. On 12 October 1898, it was dispatched to the Pacific, and served for two years on the West Coast before duty with the South Atlantic (1902–03) and North Atlantic (1903–07) Squadrons.*

TAKING FIRE

From 28 May 1898, during the Spanish–American War, USS *Iowa* was on blockade duty off Santiago de Cuba, with the Atlantic Fleet under Rear Admiral William Sampson. On 3 July it participated in battle when a squadron of six Spanish warships came out from Santiago harbour. *Iowa* was hit at least twice by Spanish shells and fire broke out on board, but was rapidly put out.

BUILDER'S TRIALS

USS *Iowa* runs builder's trials in 1897, with the house flag flying of builder, William Cramp & Sons, of Philadelphia, Pennsylvania. The ship was commissioned on 16 June 1897, four years after it was laid down.

Fuji (1897)

TYPE • *Pre-dreadnought* **COUNTRY** • *Japan*

The fifth warship of Imperial Japan's new navy to be built at a British shipyard, *Fuji* was Japan's largest warship when it was commissioned, although it was soon overtaken in size by the *Shikishima* class.

SPECIFICATIONS

Dimensions:	Length: 125.5m (412ft); Beam: 22.4m (73ft); Draught: 8.07m (26ft 6in)
Displacement:	12,734 tonnes (14,036 tons) normal load
Propulsion:	2 shafts; vertical triple-expansion engines; 10,440kW (14,000ihp)
Speed:	18 knots (33.3km/h; 20.7mph)
Range:	7730km (4170nmi)
Armament:	4 × 305mm (12in); 10 × 152mm (6in); 20 × 3-pdr; 4 × 2.6-pdr guns; 5 × 457mm (18in) torpedo tubes
Armour:	Belt 457–356mm (18–14in); barbettes 356–228mm (14–9in); casemates 152–50mm (6–2in); conning tower 356mm (14in); deck 63mm (2.5in)
Complement:	637

Left: Fuji, *as seen in 1908. From 1923 it was disarmed and used as a training ship. It remained afloat until 1945, when it capsized in harbour, and was scrapped at Sasebo, Japan, in 1948.*

Firepower
The 305mm (12in) guns were mounted in barbettes fore and aft, with light armour protection; and four 152mm (6in) guns were in sponsons placed amidships, with the others on the upper deck, protected by shields.

Pole masts
Two pole masts with broad yards were fitted, with fighting tops in which four 3-pounder guns were placed.

Powerplant
Fuji was powered by twin-shaft vertical triple-expansion engines, fired by 10 single-ended cylindrical boilers. The ship had a bunker coal capacity of 1219 tonnes (1344 tons).

BRITISH INFLUENCE

Fuji was built by Thames Ironworks at Poplar, London, laid down on 1 August 1894, launched on 31 March 1896 and completed on 17 August 1897. A companion ship, *Yashima*, was completed at Elswick on the Tyne a month later. HMS *Royal Sovereign* was the design model, although Japanese engineers incorporated many points of their own.

IMPERIAL MIGHT

Fuji photographed early in its career, before the Russo–Japanese War of 1904–05. In the Battle of Tsushima, 27 May 1905, *Fuji* was hit 12 times and one of its main guns in the after barbette was put out of action, although it sank the Russian battleship *Borodino* in the closing phase of the battle.

Argonaut (1897)

TYPE • *Submersible* **COUNTRY** • *France*

SPECIFICATIONS	
DIMENSIONS:	Length: 11m (36ft); Beam: 2.7m (9ft); Draught: 3m (10ft)
DISPLACEMENT:	59.9 tonnes (66 tons) surfaced; submerged displacement unknown
PROPULSION:	1 White & Middleton gasoline engine, 22kW (30hp); single screw
SPEED:	4 knots surfaced (7km/h; 4.5mph); 2 knots submerged (3.7km/h; 2.3mph)
RANGE:	555km (300nmi; 345 miles)
ARMAMENT:	None
COMPLEMENT:	5

Submersible designer Simon Lake's wheeled submarine can fairly be described as the precursor of all seabed-traversing vehicles, especially those used for non-military activities. It brought him riches and fame, but it also stands as a pioneering undersea craft for many other reasons.

Right: In this evocative photograph, Lake's submarine Argonaut *is prepared for launching in 1897. Press reports about the vessel were enthusiastic.*

STEERING HANDLE
The surface deck featured a steering handle, which was also replicated inside the hull.

AIR AND GAS
Air intake and gas exhaust were at first by means of canvas hoses, but two 15.2m (50ft) pipes proved more efficient. A closing valve in each tube enabled the boat to go deeper than 15m (50ft) for up to 24 hours, using air from two compressed-air cylinders.

BALLAST
Ballast tanks were at the base. *Argonaut* did not have hydroplanes for even-keel submersion. It submerged vertically, on the level, either by flooding the tanks or lowering its two 450kg (1000lb) anchor weights and reeling itself down.

DESIGN
Argonaut, built in 1897 in Baltimore at the Columbian Iron Works & Dry Dock Co., included a diver's air-lock and powered wheels for bottom crawling. The boat was driven by a 22kW (30hp) gasoline engine, even while submerged, when it used a hose supported by a surface float to supply combustion air. A supply of compressed air permitted even deeper excursions and bottoming for as long as 24 hours.

***ARGONAUT* INTERIOR**
Primarily intended for exploration and salvage, *Argonaut* travelled on the surface to designated sites, where it would descend vertically, either by ballasting down or using haul-down anchors installed in the keel. Then it would trundle along the bottom on its powered wheels, towing a surface float.

USS *Holland* (1897)

TYPE • *Submarine* **COUNTRY** • *United States*

Named after its Irish–American designer John P. Holland, this vessel was the first submarine commissioned into the US Navy. Among its many innovations, it was also the first sub to employ separate power sources for surface running and submerged operation.

SPECIFICATIONS

Dimensions:	Length: 16.4m (53ft 10in); Beam: 2.6m (8ft 6in); Draught: 2.6m (8ft 6in)
Displacement:	65 tonnes (72 tons) surfaced; 75 tonnes (83 tons) submerged
Propulsion:	Otto gasoline engine developing 34kW (45hp) with one E.D. electric motor developing 56kW (75hp); single screw
Speed:	8 knots surfaced (14.8km/h; 9mph); 5 knots submerged (9.2km/h; 5.7mph)
Range:	1150nmi (2129km; 1323 miles)
Armament:	1 × 457mm (18in) bow torpedo tube with 2 torpedoes; 2 × Zalinski pneumatic cannon (later reduced to 1)
Complement: 6	

Left: Holland 5 *was the fifth* Holland*-type submarine commissioned for the British Royal Navy, entering service in 1902.*

Aerial torpedo
The Zalinski cannon, or 'dynamite gun', fired a level shot just above the water's surface. It was known as the 'aerial torpedo'. In test firings, a 1m-long (3ft 3in), 22.5kg (50lb) wooden dummy projectile ran forward for just under 300m (980ft).

Propeller
One of J.P. Holland's fixed ideas was that the propeller of the submarine should be placed beyond the steering gear. He was eventually persuaded to have the large three-bladed screw moved inboard of the rudder.

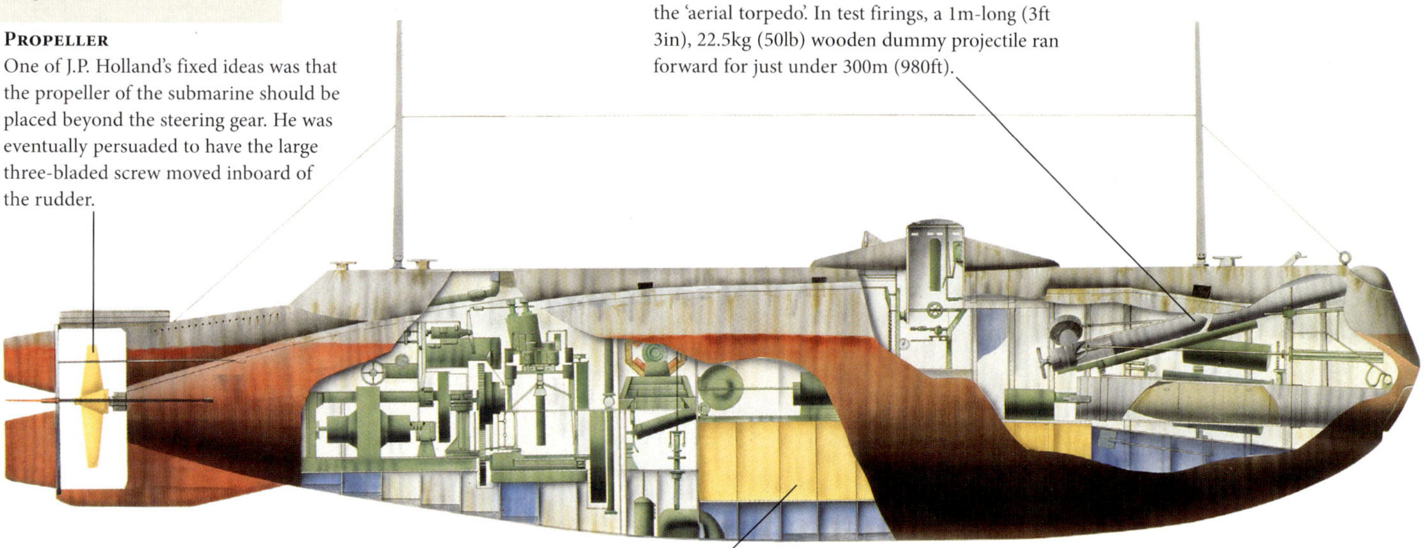

Power supply
As in subsequent submarines, the batteries in *Holland* were placed centrally close to the keel, their weight helping to maintain the boat's trim.

FAMILIARIZATION

After the launch, *Holland*'s team had to learn how to handle the boat. In March 1898, the vessel made its first successful underwater runs in open water, which required considerable adjustments of the dry ballast in order to submerge in waters of greater salinity. There was no periscope, and position of the boat and of any target had to checked by the art of 'porpoising' – bringing the craft just above the surface for a few seconds before diving again.

USS *HOLLAND* (SS-1)
The vessel was designed for attack, with a single reloadable 457mm (18in) torpedo tube and two Zalinski 210mm (8.2in) pneumatic cannon mounted fore and aft, not as deck guns, but within the hull structure (the aft one was removed in 1900).

HMS *Canopus* (1897)

TYPE • *Pre-dreadnought* **COUNTRY** • *United Kingdom*

SPECIFICATIONS

DIMENSIONS:	Length: 128m (421ft 6in); Beam: 23m (74ft); Draught: 8m (26ft)
DISPLACEMENT:	13,360 tonnes (14,723 tons) normal load
PROPULSION:	2 shafts; triple-expansion engines; 20 Belleville boilers; 11,484kW (15,400ihp)
SPEED:	18.3 knots (34km/h; 21mph)
RANGE:	N/A
ARMAMENT:	4 x 305mm (12in) guns; 12 x 152mm (6in) QF; 10 x 12-pounder QF gunsguns; 4 x 457mm (18in) torpedo tubes
ARMOUR:	Belt 146mm (5.75in); barbettes 305mm (12in); conning tower 305mm (12in); deck 50mm (2in)
COMPLEMENT:	682

Designed for long-range cruising, the *Canopus* class were the Royal Navy's first battleships to have water-tube boilers and Krupp armour. They were primarily used in the Far East to counter rising Japanese power.

Right: An essential aspect of the design was a draught able to negotiate the Suez Canal, and a saving in weight along with other advantages was obtained by the use, for the first time in a British battleship, of water-tube boilers.

ARMOUR
Another Continental borrowing was the Krupp cemented armour in the belt and barbettes; its thickness of 146mm (5.75in) equated to 190mm (7.5in) of US 'Harveyized' armour and 305mm (12in) of compound armour.

CASEMATES
The casemates on the main deck were extended out into sponsons at each end to allow end-on fire by the 152mm (6in) guns. A fire-control position was mounted on the foremast in 1907.

BOILERS
Pioneered in France, water-tube boilers worked at higher pressure and were more economical on fuel than the established cylindrical boilers.

COUNTERWEIGHT

The rapid expansion of the Imperial Japanese Navy prompted the British Admiralty to introduce a new battleship class intended for the China Station, based at Hong Kong, in order to maintain British sea power at a high level. *Canopus*, the first of six ships all placed in service between December 1899 and April 1902, was laid down at Portsmouth Dockyard on 4 January 1897, launched on 12 October in the same year and completed in December 1899.

MAIN GUNS

Sent as guard ship to the Falkland Islands in October 1914, *Canopus* was briefly in action against von Spee's German squadron in December 1914. From January to May 1915 it served in the Dardanelles campaign. In 1916, at the end of *Canopus*' effective career, the main-deck 12-pdr guns were replaced by two 152mm (6in) guns.

Kaiser Friedrich III (1898)

TYPE • *Pre-dreadnought* **COUNTRY** • *Germany*

SPECIFICATIONS	
Dimensions:	Length: 125.3m (411ft); Beam: 20.4m (67ft); Draught: 7.9m (25ft 11in)
Displacement:	11,150 tonnes (12,332 tons) normal load
Propulsion:	3 shafts; vertical triple-expansion engines; 9694kW (1300ihp)
Speed:	18 knots (33.3km/h; 20.7mph)
Range:	2980nmi (5520km; 3430 miles) at 10 knots
Armament:	4 × 238mm (9.4in) guns; 18 × 150mm (5.9in) guns; 12 × 88mm (3.5in) guns; 12 × 1-pdr guns; 6 × 457mm (18in) torpedo tubes
Armour:	Belt 305–150mm (12–5.9in); turrets 250mm (9.8in); casemates 150mm (5.9in); conning tower 254mm (10in); deck 65mm (2.6in)
Complement:	687

Lead ship of the *Kaiser* class, this vessel had relatively light main guns but a heavy secondary armament. Following the commissioning of HMS *Dreadnought*, *Kaiser*-class ships underwent considerable modification.

Right: The Kaiser *class were products of the naval rivalry between Britain and Germany that saw both countries enlarging their battle fleets at colossal expense in the early decades of the 1900s.*

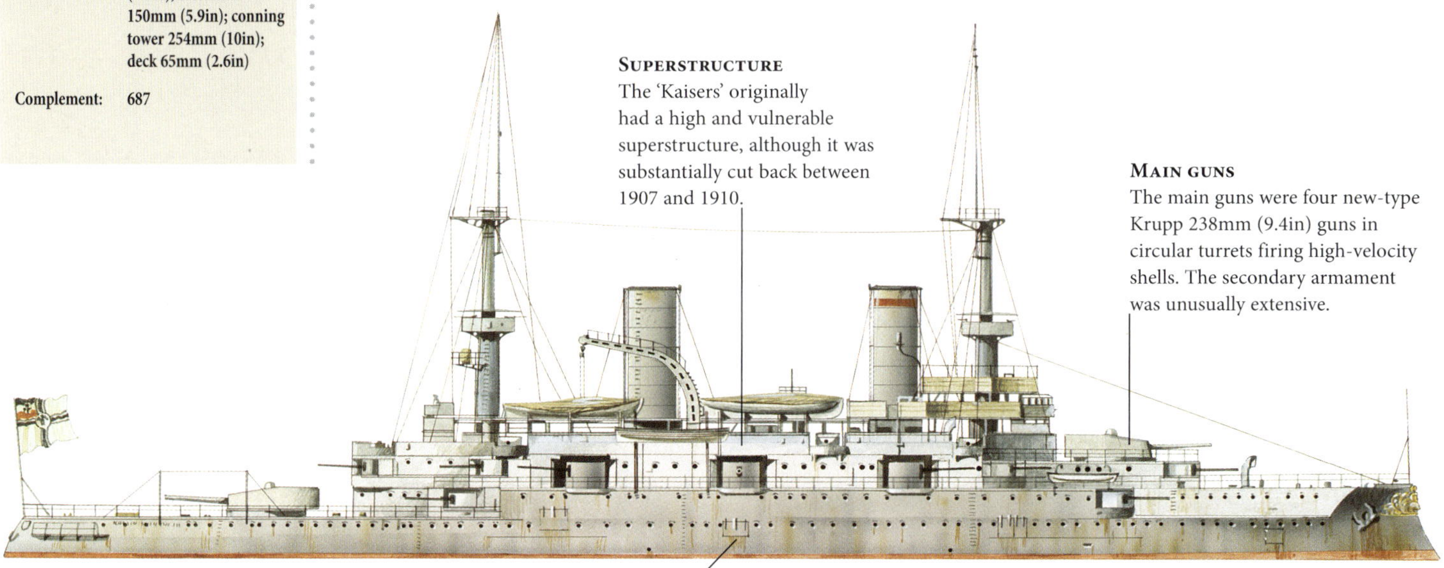

SUPERSTRUCTURE
The 'Kaisers' originally had a high and vulnerable superstructure, although it was substantially cut back between 1907 and 1910.

MAIN GUNS
The main guns were four new-type Krupp 238mm (9.4in) guns in circular turrets firing high-velocity shells. The secondary armament was unusually extensive.

FUEL
This was the first heavy German warship to use oil fuel as a supplement to coal, to help in quick steam-raising.

FLAGSHIP
The ship was assigned to the 1st Squadron of the *Heimatsflotte*, precursor of the High Seas Fleet, as flagship. On 2 January 1901, it was holed by a submerged obstacle on the eastern approach to Kiel, fire broke out and the magazines and engine rooms were flooded. Repairs were made in Wilhelmshaven. In 1910 the ship was put into reserve, but recommissioned in August 1914, when the 'Kaisers' formed the 5th Battle Squadron, operating in the Baltic Sea until February 1915.

IN KIEL, 1900
Laid down at the new German North Sea base of Wilhelmshaven in 1895, the ship was launched on 31 July 1896 and completed on 7 October 1898, the same day as *Kaiser Wilhelm II*, which had been launched a year later at the same yard. This photograph, taken in Kiel in 1900, shows the ship in its original configuration.

Charlemagne (1899)

SPECIFICATIONS

DIMENSIONS:	Length: 114m (374ft); Beam: 20.24m (56ft 5in); Draught: 8.38m (27ft 6in)
DISPLACEMENT:	11,278 tonnes (12,423 tons) normal load
PROPULSION:	3 shafts; vertical triple-expansion engines; 20 Belleville boilers; 11,185.5kW (15,000ihp)
SPEED:	18 knots (33.3km/h; 20.7mph)
RANGE:	3650nmi (6758km; 4200 miles) at 10 knots
ARMAMENT:	4 x 305mm (12in) guns; 10 x 139mm (5.5in)/ QF guns; 8 x 101mm (3.9in); 20 x 3-pdr; 4 x 1-pdr; 4 x 457mm (18in) torpedo tubes
ARMOUR:	Belt 368–203mm (14.5–8in); upper belt 102mm (4in); battery 76mm (3in); turrets 375mm (15in); conning tower 330mm (13in)
COMPLEMENT:	694

First of a class of three, with *Charlemagne* the French Navy adopted the twin-gun turrets already in wide use by other fleets. *Charlemagne* was laid down at Brest Arsenal on 14 July 1894, launched on 17 October 1895 and completed in December 1899.

Left: This illustration shows a fine depiction of the Charlemagne *under sail in Brassey's* Naval Annual *(1900).*

ARMOUR
Harveyized nickel steel armour was fitted, but although the waterline belt ran the full length of the hull, it was relatively narrow, and there was no armour protection between its upper level and the battery.

BOW
Even for its time, the ship had an exceptionally curved bow, and the forward position of the turret made it extremely wet in even a moderate head sea.

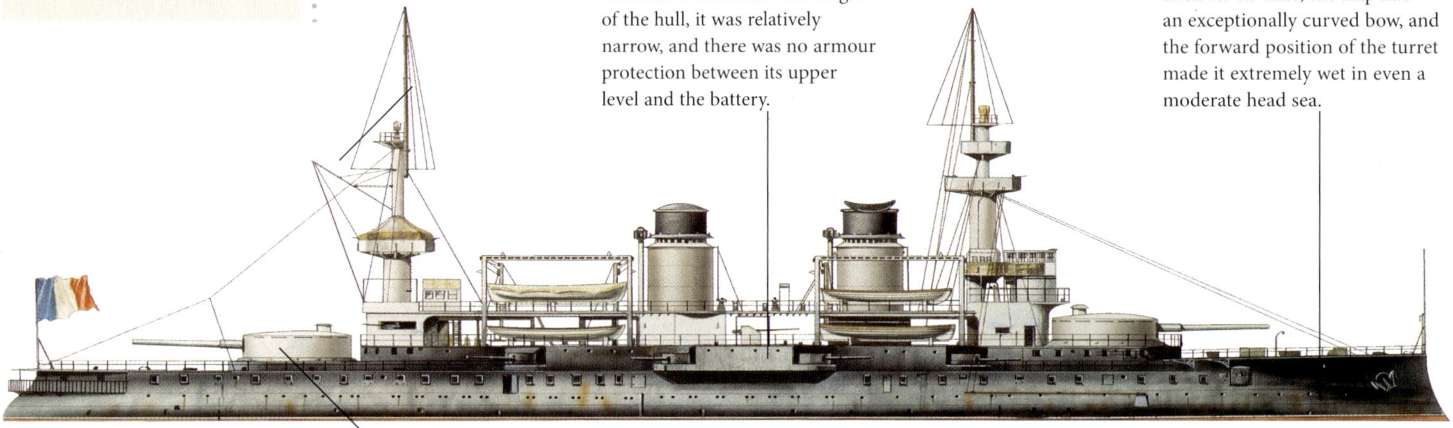

GUN TURRETS
With its two-gun turrets *Charlemagne* marked a step forward, although its 305mm (12in) M1893 guns were of 40 calibre compared to 45 calibre in the preceding, larger *Bouvet* class (1898).

WORLD WAR I

During World War I *Charlemagne* acted as a convoy escort in the Mediterranean and participated in the Dardanelles campaign between November 1914 and April 1915, bombarding Turkish shore forts and suffering some damage from return fire, then undergoing repairs at the Bizerta base. It remained in the Eastern Mediterranean until 1917, when it returned to Toulon and was disarmed. Eventually, in 1920 it was sold for breaking.

MUSTER ON DECK
The crew of the *Charlemagne* fall in for divisions on deck. The ship served with both the Northern and Mediterranean Squadrons during the 1900s and was flagship in the Mediterranean from 1901, passing into reserve in October 1912 but recommissioned as a training ship in August 1913.

USS *Kearsarge* (1900)

TYPE • *Pre-dreadnought* **COUNTRY** • *United States*

SPECIFICATIONS

DIMENSIONS:	Length: 114.4m (375ft 4in); Beam: 22.02m (72ft 3in); Draught: 7.16m (23ft 6in)
DISPLACEMENT:	10,469 tonnes (11,540 tons) normal load
PROPULSION:	2 vertical triple-expansion engines; 2 screws; 7457kW (10,000hp)
SPEED:	16 knots (29.6km/h; 18.4mph)
RANGE:	3000 nmi (5556km; 3452 miles) at 10 knots
ARMAMENT:	4 x 330mm (13in) guns; 4 x 203mm (8in) guns; 4 x 152mm (6in) guns; 20 x 6-pdr; 8 x 1-pdr guns; 4 x 457mm (18in) torpedo tubes
ARMOUR:	Main belt 419–127mm (16.5–5in); main turrets 432–381mm (17–15in); conning tower 254mm (10in)
COMPLEMENT:	553

Bearing an already famous name, *Kearsarge* became one of the best-known American warships as part of President Theodore Roosevelt's 'Great White Fleet'. It also went on to have a significant later role.

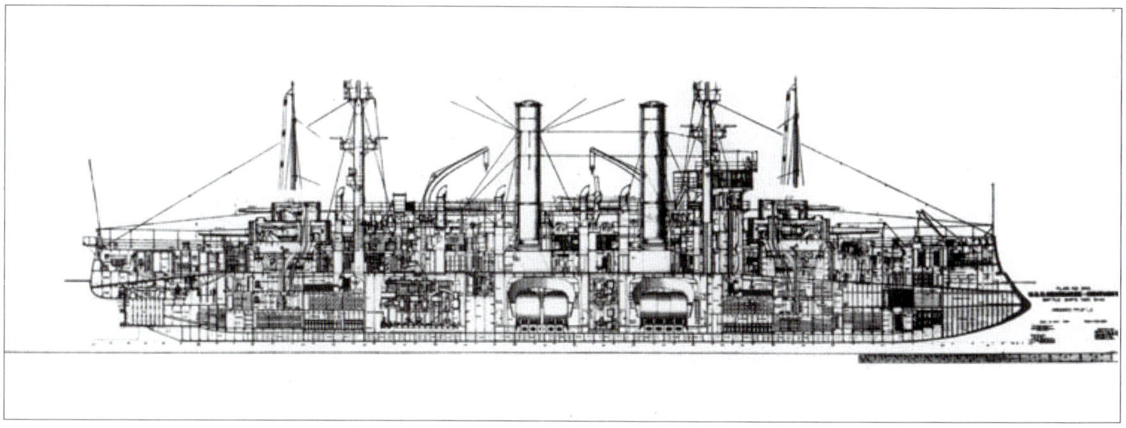

Above: A side plan of USS Kearsage *(BB-5) and* Kentucky *(BB-6).*

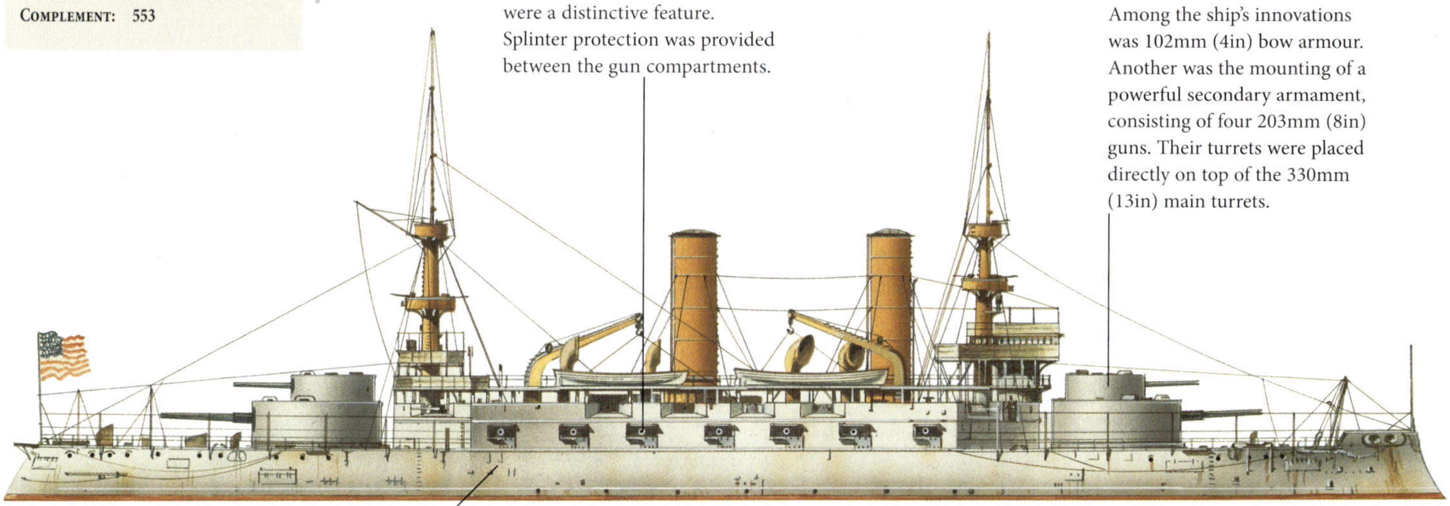

CASEMATES
The seven-a-side casemates were a distinctive feature. Splinter protection was provided between the gun compartments.

INNOVATIONS
Among the ship's innovations was 102mm (4in) bow armour. Another was the mounting of a powerful secondary armament, consisting of four 203mm (8in) guns. Their turrets were placed directly on top of the 330mm (13in) main turrets.

HULL
The wide hull, with a ratio to the ship's length of more than 20 per cent, made it highly suitable for its later conversion to a floating crane.

FLAGSHIP ROLES
Kearsarge was deployed as flagship of the North Atlantic Squadron and in the summer of 1903 it served as flagship of a special squadron on a goodwill visit to Europe. USS *Maine* took over as Atlantic flagship on 31 March 1905, but *Kearsarge* remained with the fleet until joining the 'Great White Fleet' which showed the American flag around the world between 16 December 1907 and 22 February 1909.

IN HARBOUR, C. 1900
Kearsarge was painted this way when first completed, in 1900. On 10 May 1920, it was decommissioned and work began on converting it into a crane ship, its superstructure replaced by a giant 226-tonne (250 ton) revolving crane for dockyard work and salvage.

Shikishima (1900)

TYPE • *Pre-dreadnought* **COUNTRY** • *Japan*

SPECIFICATIONS

Dimensions:	Length: 133.5m (438ft); Beam: 23m (75ft 6in); Draught: 8.29m (27ft 6in)
Displacement:	15,701 tonnes (17,310 tons) full load
Propulsion:	2 shafts; vertical triple-expansion engines; 10,812kW (14,500ihp)
Speed:	18 knots (33.3km/h; 20.7mph)
Range:	5000 nmi (9260km; 5750 miles) at 10 knots
Armament:	4 × 305mm (12in) guns; 14 × 152mm (6in) guns; 20 × 12-pdr; 6 × 3-pdr; 6 × 2.5-pdr guns; 5 × 457mm (18in) torpedo tubes
Armour:	Main belt 228–102mm (9–4in); upper belt 152mm (6in); barbettes 356–203mm (14–8in); casemates 152–50mm (6–2in); conning tower 356–76mm (14–3in); deck 102–63mm (4–2.5in)
Complement:	836

At the turn of the century, the ordering of *Shikishima* and its companion ship *Hatsuse* made clear that Japan was set on becoming a significant, perhaps even the most significant, naval power in the western Pacific.

Right: Shikishima *followed* Fuji *at the Thames Ironworks, London, laid down on 29 March 1897, launched on 1 November 1898, and completed on 26 January 1900.*

Construction
Construction was on the bracket frame system, which provided lateral spaces used for extra coal bunkers that also served as shock absorbers against shell strikes.

Appearance
Triple-funnelled, with broad yards on both masts, *Shikishima* and *Hatsuse* had a distinctly British look, hardly surprising as they were built in England and modelled on the British *Majestic* class turned out in 1895–98.

Hull
The ship had a double bottom and the new type of armoured deck designed for HMS *Renown*, sloping up from the base of the protective belt. Internally there were 261 watertight compartments.

BUILD-UP

Japan's build-up of its naval strength began as a counter to the Imperial Chinese fleet, but continued after its defeat of China in 1895, to rival Russia as a power in the western Pacific. Its construction, and that of three other battleships, was paid for by Chinese war indemnities. Although larger than the two *Fuji*-class ships, *Shikishima* and *Hatsuse* had the same main armament but a heavier battery of secondary guns.

***SHIKISHIMA*, APRIL 1947**
In 1921 *Shikishima* was classed as a coast defence ship, and under the terms of the Washington Naval Treaty was disarmed and its propellers were removed. From 1923 it was used as a floating training school, and survived until 1947, when it was scrapped at Sasebo.

Peresviet (1901)

TYPE • *Pre-dreadnought* **COUNTRY** • *Russia*

SPECIFICATIONS

Dimensions: Length: 132.4m (434ft 5in); Beam: 21.8m (71ft 6in); Draught: 7.9m (26ft)

Displacement: 13,534 tonnes (14,920 tons) normal load; 14,639 tonnes (16,125 tons) full load

Propulsion: 3 shafts; vertical triple-expansion engines; 30 Belleville boilers; 10,813kW (14,500ihp)

Speed: 16 knots (29.6km/h; 18.4mph)

Range: 6200 nmi (11480km; 7130 miles) at 10 knots

Armament: 4 x 254mm (10in) guns; 11 x 152mm (6in) guns; 20 x 76mm (3in) guns; 8 x 37mm (1.5in) guns; 5 x 381mm (15in) torpedo tubes; 45 mines

Armour: Belt 229–178mm (9–7in); turrets 229mm (9in); deck 76–51mm (3–2in)

Complement: 771

Russia's first three-funnelled battleship, *Peresviet* had the unique fate of being sunk twice, in successive wars, by enemy action. First of a class of three, it was launched on 19 May 1898 and completed in July 1901.

Left: Exceptionally high freeboard gave the ship a massive appearance, although the superstructure was relatively low. Two tubular French-type masts were fitted, with pole topmasts.

Mast
The yards and gaff on the mainmast were for signalling purposes; no rig was ever carried. Fighting tops were removed in the Japanese reconstruction of 1905–08.

Turrets
The turrets were electrically powered and the ship carried up-to-date equipment including stadiametric rangefinders, optical gunsights developed at the Obhukov Works, and Geisler electromechanical transmission between fire control and guns.

Hull
Ten transverse bulkheads divided the hull into watertight compartments and a partial centreline bulkhead separated the two forward engine rooms.

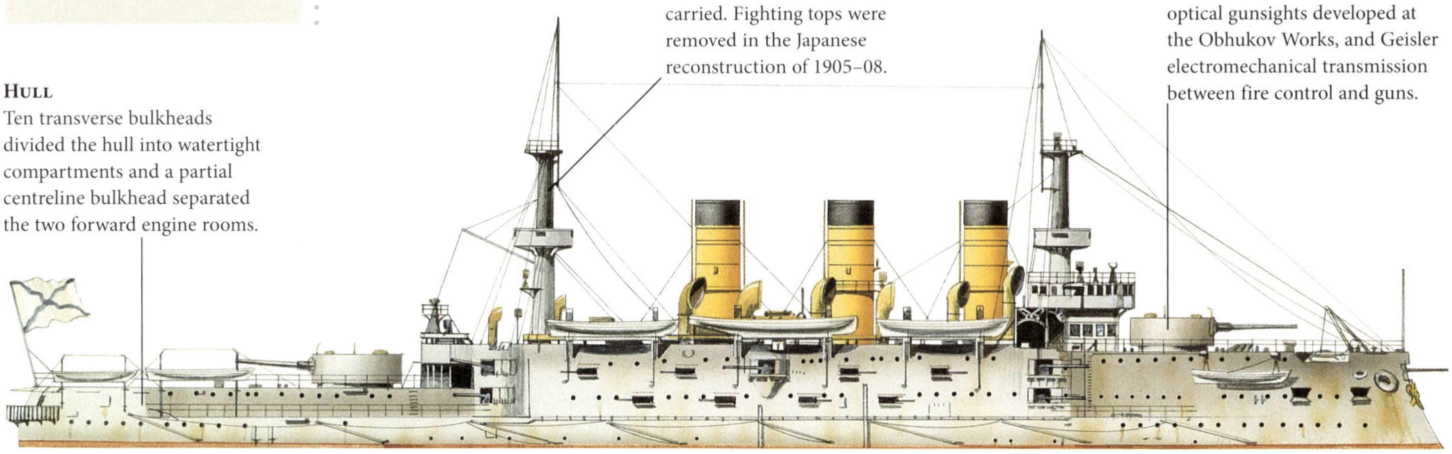

SERVICE
On completion, the ship joined the Pacific Squadron at Port Arthur as flagship of the second-in-command. In the Battle of the Yellow Sea, 10 August 1904, *Peresviet* was hit 39 times but made a safe return to Port Arthur. Following subsequent battle damage, however, the ship was scuttled in shallow water on 10 December. Repaired at Yokosuka Navy Yard in 1905–08, and fitted with Armstrong-type guns and Miyabara boilers, it was commissioned into the IJN as *Sagami* in April 1908.

PORT ARTHUR, 1904
Returned to Russia in 1916, *Sagami* was renamed *Peresviet*. On 4 January 1917, on the way from Vladivostok to join the Arctic Squadron, it hit a mine off Port Said, planted by the German *U-73*. The ship sank with the loss of 116 crew.

HMS *London* (1902)

TYPE • *Pre-dreadnought* **COUNTRY** • *United Kingdom*

SPECIFICATIONS

Dimensions:	Length: 131m (431ft 9in); Beam: 23m (75ft); Draught: 8.3m (27ft 5in)
Displacement:	15,241 tonnes (16,798 tons) normal load
Propulsion:	2 shafts; vertical triple-expansion engines; 11,558kW (15,500ihp)
Speed:	18 knots (33.3km/h; 20.7mph)
Range:	5000 nmi (9260km; 5750 miles) at 10 knots
Armament:	4 x 305mm (12in) guns; 12 x 152mm (6in) guns; 16 x 12-pdr; 6 x 3-pdr guns; 4 x 450mm (18in) submerged torpedo tubes
Armour:	Belt 229mm (9in); bulkheads 305–229mm (12–9in); turrets 254–203mm (10–8in); casemates 152mm (6in); conning tower 356mm (14in); deck 64–25mm (2.5–1in)
Complement:	760

HMS *London* was among the first capital ships to be adapted for naval aviation. It was converted to a minelayer during World War I and in this capacity it served mostly in the Mediterranean Theatre.

Right: HMS London *was a* Formidable*-class battleship in the British Royal Navy from 1902 to 1919. This broadside view shows the ship in wartime dazzle camouflage*

Deck armour
Long-range fire from high-angle guns had made decks more vulnerable, and the Londons had three armoured decks: 50mm (2in) on the main deck over the citadel, and a maximum of 64mm (2.5in) on the lower deck.

Penetration
The main guns could pierce a 305mm (12in) steel plate at a range of 4390m (14,402ft). The guns could be loaded at any position and elevation.

Minelayer
As a minelayer, the ship's main guns were removed, and storage space and launching rails for 240 mines were installed in February–April 1918.

ARMS RACE

By 1900 Britain's new battleships, now of 14,225 tonnes (14,000 tons) plus, were costing more than a million pounds. A naval arms race with Germany was under way, and between September 1901 and October 1904, 14 first-class battleships were completed for the Royal Navy. *London* was first of its class to be laid down, at Portsmouth on 8 December 1898, and launched on 18 October 1899, but HMS *Bulwark* was completed three months sooner, in March 1902.

SEAPLANE EXPERIMENTS

In June 1902 *London* was flagship at King Edward VII's Coronation Review, before being deployed with the Mediterranean Fleet until 1907. After a refit at Chatham in 1908 it served with the Atlantic Fleet as flagship. In 1912 it was used for seaplane experiments with the 5th Squadron of the 2nd Fleet, at the Nore.

Mikasa (1902)

TYPE • *Pre-dreadnought* COUNTRY • *Japan*

Admiral Heihachiro Togo's flagship and Japan's largest battleship in the Russo–Japanese War of 1905, the British-built *Mikasa* was involved in two fierce battles, in the Yellow Sea and at Tsushima. Later sunk, and raised, it is the only surviving battleship of the pre-dreadnought era.

SPECIFICATIONS

DIMENSIONS:	Length: 131.7m (432ft); Beam: 23.2m (75ft 6in); Draught: 8.28m (27ft 6in)
DISPLACEMENT:	13,789 tonnes (15,198 tons) normal load
PROPULSION:	25 Belleville water-tube boilers; 2 vertical triple-expansion engines; 11,185kW (15,000hp); 2 screws
SPEED:	18 knots (33.3km/h; 20.7mph)
RANGE:	7000 nmi (13,000km; 8080 miles) at 10 knots
ARMAMENT:	4 x 305mm (12in) guns; 14 x 152mm (6in) guns; 20 x 12-pdr; 8 x 3-pdr; 4 x 2.5-pdr guns; 4 x 457mm (18in) torpedo tubes
ARMOUR:	Belt 229–102mm (9–4in); bulkheads 305mm (12in); deck 76mm (3in); main turrets 254–203mm (10–8in);
COMPLEMENT:	830

Left: Mikasa *c. 1905. The ship was decommissioned late in 1921. Since 1925 it has been a museum display ship at Yokosuka (restored 1958–61).*

DESIGN
Mikasa was the second of three ships in the *Asahi* class, all built in British yards between 1900 and 1902. They were generally similar to the British *Majestic*-class 'pre-dreadnoughts' in design.

BRIDGES
Flying bridges fore and aft carried searchlights on each side.

ARMOUR
Krupp cemented armour, which had replaced Harveyized nickel steel as the most shell-resistant protection, was fitted to the steel plates, and the internal armoured deck sloped downwards fore and aft to meet the ram and to cover the steering gear.

RUSSO–JAPANESE WAR

Mikasa's aggressive and defensive abilities were heavily tested in the Russo–Japanese war of 1904–05, a campaign watched with intense interest by other naval powers as there had been no major naval battle for decades. On 10 August 1904 the Japanese fleet under Admiral Heihachiro Togo, with *Mikasa* as his flagship, engaged the Russian Pacific fleet in the fast-moving Battle of the Yellow Sea. No ships were sunk, but the Russians were driven into the shelter of Port Arthur.

BATTLE OF TSUSHIMA, 1905
Admiral Heihachiro Togo (holding his sword) stands on the bridge of *Mikasa* before the Battle of Tsushima, 27–28 May 1905. Although *Mikasa* took some 40 strikes in the battle, including 10 from 305mm (12in) shells, no major damage was done.

Herzogin Cecilie (1902)

TYPE • *Barque* **COUNTRY** • *Germany*

SPECIFICATIONS

Dimensions:	Length: 94.5m (310ft); Beam: 14m (46ft); Draught: 7.4m (24ft 7in)
Displacement:	2830 tonnes (3120 tons) normal load
Propulsion:	Sail only; four masts; barque rig
Speed:	20 knots (37km/h; 23mph)
Range:	N/A
Armament:	None
Armour:	None
Complement:	c. 26, plus cadets

Herzogin Cecilie **('Duchess Cecilie') was named after a German duchess and served as both a sail-training and cargo ship between 1902 and 1936. A highly admired four-masted barque of elegant design, it was wrecked off the coast of South Devon in June 1936.**

Below: Sailing mostly from Bremerhaven, Herzogin Cecilie *carried mixed cargoes of manufactured goods out to South American or Australian ports, returning with bulk loads of nitrates, grain, or timber.*

Sails
The Duchess's sail area was 3530 sq m (38000 sq ft), capable of driving at 20 knots (37km/h; 23mph).

Poop deck
In 1912 the poop deck was lengthened by 5.8m (19ft) to make room for 90 cadets.

Deckhouse
Because of its large cadet crew, the ship had a long deckhouse from quarterdeck to stern. This enabled it to carry passengers in the 1930s.

CADET SHIP

Constructed at Rickmers Reismühlen Reederei & Schiffbau, Bremerhaven, Germany, in 1902, *Herzogin Cecili* was a purpose-built sail training ship, ordered by the Norddeutscher Lloyd Steamship Company. It was designed to combine the training aspect with carrying commercial cargoes. The steel hull had shapely lines, and below the weather deck it held accommodation for a professional crew plus 63 cadets.

FULL SAIL
During World War I, *Herzogin Cecilie* was interned in Chile, leaving only as late as 1920, carrying a cargo of nitrates to Ostend. On arrival it was passed to France as part of Germany's postwar reparations.

Protector (1902)

TYPE • *Submarine* **COUNTRY** • *United States*

SPECIFICATIONS

Dimensions: Length: 19.8m (65ft); Beam: 3.35m (11ft); Draught: 3.35m (11ft)

Displacement: 117 tonnes (129 tons) surfaced; 173 tonnes (191 tons) submerged

Propulsion: Two 186kW (250hp) gasoline engines; twin 82kW (110hp) electric motors; two screws

Speed: 11 knots surfaced (20.3km/h; 12.6mph); 7 knots submerged (12.9km/h; 8mph)

Range: 1000 nmi (1852km; 1150 miles)

Armament: 3 x 381mm (15in) torpedo tubes; 1 quick-fire cannon

Complement: 6

In the early 1900s, the Imperial Russian Navy tried to accelerate the creation of a submarine arm by acquiring American technology. At the same time, Simon Lake was trying, unsuccessfully, to break the Electric Boat Company's monopoly of the US Navy market.

Left: A photograph of sea trials suggests the precarious nature of early submarine development. Lake's design was certainly superior to the Holland type in numerous specifications, but was possibly too complex for the liking of the US Navy. Multiple features also increased the potential for things to go wrong.

Exhaust vent
An exhaust funnel for the gasoline engines was mounted. Later Lake submarines would have a protective fairing built around the funnel to ease waterflow past it.

Conning tower
One photograph of *Protector* shows the top of the tower railed and with a steersman's wheel at its forward end.

Access points
Access hatches were provided forward and aft of the conning tower. The airlock and diving compartments are located just behind the bows.

MILITARY VESSEL

Simon Lake's first military vessel, *Protector*, was built speculatively in the hope of gaining sales to the US Navy. Launched at Bridgeport on 1 November 1902, it was a steel-hulled submarine, more powerful than the post-Holland A-class. *Protector*'s hull was almost ship-like, rather than cigar-shaped, although it had a flat keel. It was also the first submarine with diving planes fitted amidships, just forward of the conning tower.

SURFACE RUN

The craft could make 11 knots (20.3km/h; 12.6mph) on the surface and 7 knots (12.9km/h; 8mph) submerged, with reported underwater endurance equivalent to a radius of 80km (50 miles). It was capable of diving to 46m (150ft) and remaining submerged for 60 hours.

Preussen (1903)

TYPE • *Pre-dreadnought* **COUNTRY** • *Germany*

SPECIFICATIONS

DIMENSIONS:	Length: 147m (482ft); Beam: 16.4m (53ft 8in); Draught: 8.26m (27ft 1in)
DISPLACEMENT:	3200 tonnes (3527 tons) normal load
PROPULSION:	Sail only; five masts; full square rig
SPEED:	20.5 knots (38km/h; 23.4mph)
RANGE:	N/A
ARMAMENT:	None
ARMOUR:	None
COMPLEMENT:	46

***Preussen* ('Prussia') was one of only four five-masted square-riggers ever built. It was the second ship of the name to sail for the Reederei F. Laeisz 'Flying P' line of Hamburg, Germany, and made many fast passages between South American and northern European ports.**

Below: Preussen *was a pure sailing ship, but with a relatively small crew. Labour-saving devices were needed and numerous winches were fitted, as well as sliding shoes for the yards.*

MASTS
All five masts were fully rigged, with a claimed total sail area of 6806 sq m (73,260 sq ft), a figure that was almost certainly never exceeded.

DESIGN
Preussen's hull was built of steel, as were the lofty masts and the yards, and most of the fixed rigging was of steel cable. A central island was built between the second and third masts, providing crew accommodation.

WHEEL
Preussen's double wheel was 2m (6ft 6in) in diameter. Tacking in near-gale winds, it might take six men to hold it against the rudder's tendency to swing away from its set course.

FAST PASSAGE

All Laeisz ships had names beginning with P. *Preussen* was ordered in November 1900, laid down at the J.C. Tecklenborg Shipyard, Geestemünde, Germany, in August 1901 and launched on 7 May 1902. It was ready for service by 31 July, when it left Hamburg on its maiden voyage to the nitrate port of Iquique, Chile. *Preussen* made many rapid passages, perhaps the fastest being 55 days from Land's End, England, round Cape Horn to Iquique, in 1903, running in ballast to pick up a cargo of nitrates.

UNDER SAIL

A five-masted ship needed a larger crew than a four-master (it is interesting to compare *Preussen*'s dimensions with those of its contemporary, *Archibald Russell*), but could offer much more cargo capacity: some 8100 tonnes (8929 tons), which made a considerable difference in commercial terms.

Tsessarevich (1903)

TYPE • *Pre-dreadnought* **COUNTRY** • *Russia*

SPECIFICATIONS

DIMENSIONS:	Length: 118.5m (388ft 9in); Beam: 23.2m (76ft 1in); Draught: 8.5m (27ft 11in)
DISPLACEMENT:	13,122 tonnes (14,465 tons) normal load
PROPULSION:	20 Belleville boilers; 2 vertical triple-expansion engines; 12,155kW (16,300hp); 2 screws
SPEED:	18 knots (33.3km/h; 20.7mph)
RANGE:	5500 nmi (10186km; 6327 miles) at 10 knots
ARMAMENT:	4 × 305mm (12in) guns; 12 × 152mm (6in) guns; 16 × 75mm (3in) guns; 4 × 47mm (1.9in) 3-pdr guns; 6 × 457mm (18in) torpedo tubes
ARMOUR:	Belt 230–150mm (9–5.9in); main turrets 254mm (10in); secondary turrets 150mm (5.9in); conning tower 254mm (10in); deck 57mm (2.25in)
COMPLEMENT:	779

Tsessarevich **('Crown Prince') was built in France and closely resembled contemporary French battleships, with outward-bulging sides and large armoured fighting tops in the masts. It survived the end of the Tsarist era and was renamed 'Citizen' under the Soviet regime.**

Right: One of the four torpedo tubes was positioned higher than the ram, just above the waterline. The continuous wall of superstructure between the turrets gave Tsessarevich *a fortress-like appearance.*

MASTS
By 1914 the large fighting tops on the masts were replaced by smaller platforms carrying searchlights.

HULL
The hull had a longitudinal bulkhead dividing the armoured deck, intended for anti-torpedo protection. It separated the two engine compartments, making it possible for one to continue functioning if the other should be put out of action.

TURRETS
In typically French style, the central twin 152mm (6in) turrets were mounted on sponsons at upper deck level to give clear lines of fire.

PACIFIC SERVICE
Tsessarevich had been ordered for the Pacific fleet and was immediately deployed to Port Arthur as flagship. In the unannounced attack on Port Arthur (8 February 1904) that preceded Japan's declaration of war against Russia, it was struck by a torpedo and remained out of action for several weeks. On 10 August it led the fleet out to fight the Japanese, who were attempting to blockade the port, and fought hard in the day-long Battle of the Yellow Sea.

BALTIC FLEET
In 1905 *Tsessarevich* was transferred to the Baltic Fleet, which had lost its battleships at Tsushima, and it continued to be based at Kronstad. A courtesy visit was made to Portsmouth, England, in 1913. In World War I it was engaged on Baltic patrols.

Kniaz Potemkin (1903)

TYPE • *Pre-dreadnought* **COUNTRY** • *Russia*

Its crew staged a mutiny in 1905 and the ship was renamed, but remains famous due to the 1925 film *Battleship Potemkin*. Its entire career was spent with the Black Sea Fleet and saw considerable action against Ottoman forces during World War I.

SPECIFICATIONS

DIMENSIONS:	Length: 115.3m (378ft 6in); Beam: 22.25m (73ft); Draught: 8.2m (27ft)
DISPLACEMENT:	12,784 tonnes (14,093 tons) normal load
PROPULSION:	2 shafts; vertical triple-expansion engines; 22 Belleville boilers; 8426kW (11,300ihp)
SPEED:	16 knots (29.6km/h; 18.4mph)
RANGE:	3200 nmi (5900km; 3700 mi) at 10 knots
ARMAMENT:	4 × 305mm (12in) guns; 16 × 152mm (6in) guns; 14 × 76mm (3in) guns; 6 × 3-pdr guns; 5 × 456mm (18in) submerged torpedo tubes
ARMOUR:	Belt 229–152mm (9–6in); turrets 254mm (10in); casemates 152–125mm (6–5in); conning tower 229mm (9in); deck 76–63mm (3–2.5in)
COMPLEMENT:	741

Right: Kniaz Potemkin *pictured in the Black Sea in 1905. By the time Sergei Eisenstein was making his celebrated film, the ship was at the breakers', and other surviving pre-dreadnoughts of the Black Seas Fleet were used as stand-ins.*

DESIGN
Of similar displacement and design to *Peresviet*, *Kniaz Potemkin* had 305mm (12in) main guns mounted in small French-type turrets, but presented the same large extent of unarmoured superstructure.

TURRETS
The main turrets were operated electrically. Rate of fire was one round every four minutes, firing a 338kg (745lb) shells with a range of 12,000m (39,000ft) at an elevation of 10 degrees.

SECONDARY GUNS
The secondary battery of 16 152mm (6in)/45-calibre guns was unusually extensive.

MUTINY

On 27 June 1905, the crew mutinied against bad food and conditions, and in the fighting the captain and most of the officers were killed. The crew took the ship to Constanta, Romania, from where it was handed back to Russia. On 9 August it was renamed *Panteleimon* and carried that name until February 1917 when *Potemkin* was restored briefly, before it being renamed *Borets za Svobodu* ('Freedom Fighter').

***PANTELEIMON*, 1906**
The Russian fleet dominated the Black Sea until the outbreak of war in August 1914, and the arrival of the German battlecruiser *Goeben*, under the Turkish flag.

HMS *Dominion* (1903)

TYPE • *Pre-dreadnought* **COUNTRY** • *United Kingdom*

SPECIFICATIONS

DIMENSIONS:	Length: 138.3m (453ft 8in); Beam: 24m (78ft); Draught: 7.7m (25ft 6in)
DISPLACEMENT:	17,781 tonnes (19,595 tons) full load
PROPULSION:	2 shafts; vertical triple-expansion engines; 16 Babcock & Wilcox boilers; 13,423kW (18,000ihp)
SPEED:	18.5 knots (34.3km/h; 21.3mph)
RANGE:	5600 nmi (10,370km; 6440 miles) at 10 knots
ARMAMENT:	4 × 305mm (12in) guns; 4 × 234mm (9.2in) guns; 10 × 152mm (6in) guns; 14 × 12-pdr; 14 × 3-pdr guns; 5 × 457mm (18in) submerged torpedo tubes
ARMOUR:	Belt 229–203mm (9–8in); bulkheads, barbettes and conning tower 305mm (12in); main turrets 305–203mm (12–8in); deck 63–25mm (2.5–1in)
COMPLEMENT:	777

One of eight ships of the *King Edward VII* class, the first British battleships to carry 234mm (9.2in) guns as an 'intermediate' calibre, Dominion was laid down at the Vickers yard at Barrow-in-Furness on 23 May 1902. It launched on 25 August 1903 and completed in July 1905.

Left: The King Edwards were handsome and well-proportioned ships, but even as pre-dreadnoughts they were outclassed while still under construction by the design of the Lord Nelson *class.*

SECONDARY GUNS
The inclusion of 234mm (9.2in) guns to supplement the main armament reflected developments in other navies as well as the general increase in attackable superstructure surface.

FIRE CONTROL TOPS
As originally planned, the masts were without fighting tops, but fire-control tops on both masts were added prior to completion.

RUDDER
The balanced rudder made the ship easily manoeuvrable, but the class was notorious for its difficulty in maintaining a straight course.

BRITISH SERVICE
Dominion was commissioned into the Atlantic Fleet on 5 August 1905. In April 1909 it was assigned to the 2nd Division of the Home Fleet, then to the 2nd Battle Squadron of the First Fleet from May 1912, then from October 1913 to the 3rd Battle Squadron in the Mediterranean. From August 1914 it was part of the 3rd Battle Squadron of the Grand Fleet, serving as flagship August–September 1915.

***DOMINION* AT SEA**
Refitted at Devonport, England, in the winter of 1915–16, *Dominion* was used as a depot ship for raids on Zeebrugge and Ostend on the Belgian coast in 1917–18. Sold on 9 May 1921, it was scrapped in 1924.

Schwaben (1904)

TYPE • *Pre-dreadnought* **COUNTRY** • *Germany*

SPECIFICATIONS

DIMENSIONS:	Length: 126.8m (416ft); Beam: 22.8m (74ft 9in); Draught: 8.04m (26ft 4in)
DISPLACEMENT:	12,798 tonnes (14,114 tons) normal load
PROPULSION:	3 shafts; triple-expansion engines; 11,185.5kW (15,000ihp)
SPEED:	17.5 knots (32.4km/h; 20.1mph)
RANGE:	5000 nmi (9260km; 5750 miles) at 10 knots
ARMAMENT:	4 × 240mm (9.4in) guns; 18 × 150mm (5.9in) guns; 12 × 88mm (3.5in) guns; 12 machine guns; 6 × 450mm (17.7in) torpedo tubes
ARMOUR:	Belt 305mm (12in); deck 55mm (2.5in)
COMPLEMENT:	683

Schwaben **was last of four *Wittelsbach*-class ships, the first battleships ordered under Admiral Tirpitz's Navy Law of 1898. Modelled closely on the preceding *Kaiser* class, they had short careers in active service.**

Right: On commissioning, the ship was used for torpedo training. In May 1904 it hit a reef near Fehmarn Island in the Baltic Sea and was under repair until late in the year.

HULL
Inside the hull were 14 watertight compartments, and 70 per cent of the length was protected by a double bottom.

ARMAMENT
The limitations of the main armament of four 240mm (9.4in) guns were almost immediately apparent and it was clear that the class could be employed only on secondary duties.

LAYOUT
In appearance the Wittelsbachs resembled the *Kaiser Friedrich III* class, with the same built-up superstructure and twin funnels, and two wide-diameter pole masts, topped by tall narrow extensions. However, they were flush-decked.

WAR SERVICE
From August 1914 until 10 November 1915, *Schwaben* was involved in naval operations in the Baltic Sea. But with the other Wittelsbach, it was considered unfit for combat duties against the more heavily armed Russian *Gangut* class, and was transferred to Wilhelmshaven for engineering training, and partially disarmed from early 1916, losing its main guns and with the secondary armament reduced to six 150mm (5.9in) and four 88mm (3.5in) guns.

ON MANOEUVRES
In 1919, *Schwaben* underwent conversion as a depot ship for 12 minesweeping FM motor boats, operating in the Baltic from 1 August 1919 to 19 June 1920. It was stricken on 8 March 1921, and sold the same year for scrapping at Kiel.

B class (1904)

TYPE • *Submarine* **COUNTRY** • *United Kingdom*

SPECIFICATIONS

Dimensions:	Length: 41m (135ft); Beam: 4m (13ft 5in); Draught: 3.15m (10ft 4in)
Displacement:	292 tonnes (322 tons) surfaced; 321 tonnes (354 tons) submerged
Propulsion:	Single screw; 450kW (600hp) 16-cylinder gasoline engine and 130kW (180hp) electric motor
Speed:	12 knots surfaced (22.2km/h; 13.8mph); 6.5 knots submerged (12km/h; 7.4mph)
Range:	1110 nmi (2055km; 1277 miles) at 12 knots surfaced
Armament:	2 × 457mm (18in) torpedo tubes; 2 reload torpedoes
Complement:	15

By 1901 the British Admiralty had dropped its complacent attitude to submarines and began to take them seriously, both as a threat and as a potential new weapon. The B-class marked a real step forward in design and strategic thinking.

Left: B-2 *sails out of habour. The limited endurance of the B-class and its ability to fire only four torpedoes led to its termination, and construction ceased in 1906 with the introduction of the slightly larger C-class.*

Armament
By the start of World War I, the boats could carry the 457mm (18in) Mark VIII torpedo. This had a warhead of 150kg (330lb) of TNT and two settings for range and speed: 2300m (7545ft) at 35 knots (64.8km/h; 40.2mph), or 3700m (12,139ft) at 29 knots (53.7km/h; 33.3mph).

Power supply
Gasoline was a dangerous fuel because of its fumes and high flammability. From around 1904 the French Navy was experimenting with diesel engines, but the Royal Navy did not use diesel until the advent of D-class submarines, after the Electric Boat contract ended in 1910.

Radio communications
Unsuccessful attempts were made to use wireless transmission from the B-class in 1903. After more experiments in 1910 with *B-3* and *B-4*, most of the class had short-range radio from 1914, effective up to about 48km (30 miles).

WARTIME SUCCESS

On the outbreak of war in 1914, the B-class was already virtually obsolete, but saw considerable active service in British waters, the Mediterranean and the Adriatic. In the ill-fated Dardanelles campaign, the exploits of the submarines were among the few successful aspects. The B-class, by 1915 provided with a Maxim gun for surface defence, had several encounters with Austrian seaplanes during Adriatic patrols. *B-10* was the first submarine to be sunk by air attack, in 1916.

***B-6* IN THE SOLENT**

One the problems experienced by the B-class was the presence of salt and freshwater layers at different depths, which upset the boat's trim and required frequent alteration of the balance between water and air in the ballast tanks to keep the submarine submerged and level.

From Dreadnoughts to World War I

The launch of the British battleship HMS *Dreadnought* in 1906 instantly rendered all existing battleships obsolete with its 'all-big-gun' design, sparking an international naval arms race. Advances in fire control and gunnery soon allowed surface engagements to be fought at horizon-reaching distances by the outbreak of World War I (1914–18). The war also marked the rise of the submarine as a weapon of profound strategic influence. The German Kaiserliche Marine deployed U-boats that sank 6394 Allied ships by war's end. Consequently, submarine-hunting destroyers and frigates also rose in tactical importance, as did minesweepers, which were needed to counter the prolific use of naval mines.

USS *NEVADA*
Launched in 1914, USS *Nevada* (BB-36) exemplifies the age of the battleship, with its 10 356mm (14in) main guns, speed of 20.5 knots (38km/h; 24mph) and complement of 864 officers and men.

HMS *Dreadnought* (1906)

TYPE • *Battleship* **COUNTRY** • *United Kingdom*

SPECIFICATIONS

DIMENSIONS:	Length: 160.4m (527ft); Beam: 25m (82ft); Draught: 8.1m (26ft 6in)
DISPLACEMENT:	18,186 tonnes (20,055 tons) normal load; 22,195 tonnes (24,476 tons) full load
PROPULSION:	18 Babcock & Wilcox boilers; 4 direct-drive Parsons turbines; 17,000kW (23,000shp); 4 screws
SPEED:	21.6 knots (40km/h; 25mph)
RANGE:	6620 nmi (12,260km; 7600 miles) at 10 knots
ARMAMENT:	10 × 305mm (12in) guns; 27 × 12-pdr guns; 5 × 457mm (18in) submerged torpedo tubes
ARMOUR:	Belt 279–178mm (11–7in); bulkhead 203mm (8in); barbettes 279–102mm (11–4in); turrets 279mm (11in); conning tower 279–203mm (11–8in)
COMPLEMENT:	773

The speed of *Dreadnought*'s construction and fitting-out by the Portsmouth Navy Yard – just 14 months from laying the keel to commissioning on 2 December 1906 – was almost as remarkable as the ship itself.

Below: Dreadnought*'s guns could elevate to 13.5 degrees and fired 362kg (800lb) armour-piercing (AP) shells with a muzzle velocity of 831m (2725ft) per second, to a range of 15,040m (49,350ft).*

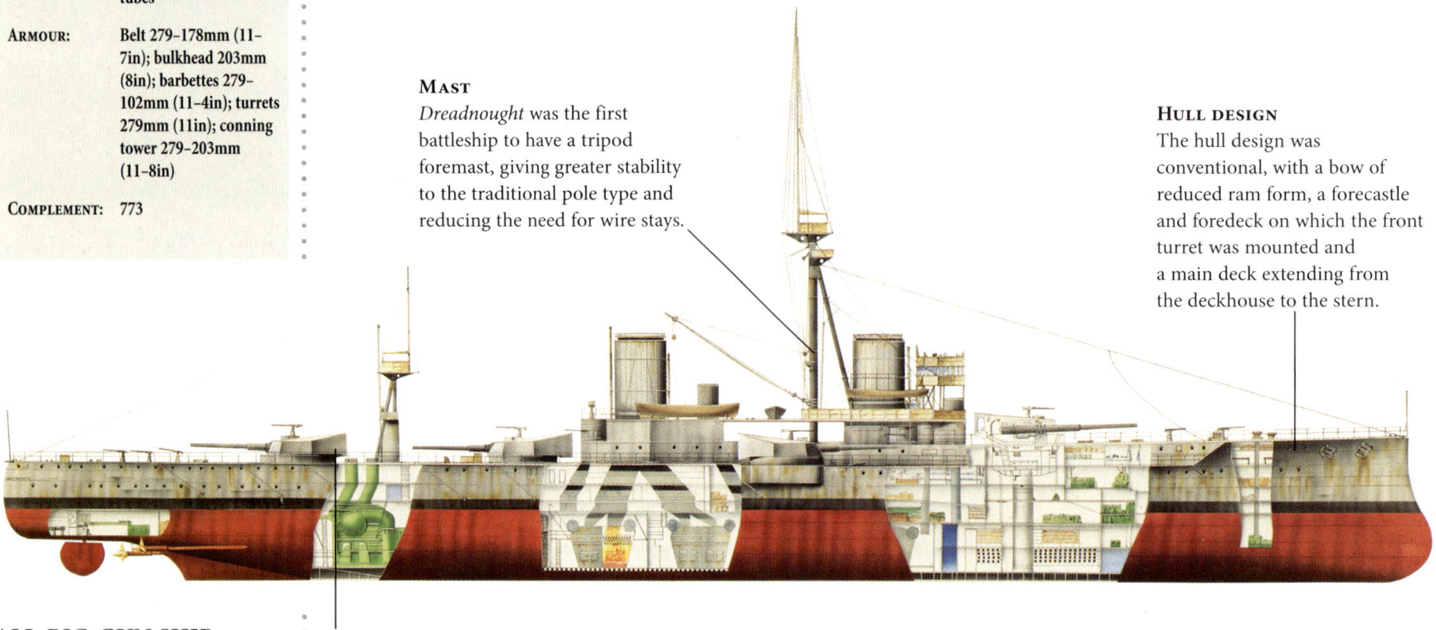

MAST
Dreadnought was the first battleship to have a tripod foremast, giving greater stability to the traditional pole type and reducing the need for wire stays.

HULL DESIGN
The hull design was conventional, with a bow of reduced ram form, a forecastle and foredeck on which the front turret was mounted and a main deck extending from the deckhouse to the stern.

MAIN GUNS
The layout of the main guns was influenced by Admiral Fisher, who considered that the ability to fire forward (or rearwards) was more important than the mounting of a full broadside.

ALL-BIG-GUN SHIP

The main design innovations were, first, the adoption of an armament composed entirely of 10 large calibre guns, with no intermediate weapons, and second, the use of turbine propulsion, which gave it greater speed. The concept of the 'all-big-gun' ship had been promoted by Italian naval engineer Vittorio Cuniberti in his 1903 article 'An Ideal Battleship for the British Fleet', and had been widely debated since. Turbine propulsion, while already used in merchant vessels, was also novel in a capital ship.

***DREADNOUGHT* UNDER STEAM**
Although *Dreadnought* had the latest fire-control systems, the limited extent of these is clear by comparison with the superstructures of post-1915 capital ships.

République (1906)

TYPE • *Battleship* **COUNTRY** • *France*

SPECIFICATIONS

DIMENSIONS:	Length: 133.8m (439ft); Beam: 24.26m (79ft 7in); Draught: 8.41m (27ft 7in)
DISPLACEMENT:	14,605 tonnes (16,099 tons) normal load
PROPULSION:	3 shafts; 3 triple-expansion engines; 13,422.6kW (18,000ihp)
SPEED:	19 knots (35.2km/h; 21.8mph)
RANGE:	4200 nmi (7778km; 4835 miles) at 10 knots
ARMAMENT:	4 × 305mm (12in) BL guns; 18 × 164mm (6.4in) QF guns; 5 × 450mm (17.7in) torpedo tubes (2 submerged)
ARMOUR:	Belt 280mm (11in); barbettes 255mm (10in); turrets 356mm (14in); conning tower 305mm (12in)
COMPLEMENT:	793

The largest warship yet built in a French yard, *République* was a pre-dreadnought battleship that marked a renewed effort to provide the navy a standardized class of capital ships. The lead ship of its class, it had only one companion ship: *Patrie*.

Left: République*'s entire service career was in the Mediterranean Sea. In World War I it patrolled the Otranto Straits between the Mediterranean and Adriatic seas in 1914–15.*

FOREMAST
A tubular foremast was fitted with two armoured control stations, with a pole topmast above and a pole mainmast. Caps and clinker screens were fitted to the funnels.

APPEARANCE
In appearance the ship was less idiosyncratic than its predecessors. The high freeboard of the forecastle deck was continued to abaft the mainmast, with the second main turret on a lower quarterdeck.

BUNKER CAPACITY
République and its companion ship *Patrie* had a maximum bunker capacity of 1854 tonnes (2044 tons), enough to make a double Atlantic crossing.

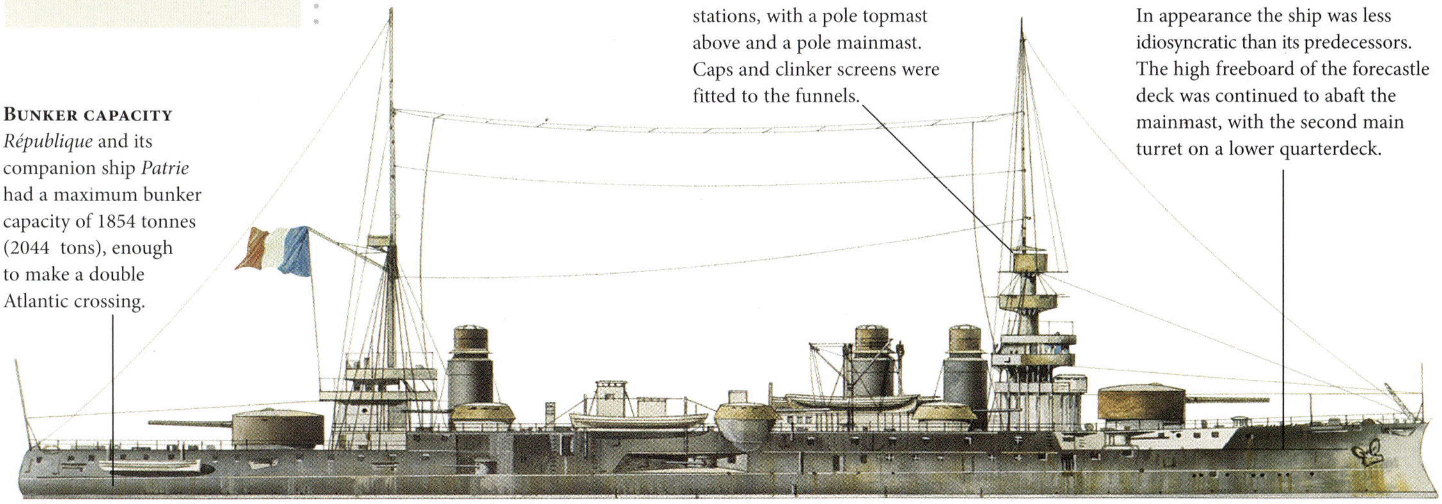

YOUNG SCHOOL
Designed by Louis-Émile Bertin, director of naval construction, *République* was laid down at the Brest Naval Arsenal on 2 December 1901, launched on 4 September 1902 and completed on 12 June 1906. Bertin had been a member of the *Jeune École* ('young school') group that viewed battleships as out of date, but since other navies continued to build them, the French felt compelled to follow suit.

EFFECTIVE DESIGN
A respectful article in a British naval journal considered *République* as almost equal to the *King Edward VII* class: it observed the increasing tendency for battleships from different navies to resemble one another. Ultimately, *République* proved to be an effective design.

Schleswig-Holstein (1908)

SPECIFICATIONS

DIMENSIONS:	Length: 127.6m (418ft 8in); Beam: 22.2m (72ft 10in); Draught: 8.2m (26ft 11in)
DISPLACEMENT:	13,190 tonnes (14,552 tons) normal load; 14,218 tonnes (15683 tons) full load
PROPULSION:	12 water-tube boilers; 3 vertical triple-expansion engines; 3 screws; 13,000kW (17,000shp)
SPEED:	19.1 knots (35km/h; 21.8mph)
RANGE:	5720nmi (10590km; 6570 miles) at 10 knots
ARMAMENT:	4 × 280mm (11in) guns; 14 × 170mm (6.7in) guns; 14 × 150mm (5.9in) guns; 4 machine guns; 6 × 450mm (17.7in) torpedo tubes
ARMOUR:	Belt 240–100mm (9.4–3.9in); deck 97–40mm (3.8–1.6in); turrets up to 280mm (9.4in); conning tower 300mm (11.8in)
COMPLEMENT:	771

The only pre-dreadnoughts to see action in both world wars were *Schlesien* and *Schleswig-Holstein* of the *Deutschland* class. Commissioned at Kiel on 6 July 1908 as the final ship of its class, *Schleswig-Holstein* exemplified the pinnacle of pre-'all-big-gun' battleship design.

Right: Schleswig-Holstein *opens fires on Polish positions with its mains guns around Danzig on 5 September 1939.*

FUNNELS
In its original form, *Schleswig-Holstein* had tall upper masts and three funnels. This is a post-1926 view. Modifications made in the interwar period included the trunking of the fore-funnel into the second funnel.

MODIFICATIONS
During the interwar period, the superstructure was extended aft to provide extra accommodation in its role as flagship (1926–35). An admiral's observation/command bridge was fitted, the navigation bridge enlarged, and searchlight platforms were mounted on the foremast.

BOILERS
Eight boilers were converted to oil firing but four remained coal-fired, which became useful in 1944–45 when oil was in short supply.

PROVOCATION

Design of the class was in 1902–03, although modifications were made to the later ships. The name was provocative: Schleswig-Holstein was a province gained by Germany from Denmark after two 19th-century wars (the crew called the ship *Sophie X*). Its dimensions were required to match the width of the Kiel Canal, to enable rapid transfer from the Baltic to the North Sea. In its original form, it had three funnels and two pole masts, with dismountable topmasts.

POSTWAR HULK

Schleswig-Holstein is remembered as the ship that fired the opening shots of World War II, at Danzig (Gdansk) on 1 September 1939, when Germany invaded Poland. Following RAF air attacks in December 1944, it sank at Gotenhafen (present-day Gdynia). Refloated by the Soviet Navy, it was towed into the Gulf of Finland,and used for target practice until the 1960s.

HMS *Indomitable* (1908)

TYPE • *Battleship* **COUNTRY** • *United Kingdom*

SPECIFICATIONS

DIMENSIONS:	Length: 173m (567ft); Beam: 23.8m (78ft); Draught: 7.3m (27ft)
DISPLACEMENT:	17,250 tonnes (19,026 tons) normal load; 21,055 tonnes (20,722 tons) full load
PROPULSION:	31 Babcock & Wilcox water-tube boilers; two paired sets of Parsons direct-drive steam turbines; 4 screws; 31,000kW (41,000shp)
SPEED:	25.5 knots (47.2km/h; 29.3mph)
RANGE:	3090nmi (5722km; 3555 miles) at cruising speed
ARMAMENT:	8 × 305mm (12in) guns; 16 × 102mm (4in) guns; 5 × 450mm (18in) torpedo tubes
ARMOUR:	Belt 152–102mm (6–4in); deck 64–38mm (2.5–1.5in); barbettes and turrets 178mm (7in); conning tower 254–152mm (10–6in)
COMPLEMENT:	784

Fast armoured cruisers had long featured in modern naval fleets, but they were relatively small and typically armed with guns no larger than 228mm (9in). HMS *Indomitable* belonged to a revolutionary new generation.

Right: In the Battle of Dogger Bank on 23 January 1915, Indomitable *was one of the ships that sank the German heavy cruiser* Blücher. *It also towed the heavily damaged HMS* Lion *back to port.*

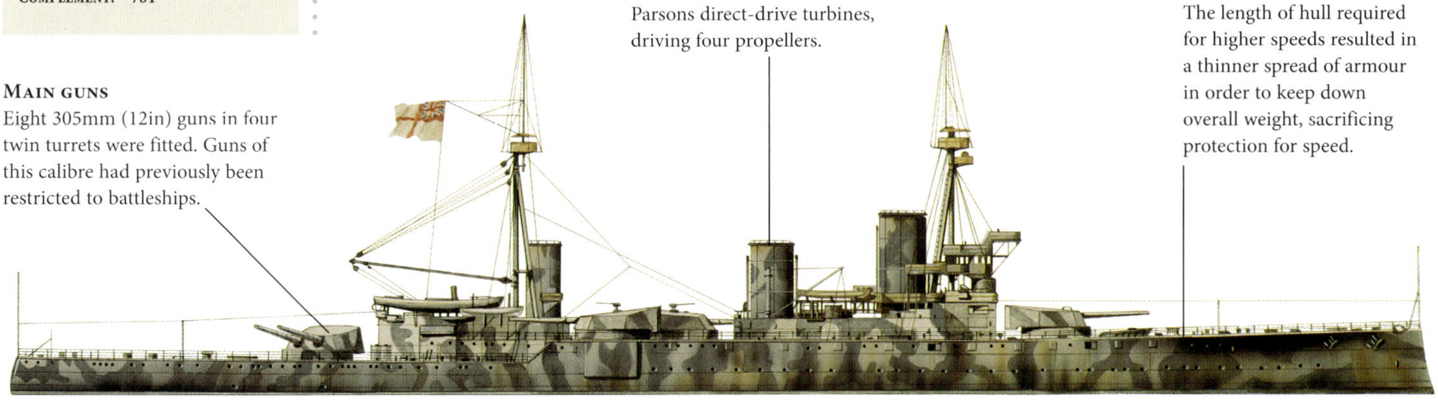

POWERPLANT
Indomitable got steam from 31 Babcock & Wilcox water-tube boilers, fitted with oil sprayers to increase burning heat. It was driven by two paired sets of Parsons direct-drive turbines, driving four propellers.

ARMOUR
The length of hull required for higher speeds resulted in a thinner spread of armour in order to keep down overall weight, sacrificing protection for speed.

MAIN GUNS
Eight 305mm (12in) guns in four twin turrets were fitted. Guns of this calibre had previously been restricted to battleships.

BATTLECRUISER
Warships were becoming larger in the 1900s, and Britain's ever-inventive Admiral Fisher sought a class of ship that would be a world-beater like *Dreadnought*, combining the speed of a cruiser with the armament of a battleship. The *Invincible* class, of which *Indomitable* was the second to enter service in June 1908, was the result. It was actually 1911 before the term 'battlecruiser' was officially bestowed on them and certain ships that followed them.

***INDOMITABLE*, C. 1908**
The long hull (to improve speed) and low superstructure of the class typified the look of battle cruisers. Their laden displacement of over 21,055 tonnes (20,722 tons) was primarily due to the heavy machinery required for speed, and to the weight of their four 305mm (12in) twin gun turrets.

Vittorio Emanuele (1908)

TYPE • *Pre-dreadnought* **COUNTRY** • *Italy*

SPECIFICATIONS

Dimensions:	Length: 144.6m (474ft); Beam: 22.4m (73ft); Draught: 8.58m (28ft 1in)
Displacement:	14,137 tonnes (15,588 tons) normal load
Propulsion:	2 shafts; 2 triple-expansion engines; 14,484kW (19,424ihp)
Speed:	21.36 knots (39.56km/h; 24.58mph)
Range:	10,000nmi (18,520km; 11,508 miles) at 10 knots
Armament:	2 × 305mm (12in) guns; 12 × 200mm (8in) guns; 16 × 76mm (3in) guns; 2 × 450mm (17.7in) torpedo tubes
Armour:	Belt 248mm (9.8in); turrets 203mm (8in); conning tower 250mm (10in); deck 38mm (1.5in)
Complement:	764

Second of the four-strong *Regina Elena* class, this pre-dreadnought warship took part in operations in the Italian–Turkish war. However, although also active throughout World War I, it saw no further combat.

Right: Vittorio Emanuele *in service. Fear of torpedo strikes below the waterline belt from Austrian submarines kept the ship out of the Adriatic Sea during World War I.*

Funnels
There were some differences within the *Regina Elena* class. *Regina Elena* had funnels 2.5m (8ft) shorter than *Vittorio Emanuele* and the last two, *Roma* and *Napoli*, had no foremast.

Armour
The armour was of Krupp-type steel, manufactured in Italy at the Terni works. At its thickest, on the belt of the ship, the armour was 248mm (9.8in) in depth.

Guns
The 203mm (8in) guns were arranged in three twin turrets on each side, with the superstructure corners angled to increase their field of fire.

DESIGN

Designed by Vittorio Cuniberti, chief constructor of the Italian Navy and a believer in the 'all-big-gun' battleship concept, *Vittorio Emanuele* was laid down at Cantieri di Castellammare di Stabia on 18 September 1901, launched on 12 October 1904 and completed on 1 August 1908. Cuniberti modified his principles considerably for the class, which carried only two heavy guns and a large array of 'semi-heavy' 200mm (8in) guns, rather than his preference for 12 305mm (12in) guns.

REGINA ELENA
Here we see the *Regina Elena* in May 1907, prior to its commissioning the following September. This class of Italian ship were, for a time, the fastest battleships in service.

HMS *Lord Nelson* (1908)

TYPE • *Battleship* **COUNTRY** • *United Kingdom*

SPECIFICATIONS

Dimensions:	Length: 135.2m (443ft 6in); Beam: 24.2m (79ft 6in); Draught: 7.9m (26ft)
Displacement:	15,604 tonnes (17,204 tons) normal load;
Propulsion:	2 shafts; vertical triple-expansion engines; 15 Babcock & Wilcox boilers; 12,490kW (16,750ihp)
Speed:	18 knots (33km/h; 21mph)
Range:	9180nmi (17,000km; 10,560 miles) at 10 knots
Armament:	4 × 305mm (12in) guns; 10 × 234mm (9.2in) guns; 24 × 12-pdr; 2 × 3-pdr guns; 5 × 450mm (17.7in) torpedo tubes
Armour:	Belt 305–50mm (12–2in); bulkhead 203mm (8in); barbettes, citadel, conning tower and main turrets 305mm (12in); decks 102–25mm (4–1in)
Complement:	817

Representing a significant improvement in firepower over its predecessors, *Lord Nelson* was almost immediately outclassed by *Dreadnought*, but gave useful wartime service in the Eastern Mediterranean and the Dardanelles Campaign.

Left: Lord Nelson *was laid down at Palmers' yard on the Tyne on 18 May 1905, launched on 4 September 1906 and completed in October 1908.*

Bridge
The deckhouse was quite narrow to provide deck space for six 234mm (9.2in) turrets on each side, the end ones carrying two guns. The ship originally had no bridge, only a high conning tower, but after trials a bridge with lateral wings, and a charthouse perched above, were added.

Hull design
Internally the hull was subdivided into self-contained sections – each with its own drainage, ventilation and pumping equipment and access lifts – by a series of solid bulkheads.

Firepower
The superiority in gun power was striking: the 305mm (12in) guns could penetrate Krupp cemented armour 305mm (12in) thick from 6950m (22,800ft), compared to 6840m (22,440ft) for HMS *Dominion*'s 40-calibre guns.

WAR SERVICE

The ship was commissioned in December 1908 in the Home Fleet's Nore Division and served in the Home Fleet until May 1912, when it was attached to the 1st Battle Squadron and, briefly, to the 4th Battle Squadron. With the outbreak of World War I, it was made flagship of the Channel Fleet until February 1915, when it was sent to the Dardanelles, and remained in the Eastern Mediterranean area until April 1919. On return to Chatham, it lay in reserve until it was sold in November 1920.

HMS *AGAMEMNON*
Lord Nelson and *Agamemnon,* the second of the two-ship class, were the last British capital ships to be fitted with reciprocating triple-expansion engines. Note here the side mounts for anti-torpedo nets.

Nassau (1909)

One of the first German big-gun ships constructed after *Dreadnought*, *Nassau* carried a substantial secondary armament alongside its 12 main guns. Powered by triple-expansion engines rather than turbines, it saw action with the High Seas Fleet at Jutland.

SPECIFICATIONS

Dimensions:	Length 146.1m (479ft 4in); Beam 26.9m (88ft 4in); Draught 8.5m (27ft 11in)
Displacement:	17,146 tonnes (18,907 tons) normal load
Propulsion:	3 vertical triple-expansion engines; 16,405kW (22,000hp); 3 screws
Speed:	19.5 knots (36.1km/h; 22.5mph)
Range:	9400 nmi (17,408km; 10,815 miles) at 10 knots
Armament:	12 × 280mm (11in) guns; 12 × 150mm (5.9in) guns; 16 × 88mm (3.5in) guns; 6 × 450mm (17.7in) torpedo tubes
Armour:	Belt 300mm (11.8in); bulkhead 200mm (7.8in); barbettes and turrets 280mm (11in); conning tower 400mm (15.6in) forward, 200mm (7.8in) aft
Complement:	963

Right: Four battleships formed the class, with Rheinland *as the first to be laid down, but* Nassau *the first to be completed. During the Battle of Jutland, on 31 May 1916,* Nassau *was hit twice by shellfire, but all damage was repaired by 10 July.*

Main guns
The main guns, 280mm (11in), were of smaller calibre than the 305mm (12in) guns being established as the British standard, but extensive testing had convinced the German navy that they were not significantly less effective.

Armour
In total, 5930 tonnes (6537 tons) of armour were carried, making up 35.2 per cent of standard displacement weight.

Secondary armament
Unlike *Dreadnought*, a large-scale secondary armament was also included, with 12 150mm (5.9in) guns mounted in casemates at a level below the port and starboard main turrets, and 16 86mm (3.4in) guns in side-mounted sponsons on the hull and superstructure.

FIRING RATE

Nassau's best firing rate was three rounds in two minutes. Comparative figures for *Dreadnought*'s guns were: barrel length identical, barrel weight 51.7 tonnes (57 tons), shell weight 385kg (849lb), range 19,000m (62,336ft) at 13 degrees of elevation and a rate of fire of two rounds a minute. The advantage would seem to have been with the British, but the German admirals believed in the armour-piercing qualities of their shells.

***NASSAU* AT ANCHOR, 1910**
The German battleship SMS *Nassau*, seen early in its career in a photograph received by the US Navy Office of Naval Intelligence on 18 May 1910. The ship entered fleet service on 1 October 1909.

Satsuma (1909)

TYPE • *Battleship* **COUNTRY** • *Japan*

SPECIFICATIONS	
DIMENSIONS:	Length: 146.9m (482ft); Beam: 25.4m (83ft 6in); Draught: 8.4m (27ft 6in)
DISPLACEMENT:	19,683 tonnes (21,708 tons) normal load
PROPULSION:	2 shafts; vertical triple-expansion engines; 20 Miyabara boilers; 12,901kW (17,300ihp)
SPEED:	18.25 knots (33.8km/h; 21mph)
RANGE:	9100 nmi (16,853km; 10,473 miles) at 10 knots
ARMAMENT:	4 × 305mm (12in) guns; 12 × 254mm (10in) guns; 12 × 120mm (4.7in) guns; 4 × 79mm (3.1in) guns; 4 × 79mm (3.1in) guns; 5 × 457mm (18in) torpedo tubes
ARMOUR:	Belt 229–102mm (9–4in); barbettes 234–178mm (9.2–7in); turrets 228–178mm (9–7in); conning tower 152mm (6in); Deck 50mm (2in)
COMPLEMENT:	887

Satsuma was the first large battleship to be built in a Japanese yard, and the last of the Imperial Navy's battleships powered by reciprocating engines. It saw no combat during World War I and was decommissioned in 1923.

Left: Disarmed at Yokosuka in 1922, Satsuma *was removed from the Navy List on 20 September 1923, and used as a target ship. On 7 September 1924, it was sunk in Tokyo Bay by shells from* Nagato *and* Mutsu.

TURBINE DRIVE
Satsuma's turbine drive, linked to the substantial armament, even if not as heavy as the original plan, earned the ship the label of 'semi-dreadnought'.

DESIGN
In its general lines and proportions, *Satsuma* retained the British look of its predecessors. Four of the 120mm (4.7in) guns were mounted in embrasures towards the bow and stern.

ENGINES
The engines were vertical triple-expansion type, with 20 Japanese Miyabara boilers, but a companion ship, *Aki*, completed almost a year later, was powered by American Curtis turbines from the Fore River Shipyard in New York.

ARMSTRONG GUNS

Satsuma was launched on 15 November 1906 and completed on 25 March 1909, the delay caused by Britain's refusal to supply its 305mm (12in) guns from Armstrong while Japan and Russia were at war. Originally designed to carry 12 of these guns, *Satsuma* could only acquire four due to Japan's limited postwar resources, supplemented by eight 254mm (10in) QF guns, also from Armstrong.

ON DEPLOYMENT

In World War I, *Satsuma* was assigned to the 1st Battleship Squadron, then the 2nd South Seas Squadron, which seized the Caroline and Palau Islands colonies from German rule in October 1914. It rejoined the 1st Battleship Squadron for the rest of the war.

U-9 (1910)

TYPE • *Submarine* **COUNTRY** • *Germany*

SPECIFICATIONS

Dimensions:	Length: 57.3m (188ft); Beam: 6m (19ft 7in); Draught: 3.5m (11ft 6in)
Displacement:	Surfaced: 431.8 tonnes (476 tons); Submerged: 610 tonnes (673 tons)
Propulsion:	4 Körting kerosene engines; electric motors; single screw
Speed:	Surfaced: 14.2 knots (26.2km/h; 16.3mph); Submerged: 8 knots (14.8km/h; 9.2mph)
Range:	3356nmi (6215km; 3862 miles) at 8.6 knots
Armament:	4 x 450mm (17.7in) torpedo tubes; 2 reloads; 1 x 37mm (1.5in) deck gun
Complement:	29

Although already outdated by the outbreak of war in 1914, the German submarine *U-9* caused a sensation in the early weeks of World War I by sinking three British cruisers in the North Sea within a single hour.

Above: A dramatic depiction of U-9 *sailing past Heligoland under the command of the victorious commander Otto Weddigen, 1914.*

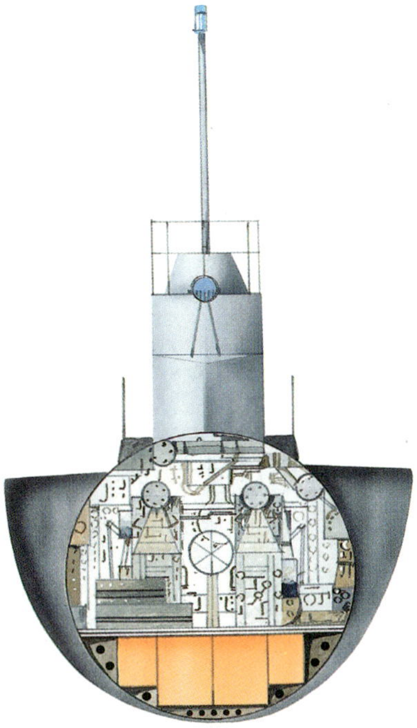

Battle bridge
The top of the conning tower was the 'battle bridge', occupied by a watch crew of officer, petty officer and seaman lookout. They had to be secured by safety lines in rough weather.

Descent speed
These early double-hulled U-boats had a slow dive-time: *U-9* required seven minutes. This was not seen as a problem until wartime conditions prevailed, especially when facing the disguised merchantmen known as Q-ships, and later air attacks.

Power supply
Batteries were placed below the living space and were constantly monitored for release of poisonous hydrogen. Ventilation tubing ran along both sides of the boat.

LANDMARK ATTACK

In August 1914, *U-9* was commanded by an exceptionally able officer, *Kapitänleutnant* Otto Weddigen. Early on 22 September 1914, *U-9* sighted the British cruisers *Cressy*, *Aboukir* and *Hogue* steaming NNE in the North Sea at 10 knots (18.5km/h; 11.5 mph) without zigzagging. In Weddigen's subsequent one-hour attack, he torpedoed and sank the three vessels. *U-9* was the only U-boat to be awarded an Iron Cross by the Kaiser. It was withdrawn from active duty in April 1916 and used for training purposes.

***U-9*, KIEL CANAL**
Laid down at the Kaiserliche Werft, Danzig (Gdansk), in 1908, *U-9* was the first of four built to a design that was already obsolete by its launch on 22 February 1910.

Minas Gerais (1910)

TYPE • *Battleship* COUNTRY • *Brazil*

SPECIFICATIONS

DIMENSIONS:	Length: 165.8m (544ft); Beam: 25.3m (83ft); Draught: 8.5m (27ft 10in)
DISPLACEMENT:	21,540 tonnes (23,744 tons)
PROPULSION:	2 shafts; vertical triple-expansion engines; 18 Babcock & Wilcox boilers; 17,524kW (23,500shp)
SPEED:	21 knots (38.9km/h; 24mph)
RANGE:	10,000nmi (18,520km; 11,500 miles) at 10 knots
ARMAMENT:	12 x 305mm (12in) guns; 22 x 120mm (4.7in) guns
ARMOUR:	Belt 230–102mm (9–4in); turrets 230–203mm (9–8in)
COMPLEMENT:	850

***Minas Gerais* was a dreadnought battleship of the Brazilian Navy. Originally planned in the pre-dreadnought years in response to powerful ships under construction for Chile, the design was later modified to make it the first dreadnought of a South American navy.**

Left: Minas Gerais *had superfiring turrets while* Dreadnought *did not, and its wing turrets were offset diagonally rather than set laterally.*

ARMOUR
The armour plate ran along the belt and encased the turrets. It was produced by Armstrong Whitworth and was of Krupp cemented steel type.

TURRETS
Hydraulic power turned the gun turrets. On completion, the ship fired the heaviest broadside of any warship in the world.

CONFIGURATION
The illustration shows *Minas Gerais* in post-1937 configuration. Originally a fore-funnel had been situated forward of the mast.

BRITISH BUILT

First proposed in a 1904 programme, *Minas Gerais* was ordered from the British Vickers-Armstrong yard at Elswick and laid down in 1907. It was a complete package deal, with the builders supplying both guns and armour. It is a curious tribute to the openness of the armament industry at that time that the design of HMS *Dreadnought*, then newly commissioned, could be adapted for a warship for another country, albeit one friendly to Britain.

MINAS GERAIS, BRAZIL, C. 1942
Launched on 10 September 1908, the ship was completed on 5 January 1910. After World War I, *Minas Gerais* underwent a refit at the New York Navy Yard and in another major refit at Rio between 1931–37 was converted to oil-firing, lost a funnel, had the bridge, superstructure, and masts modified, and acquired new fire-control systems.

USS *South Carolina* (1910)

TYPE • *Battleship* **COUNTRY** • *United States*

SPECIFICATIONS

Dimensions:	Length: 137.9m (452ft 8in); Beam: 24.5m (80ft 5in); Draught: 7.49m (24ft 7in)
Displacement:	14,515 tonnes (16,000 tons) normal load; 16,238 tonnes (17,899 tons) full load
Propulsion:	12 Babcock & Wilcox boilers; 2 vertical 4-cylinder triple-expansion engines; 12,304kW (16,500hp); 2 screws
Speed:	18.5 knots (34.3km/h; 21.3mph)
Range:	5000nmi (9260km; 5753 miles) at 10 knots
Armament:	8 × 305mm (12in) guns; 22 × 88mm (3in) QF guns; 2 × 533mm (21in) underwater torpedo tubes
Armour:	Belt 305–228mm (12–9in); casemates and barbettes 254–203mm (10–8in); turrets 304mm (12in); deck 38mm (1.5in); conning tower 305mm (12in)
Complement:	869

The US Navy placed greater emphasis on striking power and endurance than on speed. This two-ship class mounted a formidable eight-gun 305mm (12in) broadside, and their superfiring turrets arrangement would go on to become the standard for capital ships.

Right: The gun layout did away with the need for wing turrets, making better use of magazine space and improving magazine protection, and still enabling an eight-gun broadside to be fired.

Basket masts
BB-25 and -26 were the first US ships to carry the so-called lattice masts, often referred to as basket masts. It was felt that the nature of the structure would protect fire-control positions from the shocks and vibrations that were transmitted through the traditional form of mast.

Turret design
Drawn up before HMS *Dreadnought*, this was the first battleship planned with all its big guns in centreline turrets (although the first design incorporated two single turrets to port and starboard).

Propulsion
Turbine propulsion had been considered, but price considerations prevailed. The US Navy command considered speed to be less vital than firepower and armour protection.

WAR SERVICE

From the US entry into World War I in April 1917, *South Carolina* was used as a gunnery training ship. After the Armistice, from February to July 1919, it made four Atlantic crossings to repatriate American troops from France. With *Michigan* it made a training cruise to Honolulu in 1920 and went on further training and goodwill cruises until paid off at Philadelphia on 15 December 1921.

SOUTH CAROLINA
Although *Michigan* was commissioned two months earlier, *South Carolina* had already been allocated the pennant number BB-25, and its companion ship was BB-26. Ordered as first-class battleships, their standard displacement was restricted to 14,515 tonnes (16,000 tons) by Congressional mandate, but fully loaded they were closer to 16,329 tonnes (18,000 tons).

Von der Tann (1910)

TYPE • *Battlecruiser* **COUNTRY** • *Germany*

SPECIFICATIONS

Dimensions:	Length: 171.7m (563ft 4in); Beam: 26.6m (87ft 3in); Draught: 9m (26ft 7in)
Displacement:	19,370 tonnes (21,344 tons) normal load; 22,048 tonnes (24,314 tons) full load
Propulsion:	4 shafts; Parsons turbines; 18 Schulz-Thornycroft boilers; 32,512kW (43,600shp)
Speed:	24.75 knots (45.8km/h; 28.5mph)
Range:	4400nmi (8149km; 5063 miles) at 14 knots
Armament:	8 × 280mm (11in) guns; 10 × 150mm (5.9in) guns; 16 × 88mm (3.5in) guns; 4 × 450mm (17.7in) torpedo tubes
Armour:	Belt 250–80mm (10–3.2in); bulkheads 180–100mm (7–4in); battery 152mm (6in); barbettes 230–30mm (9–1.2in); turrets 230–60mm (9–2.4in)
Complement:	1174

The third ship of Germany's 1907 building programme, *Von der Tann* was Germany's first battlecruiser. Matched against its British counterparts, it showed considerable superiority in combat at the Battle of Jutland in June 1916.

Right: Constructed by Blohm & Voss in Hamburg, the ship was laid down on 21 March 1908, launched on 20 March 1909 and commissioned on 1 September 1910, at a cost of 36.5 million Goldmarks.

Guns
The SK designation of the guns refers to their Krupp manufacture (Siegfried Krupp); it could also stand for *schwere Kanone*, meaning 'heavy gun'.

Crew quarters
Unlike the British battlecruisers built around this time, *Von der Tann* was not planned for long-range missions, and crew quarters were limited.

Turret designation
The turrets were designated 'A', 'B', 'C' and 'D' (Anton, Bruno, Cäsar and Dora), with the rear turret as 'C' and the port central turret as 'D'.

JUTLAND

At the start of the Battle of Jutland, 31 May–1 June 1916, the 1st Scouting Group opened fire on the British battlecruisers at a range of 14,000m (45,932ft), and shells from *Von der Tann* sank HMS *Indefatigable*. Six minutes later a 375mm (15in) shell from HMS *Barham* pierced the armour and flooded the steering room. Return fire by *Von der Tann* caused serious damage to *Barham*. *Von der Tann* survived the battle and was subsequently repaired.

***VON DER TANN*, 1910**
The great battleship *Von der Tann*, photographed shortly after its commission. During the war, it remained active until Germany's capitulation. Interned with the rest of the Fleet in Scapa Flow, in the Orkney Islands, it was scuttled by its crew on 21 June 1919.

HMS *Hercules* (1911)

TYPE • *Battleship* **COUNTRY** • *United Kingdom*

SPECIFICATIONS

DIMENSIONS:	Length: 166.3m (545ft 9in); Beam: 26m (85ft 2in); Draught: 8.2m (27ft)
DISPLACEMENT:	20,350 tonnes (22,432 tons) normal load
PROPULSION:	18 Yarrow boilers; 4 sets Parsons direct-drive steam turbines; 2 screws; 19,000kW (26,000shp)
SPEED:	21 knots (38.5km/h; 24mph)
RANGE:	6680nmi (12,370km; 7677 miles) at 10 knots
ARMAMENT:	10 × 305mm (12in) guns; 16 × 102mm (4in) guns; 3 × 533mm (21in) torpedo tubes
ARMOUR:	Belt 279–203mm (11–8in); Turrets 280mm (11in); Barbettes 279–102mm (11–4in); Conning tower 280mm (11in)
COMPLEMENT:	791

When *Hercules* was laid down on 30 July 1909, British designers were being required to maximize the broadside capacity of battleships. In the Battle of Jutland in June 1916, *Hercules* fired 98 305mm (12in) shells at enemy ships and claimed hits on the *Seydlitz*.

Left: Several alterations were made before the ship saw action: a searchlight platform was installed, the fore funnel was heightened and a 'clinker cowl' fitted, to stop or reduce the amount of ash that fell on the navigating bridge.

ENGINE ROOMS
There were three engine rooms, with the two inner shafts operated from a central room and each outer shaft from separate rooms to port and starboard.

TURRETS
The three ships of the *Colossus* class had two 305mm (12in) turrets centre-mounted en echelon, enabling all 10 guns to be fired to either port or starboard.

COAL FIRED
Hercules was coal-fired, with 18 Yarrow boilers working at 16.1kg/sq cm (230psi), serving 4 Parsons direct-drive steam turbines. Oil sprayers were fitted to speed up the process of raising steam.

IMPROVEMENTS

In late 1912, *Hercules* was one of 12 ships fitted with short distance 8km (5-mile) radio. Gun direction equipment was installed in 1915. Hercules was also among the ships trying out new devices such as the Waymouth-Cooke rangefinder, a handheld instrument combining a sextant with a telescope, which was used for keeping a ship in station with the rest of a squadron as well as for fire control.

ADVANCED WARSHIP

A variety of range-finding and fire-control instruments were tested on ships throughout World War I. Another set tested on Hercules was 'Walker's instruments' – a triple set intended to establish the deflection rate for a torpedo director with the ship in motion.

USS *Utah* (1911)

TYPE • *Battleship* **COUNTRY** • *United States*

USS *Utah* and USS *Florida* were the United States Navy's most powerful battleships when commissioned in 1911, although by then the USS *Wyoming*, armed with 12 305mm (12in) guns, had already been launched.

SPECIFICATIONS

Dimensions:	Length: 159m (521ft 8in); Beam: 26.9m (88ft 3in); Draught: 8.6m (28ft 3in)
Displacement:	23,033 tonnes (25,375 tons) normal load
Propulsion:	4 shafts; Parsons turbines; 12 Babcock & Wilcox boilers; 20,880kW (28,000shp)
Speed:	21 knots (39km/h; 24.2mph)
Range:	6720nmi (12,450km; 7730 miles) at 10 knots
Armament:	10 × 305mm (12in) guns; 16 × 127mm (5in) guns; 2 × 533mm (21in) torpedo tubes
Armour:	Belt 279–229mm (11–9in); turrets 305mm (12in); barbettes 254–203mm (10–8in); conning tower 292mm (11.5in); deck 38mm (1.5in)
Complement:	1001

Above: Officers and crew of the engineer force present on the deck of the USS Utah *(BB-31).*

Funnels
Unlike the *Wyoming* class, the funnels were both placed between the narrow basket masts, all concentrated amidships, in order to make room for the triple set of turrets carried aft of the mainmast.

Design
The image shows *Utah* as it was around 1923. Several alterations were made to control and signalling positions on the masts during World War I.

Gun protection
Following criticism of secondary gun protection on *Delaware*, a lightly armoured upper casemate protected the secondary 127mm (5in) guns on each side.

WAR AND PEACE

Utah's war service began with the US occupation of Veracruz, Mexico, where along with *Florida* it landed marines on 21 April 1914. From 6 April 1917, it was based at Chesapeake Bay for engineering and gunnery training until 30 August 1918, when it crossed the Atlantic to Bantry Bay, Ireland, as flagship of Battleship Division 6, patrolling the Western Approaches to the British Isles. After the war, it escorted President Woodrow Wilson's ship to the Versailles negotiations.

PUGET SOUND, 1941
USS *Utah* undergoes a paint job at Puget Sound Navy Yard in 1941. In December 1941, the ship was at Pearl Harbor. In the Japanese surprise attack of 7 December, it was struck by two aerial torpedoes and capsized, with the loss of 64 men.

HMS *Lion* (1912)

TYPE • *Battlecruiser* **COUNTRY** • *United Kingdom*

SPECIFICATIONS

Dimensions:	Length: 213.4m (700ft); Beam: 27m (88ft 6in); Draught: 9.9m (32ft 5in)
Displacement:	26,925 tonnes (29,680 tons) full load
Propulsion:	42 Yarrow water-tube boilers; 2 sets Parsons direct-drive steam turbines; 4 screws; 52,199kW (70,000shp)
Speed:	28 knots (51.9km/h; 33.4mph)
Range:	5610nmi (10,390km; 6451 miles) at 10 knots
Armament:	8 × 343mm (13.5in) guns; 16 × 102mm (4in) guns; 2 × 533mm (21in) torpedo tubes
Armour:	Belt 229–102mm (9–4in); bulkheads 102mm (4in); barbettes 229–203mm (9–8in); turrets 229mm (9in); deck 64–25mm (2.5–1in); conning tower 254mm (10in)
Complement:	997

First of a new class of battlecruisers, and the first Royal Navy warship to cost over £2 million, *Lion* was a developed version of the *Indefatigable* class, faster by 3.7km/h (2 knots) and carrying eight 343mm (13.5in) guns.

Right: At the Battle of Dogger Bank, 24 January 1915, a shot from Lion *knocked out the rear turret of Admiral Hipper's flagship* Seydlitz. *However,* Lion *sustained 17 hits, narrowly avoided flooding of the engine room, and fell out of the action.*

Rangefinders
Rangefinders were located in 'B' and 'Y' turrets and in the conning tower with the fire-control position (this was later transferred to the mast, which was reinforced by struts to support it).

Searchlights
By 1918 *Lion* had searchlight towers mounted on the after funnel, and flying-off platforms mounted on 'Q' and 'X' turrets.

Fore-funnel
As in *Dreadnought* and *Orion*, the fore-funnel was in front of the mast, making the masthead installations often uninhabitable through smoke and heat.

SUPERFIRING
The British Navy adopted the superfiring turret arrangement in HMS *Neptune* (1911) and with the *Orion* class (1912) introduced the 343mm (13.5in) gun. Both were combined in *Lion*, laid down at Devonport Naval Dockyard on 29 November 1909, launched on 6 August 1910 and commissioned on 4 June 1912. The three ships of the *Lion* class were the largest, fastest capital ships yet built and also the most expensive. However, they had serious defects.

HMS *LION*, C. 1919
HMS *Lion* survived World War I, despite being involved in the Battles of Dogger Bank and Jutland, the ship receiving enemy shell strikes and sustaining serious damage in each engagement.

USS *Wyoming* (1912)

TYPE • *Battleship* **COUNTRY** • *United States*

SPECIFICATIONS

DIMENSIONS:	Length: 171m (562ft); Beam: 28.4m (93ft 2in); Draught: 8.7m (28ft 6in)
DISPLACEMENT:	26,416 tonnes (29,129 tons) normal load
PROPULSION:	12 Babcock coal/oil boilers; 4 Parsons direct-drive turbines; 4 screws; 20,880kW (28,000shp)
SPEED:	20.5 knots (37.9km/h; 23.7mph)
RANGE:	8000nmi (14,800km; 9206 miles) at 10 knots
ARMAMENT:	12 × 305mm (12in) guns; 21 × 130mm (5in) guns; 2 × 533mm (21in) torpedo tubes
ARMOUR:	Bulkhead 279–229mm (11–9in); belt 279–127mm (11–5in); decks 76–25mm (3–2in); turrets 305mm (12in); barbettes 279mm (11in);
COMPLEMENT:	1063

***Wyoming* served in both world wars, in both the Atlantic and Pacific oceans, and went through a series of changes in appearance and function, ending up as the 'Chesapeake Raider' anti-aircraft training ship.**

Left: USS Wyoming *passes through the Galliard Cut, Panama Canal, on 26 July 1919. From 19 July 1919 until August 1921 the ship was deployed in the Pacific.*

MAST
In 1926 the aft basket mast was replaced by a tripod structure, and a catapult was installed to launch a float plane off 'C' turret.

MAIN GUNS
The main guns were 305mm (12in) 50-calibre Mk 7, introduced for the Wyoming class and made at the US Naval Gun Factory. Maximum elevation was 150 degrees; they fired 390kg (870lb) AP shells to a range of 18,000m (59,055ft).

DECK ARRANGEMENT
The flush deck arrangement placed the midships battery about 1.2m (4ft) higher than on previous ships: a much more effective position.

DISARMED
In 1933, to conform with the Washington naval treaty, the ship was partially disarmed, losing the guns of III, IV and V turrets and the fantail 152mm (6in) gun. It was reclassified as a training ship, AG-17 (AG denoted 'miscellaneous auxiliaries'), and used for this purpose until November 1941 when it became a gunnery training ship, operating mainly in the Chesapeake Bay area, primarily with AA guns.

USS *WYOMING*, C. 1920
In World War I, *Wyoming* was one of the four ships in BatDiv9, sent across the Atlantic to join the British Grand Fleet during 1917–18. It was also among the ships escorting the surrendered German High Seas Fleet to Scapa Flow in November 1918.

Viribus Unitis (1912)

TYPE • *Battleship* **COUNTRY** • *Austria-Hungary*

SPECIFICATIONS

Dimensions:	Length: 152.2m (499ft 3in); Beam: 27.3m (89ft 8in); Draught: 8.9m (29ft)
Displacement:	22,032 tonnes (24,285 tons) normal load; 21,898 tonnes (24,117 tons) full load
Propulsion:	12 Yarrow coal/oil boilers; 4 Parsons turbines; 4 screws; 19,700kW (26,400shp)
Speed:	20.3 knots (37.6km/h; 23mph)
Range:	4200nmi (7780km; 4831 miles) at 10 knots
Armament:	12 x 305mm (12in) guns; 12 x 150mm (5.9in) guns; 18 x 11-pdr guns; 4 x 533mm (21in) torpedo tubes
Armour:	Belt 280–152mm (11–5in); decks 25–19mm (2–1.5in); turrets 280mm (11in); casemates 180mm (7.1in)
Complement:	1087

The first triple-turret battleship and the first Austro-Hungarian dreadnought of the *Tegetthoff* class, *Viribus Unitis* carried the bodies of Archduke Franz Ferdinand and Archduchess Sophie back to Trieste after their fateful assassination on 28 June 1914.

Right: The huge weight of the four turrets required close attention to the metacentric height in design. They were originally conceived for a larger hull, but financial constraints resulted in a reduced freeboard and shorter length. In profile Viribus Unitis *was a typical dreadnought type.*

Double bottom
A double bottom was fitted, coming up to meet the base of the armour plate, but of relatively thin plates, the inner torpedo bulkhead being only 50mm (2in).

Main guns
The main guns were Škoda 305mm (12in) 45-cal, Model K10, with a maximum elevation of 200 degrees, firing shells of 450kg (990lb) to a maximum range of 22,000m (72,178ft) and with a claimed rate of fire of three rounds per minute.

Superstructure
The superstucture was a simple round-fronted tower with a closed navigating deckhouse above, on a platform built around the pole foremast and partially supported by lateral lattice frames. The two funnels were set close behind and between the two tall pole masts.

UNITED STRENGTH

Austria-Hungary's only dreadnought class was named for Austrian Admiral Tegetthoff. *Viribus Unitis* ('With United Strength') was first to be laid down, at Stabilimento Tecnico Triestino on 24 July 1910, launched on 24 June 1911 and completed on 5 December 1912. The class was designed to counter Italy's triple-turreted *Dante Alighieri*, but in fact *Viribus Unitis* was the first 305mm (12in) triple-turret battleship to be commissioned into service.

AERIAL VIEW
Apart from delivering shore bombardment against the Italian city of Ancona, and the Montenegrin coast, the ship saw little action in 1914–18. Following the Austrian surrender, it was briefly handed over to the new state of Yugoslavia, but was sunk by limpet mines set by Italian frogmen on 1 November 1918.

Derfflinger (1913)

TYPE • *Battleship* **COUNTRY** • *Germany*

SPECIFICATIONS

Dimensions:	Length: 210.4m (690ft 3in); Beam: 29m (95ft 2in); Draught: 9.2m (30ft 3in)
Displacement:	26,600 tonnes (29,326 tons) normal load
Propulsion:	18 boilers; Parsons turbines; 4 screws
Speed:	26.5 knots (49.1km/h; 30.7mph)
Range:	5600nmi (10,400km; 6440 miles) at 14 knots
Armament:	8 × 305mm (12in) guns; 12 × 150mm (5.9in) guns; 4 × 88mm (3.4in) guns; 4 × 88mm (3.4in) AA guns; 4 × 500mm (19.6in) torpedo tubes
Armour:	Side belt 305–150mm (12–6in); bulkhead 250–100mm (9.8–3.9in); barbettes 260mm (10.2in); turrets 270mm (11in); deck 30mm (1.2in)
Complement:	1112–1182

Unlike any previous German capital ship in appearance, with a flush deck rising gently towards a straight bow, *Derfflinger* was widely considered one of the most handsome warships of its time.

Left: Derfflinger *unleashes a full broadside. The normal barrel life of the main guns was 200 rounds. The magazines held 720 shells.*

Torpedo nets
Originally, extendable torpedo nets were stowed along the sides. These were removed during 1916, as with most other capital ships, following the introduction of wire-cutting torpedoes.

Gun turrets
The 305mm (12in) gun turrets traversed by electric power, but the guns elevated hydraulically (maximum 13.5 degrees, increased to 16 degrees in 1916), firing 405.5kg (894lb) AP shells at a rate of up to three per minute and with a maximum range from 1916 of 20,400m (69,929ft).

Hull design
The hull design showed a marked change from older practice, with a full-length flush deck. Internally it did away with transverse frames and relied on longitudinal supports, although still compartmentalized against shell or torpedo strikes.

JUTLAND

At Jutland on 31 May 1916, with the battlecruiser *Seydlitz*, *Derfflinger* sank the British battleship *Queen Mary*, and with *Lützow*, sank HMS *Invincible*, but itself took 17 heavy hits and four lesser ones. By the end of the battle, with only two guns still operational, it returned to Wilhelmshaven, Germany, under its own steam, with some 3050 tonnes (3362 tons) of water inside the hull. Repairs were completed at Kiel by mid-October, but it remained in port except for a sortie into the northern North Sea, with other ships, on 23 April 1918.

COMBAT DAMAGE
A view of combat damage aboard *Derfflinger*, photographed in June 1916 after the Battle of Jutland. One of the 150mm (5.9in) guns has had its muzzle shot away; also visible are the 88mm (3.4in) anti-aircraft guns, and damaged torpedo nets.

Kongō (1913)

TYPE • *Battleship* COUNTRY • *Japan*

SPECIFICATIONS

DIMENSIONS:	Length: 222m (728ft 4in); Beam: 31m (101ft 8in); Draught: 9.7m (31ft 10in)
DISPLACEMENT:	37,187 tonnes (41,004 tons) normal load
PROPULSION:	36 Yarrow water-tube boilers; 2 sets Parsons direct-drive turbines; 4 screws
SPEED:	27.5 knots (50.9km/h; 31.6mph)
RANGE:	10,000nmi (19,000km; 11,875 miles) at 14 knots
ARMAMENT:	8 × 356mm (14in) guns; 16 × 152mm (6in) guns; 8 × 76mm (3in) guns; 8 × 533mm (21in) torpedo tubes
ARMOUR:	Belt 279–203mm (11–8in); deck 58–38mm (2.3–1.5in); turrets 229mm (9in); barbettes 254mm (10in)
COMPLEMENT:	1360

The last Japanese capital ship built outside Japan, *Kongō* was the first warship to carry 356mm (14in) guns. It was the first battlecruiser of the *Kongō* class and was the only Japanese battleship to be sunk by a submarine.

Right: In many ways Kongō *was an improved version of the British* Lion *class. It was the first of a class of four, with various differences in appearance.*

MODERNIZATION
Another drastic modernization was made in 1935–37. The stern was extended and reshaped. Sixteen new Kampon oil-fired boilers and Brown-Curtis direct-drive turbines doubled the power output, with speed increased to 30.5 knots (56km/h).

ARMAMENT
Kongō's 356mm (14in) guns had the longest range of any at the time, up to 35.45km (19.14nmi). Its eight 533mm (21in) submerged torpedo tubes were the most ever fitted to a major warship.

MODIFICATIONS
Between 1929 and 1931, *Kongō* and three companion ships were rearmed and reconstructed as fast battleships, although increased armour and anti-torpedo bulges actually slowed them down. The funnels were reduced to two.

SINKING

In World War II, *Kongō* was involved in many Pacific operations, from the Malayan and Philippines invasions to the battle of Midway and the Guadalcanal campaign. With reinforced AA armament, it participated in the vast battle of Leyte Gulf in October 1944. On 21 November 1944 it was passing through the Formosa Strait when it was hit by two torpedoes from USS *Sealion*. For two hours it struggled on, before sinking with the loss of 1200 crew.

STARBOARD MIDSECTION
Kongō was laid down at the Vickers yard at Barrow-in-Furness, England, on 17 January 1911, launched on 18 May 1912 and completed on 16 August 1913, to plans drawn up under the direction of Sir George Thurston, chief naval architect of Vickers.

Courbet (1913)

TYPE • *Battleship* **COUNTRY** • *France*

SPECIFICATIONS

DIMENSIONS:	Length: 168m (551ft 2in); Beam: 27.9m (91ft 6in); Draught: 9m (29ft 6in)
DISPLACEMENT:	22,545 tonnes (24,846 tons) normal load; 26,427 tonnes (29,143 tons) full load
PROPULSION:	24 Belleville boilers; 4 sets Parsons direct-drive turbines; 4 screws; 32,065kW (43,000shp)
SPEED:	20 knots (39km/h; 24mph)
RANGE:	4200nmi (7800km; 4830 miles)
ARMAMENT:	12 × 305mm (12in) guns; 22 × 138.6mm (5.4in) guns; 4 × 47mm (1.8in) guns; 4 × 450mm (18in) torpedo tubes
ARMOUR:	Belt 270–140mm (10.6–5.5in); deck 70–40mm (2.8–1.6in); turrets 320mm (12.6in); conning tower 266mm (10.5in)
COMPLEMENT:	1085

France's first dreadnought battleship, *Courbet* was leader of a class of four, all of which served in the Mediterranean Sea during World War I. *Courbet* was laid down at the Brest Naval Arsenal in 1910, launched in 1911 and completed on 19 November 1913.

Left: Restored to operational status in May 1940, as part of the 3rd Line Division, in June Courbet *supported troop evacuations from Cherbourg, France, and then steamed to Portsmouth, England. Briefly taken over by the Royal Navy, it was handed to the Free French forces and used by them as an accommodation ship and stationary AA platform.*

MAIN GUNS
Twelve 305mm (12in) guns were mounted in double turrets, eight in superfiring turrets on the centreline and four in parallel turrets to port and starboard between the second and third funnels.

REFIT
A refit from 9 July 1923 to 16 April 1924, after a boiler fire, installed some oil-fired boilers and combined the two forward funnels into one.

ARMOUR
Almost half the length was protected by side armour, to 2.35m (7ft 7in) above and 2.4m (7ft 9in) below the waterline, and with a maximum thickness of 270mm (10.6in), comparable with *Dreadnought*.

FIRST KILL
In August 1914, a French and British fleet was in the Adriatic, hoping to draw the Austro-Hungarian fleet into battle, but the Austro-Hungarians stayed in port. The only victim was the light cruiser *Zenta*, caught off Antivari (Montenegro) and sunk by salvoes from *Courbet* and others on 16 August. *Courbet* was the flagship of the French squadron engaged in watching the Otranto Straits.

FORWARD TURRETS
The forward turrets of *Courbet* photographed in Toulon, France, c. 1919. Note the triplex rangefinder on the conning tower. The 305mm (12in) 45-calibre guns were an improved version of a 1906 design.

Dante Alighieri (1913)

SPECIFICATIONS

DIMENSIONS:	Length: 168.1m (551ft 6in); Beam: 26.6m (87ft 3in); Draught: 8.8m (28ft 10in)
DISPLACEMENT:	19,866 tonnes (21,905 tons) normal load; 21,947 tonnes (24,177 tons) full load
PROPULSION:	4 shafts; Parsons geared turbines; 23 Blechynden boilers; 24,004kW (32,190shp)
SPEED:	22.8 knots (42.2km/h; 26.2mph)
RANGE:	4800nmi (8890km; 5526 miles) at 10 knots
ARMAMENT:	12 × 305mm (12in) guns; 20 × 120mm (4.7in) guns; 13 × 76mm (3in) guns; 3 × 450mm (17.7in) torpedo tubes
ARMOUR:	Belt and main turrets 254mm (10in); secondary turrets and battery 98mm (3.8in); conning tower 305mm (12in); deck 38mm (1.5in)
COMPLEMENT:	981

In typical fashion, Italy's naval architects stamped their own mark on the dreadnought concept. *Dante Alighieri* was the first battleship to be designed with triple turrets for the heavy guns.

Right: Dante Alighieri *photographed sometime between 1923 and 1927, following a refit with a tripod foremast.*

DESIGN
The ship's ratio of length to beam, 6.32 times, was unusually large for a battleship.

LIGHT GUNS
The 76mm (3in) guns mounted on the turret roofs could not be fired if the main guns were in use.

SECONDARY ARMAMENT
Dante's secondary armament did not match the 150mm (5.9in) guns of the *Tegetthoff* class, but consisted of rapid-firing 120mm (4.7in) guns capable of six rounds a minute.

DESIGN PRINCIPLES

Designed on principles set by Vittorio Cuniberti, *Dante Alighieri* resembled no other big-gun ship of 1912 (although the Russian *Gangut*-class battleships of 1914 later bore a strong resemblance). Laid down at Castellammare on 6 June 1909, launched on 20 August 1910 and completed on 15 January 1913, it had a long, lowset look. It was 52.1m (57ft) longer than the Austrian *Viribus Unitis*, which also carried 12 305mm (12in) guns, but the Austrian ship made use of superfiring turrets.

VENICE, C. 1919
Dante Alighieri is shown here at Venice after the end of World War I and probably before May 1919, when the ship deployed to Fiume across the Adriatic. By 1928 the ship was obsolete and expensive to maintain in seagoing condition. It was stricken on 1 July that year, and disposed of for scrap.

HMS *Iron Duke* (1914)

TYPE • *Battleship* **COUNTRY** • *United Kingdom*

SPECIFICATIONS

Dimensions:	Length: 189.8m (622ft 9in); Beam: 27.4m (90ft); Draught: 9m (29ft 6in)
Displacement:	25,401 tonnes (28,001 tons) normal load; 30,034 tonnes (33,104 tons) full load
Propulsion:	4 shafts; Parsons turbines; 18 Babcock & Wilcox boilers; 21,625kW (29,000shp)
Speed:	21.25 knots (39.3km/h; 24.45mph)
Range:	7870nmi (14,580km; 9063 miles) at 10 knots
Armament:	10 × 343mm (13.5in) guns; 12 × 152mm (6in) guns; 2 × 76mm (3in) AA guns; 4 × 3-pdr guns
Armour:	Belt 305–102mm (12–4in); bulkheads 203–40mm (8–1.6in); barbettes 254–76mm (10–3in); turret faces 279mm (11in); deck 65–25mm (2.5–1in)
Complement:	1022

The four ships of the *Iron Duke* class were Britain's last coal-burning battleships, and also marked a return to the 152mm (6in) gun as a useful supplement to the main armament. *Iron Duke* was laid down at Portsmouth on 12 January 1912 and completed in March 1914.

Left: The front turrets of Iron Duke *receive a pounding from a heavy sea. At Jutland in May 1916, it was the flagship of Admiral John Jellicoe, making it the command centre of the Grand Fleet in the largest dreadnought battle in history.*

Tripod mast
A single heavy tripod mast was fitted, to carry director and spotting stations, and the superstructure was built up around its base to carry the 152mm (6in) gun directors.

Conning tower
The use and value of the conning tower was much debated at the time. Its limited lookout capacity and poor ventilation against shell-fumes and smoke discouraged its use and officers from admirals down preferred to use the bridge deck.

AA guns
Iron Duke was among the first battleships to carry AA guns, two 3-pounders, although its AA defences were improved during the war by two 76mm (3in) AA guns on the after superstructure.

ACTIVE DUTY

On completion *Iron Duke* was made flagship of the Home Fleet and then the Grand Fleet. At Jutland it was the flagship of Admiral John Jellicoe, and sustained no casualties in the battle. From 1916–19 it was with the 2nd Battleship Squadron, and from 1919 to 1926 it was in the Mediterranean and also the Black Sea, where it was engaged in operations against Bolshevik Russian units. Its final active spell was with the Atlantic Fleet from 1926 to 1929.

COAL-BURNING BATTLESHIP
HMS *Iron Duke* in profile, steam billowing from its rear funnel.

HMS *Royal Oak* (1914)

TYPE • *Battleship* **COUNTRY** • *United Kingdom*

SPECIFICATIONS

Dimensions:	Length: 189m (620ft 6in); Beam: 31.2m (102ft 2in); Draught: 8.7m (28ft 6in)
Displacement:	29,110 tonnes (32,097 tons) normal load; 34,037 tonnes (37,535 tons) with bulges
Propulsion:	18 Yarrow boilers; Parsons geared turbines; 4 screws; 30,000kW (40,000shp)
Speed:	23 knots (42.6km/h; 26.5mph)
Range:	4200nmi (7400km; 4830 miles) at 10 knots
Armament:	8 × 381mm (15in) Mk I guns; 14 × 152mm (6in) guns; 8 × 102mm (4in) guns; 4 × 533mm (21in) torpedo tubes
Armour:	Belt 330mm (13in); turret faces 330mm (13in); barbettes 254mm (10in); bulkheads 152mm (6in); deck 51–25mm (2–1in)
Complement:	997

***Royal Oak*, a *Revenge*-class battleship, was the first capital ship sunk in World War II, torpedoed at anchor in Scapa Flow in October 1939 by an audacious U-boat attack. The ship was launched on 29 April 1915 and completed in May 1916.**

Right: Royal Oak *seen in September 1937. From 24 January that year, it was with the Mediterranean Fleet at Gibraltar, and was involved in neutrality patrols off the Spanish coast during the Spanish Civil War, taking two minor hits from incidental firing. From June 1937 it was again deployed with the Home Fleet.*

Mainmast
Up to 1938, the mainmast was of pole type, supporting a derrick, but when gunnery control instrumentation was installed at the crosstrees it was fitted with additional supports as a tripod.

Launcher
This view shows *Royal Oak* as it appeared on the outbreak of war in 1939, with an aircraft launcher fitted on 'X' turret plus the addition of a luffing crane.

Armour
The blister shape of the torpedo bulge is apparent. The total armour weight for the ship was an impressive 8382 tonnes (9240 tons).

TRAGEDY
In 1939 the extent of modernization on board *Royal Oak* was quite limited. It still had the original engines and its speed was no longer sufficient to maintain station with the fleet. With the outbreak of war, it was moved to the Scapa Flow anchorage, where it was torpedoed in a bravura operation on the night of 14 October 1939 by *U-47*. A first torpedo hit the bows, followed by a second salvo, which scored three hits, exploding beneath the ship's bottom. The ship capsized and sank within 10 minutes, with the loss of 833 lives.

1936 VIEW
Between 1917 and 1923 all ships of the class had aircraft take-off platforms fitted to the superfiring turrets; and in 1934–35, *Royal Oak* had a catapult mounted on X turret, and an aircraft crane was placed on the port side of the mainmast.

USS *Texas* (1914)

TYPE • *Battleship* **COUNTRY** • *United States*

SPECIFICATIONS

DIMENSIONS:	Length: 174.5m (572ft 7in); Beam: 29m (95ft 3in); Draught: 8.7m (28ft 5in)
DISPLACEMENT:	27,433 tonnes (30,235 tons) normal load
PROPULSION:	14 boilers; 2 vertical triple-expansion engines; 2 screws; 20,954kW (28,100shp)
SPEED:	21 knots (39km/h; 24.4mph)
RANGE:	8000nmi (14,816km; 9206 miles) at 10 knots
ARMAMENT:	10 × 356mm (14in) guns; 21 × 127mm (5in) guns; 4 × 533mm (21in) torpedo tubes
ARMOUR:	Belt 304–254mm (12–10in); deck 76mm (3in); turrets 356mm (14in); barbettes 305mm (12in); conning tower 305mm (12in)
COMPLEMENT:	1530

The first US warship to launch an aircraft, USS *Texas* saw service in both world wars, and still survives as a museum ship at La Porte, Texas, in its name state. Assigned pennant number BB-35, it was commissioned in March 1914.

Left: The basket masts indicate that here Texas *pre-dates its 1925–27 modernization.* Texas *was the first of several battleships to be transferred by Act of Congress from the Navy to become a museum ship, in 1948.*

AA GUNS
Quick-firing high-elevation 76mm (3in) tower-mounted AA guns, capable of 30 shots a minute, were tested on *Texas* in August 1916 and 76mm (3in) guns were fitted on turrets 3 and 4 in 1921.

RADAR
Texas was fitted with radar from 1938, initially an experimental bridge-mounted CXZ antenna, replaced by 1943 with SRa forward, and SRa and SC aft. In 1948 SK (aft) was also fitted.

MODERNIZATION
A major modernization between 1925 and 1927 saw tripod masts installed, conversion to oil fuel with six new Bureau Express boilers and a single funnel, an aircraft catapult on the midships turret, with adjacent crane, and anti-torpedo blisters.

DEPLOYMENTS
In 1918, *Texas* crossed the Atlantic to join the British Grand Fleet in the 6th Battle Squadron, and it became the first US battleship to launch an aircraft, on 19 March 1919. Following a major modernization in the 1920s, and service with the Atlantic and Pacific fleets in the 1930s, during World War II the ship was employed as a convoy escort and also gave support to the shore landings of Operation *Torch* (1942) and *Overlord* (1944). Deployed to the Pacific in 1945, it fought in the Battle of Iwo Jima and gave support to the landings on Okinawa.

WWII UPGRADES
In 1942 an SG radar aerial was fitted to a platform on the foremast, and SK radar was added on the aftermast in 1943. Also from 1943 six 20mm (0.79in) AA guns were mounted on the top of 'B' turret.

HMS *E-11* (1914)

TYPE • *Submarine* **COUNTRY** • *United Kingdom*

SPECIFICATIONS

Dimensions:	Length: 55.17m (181ft); Beam: 6.91m (22ft 8in); Draught: 3.81m (12ft 6in)
Displacement:	Surfaced – 677 tonnes (747 tons); Submerged – 820 tonnes (904 tons)
Propulsion:	2 Vickers diesel engines 1600hp (1192kW); two electric motors 420hp (313kW); two screws
Speed:	Surfaced: 14 knots (25.9km/h; 16.1mph); Submerged: 9 knots (16.6km/h; 10.3mph)
Range:	3579nmi (6628km; 4118 miles) at 10 knots
Armament:	5 × 457mm (18in) torpedo tubes; 1 × 6pdr QF deck gun
Complement:	30

The E-class submarines were the main element in the Royal Navy's attack submarine force during World War I, operating in the North and Baltic Seas, the Mediterranean and the Dardanelles. In some respects inferior to U-boats, they still gave distinguished service.

Below: Twelve-pounder QF guns were installed on the walking deck, either immediately aft of the tower or on the extreme end of the deck.

Layout
The appearance was typical of British design at the time, with bulbous sides, a flat surfaced 'walking deck' and a low, squat tower giving minimal protection against waves and wind when running on the surface. Diving planes were fitted at the bow and stern.

Beam tubes
The beam tubes fired to port (forward tube) and starboard (after tube) respectively. They were opened laterally to load the torpedoes. Impracticable in action, they were hard to aim and required the boat to come to a stop for firing.

Bulkheads
The E-class was the first British submarine class to be provided with transverse bulkheads, dividing the boat into three compartments and increasing its survivability.

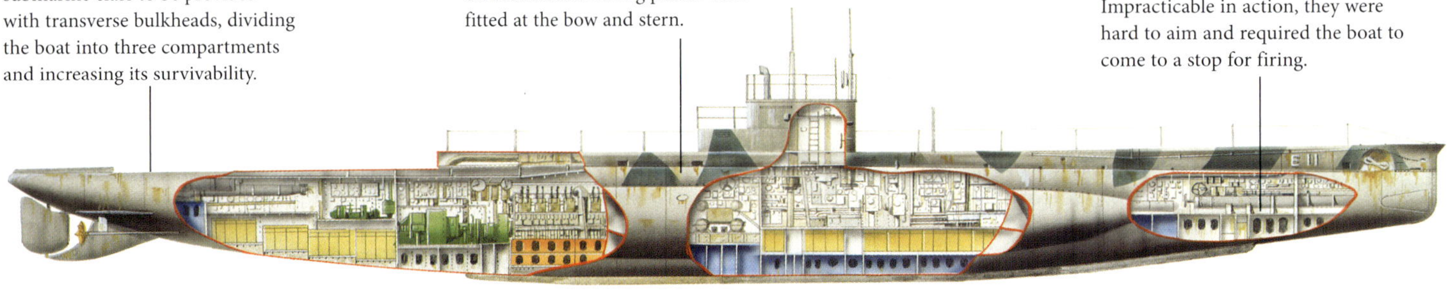

DIVERTED FUNDS

An improved version of the previous D-class submarine, with higher freeboard and increased reserve buoyancy, the first of the class, *E-1* was launched in November 1912. By 1914 the British Admiralty was well aware of the strategic possibilities of the ocean-going submarine – and of Britain's lack of such vessels. *E-9* to *E-24* were therefore built with funds originally earmarked for a battleship, HMS *Resistance*, that was never laid down.

GREECE, 1915
E-11 and its crew, probably at Mudros port, Lemnos, Greece, in 1915. Wartime fittings include the wave-deflector and the extended upper platform on the tower. In May 1915, *E-11* joined in operations at the Dardanelles and in the Sea of Marmara. Entering the Dardanelles Straits on 19 May in the hope of encountering the German battleships *Goeben* and *Breslau*, the boat had to dive below 30m (98ft) to pass below three minefields.

UC-II class (1914)

TYPE • *Submarine* **COUNTRY** • *Germany*

SPECIFICATIONS

DIMENSIONS:	Length: 49.45m (162ft 3in); Beam: 5.28m (17ft 4in); Draught: 4m (13ft 1in)
DISPLACEMENT:	Surfaced – 400 tonnes (441 tons); Submerged – 480 tonnes (529 tons)
PROPULSION:	2 6-cylinder diesel engines 500hp (370kW); two Siemens-Schukert electric motors 460hp (340kW); 2 screws
SPEED:	Surfaced: 11.5 knots (21.2km/h; 13.2mph); Submerged: 7.2 knots (13.3km/h; 8.2mph)
RANGE:	9260nmi (17,149km; 10,656 miles) at 7 knots
ARMAMENT:	3 × 500mm (20in) torpedo tubes; 7 × torpedoes; 18 × UC200 mines; 1 × 88mm (3.5in) deck gun
COMPLEMENT:	26

The Imperial German Navy's Type UC-II consisted of small but powerfully armed submarines. A total of 64 were manufactured and they turned out to be among the most versatile and effective U-boats of World War I.

Below: UC-II *class* UC-74, *commissioned on 26 November 1916 and based at Pola from 7 March 1917, was at sea when Germany surrendered. It made for Kiel, but ran out of fuel and was interned at Barcelona on 21 November 1918. Later handed over to France, it was broken up in 1921.*

CONNING TOWER
The least satisfactory aspect of the design was that the conning tower was not armoured and there was no means of sealing it off from the control room below.

FOREDECK
The front portion of the foredeck is raised above the level of the gun platform, as a fairing over the mine tubes. These discharged through the boat's bottom.

MASS PRODUCTION
UC-II could be considered the first mass-produced submarine type, with five yards involved in its construction. *UC-25* was first of a batch of nine built at the Vulcan Works in Hamburg. Launched on 10 June 1916, it was commissioned only 18 days later.

BACKGROUND

Drawn up by the Development Section of the U-boat Inspectorate as Project 41, the class was a developed version of UC-I, the first operational minelaying submarines (Russia's *Krab*, although an earlier design, was later in service). The UC-Is were single-hulled vessels designed for coastal operations, with limited sea-keeping abilities. UC-II, designated Project 41, were larger, double-hulled and, although still considered 'coastal', had improved range and seakeeping.

***UC-30* BEACHED**
The *UC-II* submarines contributed to huge Allied merchant ship losses. In 1916 alone, the Allies lost 415 ships in the Mediterranean, the main theatre for the *UC-IIs*, totalling half the tonnage of all Allied ships sunk anywhere.

Andrea Doria (1915)

TYPE • *Battleship* COUNTRY • *Italy*

SPECIFICATIONS

Dimensions:	Length: 186.9m (613ft 2in); Beam: 29.2m (95ft 3in); Draught: 8.6m (28ft 3in)
Displacement:	29,100 tonnes (32,088 tons) normal load; 29,400 tonnes (32,401 tons) full load
Propulsion:	8 Yarrow superheated boilers; 2 Belluzzo reduction-gear turbines; 2 screws; 67,113kW (90,000shp)
Speed:	28 knots (48km/h; 31mph)
Range:	6700nmi (12408km; 7707 miles)
Armament:	13 × 305mm (12in) guns in 2 triple and 2 twin turrets; 12 × 135mm (5.5in) guns; 10 × 90mm (3.5in) guns; 12 × 37mm (1.45in) guns; 15 × 20mm (0.79in) AA guns
Armour:	Belt 250–130mm (10–5.1in); barbettes 280–130mm (11–5.1in); deck 166–135mm (6.5–5.3in); conning tower 250mm (10in)
Complement:	1300

Ordered in 1912 and commissioned in 1915, the Italian Navy's *Andrea Doria* was not involved in combat until the 1940s. An improved version of the preceding *Cavour* class, it had similar dimensions and the same armament.

Above: Andrea Doria *photographed in late 1940 after the completion of its 1937–40 reconstruction.*

Modified design
Andrea Doria as it appeared after 1937, with the forecastle deck extended as far as the mainmast and the central turret removed.

Firepower
Ten 317mm (12.6in) guns replaced the original main guns, and 12 135mm (5.3in) guns, mounted amidships in triple turrets replaced the original 152mm (6in) guns; 41 AA guns were also installed.

Turbine
New turbine machinery was installed, with twin shafts replacing the original four.

WORLD WAR II
From 8 April 1937 to 26 October 1940, *Andrea Doria* underwent a major reconstruction at Trieste. It joined the 5th Division of the Italian Navy at Taranto on 26 October 1940, surviving the British attack of 11–12 November unscathed, and moved to Naples on 12 November. Movements to intercept British warships and convoy escort, were its main wartime duties. After the fall of Italy it was interned at Malta until 8 June 1944, and ultimately returned to Italy.

MODERNIZED SHIP
Andrea Doria, designed by Vice-Admiral Giuseppe Valsecchi and built at La Spezia, retained its form as commissioned on 30 June 1916, until a virtual rebuilding in 1937–40. The ship that emerged from the Trieste yard was longer and heavier, with a redesigned bow and stern and an extended forecastle deck.

HMS *Queen Elizabeth* (1915)

TYPE • *Battleship* **COUNTRY** • *United Kingdom*

SPECIFICATIONS

DIMENSIONS:	Length: 195.34m (640ft 11in); Beam: 27.6m (90ft 6in); Draught: 9.1m (30ft)
DISPLACEMENT:	29,995 tonnes (33,020 tons) full load
PROPULSION:	24 Babcock & Wilcox boilers; 4 Parsons geared turbines; 55,927kW (75,000hp); 4 screws
SPEED:	25 knots (46.3km/h, 28.9mph)
RANGE:	7500nmi (13,840km; 8630 miles) at 12.5 knots
ARMAMENT:	8 × 381mm (15in) guns; 16 × 152mm (6in) guns; 2 × 76mm (3in) AA guns; 4 × 533mm (21in) torpedo tubes
ARMOUR:	Belt 330–102mm (13–4in); bulkheads 152–102mm (6–4in); barbettes 254–102mm (10–4in); turrets 330–127mm (13–5in); deck upper 45–32mm (1.7–1.2in)
COMPLEMENT:	951

***Queen Elizabeth* was the lead ship of a class of five 'super-dreadnought' battleships that served in both world wars. In World War I, their speed and firepower set them apart from the rest of the Grand Fleet, operating as a separate battle squadron.**

Right: Queen Elizabeth *in 1943, with augmented AA armament and radar equipment after a refit in the United States. The aircraft and catapult were soon removed.*

RADAR
From 1941 Type 273 (sea search) radar was carried on the foremast, the aerial rotating within a protective plastic cylinder.

AA DEFENCES
AA defences included 20mm (0.79in) guns fitted on the superfiring turrets; up to 52 of these guns were installed during World War II in addition to 32 40mm (1.5in) eight-barrelled AA guns and 16 four-barrelled heavy machine guns.

AIRCRAFT CATAPULT
From 1939 to 1943 an aircraft catapult was installed on the boat deck. Four float planes could be carried, but the British soon dispensed with aircraft on board battleships, relying instead on carrier-borne planes for scouting.

REFITTING
One rather anachronistic feature of *Queen Elizabeth* was a stern gallery, reminiscent of the old wooden men-of-war. Over the years, and especially between 1937and 1941, ships of the class underwent many alterations and additions that substantially changed their profile. By 1942 only the hulls and the heavy guns remained unchanged, although the latter could now elevate to 30 degrees. New machinery was installed and a single large funnel mounted with a searchlight placement at its after edge.

ROYAL VISIT
King George V stands in front of the main guns of HMS *Queen Elizabeth*. The British 381mm (15in) gun was developed with remarkable speed in response to Germany's reported adoption of the same size in SMS *Bayern*.

HMS *Warspite* (1915)

TYPE • *Battleship* **COUNTRY** • *United Kingdom*

Laid down in 1912 as fast battleships to replace battlecruisers as the offensive element of the British battle fleet, the *Queen Elizabeth* class of dreadnoughts included the *Barham*, *Malaya*, *Valiant* and *Warspite*. A sixth vessel, *Argonaut*, was never built.

SPECIFICATIONS

Dimensions:	Length: 196.2m (643ft 8in), Beam: 27.6m (90ft 7in), Draught: 10.1m (33ft 2in)
Displacement:	31,818 tonnes (36,497 US tons)
Propulsion:	4 screws, four Parsons geared turbines, 24 Yarrow boilers, 55,927kW (75,000shp)
Speed:	23 knots (26.5mph; 46.3km/h)
Range:	9300km (5800 miles; 5000nm) at 12 knots
Armament:	8 × 381mm (15in), 16 × 152mm (6in) guns; 2 × 76mm (3in) AA guns; 4 × 533mm (21in) torpedo tubes
Armour:	Belt 330–102mm (13–4in); bulkheads 152–102mm (6–4in); barbettes 254–102mm (10–4in); turrets 330–127mm (13–5in)
Complement:	1025

Right: HMS Warspite *in July 1937. Although the* Queen Elizabeths *were handsome vessels, their handling was not as good as it might have been. They were seriously overweight because the designers attempted to incorporate too many innovations.*

Superstructure
A distinguishing feature of this class was the pole-type mainmast. The conning tower was redesigned and enlarged during reconstruction.

Fire control
The *Queen Elizabeth* class had the benefit of an accurate fire-control system.

Firepower
The *Queen Elizabeth* class were the first battleships to mount 381mm (15in) guns. Their firepower was designed to overwhelm any existing enemy battleship.

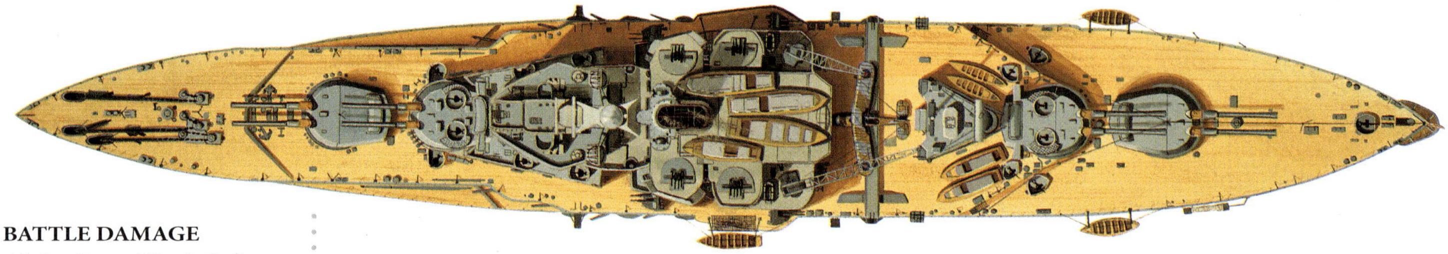

BATTLE DAMAGE
All the *Queen Elizabeth* class survived World War II except *Barham*, which exploded after being hit by three torpedoes off Sollum on 25 November 1941. Like *Queen Elizabeth*, *Valiant* was badly damaged by Italian frogmen and was further damaged in a series of accidents in 1944, being broken up in 1948. *Malaya* was also broken up in 1948, having survived a torpedo attack in 1941. *Warspite*, having suffered severe damage in the Mediterranean in 1941 and again in 1943, was scrapped in 1947.

***WARSPITE*, MALTA, 1938**
The arrangement for accommodating aircraft, with a hangar each side of the forward funnel and rails to a catapult set athwartships, began with HMS *Warspite* in the mid-1930s. It was admired by the German Navy, which adapted it for the *Bismarck* class.

HMS *Canada* (1915)

TYPE • *Battleship* **COUNTRY** • *United Kingdom*

Originally built and launched as a vessel for the Chilean government, but taken into British service in 1914, HMS *Canada* was a one-off, the only British battleship of the time with 356mm (14in) guns.

SPECIFICATIONS

Dimensions:	Length: 191m (625ft); Beam: 28.2m (92ft 6in); Draught: 10m (33ft)
Displacement:	25,400 tonnes (27,998 tons) normal load; 32,514 tonnes (35,836 tons) full load
Propulsion:	21 Yarrow boilers; 2 Brown-Curtis HP and 2 Parsons LP turbines; 27,591kW (37,000shp); 4 screws
Speed:	22.75 knots (42.1km/h; 26.2mph)
Range:	4350nmi (8153km; 5006 miles)
Armament:	10 × 356mm (14in) guns; 16 × 152mm (6in) guns; 2 × 76mm (3in) guns; 4 × 3-pdr guns; 4 × 533mm (21in) torpedo tubes
Armour:	Belt 230mm (9in); barbettes and turrets 254mm (10in); conning tower 280mm (11in)
Complement:	834

Left: The Almirante Latorre*, before transfer to the Royal Navy. While conforming to the requirements of the Chilean Navy, this was essentially a modern British-style battleship, and there were broad similarities with the* Iron Duke *class of 1914.*

Modifications
Following the handover modifications were made, including removal of the bridge and the placing of two open platforms on the foremast, and a boat derrick between the funnels.

Armour
Armour was comparatively light, comparable to that of a battlecruiser rather than a battleship, perhaps anticipating fighting conditions in the eastern Pacific.

Machinery
The machinery comprised 21 mixed-firing Yarrow boilers, supplying steam to two Brown-Curtis high pressure and two Parsons direct low-pressureturbines, giving 27,591kW (37,000shp).

ADOPTION

Ordered by the Chilean government from the Tyneside commercial shipyard of Armstrong Whitworth (specialists in warship building) in 1911 and launched as *Almirante Latorre* on 27 November 1913, the ship was still fitting out when Britain declared war on Germany on 4 August 1914. By agreement between the two governments, it was taken over by the Royal Navy and renamed HMS *Canada*. It was commissioned on 15 August 1915.

HMS *CANADA*
In 1921, the ship was finally purchased by Chile and regained its original name. In 1929–31, a modernization was carried out at Plymouth. With new oil-fired boilers and turbines, it also acquired anti-torpedo bulges, a higher mainmast (18.3m/60ft) and AA guns mounted on the superstructure.

Fuso (1915)

TYPE • *Battleship* COUNTRY • *Japan*

The first 'super-dreadnought' of the Imperial Navy designed and built in Japan, *Fuso* was the world's largest battleship when commissioned. Originally, it showed a mix of British and Japanese influences, but later modifications changed its appearance entirely.

SPECIFICATIONS

DIMENSIONS:	Length: 205m (672ft 6in); Beam: 10.08m (33ft 1in); Draught: 8.6m (28ft)
DISPLACEMENT:	35,512 tonnes (39,145 tons) full load
PROPULSION:	6 Kampon boilers; 4 Kampon steam turbines; 55,927kW (75,000hp); 4 shafts
SPEED:	27.74 knots (51.4km/h; 31.6mph)
RANGE:	11,800nmi (21,854km; 13,576 miles) at 16 knots
ARMAMENT:	12 × 356mm (14in) guns; 14 × 152mm (6in) guns (1938); 8 × 127mm (5in) guns; 16 × 132mm (5.2in) AA guns in quadruple mountings
ARMOUR:	Side belt 305–102mm (12–4in); main turrets 297–114mm (11.7–4.5in); casemate 152mm (6in); deck 132–51mm (5.2–2in)
COMPLEMENT:	1396

Right: The Yamashiro *was the other warship in the* Fuso *class. Like* Fuso, *the* Yamashiro *was also sunk during the Battle of Surigao Strait in October 1944.*

TOWER
During the modernization, a lofty 'pagoda' tower was erected, reaching 44m (144ft) above the waterline, with navigation bridges, lookout posts and fire-control stations.

1915 CONFIGURATION
The original 356mm (14in) guns (seen here) elevated from 0 degrees to 30 degrees. Post-1933 elevation was between -50 and 43 degrees, with a maximum range of 35,450m (116,306ft).

POWERPLANT
From 12 March 1930, the ship underwent a major modernization programme at Yokosuka Naval Yard. Completed in several phases, with service periods between, it included six new oil-fired Kampon boilers replacing the original eight double-ended and 16 single-ended Mijabara coal-fired boilers.

ARMS RACE
While the battleship-building race between the two main North Sea powers was intensely public, a quieter but equally dramatic race was taking place in the Pacific Ocean, between Japan and the United States. *Fuso* was built to rival USS *Texas* and *New York*. Designed in Japan, its builders drew heavily on British and European practice. Laid down in 1912, *Fuso* was not commissioned until 1917, and played no more than a patrolling role in World War I.

***FUSO* TRIALS, 1933**
Fuso served throughout World War II, supporting the Pearl Harbor attack in December 1941 and participating in the Battle of Midway in June 1942. The ship was ultimately sunk by US Navy firepower during the Battle of Surigao Strait in October 1944.

Provence (1915)

TYPE • *Battleship* **COUNTRY** • *France*

SPECIFICATIONS

DIMENSIONS:	Length: 166m (544ft 7in); Beam: 26.8m (88ft 3in); Draught: 8.9m (29ft 2in)
DISPLACEMENT:	23,936 tonnes (26,380 tons)
PROPULSION:	18 Guyot du Temple boilers; 2 sets Parsons direct-drive turbines; 4 screws; 22,000kW (29,000shp)
SPEED:	20 knots (37km/h; 23mph)
RANGE:	4600nmi (8500km; 5315 miles) at 10 knots
ARMAMENT:	10 × 340mm (13.4in) guns; 22 × 138mm (5.4in) guns in casemates; 4 × 47mm (1.8in) guns; 4 × 450mm (17.7in) torpedo tubes
ARMOUR:	Belt 270–160mm (10.6–6.25in); barbettes 248mm (9.75in); deck 77–40mm (3.2–1.6in); conning tower 340mm (13.4in)
COMPLEMENT:	1174

Built at the Arsenal de Lorient, with its keel laid on 1 August 1912, the French battleship *Provence* was first of the *Bretagne* class of three ships commissioned, on 1 March 1916. It spent much of its career in the Mediterranean.

Left: During its 1932–35 refit, the superstructure was rebuilt, a massive tripod foremast was fitted, modern fire control was installed and the central side armour was strengthened.

PROPULSION
Provence had 18 Guyot du Temple coal boilers, powering two sets of Parsons turbines. Fuel capacity was 2680 tonnes (2954 tons) of coal and 300 tonnes (330 tons) of oil, giving an operating range of 8500km (4600nmi).

AA GUNS
During the refit, AA defences were increased: eight 75mm (2.95in) guns replaced older 138mm (5.4mm) guns in the casemates, and 12 13mm (0.51in) machine guns mounted on the superstructure.

CENTRE TURRET
A centre turret, firing to port or starboard, was placed between the two funnels, and all heavy guns were placed on the centreline.

WAR SERVICE
Provence served with the Mediterranean Fleet in 1914–18, participating in the blockade of the Adriatic, which bottled up the Austro-Hungarian fleet, although its activity was heavily limited by inability to obtain coal for the boilers. In Operation *Catapult* at Mers-el-Kebir on 3 July 1940, *Provence* was the first ship to return fire on the British attackers. The ship was scuttled at Toulon in 1942.

***PROVENCE*, C. 1930**
The original foremast shows *Provence* before its major refit in 1932–35. After the ship was scuttled at Toulon in November 1942, the Germans removed the two aft turrets and set them up as coastal defence. In August 1944, they exchanged fire with *Provence*'s companion battleship *Lorraine* and USS *Nevada*.

Deutschland (1916)

TYPE • *Submarine* **COUNTRY** • *Germany*

SPECIFICATIONS

Dimensions:	Length: 65m (213ft 3in); Beam: 8.8m (28ft 10in); Draught: 5.3m (17ft 5in)
Displacement:	Surfaced 1536 tonnes (1693 tons); submerged 1905 tonnes (2101 tons)
Propulsion:	2 six-cylinder Viertakt diesel engines; 588kW (800hp); electric motors; twin screws
Speed:	Surfaced: 12.4 knots (22.9km/h; 14.2mph); Submerged: 5.2 knots 9.6km/h; 5.9mph)
Range:	13,130nmi (24,316km; 15,109 miles) at 9 knots
Armament:	6 × 500mm (20in) torpedo tubes; 18 torpedoes; 2 × 150mm (5.9in) deck guns
Complement:	29 (merchant); 56 (military)

First as a cargo-carrier that made two voyages across the Atlantic, then as an armed submarine cruiser *U-155*, *Deutschland* demonstrated the rapid development of German U-boat design and technology during World War I.

Right: The U-155/ Deutschland *– the addition of two 150mm (5.9in) deck guns radically changed the external appearance of the submarine.*

Propellers
New larger propellers were fitted as part of the U-boat conversion in the hope of improving speed. The diesel and electric motors could drive the shaft simultaneously, although only for limited periods, to obtain maximum speed.

Armament
When converted to a fighting vessel, six torpedo tubes were fitted beneath the deck and outside the pressure hull. Although two internal bow tubes were later installed, little use was made of them.

Mercantile form
The profile shows the boat in its original unarmed mercantile form. Three were built like this and four were converted to combat boats while under construction.

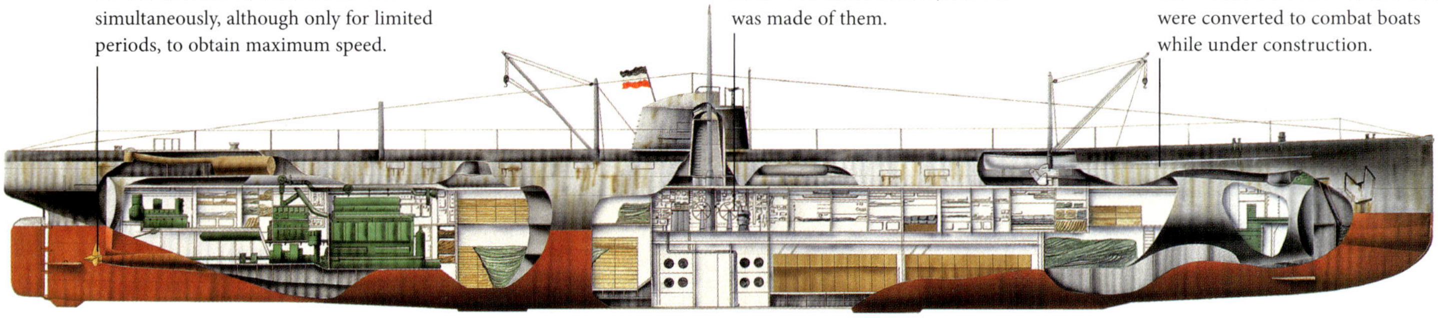

DIPLOMATIC INCIDENT
On 23 June 1916, *Deutschland* left Wilhelmshaven carrying around 700 tonnes (772 tons) of high value cargo destined for still-neutral USA. Its arrival at Baltimore on 9 July caused an international dispute. The British and French demanded it should be impounded as a warship, while the Germans insisted it was a cargo vessel. The American government concurred with the Germans and *Deutschland* was allowed to discharge and load cargo.

U-155
Despite its slow speed, the exploits of *U-155* were sufficient to promote the idea of the U-boat cruiser. One of its commanders, *Korvettenkapitän* Ackermann, noted that 'even such clumsy and slow vessels as the former merchant submarines, when handled correctly, are equal to their tasks both surfaced and submerged, all of which justifies the construction of even larger U-cruisers than are now in construction.'

USS *Nevada* (1916)

TYPE • *Battleship* **COUNTRY** • *United States*

SPECIFICATIONS

Dimensions:	Length: 177.8m (583ft 4in); Beam: 29m (95ft 3in); Draught: 8.7m (28ft 5in)
Displacement:	24,947 tonnes (27,500 tons) normal load; 26,217 tonnes (28,900 tons) full load
Propulsion:	12 Yarrow boilers; 2 Curtis turbines plus 2 geared cruising turbines; 19,761kW (26,500hp); 2 screws
Speed:	20.5 knots (38km/h; 23.6mph)
Range:	10,000nmi (18,520km; 11,432 miles) at 10 knots
Armament:	10 × 356mm (14in) guns; 21 × 127mm (5in) guns; 2 × 76mm (3in) AA guns; 2 × 533mm (21in) torpedo tubes
Armour:	Belt 343–203mm (13.5–8in); bulkheads 38mm (1.5in); deck upper 76mm (3in), barbettes 343mm (13.5in); turrets 457–127mm (18–5in)
Complement:	864

The United States' first 'super-dreadnoughts', the two ships of the Nevada class – *Nevada* and *Oklahoma* – incorporated a number of innovations, including a revised armour arrangement, oil-firing and triple turrets.

Left: Nevada *swings all guns to starboard to deliver a mighty broadside. The guns were the same as those of USS* Texas, *although the triple sets were mounted in common cradles, which meant that they could not be individually elevated.*

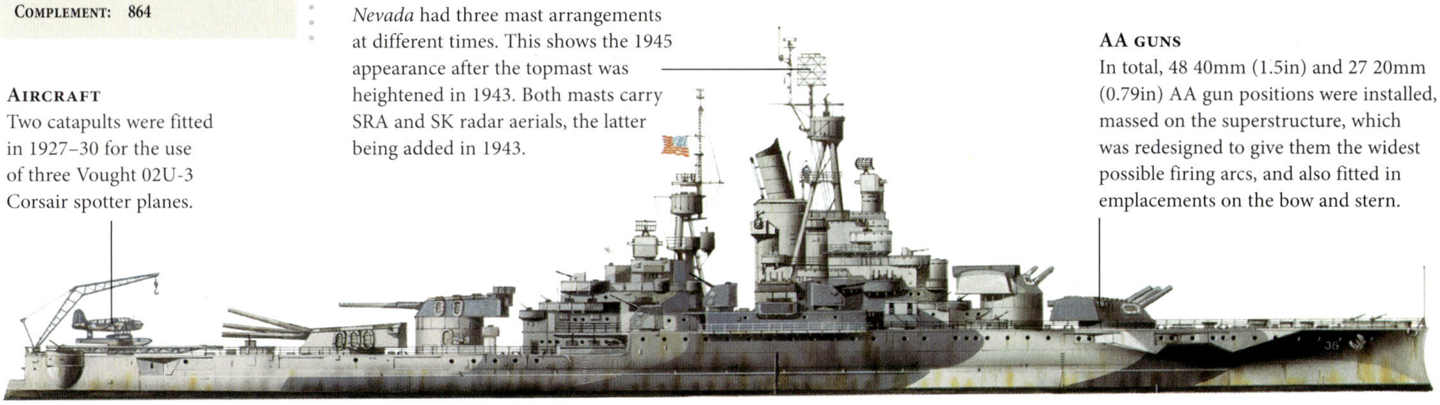

MAST
Nevada had three mast arrangements at different times. This shows the 1945 appearance after the topmast was heightened in 1943. Both masts carry SRA and SK radar aerials, the latter being added in 1943.

AA GUNS
In total, 48 40mm (1.5in) and 27 20mm (0.79in) AA gun positions were installed, massed on the superstructure, which was redesigned to give them the widest possible firing arcs, and also fitted in emplacements on the bow and stern.

AIRCRAFT
Two catapults were fitted in 1927–30 for the use of three Vought 02U-3 Corsair spotter planes.

OIL FIRING
The transition to oil firing had a major influence on warship design. Lateral coal bunkers had been considered as part of the defensive system, but this aspect was completely abandoned. *Nevada*'s oil fuel weighed 1848 tonnes (2037 tons), compared to *Texas*'s 2623 tonnes (2892 tons) of coal, and was carried in double-bottom tanks. Oil-fired boilers were more compact and lighter, enabling more weight to be given to armour and more space to be given to stores, accommodation and protective compartments.

LATTICE MASTS
The USS *Nevada* initially had lattice mast structures, a type of cage-like mast used on many early 20th-century American warships. Alterations in 1927–29 replaced these masts with tripod types.

Bayern (1916)

TYPE • *Battleship* **COUNTRY** • *Germany*

The 'super-dreadnoughts' *Bayern* and *Baden* – the only two completed of a projected four – have always been compared with the contemporary British *Queen Elizabeth* class, sharing broadly similar tonnage and dimensions, as well as the same main armament of eight 381mm (15in) guns.

SPECIFICATIONS

Dimensions:	Length: 180m (590ft 7in); Beam: 30m (98ft 5in); Draught: 9.4m (30ft 10in)
Displacement:	32,200 tonnes (31,700 tons) normal load
Propulsion:	14 marine boilers; 3 Schichau geared turbines; 38,776kW (52,000shp); 3 screws
Speed:	22.25 knots (41.2km/h; 25.6mph)
Range:	7,800nmi (14,500km; 8,991 miles) at 10 knots
Armament:	8 × 381mm (15in) guns; 16 × 152mm (6in) guns; 2 × 88mm (3.5in) guns; 5 × 600mm (24in) torpedo tubes
Armour:	Belt 350–170mm (13.8–6.7in); bulkheads 200–170mm (7.9–6.7in); deck 150–30mm (5.9–1.2in); barbettes 350mm (13.8in); turrets 350–250mm (13.8–9.8in);
Complement:	1171

Right: The main guns were mounted in twin superfiring turrets fore and aft, the turrets projecting backwards on the barbettes, with no two turrets at the same level. The forecastle deck extended aft as far as X turret, above a long casemate with eight 152mm (6in) guns on each side.

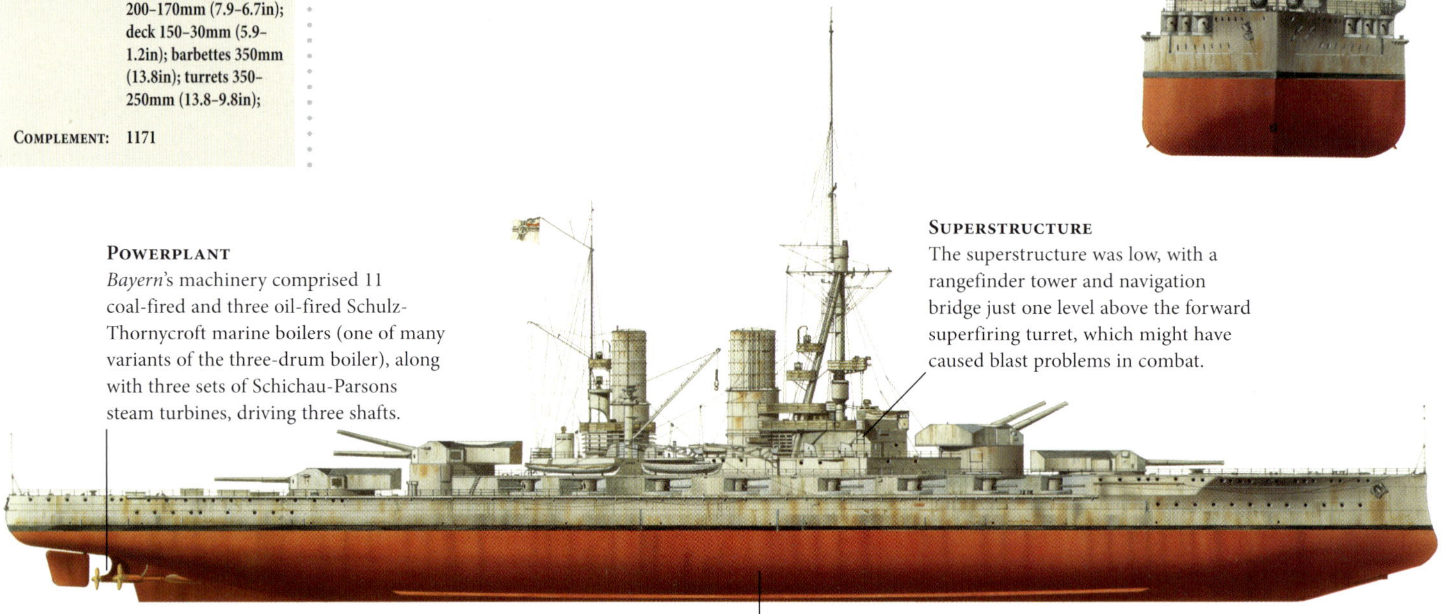

Powerplant
Bayern's machinery comprised 11 coal-fired and three oil-fired Schulz-Thornycroft marine boilers (one of many variants of the three-drum boiler), along with three sets of Schichau-Parsons steam turbines, driving three shafts.

Superstructure
The superstructure was low, with a rangefinder tower and navigation bridge just one level above the forward superfiring turret, which might have caused blast problems in combat.

Electricity and fuel
Eight small diesel generators provided electric power (2400kW/3220hp capacity). Fuel bunkerage was 3400 tonnes (3748 tons) of coal and 620 tonnes (683 tons) of fuel oil.

CONSTRUCTION

Built at the Howaldtswerke, Kiel and commissioned in July 1916, *Bayern* was constructed of riveted steel plates attached to longitudinal and transverse bulkheads, with a cellular double bottom along almost 90 per cent of the hull. Internally, it was divided into 17 watertight compartments. Alongside the boiler and engine rooms, there was an inner torpedo bulkhead, flanked on the inboard side by coal bunkers.

SEA TRIALS, 1915
While completed in March 1916, *Bayern* was still running trials the following May and was not involved in the Battle of Jutland. On 19 August and 19 October 1916 it made unsuccessful sorties into the North Sea. In 1917 it was in action in the Baltic Sea against Russian units.

HMS *Royal Sovereign* (1916)

TYPE • *Battleship* **COUNTRY** • *United Kingdom*

SPECIFICATIONS

DIMENSIONS:	Length: 190.3m (624ft 3in); Beam: 27m (88ft 6in); Draught: 8.7m (28ft 6in)
DISPLACEMENT:	28,449 tonnes (31,363 tons) normal load; 31,497 tonnes (34,741 tons) full load
PROPULSION:	18 oil-fired Babcock & Wilcox boilers; 2 sets Parsons direct-drive turbines; 4 screws; 29,828kW (40,000shp)
SPEED:	23 knots (42.6km/h; 26.5mph)
RANGE:	7000nmi (13,000km; 8,100 miles)
ARMAMENT:	8 x 381mm (15in) guns; 14 x 152mm (6in) guns; 2 x 76mm (3in) AA guns; 4 x 3-pdr guns; 4 x 533mm (21in) torpedo tubes
ARMOUR:	Belt 330–25mm (13–1in); bulkheads 152–102mm (6–4in); barbettes 254–102mm (10–4in); turret faces 330mm (13in); conning tower 280mm (11in)
COMPLEMENT:	997

Completed at Portsmouth Naval Dockyard in May 1916, *Royal Sovereign* just missed the Battle of Jutland, where its eight 381mm (15in) guns and up-to-date fire control equipment might have been useful to the Royal Navy.

Left: HMS Royal Sovereign *in Cape Town, South Africa, October 1940. In May 1944, the ship was secretly transferred to the Soviet Navy as* Archangelsk *and used for escorting convoys to Murmansk. Returned to the Royal Navy in February 1949, it was found to be in poor condition and was sent for scrapping that year.*

FIRE CONTROL
Unlike earlier ships, *Royal Sovereign* was equipped with fire-control systems from the start. A forward one was placed above the conning tower and the aft one was on the mainmast (both masts were tripods) with 4.6m (15ft) rangefinders. In the event of both being knocked out, there was a further control point in 'X' turret.

TURBINES
The turbines were in three separate compartments; the outer (high pressure) set drove the outer shafts and the low-pressure inner pair in the central room drove the inner two propellers.

QUARTERDECK
In its 1933–36 refit *Royal Sovereign* had a catapult and crane placed on the quarterdeck and could launch and recover float planes such as the Fairey III, but these were removed in 1937.

MODERN LOOK
Although the design period was pre-war, the ship, like others in the R-class, took on a more modern appearance, helped by the single funnel and the more substantial tower that now held control stations as well as the navigating deck. The main illustration appears to show how it looked after a refit in 1927–29 when, among other modifications, new rangefinders and eight searchlights were fitted, and the tipped funnel cap was added.

REFIT, PORTSMOUTH, 1920s
As with all new battleship classes, the ultimate design was a compromise between what the designers wanted and what the government was willing to spend The R-class emerged as a scaled-down version of the *Queen Elizabeth* class – smaller, slower, but with good protection and endurance.

K-4 (1916)

TYPE • *Submarine* COUNTRY • *United States*

SPECIFICATIONS

Dimensions:	Length: 47m (153ft 10in); Beam: 5m (16ft 9in); Draught: 4m (13ft 2in)
Displacement:	Surfaced: 398 tonnes (439 tons); Submerged: 530 tonnes (584 tons)
Propulsion:	2 NELSECO diesel engines; 354kW (475bhp) 2 electric motors; 2 screws
Speed:	Surfaced: 14 knots (26km/h; 16mph); Submerged: 10.5 knots (19km/h; 12mph)
Range:	4500nmi (8334km; 5182 miles) at 10 knots
Armament:	4 × 457mm (18in) torpedo tubes
Complement:	31

Serving between 1914 and 1923, the US K-class consisted of eight submarines, in many ways a slightly scaled-up version of the preceding H-class. Although they served in World War I, they quickly became obsolete in the interwar years.

Right: K-4 photographed circa 1918. In 1918, the US Navy had a total of 79 submarines. None had been lost to enemy action; F-4 had foundered and H-1 had been wrecked in the Gulf of Mexico. The L-class boats based at Berehaven in Ireland returned home, as did the K-class from the Azores.

Hull form
The K-class class of eight boats were relatively small in size compared to postwar American submarines. The hull form is still reminiscent of earlier US types.

K-4
K-4 never carried its proposed name, *Walrus*. It was commissioned in October 1914 and based first at Hawaii, then at Key West, Florida.

Propulsion
For surface running, K-class boats were powered by two 354kW (475bhp) NELSECO diesel engines, each driving one propeller shaft. When submerged each propeller was driven by a 127kW (170bhp) electric motor.

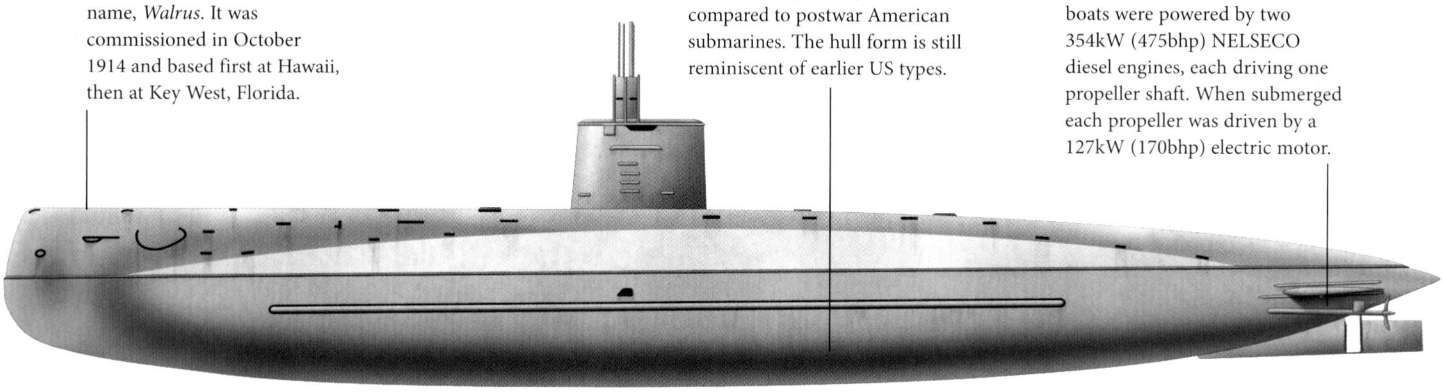

STORM LOSS

At the outbreak of the war, the United States was lagging behind the great powers of Europe in submarine development. Nevertheless, when the United States declared war on Germany in April 1917, the US Navy wanted to get its submarines involved. K-class boats were not designed for blue-water operations, but the Navy decided to tow the eight *K-1* to *K-8* boats to the Mediterranean. This proved to be a disaster, with *K-3, K-4, K-6* and *K-7* lost in a storm, while the survivors were kept in the Azores for the rest of the war.

***K-8* UNDER CONSTRUCTION, 1920s**
America's first diesel-powered submarine was *E-1*, commissioned at Groton in February 1912. The US Navy made relatively little use of its submarines in the war; their operational range was limited and the Navy, mistakenly, did not anticipate hostile action close to the American coast.

HMS *Courageous* (1916)

TYPE • *Battlecruiser/Aircraft carrier* **COUNTRY** • *United Kingdom*

SPECIFICATIONS

DIMENSIONS:	Length: 239.6m (786ft); Beam: 24.7m (81ft); Draught: 6.78m (22ft 3in)
DISPLACEMENT:	20,829 tonnes (22,959 tons) full load
PROPULSION:	18 Yarrow boilers; 4 Parsons geared turbines; 4 screws; 69,932kW (93,780hp)
SPEED:	31 knots (57.4km/h; 35.7mph)
RANGE:	3200nmi (5920km; 3200 miles)
ARMAMENT:	4 × 381mm (15in) guns; 18 × 102mm (4in) guns; 2 × 76mm (3in) AA guns; 2 × 533mm (21in) torpedo tubes (underwater)
ARMOUR:	Armoured deck 76–25mm (3–2in); belt 76–51mm (3–2in); bulkhead 76–37mm (3–1.5in); turrets 330–114mm (13–4.5in)
COMPLEMENT:	842

Built for an invasion that never happened and labelled as 'battlecruisers', the *Courageous* class was completely unsuited to the role. All three were eventually converted to aircraft carriers and gave valuable service in that form.

Left: Courageous *and* Glorious *functioned effectively as carriers, but* Courageous *was sunk by torpedoes from* U-29 *on 17 September 1939, and* Glorious *was sunk by gunfire from* Scharnhorst *and* Gneisenau *on 8 June 1940.* Furious *survived World War II and was scrapped in 1948.*

TRIPOD MAST
Courageous in 1918 after fitting of the tripod mast. It also included flying-off platforms on both main turrets.

ENGINES
The best aspect of *Courageous* was the geared turbine engines, which gave it a 30-knot speed.

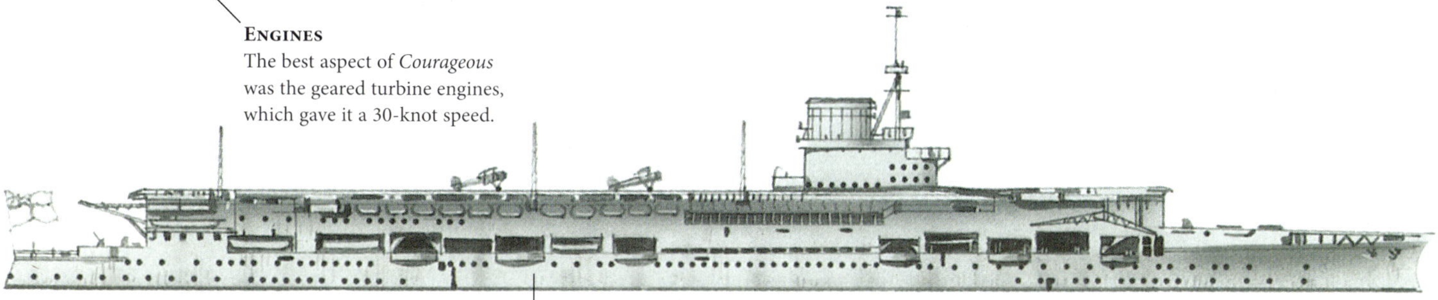

***COURAGEOUS* IN 1938**
Courageous and *Glorious* were given what would become the standard appearance of the aircraft carrier, with deckhouse and funnel set on the extreme starboard to give maximum flight-deck space.

LIGHT CRUISERS

Whether the 'battlecruiser' truly was a viable concept will always be a source of debate. HMS *Courageous* and its two companion ships, *Furious* and *Glorious*, exhibited all the weaknesses and few of the strengths of the type. Planned for service in the Baltic Sea, to support a Russian invasion of northern Germany, they were first designated as 'large light cruisers', but later were officially listed as battlecruisers.

CARRIER CONVERSION

Courageous was converted into an aircraft carrier in 1924–28, with an original complement of 48 aircraft. A leading warship historian described the *Courageous* trio as 'the largest tin-clad targets afloat'. Another writer called them 'Fisher's folly'. But converted to aircraft carriers, they gave effective service.

Ise (1917)

TYPE • *Battleship carrier* **COUNTRY** • *Japan*

SPECIFICATIONS

DIMENSIONS:	Length: 219.68m (720ft 6in); Beam: 31.75m (104ft 2in); Draught: 9.45m (31ft)
DISPLACEMENT:	40,444 tonnes (44,593 tons) full load
PROPULSION:	8 oil-fired Kampon boilers; 4 sets Kampon turbines; 4 screws; 60,000kW (80,000shp)
SPEED:	24.5 knots (45.4km/h; 28.2mph)
RANGE:	7870nmi (14,580km; 7870 miles) at 16 knots
ARMAMENT:	4 × 356mm (14in) guns; 8 × 127mm (5in) guns; 31 × 25mm (1in) AA guns; 11 × 25mm (1in) AA guns; 6 × 127mm (4.5in) AA rocket launchers
ARMOUR:	Deck 152mm (6in); 200mm (7.8in) of concrete on flight deck
COMPLEMENT:	1463

Built by Kawasaki at Kobe, and commissioned on 1 December 1917, *Ise*'s first incarnation was as a battleship armed with 12 356mm (14in) guns in twin turrets, a direct development of the preceding *Fuso* class.

Right: Ise *undergoes sea trials after its conversion to a hybrid battleship carrier. The ship took part in the Leyte Gulf campaign and the Battle of Cape Engaño, shooting down five US aircraft but taking three hits.*

AIRCRAFT
Maximum aircraft capacity was 22, with 11 stationed on the flight deck and the remainder in the hangar.

HANGAR
The hangar below the flight deck was fully fitted with firefighting sprays. *Ise* could carry 11 Yokosuka D4Y dive bombers plus Aichi reconnaissance planes, but never carried a full air group.

MODIFICATIONS
In 1935–37, at Kure, the original 24 coal/oil boilers were replaced by eight Kampon oil boilers and new Kampon turbines. Anti-torpedo bulges were fitted and its six torpedo tubes were removed.

CONVERSION

After Japan's disastrous losses at the Battle of Midway, the decision was made to convert *Ise* into a hybrid battleship carrier to support the depleted carrier fleet. Between February-August 1943 it was remodelled, retaining the front end and guns of a battleship, but with the two aft turrets removed along with their barbettes (though not the magazines that now stored AA ammunition). A 70m (300ft) flight deck was installed between the reduced mainmast and the stern.

WRECK OF THE *ISE*
Following intensive combat around the Philippines, *Ise* returned to Kure in February 1945, where it was used as a stationary gun platform. Towed to the base of Ondo Seto, it was under constant air attack until it was finally sunk on 28 July 1945.

USS *Mississippi* (1917)

TYPE • *Battleship* **COUNTRY** • *United States*

SPECIFICATIONS (Post-1933)

DIMENSIONS:	Length: 9.7m (97ft 5in); Draught: 9.1m (30ft)
DISPLACEMENT:	33,960 tonnes (37,422 tons) normal load; 36,737 tonnes (40,494 tons) full load
PROPULSION:	6 Bureau Express boilers; 2 sets Westinghouse geared turbines; 4 screws; 30,000kW (40,000shp)
SPEED:	22 knots (41km/h; 25mph)
RANGE:	6400nmi (11,900km; 6400 miles) at 10 knots
ARMAMENT:	12 × 356mm (14in) guns; 12 × 127mm (5in) guns; 8 × 127mm (5in) AA guns; 4 × 76mm (3in) guns
ARMOUR:	Belt 343–203mm (13.5–8in); turret faces 457mm (18in); conning tower 406mm (16in); deck 139mm (5.5in)
COMPLEMENT:	1076

For the *New Mexico* class, which included the USS *Mississippi* (BB-41), the US Navy wanted a completely new design that incorporated the lessons learned from its own and other navies' experience of war.

Left: USS Mississippi *ploughs through rough Atlantic seas in 1941. The great ship served the US Navy from the last year of World War I until t he mid-1950s.*

MAIN GUNS
The 356mm (14in) 50-calibre gun was a developed version of the same 45-calibre guns fitted on *Nevada*. Each gun could elevate independently. In 1931 *Mississippi* was rearmed with the Mark 11 version.

1944 PROFILE
Modifications in the 1931–33 refit included raising of maximum elevation of the main guns, building up the superstructure, installing gun directors for the main and 127mm (5in) guns, heightening the funnel, taking out the foremast and replacing the mainmast with a short pole.

ARMOUR INCREASE
Armour protection was increased in the 1930s, with deck armour now reaching 140mm (5.5in), capable of withstanding a 725kg (1600lb) bomb dropped from up to 1830m (6000ft), and the splinter protection above the machinery spaces was raised to 70mm (2.75in).

ORIGINAL FEATURES

Most noticeable in the original ship was the raked clipper-type bow that gave the ship the racy look of a hunter. The angle of the stem changed to vertical at the waterline. Oil-fired from the start, the boiler take-ups were trunked into a single funnel. The casemate arrangement for secondary guns was at last being abandoned, and during construction most of the 127mm (5in) guns were mounted at the same deck level as 'A' turret. The remaining casemate guns were removed in the 1920s.

ON THE MISSISSIPPI
Here the USS *Mississippi* sails along its namesake river, en route to take part in Navy Day celebrations at New Orleans, Louisiana, 16 October 1945. Note the anchors suspended below their normal stowed position at the bow.

HMS *Furious* (1917)

TYPE • *Aircraft carrier* **COUNTRY** • *United Kingdom*

SPECIFICATIONS

DIMENSIONS:	Dimensions: Length: 224.1m (735ft 2in); Beam: 26.8m (88ft); Draught: 8.3m (27ft 3in)
DISPLACEMENT:	22,900 tonnes (25,245 tons) normal load; 26,900 tonnes (29,652 tons) deep load
PROPULSION:	18 Yarrow boilers; 4 Brown-Curtis geared turbines; 4 shafts; 67,113kW (90,000shp)
SPEED:	30 knots (56km/h; 34.5mph)
RANGE:	7480nmi (13,850km; 8610 miles) at 10 knots
ARMAMENT:	10 × 140mm (5.5in) BL Mk I guns; 6 × 102mm (4in) QF AA guns
AIRCRAFT:	36
COMPLEMENT:	795

Originally intended as a cruiser, *Furious* became the first warship in regular use as an aircraft carrier, undergoing successive conversions from partial to full flight deck. With HMS *Argus*, it was one of only two ships to serve as a carrier in both world wars.

Right: Furious *with a barrage balloon on its short flight deck. Although Squadron Commander E.H. Dunning made the first successful landing on a carrier steaming into the wind in a Sopwith Pup on* Furious *on 2 August 1917, he was killed in his second attempt. The ship at that time had no arrestor gear, and deck landings were abandoned as too risky.*

EXHAUST
Exhaust gases were ducted out at the stern in the space between the flight deck and the hangar roof, a bad location for planes landing, flying into the wind.

PROFILE
The profile shows the carrier's appearance in World War II, with guns and direction-finding equipment forward of the flight deck.

FLY-OFF DECK
Initially the lower fly-off deck was laid over the forecastle, tapering with the bow. *Furious* was also the first carrier to have the after edge of the flight deck rounded down rather than presenting a sharp edge, which smoothed airflow as the ship steamed into the wind

CARRIER CONVERSION

In 1917 *Furious* was given a short flight deck with a small hangar and lift instead of a forward gun turret, and later the aft turret was removed to provide a further short flight deck. It underwent further alteration between 1921 and 1925 to become a full flush-deck carrier, although at first the flush deck was used only for landings and a short lower-level launch deck was retained over the bow (removed in 1932). More changes were made in 1931–32, when the stern was built up to the level of the flight deck, and in 1939 when a small island was added on the starboard side.

PRE-CARRIER
One of three companion ships planned as heavy-gun cruisers for coastal bombardment in the Baltic Sea, all of which ended up as carriers, *Furious* was laid down at Armstrong Whitworth's Elswick Yard on the Tyne on 8 June 1915.

U-139 (1917)

SPECIFICATIONS

DIMENSIONS:	Length: 94.8m (311ft); Beam: 9m (29ft 9in); Draught: 5.2m (17ft)
DISPLACEMENT:	Surfaced: 1961 tonnes (2162 tons); Submerged: 2523 tonnes (2781 tons)
PROPULSION:	2 diesel engines delivering 2458kW (3300hp); electric motors 1326kW (1780hp); 2 screws
SPEED:	Surfaced: 15.8 knots (29.2km/h; 18.1mph); Submerged: 7.6 knots (14km/h; 8.7mph)
RANGE:	12,630nmi (23,390km; 14,534 miles) at 8 knots
ARMAMENT:	6 × 508mm (20in) torpedo tubes; 2 × 150mm (5.9in) guns
COMPLEMENT:	62

The first of a class of three long-range attack submarines, this was the last German submarine to sink an enemy ship before the war ended in 1918. Its design influenced the postwar submarine designs of several other navies.

Left: U-139*'s companion ship* U-140 *here sits alongside an American warship. This submarine was transferred to the United States in 1919 and its fate was to be sunk as a target ship in 1920.*

TARGET OPTICS
The attack periscope used a system known as UZO (U-Boot Ziel Optik, 'U-boat target optics') incorporating a Stadimeter split-prism rangefinder that gave a double image of the target and a range calculation up to 6000m (19,685ft).

AMMUNITION TRUNKING
The ammunition trunking was protected by 250mm (1in) armour. In total 980 rounds of gun ammunition were carried.

ARMAMENT
As the French *Halbronn*, *U-139* kept its deck guns, but the fairing round the conning tower was removed, leaving the armoured 90mm (3.5in) tower form.

COMBAT ACTION

Encountering an Allied Atlantic convoy on 1–2 October 1918, *U-139* sank three small merchant ships and damaged the armed escort *Perth*. From this engagement comes an unsubstantiated story that one ship came down on top of the submarine, forcing it towards the seabed. All tanks were blown empty by compressed air in the effort to wrest *U-139* free, and it disengaged and shot to the surface. Damaged, its periscopes bent, it managed to make its escape.

***U-139*, BREST, FEBRUARY 1919**
In postwar reparations, *U-139* was handed over to France, and served until 1935 in the Marine Nationale as Halbronn, its design influencing the French interest in large submarine cruisers. Here the French vessel beyond it is a *Gustave Zédé*-class submarine.

HMS *Walker* (1918)

TYPE • *Destroyer* **COUNTRY** • *United Kingdom*

SPECIFICATIONS

DIMENSIONS:	Length: 91m (300ft); Beam: 8.15m (26ft 9in); Draught: 2.7m (8ft 10in)
DISPLACEMENT:	1117 tonnes (1231 tons) normal load
PROPULSION:	3 Yarrow water-tube boilers; Brown-Curtis steam turbines; 2 shafts; 20,134kW (27,000shp)
SPEED:	34 knots (63km/h; 39mph)
RANGE:	3500nmi (6000km; 4039 miles) at 15 knots
ARMAMENT:	4 × 102mm (4in) QF Mk V guns; 2 × 40mm (2-pdr) QF Mk II pom-pom or 1 × 76mm (3in) QF 20cwt HA gun; 2 × 3 × 533mm (21in) torpedo tubes
ARMOUR:	Splinter protection only
COMPLEMENT:	110

Launched in November 1917 and commissioned the following February, HMS *Walker* survived all six years of World War II, principally in the role of troopship and convoy escort.

Right: HMS Wishart *(D67) was a Modified W-class destroyer of the British Royal Navy that saw service in World War II.*

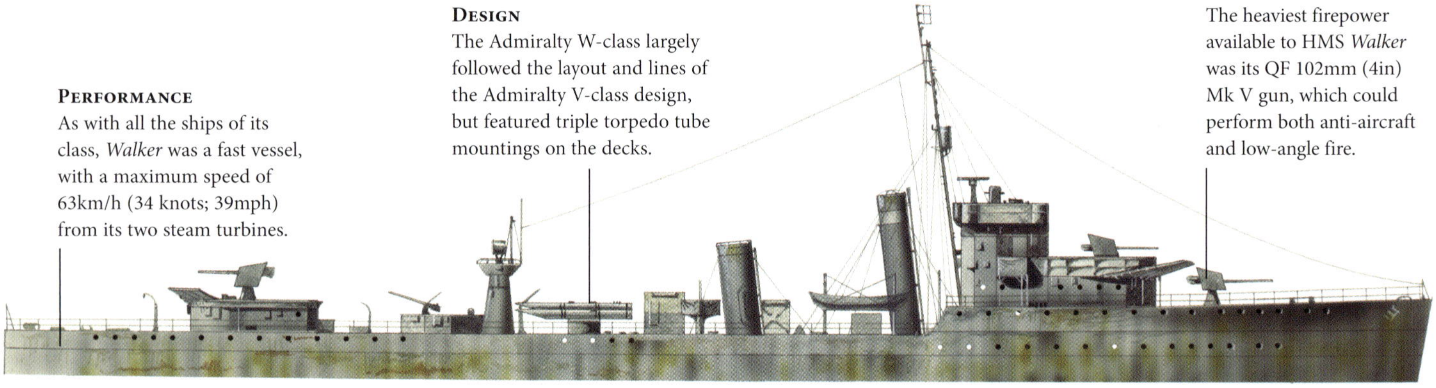

PERFORMANCE
As with all the ships of its class, *Walker* was a fast vessel, with a maximum speed of 63km/h (34 knots; 39mph) from its two steam turbines.

DESIGN
The Admiralty W-class largely followed the layout and lines of the Admiralty V-class design, but featured triple torpedo tube mountings on the decks.

ARMAMENT
The heaviest firepower available to HMS *Walker* was its QF 102mm (4in) Mk V gun, which could perform both anti-aircraft and low-angle fire.

D-DAY SERVICE

HMS *Walker* performed regular and arduous convoy escort duties during World War II, guiding merchant ships through the Western Approaches or making the perilous run through Arctic waters to northern Russia. During the Normandy landings of June 1944, however, the warship was tasked with escorting troopships to the invasion beaches, as well as supporting logistics vessels once the beachhead had been established.

W-CLASS
HMS *Walker* (D-27) was specifically of the Admiralty W-class. In total, British shipyards built 67 V- and W-class destroyers, making them among the most prolific Royal Navy warship types of the conflict.

HMS *Argus* (1918)

TYPE • *Aircraft carrier* **COUNTRY** • *United Kingdom*

SPECIFICATIONS

DIMENSIONS:	Length: 172.2m (565ft); Beam: 20.7m (68ft); Draught: 7m (23ft 2in)
DISPLACEMENT:	14,935 tonnes (14,700 tons) normal load; 16,028 tonnes (15,775 tons) deep load
PROPULSION:	12 Scotch boilers; 4 Parsons turbines; 2 shafts; 15,000kW (20,000shp)
SPEED:	20 knots (37km/h; 23mph)
RANGE:	3563nmi (6598km; 4100 miles)
ARMAMENT:	4 × 102mm (4in) guns; 4 × 102mm (4in) AA guns
AIRCRAFT:	48
COMPLEMENT:	495

Commenced as a passenger liner in 1914, HMS *Argus* became the first carrier with a full-length flight deck, but was commissioned too late for action in World War I. It was first a plane ferry then, for a time, a fighting carrier in World War II.

Left: Argus *in British waters, c. late 1918. The aircraft carrier is painted in wartime dazzle camouflage –* Argus *was initially painted in camouflage style before assuming peacetime naval grey.*

FLIGHT DECK
Only 17.7m (58ft) wide until 1938, the flight deck was barely wider than the aircraft carried, and three planes were blown off the ship by wind on the ship's first cruise.

SUPERSTRUCTURE
A small retractable pilot house was placed behind the forecastle, for use when there were no launchings planned. Exhaust smoke and fumes from the boilers were trunked to the stern through tubes set between the flight deck and the roof of the hangar deck, with some of the heat dissipated by electric fans.

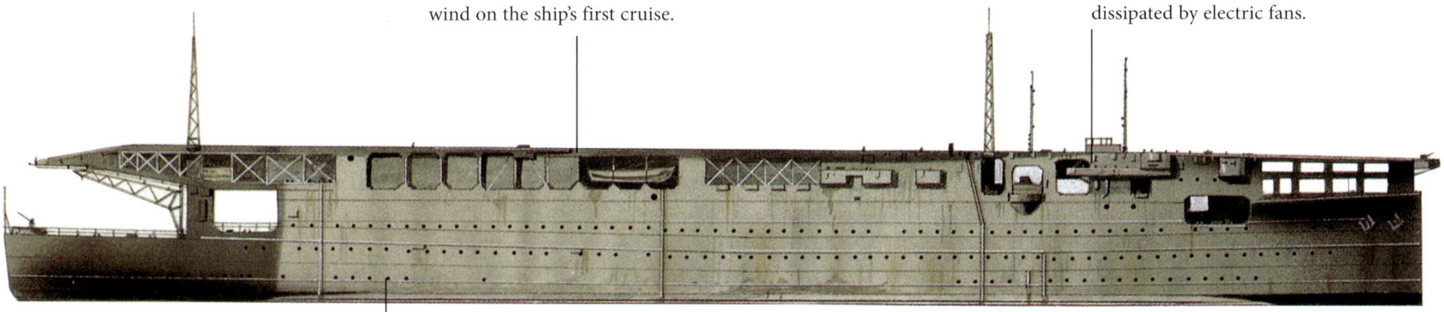

HULL
The hull was that of a merchant ship, and armour was applied as a 50mm (2in) box round the after magazine and torpedo room, while the forward magazine was protected only by a 50mm (2in) deck above it.

WORLD WAR II

In the early years of World War II, carrying Fairey Swordfish planes, *Argus* was engaged in convoy escort and aircraft-ferrying duties to Iceland, Malta, the Gold Coast (Ghana), Gibraltar and Murmansk. In November 1942 it supported the North African landings and sustained a bomb hit, which killed four of the crew and required a return to the UK for repairs. It was reclassified as an escort carrier in May 1943, but by this time more up-to-date ships were becoming available, and it continued to be used for pilot training.

ARGUS FLYOVER
Commissioned on 16 September 1918, *Argus* was the first carrier in the world to be fitted with a full-length flight deck, enabling wheeled aircraft to land on board, and earning the nickname 'flat iron'.

47
52

Interwar Period

The interwar period (1918–39) was a time for navies to reflect on the tactical lessons learned from four years of intense warfare. The focus remained on maintaining a powerful, mixed-type surface fleet, leading to successive generations of more capable battleships and heavy cruisers, supported by fast-moving destroyers, frigates, torpedo-boats and other smaller craft. Submarines had demonstrated their worth, but in many navies they were still secondary assets, constrained by limited range and the need to spend much of their operational time on the surface. By 1939, Italy, Japan and the USSR had the largest submarine fleets, although Germany's 57 U-boats were tactically more advanced. What Germany did not develop during this period were aircraft carriers – the first purpose-built carriers were constructed by Japan, the United Kingdom and the United States.

U-47
The famous Kapitänleutnant Günther Prien and his officers gather on the deck of *U-47*, a Type VIIB U-boat, alongside *U-46*, *U-52* and *U-51* at the Germaniawerft shipbuilder, Kiel, on 1 August 1939.

Hood (1920)

TYPE • *Battleship* COUNTRY • *United Kingdom*

SPECIFICATIONS

DIMENSIONS:	Length: 262m (860ft 7in); Beam: 31.8m (104ft 2in); Draught: 9.8m (32ft)
DISPLACEMENT:	45,923 tonnes (50,617 tons) full load
PROPULSION:	24 Yarrow small-tube boilers; 4 Brown-Curtis geared turbines; 107,381kW (144,000shp); 4 screws
SPEED:	31 knots (57.4km/h; 35.9mph)
RANGE:	5000nmi (9260km; 5753 miles) at 18 knots
ARMAMENT:	8 × 381mm (15in) guns; 12 × 140mm (5.5in) guns; 4 × 102mm (4in) AA guns; 6 × 533mm (21in) torpedo tubes
ARMOUR:	Belt 305–152mm (12–6in); bulkheads 127–102mm (5–4in); main deck 76–19mm (3–0.75in); turrets 381–279mm (15–11in); conning tower 279–229mm (11–9in)
COMPLEMENT:	1433

***Hood* was described by one naval historian in 1968 as 'a ship far in advance of its time'. He viewed it as a precursor of the fast, heavily armed battleships that would dominate World War II. It was the Royal Navy's largest warship, intended as the first of five.**

Below: Hood *was the last battlecruiser to be built for the Royal Navy. A postwar generation of even larger battlecruisers was planned but cancelled under the terms of the 1921 Washington Treaty.*

BOILERS
The new small-tube boilers provided 107,381kW (144,00hp) for the same weight as HMS *Renown*'s boilers that produced 83,518kW (112,000hp).

AIRCRAFT
In 1932 an aircraft crane and telescopic catapult were experimentally mounted on the aft deck, but not retained.

ARMOUR
Total weight of armour carried was 12,292 tonnes (13,550 tons), 30.3 per cent of design weight. This compared favourably to the armour weight of British World War I battleships: *Royal Oak*'s armour weight, 7484 tonnes (8250 tons), was 32 per cent of design displacement

SINKING

Hood's wartime operations included the chase of *Scharnhorst* and *Gneisenau* in the English Channel in November 1939, and in June–July 1940 it was flagship of 'Force H' in the Mediterranean, On 22 May 1941 it was flagship of Vice Admiral Holland's Battle Cruiser Force, sent with HMS *Prince of Wales* to intercept *Bismarck* and *Prinz Eugen*, south of Greenland. While exchanging gunfire with the German ships at a range of around 22,860m (75,000ft), *Hood* exploded at 06:00 on 24 May and sank a few minutes later. Only three of the crew survived.

***HOOD*, PANAMA CANAL, 1924**
From 1920 to 1929, *Hood* served with the Home Fleet on exercises and ceremonial duties, including a 10-month world cruise with HMS *Repulse* in 1923. A modernization programme at HM Dockyard Portsmouth lasted from June 1929 to May 1931, after which it rejoined the Home Fleet.

Nagato (1920)

TYPE • *Battleship* **COUNTRY** • *Japan*

SPECIFICATIONS

Dimensions:	Length: 215.8m (708ft); Beam: 29m (95ft 3in); Draught: 9.1m (29ft 9in)
Displacement:	32,720 tonnes (36,067 tons) standard; 39,116 tonnes (43,119 tons) full load
Propulsion:	21 boilers (15 oil, 6 coal); 4 Gihon turbines; 59,656kW (80,000shp); 4 screws
Speed:	26.75 knots (49.6km/h; 31mph)
Range:	5700nmi (10,550km; 6570 miles) at 16 knots
Armament:	8 × 406mm (16in) guns; 20 × 140mm (5.5in) guns; 4 × 76mm (3in) AA guns; 8 × 533mm (21in) torpedo tubes
Armour:	Belt 305–102mm (12–4in); deck upper 44–25mm (1.7–1in), lower 75–50mm (2.9–2.7in); barbettes 300mm (11.8in); turrets 356mm (14in); conning tower forward 371mm (14.6in), aft 97mm (3.8in)
Complement:	1333

The war of 1904–05 with Russia had confirmed Japan as a major naval power, able to combine advanced warships with new tactics. With the *Nagato* class, Japan possessed two of the largest, most powerful and fastest battleships in the world.

Below: A deck plan of the 1924–34 layout. Note the prototype Heinkel catapult fitted on 'B' turret.

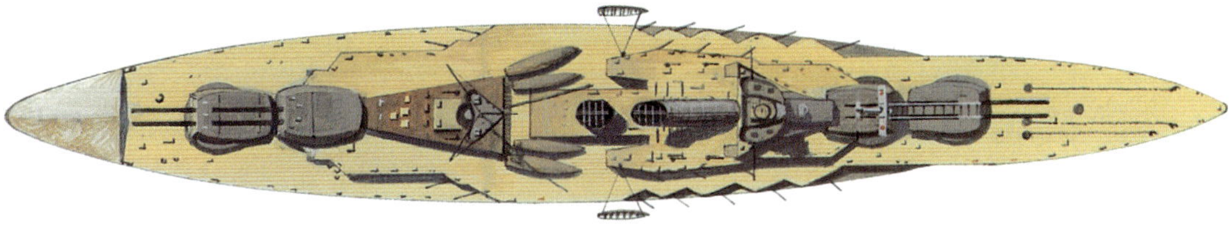

'Pagoda' structure
The tower-foremast structure went through two major changes, resulting in an increasingly higher 'pagoda' form that was typical of large Japanese battleships in the 1930s.

Gun turrets
The 406mm (16in) gun turrets weighed 930 tonnes (1025 tons) each, including rotary armour of 498 tonnes (549 tons).

Displacement
From 1936 maximum displacement was 38,814 tonnes (42,785 tons), length was 224.5m (736ft 6in), Beam 34.6m (113ft 5in) and Draught 9.5m (31ft 2in).

LONG-RANGE FIRING

From 1931, the IJN used the Type 91AP shell for long-range gunnery – 'boat-tailed', with a longer-delay fuse, anti-premature detonation device, streamlined nose cone incorporating a blunt cap-head that detached if the shell hit water. Its Anisol explosive charge was only 1.46–1.65 per cent of total weight. With maximum elevation, a range of 35,000m (114,829ft) could be achieved with 356mm (14in) guns.

***NAGATO* ILLUSTRATION**
A Japanese artwork celebrates *Nagato*'s glory days. In 1944, *Nagato* was engaged in the battles of the Philippine Sea and Leyte Gulf, incurring some damage in the latter. Handed over to the USA in 1945, it was sunk in nuclear weapons tests in 1946.

Jaime I (1922)

TYPE • *Battleship* COUNTRY • *Spain*

Third ship in Spain's last class of battleships, *Jaime I*'s construction was delayed by World War I. While well-armed, its endurance under fire was limited. As the smallest dreadnought-type ships ever built, the *España* class can be seen as anticipating the pocket battleships of the 1930s.

SPECIFICATIONS

Dimensions:	Length: 140m (459ft 2in); Beam: 24m (78ft 9in); Draught: 7.8m (25ft 6in)
Displacement:	15,700 tonnes (17,321 tons) standard; 15,952 tonnes (17,586 tons) full load
Propulsion:	4 shafts; Parsons turbines; 12 Yarrow boilers; 11,558kW (15,500shp)
Speed:	16.75 knots (31km/h; 19.3mph)
Range:	9260km (5000nmi; 5752 miles) at 10 knots
Armament:	8 x 305mm (12in) guns; 20 x 102mm (4in) guns; 4 x 47mm (3-pdr) guns
Armour:	Belt 203–102mm (8–4in); upper belt 152mm (6in); barbettes 254mm (10in); turrets 203mm (8in); conning tower 256mm (10in); deck 38mm (1.5in)
Complement:	854

Right: Jaime I *saw action against the Riff revolt in Spanish Morocco in 1922–23 and during the Spanish Civil War (1936–39). While in Cartagena dockyard on 17 June 1937, an accidental explosion and fire caused major damage, killing around 300 men and leaving the ship unusable.*

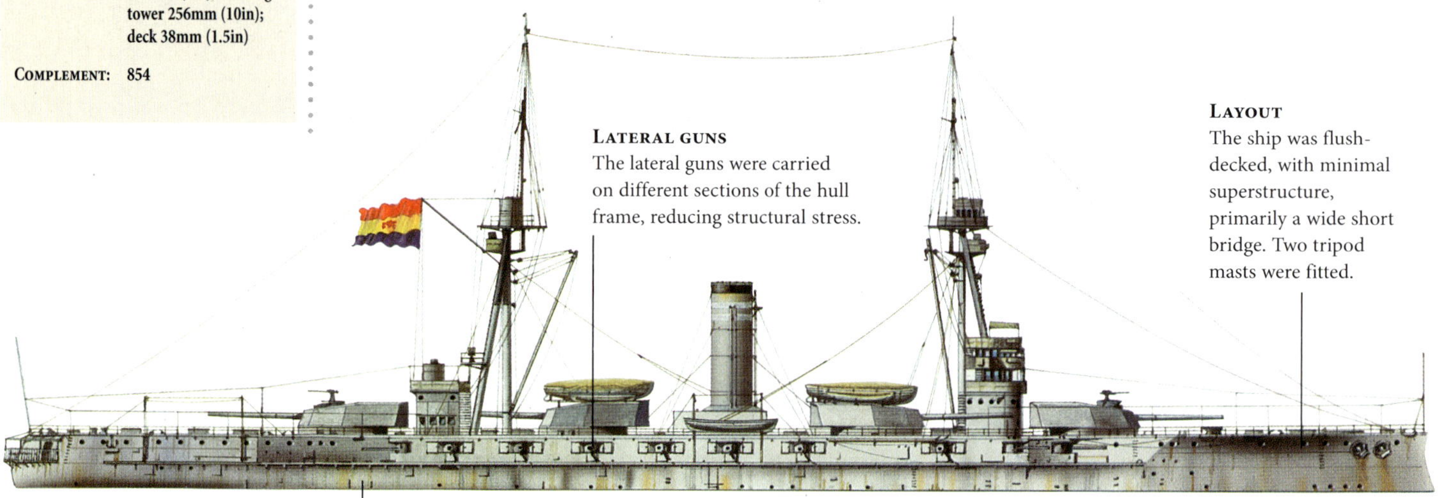

Lateral guns
The lateral guns were carried on different sections of the hull frame, reducing structural stress.

Layout
The ship was flush-decked, with minimal superstructure, primarily a wide short bridge. Two tripod masts were fitted.

Armour
The ship did not lack armour, carrying 4826 tonnes (5320 tons), 29.9 per cent of displacement, but this was considerably thinner than that of larger dreadnoughts.

CONSTRUCTION

The type's dimensions were restricted by the size of the docks at Spanish naval bases; funds did not allow for new docking facilities and new capital ships. Designed on the 'all big-gun' principle, *Jaime I* was laid down at Ferrol, Spain, on 5 February 1912 and launched on 21 September 1914. Construction was in partnership with Vickers-Armstrong of England, and the supply of material and parts was interrupted by the war years. Consequently, Jaime I was not completed until 1922.

***JAIME I*, 1931**
Jaime I was the last survivor of the class. *España* was wrecked on 26 August 1923, and *Alfonso XIII* (renamed *España* in 1931) was sunk by a mine on 30 April 1937.

Chacal (1922)

TYPE • *Destroyer* COUNTRY • *France*

SPECIFICATIONS

DIMENSIONS:	Length: 126.8m (416ft); Beam: 11.1m (36ft 5in); Draught: 4.1m (13ft 5in)
DISPLACEMENT:	2126 tonnes (2343 tons) standard; 3075 tonnes (3388 tons) full load
PROPULSION:	5 du Temple boilers; 2 geared steam turbines; 2 shafts; 37,000kW (49,000shp)
SPEED:	35.5 knots (65.7km/h; 40.9mph)
RANGE:	3000nmi (5600km; 3500 miles) at 15 knots (28km/h; 17mph)
ARMAMENT:	5 × 130mm (5.1in) guns; 2 × 75mm (3in) AA guns; 2 × triple 550mm (21.7in) torpedo tubes; 46 depth charges (2 chutes, 4 throwers)
COMPLEMENT:	221 (wartime)

The *Chacal* was a member of the French *Jaguar*-class destroyers (also known as the *Chacal* class) launched and commissioned during the 1920s. Of the six vessels constructed, only *Tigre* survived World War II.

Left: Tigre *(seen here) spent most of its operational career in the Mediterranean. During World War II it escaped destruction on several occasions, and passed through French, German, Italian and Free French hands.*

DEPTH CHARGES
The stern of the ship was fitted with two chutes for deploying heavy 200kg (440lb) depth charges, as well as four throwers for lighter ordenance.

MAIN GUNS
The principal deck armament was five Canon de 130mm (5.1in) Modèle 1919 guns set in single mounts, two forward, two aft and one abaft the rear funnel.

CREW
The *Jaguar* class had a peacetime complement of 10 officers and 187 crewmen, rising to 12 officers and 209 enlisted men in wartime conditions.

CONTRE-TORPILLEUR

The *Jaguar*-class vessels belonged to a category of French naval vessel known as a *contre-torpilleur*. The French term literally means 'counter-torpedo boat', but the ship type has also been regarded as a destroyer or a light cruiser. The ships were designed around high speed plus an armament sufficient enough to take on enemy destroyers and cruisers. The primary role of the *contre-torpilleur* was aggressive scouting.

***CHACAL*, 1924**
Chacal was launched on 27 September 1924 and completed nearly two years later, on 12 June 1926. It was bombed and sunk by German air attack on 24 May 1940; another ship of the class, *Jaguar*, had been similarly lost the previous day.

HMS *Eagle* (1924)

TYPE • *Aircraft carrier* **COUNTRY** • *United Kingdom*

SPECIFICATIONS

Dimensions:	Length: 203.5m (667ft 6in); Beam: 35.1m (115ft); Draught (max.): 8.1m (26ft 8in)
Displacement:	22,200 tonnes (24,469 tons) standard
Propulsion:	32 Yarrow boilers; 4 geared turbines; 37,000kW (50,000shp)
Speed:	24 knots (44km/h; 28mph)
Range:	8900km (4800nmi; 5500 miles) at 16 knots
Armament:	9 x 152mm (6in) guns; 5 x 102mm (4in) AA guns
Aircraft:	24
Complement:	791

Like all first-generation carriers to emerge in the 1920s, HMS *Eagle* was not originally intended for the role, and retained much of its battleship appearance even after conversion to an aircraft carrier.

Right: HMS Eagle *and HMS* Malaya *in the Mediterranean during Operation* Spotter*, which delivered 16 RAF Spitfire Mk V fighters to Malta on 7 March 1942.*

Flight deck
Eagle was the first carrier to have an island structure on the flight deck, establishing the convention that it would be positioned on the starboard side of the ship.

Arms and armour
Nine 152mm (6in) guns were mounted on the main deck. Armour consisted of a 114mm belt (4.5in), deck armour of 25–38mm (1–1.5in) and 102mm (4in) transverse bulkheads

Engine room
The engine room instruments were ordered prior to purchase by the British, and uniquely for a British warship were calibrated in metric units and labelled in Spanish.

ORIGINS

The Chilean Navy ordered the dreadnought-type battleship *Almirante Cochrane* from Armstrong Whitworth's yard on the Tyne in July 1912, and it was laid down on 20 February 1913, but work ceased with the outbreak of war in August 1914. On 13 February 1918 the British Admiralty completed purchase of the unfinished hull for conversion into a seaplane carrier, later amended to be a standard aircraft carrier, and was first commissioned as HMS *Eagle* at Devonport on 6 April 1920.

DECK VIEW, 1931
The dreadnought-type hull was evident, with the hangar decks built up over it, and a full-length armoured flight deck. A substantial island structure was mounted to starboard.

V-boat (1924)

TYPE • *Submarine* **COUNTRY** • *United States*

SPECIFICATIONS

Dimensions:	Length: 99.4m (326ft); Beam: 8.3m (27ft 3in); Draught: 4.5m (14ft 9in)
Displacement:	Surfaced: 2032 tonnes (2240 tons); Submerged: 2662 tonnes (2935 tons)
Propulsion:	Twin-shaft diesel engines; electric motors
Speed:	Surfaced: 18 knots (33.3km/h; 20.7mph); submerged: 11 knots (20.4km/h; 12.7mph)
Range:	11,118km (6000nmi; 6893 miles) at 11 knots
Armament:	6 × 533mm (21in) torpedo tubes; 1 × 76mm (3in) gun
Complement:	85

Drawing on U-boat designs, the United States built nine V-boats between 1919 and 1934. The striking ability of the *U-151* and similar submarines against the American coast left a deep impression. These nine boats were organized into five separate types.

Below: V-5 and V-6, Narwhal *and* Nautilus, *were commissioned in 1930. They carried two 150mm (6in) guns, as did* Argonaut. *Plans for accommodating a seaplane were dropped.*

USS *Bass*
Originally *V-2*, *Bass* was twice the size of the S-boats, with only two more torpedo tubes. It served with Submarine Division 20 in the Caribbean and Pacific. Put on reserve in 1937, recommissioned in 1940, it was part of Submarine Squadron 3, Submarine Division 31, Atlantic Fleet. A fire on 17 August 1942 killed 25 crewmen.

Improvements
Later V-class submarines incorporated numerous innovations, from air conditioning to partially welded hulls and sophisticated fire-control. Ineffective in wartime conditions, they were relegated to training activities.

Repurposed
Decommissioned in 1937, *V-1* to *V-3* were re-entered and overhauled in 1942–43, converted to cargo carrying, but never used in this role. *Bass* was scuttled as a sonar target in 1945.

FLEET BOAT

Originally designated only by the V and a number, the V-boats received names in 1931. The basic aim was to find an acceptable 'fleet boat' design, on the lines of the German cruisers. *V-1* to *V-3*, later named *Barracuda*, *Bass* and *Bonita*, were not considered successes, partly because of unreliable and underpowered motors, and because they failed to make the required surface and submerged speeds of 21 knots (38.9km/h; 24.2 mph), and 9 knots (16.7km/h; 10.4mph).

V-BOATS AND TENDER
Five V-boats dock alongside the submarine tender USS *Holland*. From left to right the submarines are: USS *Cachalot* (SS-170), USS *Dolphin* (SS-169), USS *Barracuda* (SS-163), USS *Bass* (SS-164), USS *Bonita* (SS-165), USS *Nautilus* (SS-168) and USS *Narwhal* (SS-167).

HMS *Rodney* (1927)

TYPE • *Battleship* **COUNTRY** • *United Kingdom*

SPECIFICATIONS

Dimensions:	Length: 216.8m (710ft); Beam: 32.4m (106ft); Draught: 9.1m (30ft)
Displacement:	30,799 tonnes (33,950 tons) standard; 34,473 tonnes (38,000 tons) full load
Propulsion:	8 three-drum boilers, Brown-Curtis geared turbines; 33,556kW (45,000shp), 2 screws
Speed:	23 knots (42.6km/h; 26.6mph)
Range:	14,300nmi (26,500km; 16,460 miles) at 12 knots
Armament:	9 x 406mm (16in) guns; 12 x 152mm (6in) guns; 6 x 119mm (4.7in) AA guns; 8 x 2-pdr pom-poms; 2 x 622mm (24.5in) underwater torpedo tubes
Armour:	Belt 356mm (14in); bulkheads 356–76mm (14–3in), barbettes 381mm (15in), turret faces 406mm (16in),
Complement:	1314

Unusually designed with all the heavy guns mounted forward, *Rodney* was one of the two Royal Navy battleships built under the terms of the Washington Naval Treaty of February 1922, and was the only British warship to carry 406mm (16in) guns.

Above: Rodney *steamed 251,000km (156,000 miles) on war service. From December 1944, with engines badly needing an overhaul, it remained at Scapa Flow then passed into reserve after the end of the war.*

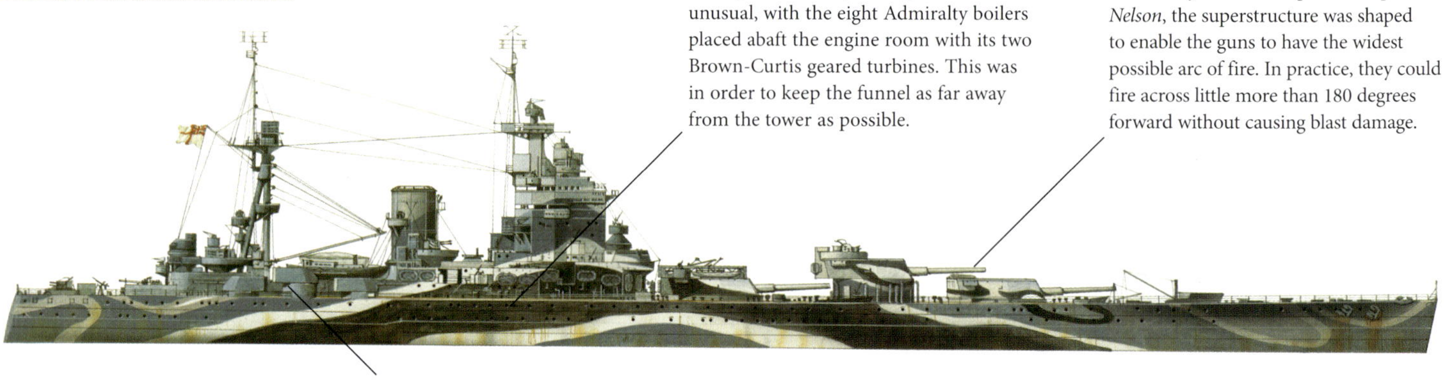

MACHINERY
Rodney's machinery arrangement was unusual, with the eight Admiralty boilers placed abaft the engine room with its two Brown-Curtis geared turbines. This was in order to keep the funnel as far away from the tower as possible.

FIRING ARC
On *Rodney* and its companion ship *Nelson*, the superstructure was shaped to enable the guns to have the widest possible arc of fire. In practice, they could fire across little more than 180 degrees forward without causing blast damage.

AA GUNS
The final AA armament fitted was 16 40mm (1.6in), 48 2-pounder pom-poms and 61 20mm (0.79in) guns.

ARMOUR SYSTEM

With these ships, the Royal Navy opted for the American 'all or nothing' armour system. The horizontal armour above the engine room was 76mm (3in) thick (it was pierced by a bomb in 1940, which failed to explode) and above the 406mm (16in) magazines it was 159mm (6.25in). For the first time, side armour was not fitted as an external belt, but on the inner side of the steel hull, angling inwards from the armoured deck: this was intended to reduce the penetrative power of an incoming shell and also to deflect plunging shells.

SHELL LOADING
Here shells for the 406mm (16in) guns are being loaded from a supply vessel during World War II. Each one weighed 1198kg (2641lb). Details of rangefinding and radar equipment can be seen.

Chervona Ukraina (1927)

TYPE • *Cruiser* **COUNTRY** • *Soviet Union*

SPECIFICATIONS

DIMENSIONS:	Length: 166.7m (546ft 11in); Beam: 15.7m (51ft 6in); Draught: 6.2m (20ft 4in)
DISPLACEMENT:	8400 tonnes (9260 tons) deep load
PROPULSION:	4 shafts; 4 geared steam turbines; 14 Yarrow boilers; 41,000kW (55,000shp)
SPEED:	29.5 knots (54.6km/h; 33.9mph)
Range:	3700nmi (6900km; 4300 miles)
ARMAMENT:	15 × 130mm (5.1in) guns; 3 × 100mm (3.9in) AA guns; 4 × 45mm (1.8in) AA guns; 2 × 457mm (18in) triple torpedo tubes; 60–100 mines
ARMOUR:	Belt lower 76mm (3in); belt upper 25mm (0.98in); upper and lower armoured decks 20mm (0.79in) each; gun shields 25mm (0.98in)
COMPLEMENT:	830

Chervona Ukraina's wartime service came to an abrupt end in November 1941 when it was sunk off Sevastopol by German dive-bombers. The ship was nevertheless raised in 1947 and thereafter used as a training hulk.

Left: Chervona Ukraina, *as it appeared in the 1930s. In November 1941, the ship was sunk at Sevastopol by a dive-bombing attack from German Ju-87 aircraft.*

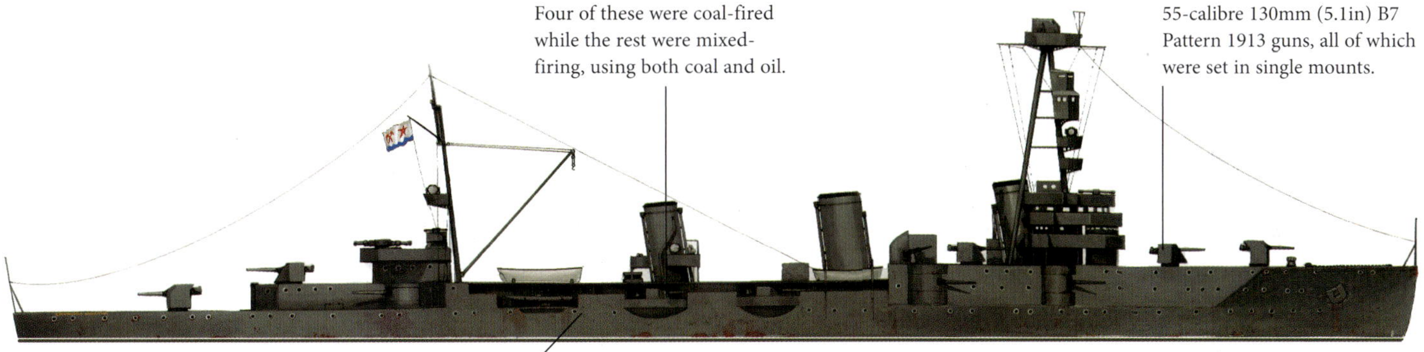

POWER
The ship's power came from 14 Yarrow water-tube boilers. Four of these were coal-fired while the rest were mixed-firing, using both coal and oil.

MAIN ARMAMENT
Chervona Ukraina's main armament consisted of 15 55-calibre 130mm (5.1in) B7 Pattern 1913 guns, all of which were set in single mounts.

CREW
As designed, the warship had a complement of 630 officers and men, but by the time it reached service the actual figure had swelled to more than 800.

SHIP CLASS
The *Chervona Ukraina* belonged to the *Admiral Nakhimov* class of light cruisers. Four were intended, but only two actually proceeded to the shipyards. These were laid down between 1913 and 1932, but as with many pre-World War I Russian ship types, the Russian Revolution interrupted their construction and they did not enter service until the 1920s and 1930s.

ISTANBUL, 1935
Chervona Ukraina was laid down in October 1913, originally as the *Admiral Nakhimov*, but was commissioned into service under its final name in March 1927.

Akagi (1927)

TYPE • *Aircraft carrier* **COUNTRY** • *Japan*

SPECIFICATIONS

Dimensions:	Length: 260.6m (855ft); Beam: 31.4m (103ft); Draught: 8.8m (29ft)
Displacement:	37,084 tonnes (40,891 tons) standard; 41,961 tonnes (46,281 tons) full load
Propulsion:	4 shafts; 4 turbines; 19 Kampon boilers; 41,013kW (55,000shp)
Speed:	31 knots (57.3km/h; 35.6mph)
Range:	10,000nmi (18,520km; 11,510 miles) at 14 knots
Armament:	6 × 203mm (8in) guns; 12 × 120mm (4.7in) guns; 14 × 25mm AA twin guns
Aircraft:	91
Complement:	1630

Flagship of Japan's Pearl Harbor attack fleet and again at the start of the Battle of Midway – where it was sunk in a devastating US carrier attack– the aircraft carrier *Akagi* participated in every major action during the early stages of the Pacific War.

This page: Three views showing Akagi *after the 1935 refit that gave the ship a full-length flight deck, setting the pattern for further carriers of the Imperial Navy, although the port-side island was repeated only on* Hiryu.

Hangar decks
The hangar decks were fully enclosed, which contributed to the ship's destruction when accumulated gas and vapour exploded in the unventilated space.

Island
The position of the island was on the port side to keep it away from smoke and gases from the engine exhausts. This was also done on *Hiryu*, but not repeated on subsequent carriers. It is shown here with external protective wrapping against shells and fragments.

Flight deck
From 1938 the flight deck was 249.2m (817ft 6in) long. The flight decks of Japanese carriers were wood-laid, with the planks running longitudinally rather than transversely as in US carriers.

PEARL HARBOR
On 7 December 1941, *Akagi* launched two waves of aircraft, the first including 15 dive-bombers, 12 torpedo planes and 10 fighter planes; the second formed of eight dive-bombers and nine fighters. Apart from the loss of five battleships and damage to many other warships, the United States lost 239 aircraft in simultaneous raids on the airfields. *Akagi*'s torpedo bombers were responsible for the destruction of USS *Oklahoma* and *West Virginia*.

UNDER CONSTRUCTION, 1925
Laid down as a battlecruiser at Kure Naval Yard on 6 December 1920, *Akagi* was finally launched as an aircraft carrier at a construction cost of 53 million yen.

USS Lexington (1927)

TYPE • *Aircraft carrier* **COUNTRY** • *United States*

SPECIFICATIONS

Dimensions:	Length: 270.66m (888ft); Beam: 32.12m (105ft 5in); Draught: 10.15m (33ft 4in)
Displacement:	38,284 tonnes (42,216 tons) normal load; 43,744 tonnes (48,197 tons) full load
Propulsion:	4 shafts; turbo-electric drive; 16 water-tube boilers; 134,226kW (180,000shp)
Speed:	33.25 knots (61.5km/h; 38mph)
Range:	10,500nmi (19,456km; 12,075 miles) at 15 knots
Armament:	8 × 203mm (8in) guns; 12 × 127mm (5in) guns; 48 28mm (1.1in)
Aircraft:	63
Complement:	2791

Laid down first as a battlecruiser, *Lexington* (CV-2) was redesigned as a carrier and taught the US Navy much about sea flying in the 1930s. In 1942 it played a key part in the Battle of the Coral Sea before succumbing to fire after heavy attacks.

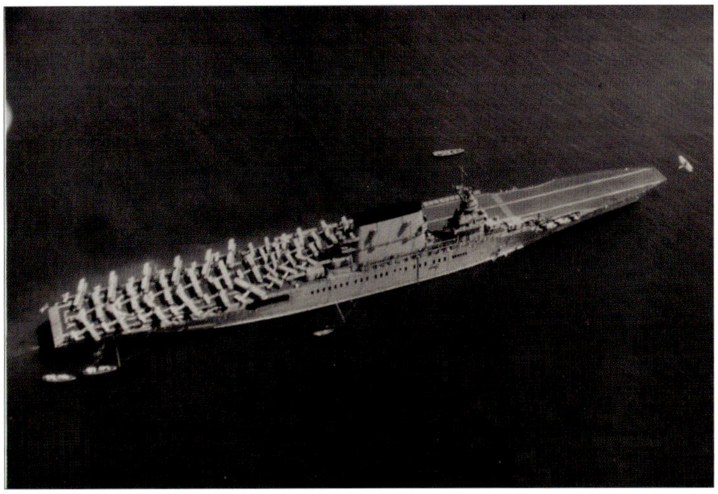

Left: USS Lexington *(CV-2) seen from above, its aircraft stacked thickly along the aft half of the flight deck.*

Funnel
Lexington's massive funnel structure, separate from the island, gave it a unique profile. A gallery was built below the funnel cap in the 1937–38 refit and a CXAM radar grid was mounted on its top forward edge in October 1941.

Bow
The *Lexington* class featured a bulbous bow, intended to reduce water resistance and give better support to the forecastle area beneath the forward flight deck.

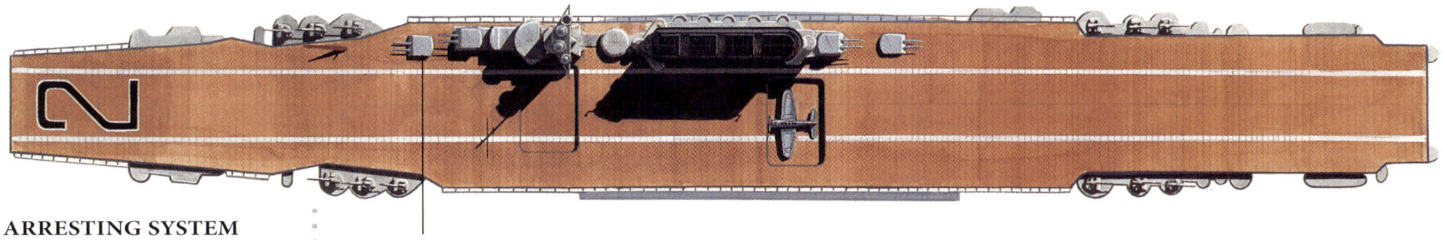

ARRESTING SYSTEM
Lexington was the first carrier to be fitted with an effective transverse-wire arresting system to allow faster and heavier planes to land. In August 1931, the original Mark 1 system was replaced by a hydraulically powered transverse-only system of greater power and elasticity. Further improvements were made in 1934, and 1936 modifications included fitting a four-wire arrestor system on the forward end of the flight deck in case damage impeded normal stern landings.

Armament
Original armament was eight 203mm (8in) guns in quadruple turrets, 12 127mm (5in) and 48 28mm (1.1in) guns. It is probable that if they were fired to port (across the deck) the blast would have damaged the flight deck.

LAUNCHED
USS *Lexington* is launched on 3 October 1925. *Lexington* was laid down at the Fore River yard at Quincy, Massachusetts, on 8 January 1921, launched on 3 October 1925 and commissioned on 14 December 1927, designated CV-2.

Argonaute '600' and '630' classes (1927)

TYPE • *Submarine* **COUNTRY** • *France*

SPECIFICATIONS (Group I)

Dimensions:	Length: 63.4m (208ft); Beam: 6.4m (21ft); Draught: 4.24m (13.9ft)
Displacement:	630 tonnes (694 tons) surfaced; 798 tonnes (880 tons) submerged
Propulsion:	2 screws; two Schneider-Carel diesels; 485kW (650hp) total; two electric motors, 373kW (500hp)
Speed:	14 knots (26km/h; 16mph) surfaced; 9 knots (17km/h; 10mph) submerged
Range:	4000nmi (7408km; 4603 miles) surfaced; 82nmi (152km; 84.4 miles) submerged
Armament:	6 × 550mm (21.45in) torpedo tubes with 13 torpedoes; 2 × 400mm (15.7in) torpedo tubes with 9 torpedoes; 1 × 76mm (3in)/35cal deck gun; 1 × 8mm (0.3in) machine gun
Complement:	41

French naval policy required an effective fleet of coastal submarines, particularly in the Mediterranean Sea, where Italy was considered a potential enemy. This resulted in two main classes of broadly similar type, defined by their displacement tonnage.

Right: Pictured here at Cherbourg in 1932, the Ariane, *a 600 series submarine, rests in the harbour.*

Dimensions
Although the dimensions of the 630 class were slightly less than the 600s, the displacement was marginally higher and habitability for the crew of 41 was improved. Operating depth was 80m (262ft).

Layout
All were double-hulled, with a sharply raked stem, some with a pronounced downwards-sloping foredeck. The tower was oblong, with varying profiles, but all had the bridge forward and a space behind for machine guns.

Argonaute
Argonaute was a Schneider-Laubeuf design. Assigned to the Mediterranean squadron, it was sunk by British ships at Oran in November 1942.

CLASS TYPES

The '600' class, built between 1923 and 1927, numbered 12, completed by 1930. Four were built in each of three yards, AC de la Loire, Nantes, under the direction of Jean Ernest Simonot; Chantiers Augustin Normand, Le Havre, by Normand and Fernand Fenaux; and at Chantiers Schneider, Bordeaux by Maxime Laubeuf. The '630 tonne' class was formed of four distinct groups. Group I was five boats of the Schneider-Laubeuf type, including *Argonaute*; Group II, Normand-Fenaux type, nine boats; and Group III, Loire-Simonot type, two boats.

SUB LAUNCH
The French submarine *Vénus* launches on 6 April 1935. The *Minerve* class was based on the 630 series. By 1939 the 600/630 class, especially the 600s, were obsolescent although most did wartime service.

Oberon and *Odin* class (1927)

TYPE • *Submarine* **COUNTRY** • *United Kingdom*

HMS *Oberon* marked the beginning of a new generation of post-World War I British submarines. Built for long-range patrol duties, and commissioned in 1927, *Oberon* was the first Royal Navy submarine to be named rather than simply numbered.

SPECIFICATIONS (HMS *Odin*)

DIMENSIONS:	Length: 86.4m (283ft 6in); Beam: 9.1m (29ft 10in); Draught: 4.8m (15ft 9in)
DISPLACEMENT:	1810 tonnes (1996 tons) surfaced; 2060 tonnes (2271 tons) submerged
PROPULSION:	2 screws; 2 Admiralty diesels, 1640kW (2200hp) total; two electric motors, 492kW (660hp)
SPEED:	17.5 knots (32.4km/h; 20.1mph) surfaced; 9 knots (17km/h; 10mph) submerged
RANGE:	8400nmi (15,557km; 9667 miles) at 10 knots surfaced; 70nmi (130km; 81 miles) at 4 knots submerged
ARMAMENT:	8 × 533mm (21in) torpedo tubes with 14 torpedoes; 1 × 102mm (4in) QF deck gun; 2 × 7.7mm (0.303in) Lewis guns
COMPLEMENT:	56

Left: Completed in 1927, HMS Oberon *featured an advanced design for the period, along with up-to-date instrumentation. One defect was the riveted saddle tanks, which contained additional fuel, but tended to leave a trail of leakage. Placed in reserve between 1937 and 1939, it was used in World War II primarily for training, and decommissioned in July 1944.*

TANKAGE
Tankage of 161 tonnes (177 tons) enabled a range of 8400nmi (15,557 km; 9667 miles) at 10 knots (18.5 km/h; 11.5 mph).

HMS *PARTHIAN*
New higher-capacity batteries gave the *Parthian* class greater underwater endurance. If not carrying mines, they took 14 torpedoes of the standard British Mark VIII type.

***PARTHIAN* CLASS**
Similar to the *Odin* class and also designed for long-range patrol in the Far East, the six *Parthian*-class boats (1929) had a more bulbous bow, were slightly longer and with diesels uprated to 3430kW (4600hp).

CLASS DESIGN

Oberon's design drew on the L-class of World War I, but a range of improvements made it a starkly different craft. Although it has been regarded as the lead ship of the O-class, there were significant differences between it and its successors. The O-class is more properly named after HMS *Odin*, which was shorter, lighter and less powerful than the subsequent boats. Apart from *Odin*, *Orpheus* and *Oswald* were also sunk by Italian destroyers, and *Olympus* struck a mine off Malta.

OTWAY* AND *OSIRIS
HMS *Otway* (left, originally HMAS *Otway* of the Royal Australian Navy) and HMS *Osiris* (N67), both *Odin*-class submarines of the Royal Navy, tie up at the Shadwell Basin London Dock, Wapping, on 5 July 1938.

Pisani and *Mameli* classes (1928)

SPECIFICATIONS
(*Pisani* class)

DIMENSIONS:	Length: 68.2m (223.9ft); Beam: 6m (20ft); Draught: 4.9m (16.2ft)
DISPLACEMENT:	894 tonnes (986 tons) surfaced; 1075 tonnes (1185 tons) submerged
PROPULSION:	2 diesels; 1118.5kW (1500shp); two 410kW (550hp) electric motors; 2 screws
SPEED:	15 knots (27.8km/h; 17.3mph) surfaced; 8.2 knots (15.2km/h; 9.4mph) submerged
RANGE:	5000nmi (9260km; 5755 miles) at 8 knots; 61nmi (113km; 70 miles) at 4 knots
ARMAMENT:	6 × 533mm (21in) torpedo tubes; 9 torpedoes; 1 × 102mm (4in) deck gun; 2 × 13.2mm (0.5in) MGs
COMPLEMENT:	49

Unlike most other navies, the Regia Marina had a large and relatively up-to-date submarine arm in 1939. These two similar classes, although from separate design studios, set the pattern for future development of the submarine force.

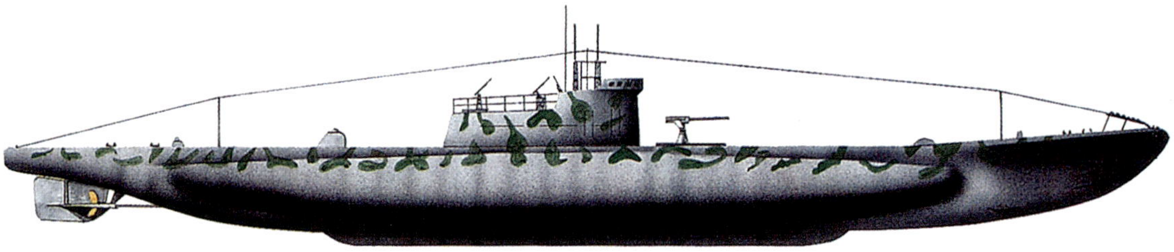

Above: Despite its age and obsolescence, Giovanni da Procida *(*Mameli *class) was on active service until January 1942 and was stricken only in 1948.*

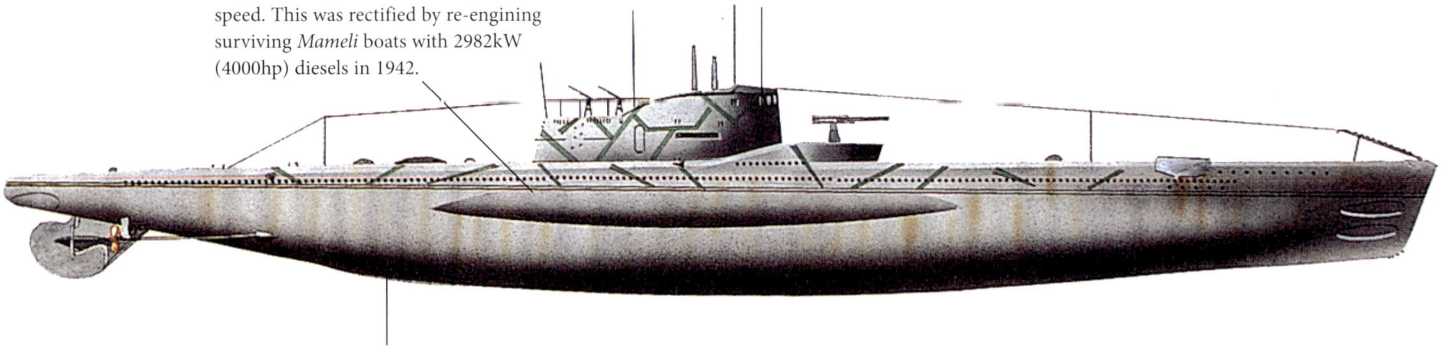

TOWER
A typical feature was the large tower, which persisted on Italian boats despite its prominence as a target and its contribution to surface instability and underwater resistance.

PERFORMANCE
Bulges fitted to the outer hull sections helped stability but slowed surface speed. This was rectified by re-engining surviving *Mameli* boats with 2982kW (4000hp) diesels in 1942.

GIOVANNI BAUSAN
The double-hull concept had originated in Italy, and both classes had this feature. This *Pisani*-class vessel was converted to an oil barge in spring 1942.

PISANI ISSUES

The *Pisani* class suffered from instability problems, which were partly rectified by fitting bulges to the bows. However this, and their obsolescent design, meant that little use was made of them as combat boats in World War II. *Des Geneys* was used as a battery-charging hulk and *Marcantonio Colonna* was broken up in 1943. Only *Vettor Pisani* survived the war. *Giovanni Bausan* was converted to an oil barge in spring 1942.

SS *MAMELI*, AUGUST 1944
The main external difference between the *Mameli* and *Pisani* classes was that the former's102mm (4in) gun was mounted on a low breakwater platform. This class, built at the Tosi yard in Taranto, was powered by two Franco Tosi four-stroke S8 eight-cylinder diesels developing 1156kW (1550hp).

Kaga (1928)

TYPE • *Aircraft carrier* **COUNTRY** • *Japan*

SPECIFICATIONS

DIMENSIONS:	Length: 247.5m (812ft); Beam: 32.6m (107ft); Draught: 9.45m (31ft)
DISPLACEMENT:	38,811 tonnes (42,773 tons) normal load; 43,222 tonnes (47,646 tons) full load
PROPULSION:	8 Kampon boilers; 4 Kampon geared turbines; 4 shafts; 95,002kW (127,400shp)
SPEED:	28 knots (52km/h; 32mph)
RANGE:	10,000nmi (18,500km; 11,500 miles) at 16 knots
ARMAMENT:	10 × 203mm (8in) guns; 16 × 127mm (5in) guns; 22 × 25mm AA guns
AIRCRAFT:	90
COMPLEMENT:	1708

Planned as a battleship, in aircraft carrier form *Kaga* was part of the Japanese fleet that attacked Pearl Harbor. It went on to engage in other operations in the Pacific before being sunk in the Battle of Midway in 1942.

Above: Kaga's size and speed made it highly suitable for transformation into Japan's first fleet carrier.

MODERNIZATION
The profile shows the ship in post-1934 modernized form, with the lower flight decks removed. The small island, essentially a conning tower, was added at that time.

GUNS
The 203mm (8in) guns were mounted so low, beneath main deck level, as to be of little use except in calm conditions.

ARRESTOR WIRES
Japanese carriers could have six arrestor wires. Where only three were fitted, hooking on to the second wire was the ideal, and pilots who achieved this were given a pay bonus of 5 yen.

BATTLE OF MIDWAY

Kaga was one of four carriers in the Japanese fleet at the Battle of Midway. On 4 June 1942 at 10.22 it was attacked by dive-bombers from USS *Enterprise* and hit by at least four bombs, three 225kg (500lb) and one 450kg (1000lb). Two broke through to the upper hangar, causing explosions and fires. By 17.00 the surviving crew was evacuated, and the still-burning hulk was sunk at 19.25 by two torpedoes from the escort destroyer *Hagikaze*. In total 811 men were lost.

FLIGHT DECK
Aircraft crowd the deck of *Kaga* in 1937. A thorough reconstruction of the upperworks had been made in 1934–35. *Kaga* emerged with a lengthened hull, a single flight deck of 250 × 32.6m (812 x × 106ft) laid with longitudinal wood planking above a thin steel base, and an increased displacement.

Köln (1928)

TYPE • *Cruiser* **COUNTRY** • *Germany*

SPECIFICATIONS

DIMENSIONS:	Length: 174m (571ft); Beam: 15.29m (50ft 2in); Draught: 5.56m (18ft 3in)
DISPLACEMENT:	6500 tonnes (7165 tons) standard
PROPULSION:	6 navy boilers; 2 double-acting 4-stroke 10-cylinder MAN diesel engines; 1300kW (1800shp) diesel power; 48,470kW (65,000shp) turbine power; 2 screws
SPEED:	32 knots (59km/h; 37mph)
RANGE:	5700nmi (10,600km; 6600 miles) at 19 knots
ARMAMENT:	9 x 152mm (6in) guns; 4 x 88mm (3.4in) guns; 8 x 37mm (1.5in) guns; 8 x 20mm (0.79in) AA guns; 8 x 500mm (19.7in) torpedo tubes; 2 aircraft
ARMOUR:	Belt 50–70mm (2–2.75in); torpedo bulkhead 12mm (0.5in); turrets 31mm (1.25in)
COMPLEMENT:	850

The Kriegsmarine's three K-class cruisers were the public face of the resurgent German Navy during the 1930s. They were a bold new design that used modern techniques such as electric welding in their construction.

Above: The German cruiser Köln *lies damaged at Wilhelmshaven on 9 May 1945, as a British officer inspects one of its main gun turrets.*

REAR TURRETS
The two rear turrets were positioned offset to the centreline and could fire at 40 degrees elevation in all directions.

MAIN ARMAMENT
The nine 152mm (6in) main guns were fitted in three triple turrets, one fore and one aft.

POWERPLANT
A combination of steam and diesel powerplants was fitted, although the two could only function separately. This still allowed for a diesel cruising speed of 19km/h (10 knots; 11.5mph).

BATTLE OF NORWAY

Once the war had started, it was no longer plain sailing for the K-class. Both *Königsberg* and *Karlsruhe* were diverted from mine-laying duties in the Atlantic to participate in the invasion of Norway, as part of the ill-fated Task Forces 3 and 4. Both were so badly damaged – on 10 and 9 April 1940 respectively – that they were put out of action for the rest of the war. *Köln* had joined *Königsberg* as part of Task Force 3 but managed to escape the devastation inflicted upon the German ships.

***KARLSRUHE*, *KÖNIGSBERG* CLASS**
Karlsruhe, the second ship of the three-ship *Königsberg* class (or K-class), had a relatively short wartime career. It was torpedoed by a British submarine off Norway on 9 April 1940 and had to be scuttled.

Surcouf (1929)

TYPE • *Submarine* **COUNTRY** • *France*

SPECIFICATIONS

Dimensions:	Length: 110m (361ft); Beam: 9m (29ft 6in); Draught: 7.25m (23ft 9in)
Displacement:	3300 tonnes (3630 tons) surfaced; 4373 tonnes (4820 tons) submerged
Propulsion:	2 diesels; 2834kW (3800hp); 2 electric motors; 1250kW (1700hp); 2 screws
Speed:	18.5 knots (34.3km/h; 21.3mph) surfaced; 10 knots (19km/h; 12mph) submerged
Range:	10,000nmi (18,520km; 11,508 miles) at 10 knots surfaced; 59nmi (130km; 81 miles) at 4.5 knots submerged
Armament:	8 × 550mm (21.45in) torpedo tubes; 4 × 400mm (15.7in) torpedo tubes; 2 × 203mm (8in) guns; 2 × 37mm (1.5in) AA guns; 4 × 13.2mm (0.5in) machine guns
Complement:	118

Until Japan built even bigger ones, *Surcouf* was the largest submarine in the world. The large 'cruiser-submarine' was an idea that took hold in several navies. However, the great French experiment was not a military success.

Left: Surcouf's *double hull, an enlarged* Redoutable, *was a bold attempt at combining the stealth qualities of a submarine with the surface fighting and scouting ability of a cruiser.*

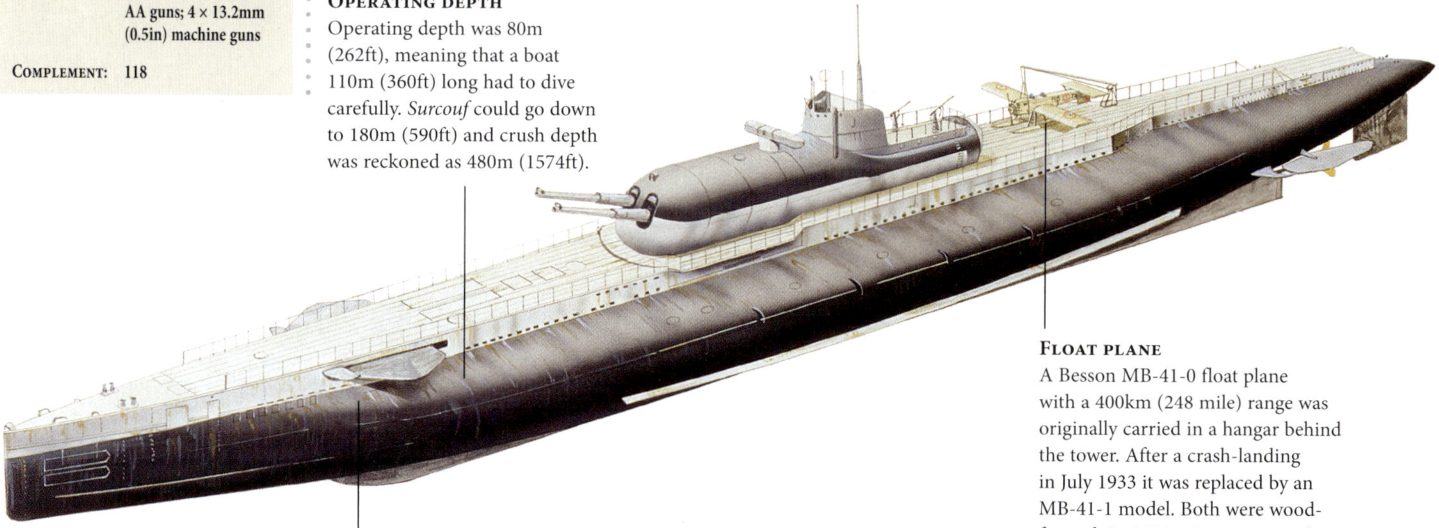

Operating depth
Operating depth was 80m (262ft), meaning that a boat 110m (360ft) long had to dive carefully. *Surcouf* could go down to 180m (590ft) and crush depth was reckoned as 480m (1574ft).

Float plane
A Besson MB-41-0 float plane with a 400km (248 mile) range was originally carried in a hangar behind the tower. After a crash-landing in July 1933 it was replaced by an MB-41-1 model. Both were wood-framed. In 1938 a Breguet gyroplane was tested as a possible replacement.

Colour scheme
Originally painted in French naval grey, *Surcouf* was repainted in 'dark Prussian blue' from 1934 until 1940. From then it was painted in two-tone grey.

CORSAIR-SUBMARINE

Surcouf's design began as 'Project Q' in 1926, under Jean-Jacques Roquebert, who also designed the *Redoutable* class. Described as a 'corsair-submarine', *Surcouf* was intended as a commerce raider, equipped with a Besson MB411 float plane and a 4.5m (15ft) motorboat, and with room to hold around 40 prisoners. Its cruising endurance was a maximum of 90 days. Its designers did not anticipate attacks from the air, as it took two minutes to dive.

DECK GUNS
The 203mm (8in) guns and gun-laying equipment were not new, as many guns in good condition were available from decommissioned surface ships. The twin turret weighed 185 tonnes (204 tons) and could be ready to fire within three minutes of surfacing. This photograph shows the gun turret without the streamlined casing that was fitted for seagoing service.

USS *Pensacola* (1930)

SPECIFICATIONS

DIMENSIONS:	Length: 178.46m (585ft 8in); Beam: 19.89m (65ft 3in); Draught: 6.7m (21ft 11in)
DISPLACEMENT:	9246 tonnes (10,191 tons) standard
PROPULSION:	4 Parsons reduction steam turbines; 12 White-Forster boilers; 80,000kW (107,000shp); 4 screws
SPEED:	32.7 knots (60.6km/h; 37.6mph)
RANGE:	10,000nmi (19,000km; 12,000 miles) at 15 knots
ARMAMENT:	10 × 203mm (8in) guns; 8 × 127mm (5in) AA guns; 2 × 47mm (1.9in) 3-pdr saluting guns; 6 × 533mm (21in) torpedo tubes
ARMOUR:	Belt 64–102mm (2.5–4in); deck 25–44mm (1–1.75in); barbettes 19mm (0.75in); turrets 19–64mm (0.75–2.5in)
COMPLEMENT:	530

The two *Pensacola* class heavy cruisers were the first warships laid down in the United States as 'treaty cruisers', reflecting the 10,000-ton displacement limit imposed by the 1922 Washington Naval Treaty.

Right: Pensacola *photographed at anchor during the 1930s. The ship was laid down on 27 October 1926, launched on 25 April 1929 and commissioned on 6 February 1930.*

FIREPOWER
The USS *Pensacola* was geared towards firepower and armed with 203mm (8in) guns at a time when most rival cruisers had only 152mm (6in) guns.

LAYOUT
Some security from the ship's lack of heavy armour came from the layout of the fire and engine rooms, which alternated through the ship, and thus were more likely to survive single shell or torpedo hits.

FLOAT PLANES
Pensacola could carry four float planes for patrol and reconnaissance purposes, with two launch catapults fitted amidships.

PACIFIC SERVICE

By February 1942, *Pensacola* was operating as part of Task Force 11 with the carrier *Lexington*. It provided AA protection for the *Lexington* against waves of Japanese bombers during the fighting off Bougainville, in the Solomons. *Pensacola* continued duty as a carrier escort in the Coral Sea, New Caledonia, and at the Battle of Midway. At Midway, it escorted *Enterprise* and defended the stricken carrier *Yorktown*, preventing it from being attacked by Japanese torpedo bombers.

BATTLE TRAINING
Pensacola is here seen in Hawaiian waters, conducting battle training with other units of the Pacific Fleet in early August 1942, shortly before the ship deployed to the contested waters of the South Pacific.

HMS *Glorious* (1930)

TYPE • *Aircraft carrier* **COUNTRY** • *United Kingdom*

SPECIFICATIONS

Dimensions:	Length: 239.8m (786ft 9in); Beam: 27.6m (90ft 6in); Draught: 8.5m (27ft 9in)
Displacement:	25,370 tonnes (27,967 tons) normal load; 27,859 tonnes (30,706 tons) full load
Propulsion:	18 Yarrow boilers; 4 geared turbines; 4 shafts; 67,113kW (90,000shp)
Speed:	30 knots (55.5km/h; 34.5mph)
Range:	6000nmi (11,000km; 6900 miles) at 20 knots
Armament:	16 × 120mm (4.7in) DP guns
Aircraft:	48
Complement:	1283

Companion ship to HMS *Furious* and *Courageous*, *Glorious* became one of the few carriers to be sunk by gunfire from a surface ship, rather than by aircraft or submarines, in an encounter with *Scharnhorst* that remains controversial.

Left: HMS Glorious *photographed off Gibraltar in 1931. Relieving* Courageous *in the Mediterranean Fleet in 1930,* Glorious *remained in the Mediterranean until 1939, and was based at Alexandria when war was declared on 3 September.*

Deck arrangement
The upper hangar deck initially opened on to a flyoff deck built above the forecastle, enabling planes to take off and land at the same time. The main flight deck incorporated an island with a tall funnel and pole mast. Two aircraft lifts were installed forward and aft.

Aircraft
Like *Courageous*, *Glorious* could carry and service four squadrons of aircraft, up to 36 planes in total. For defence it carried 16 120mm (4.7in) dual-purpose guns mounted in single turrets.

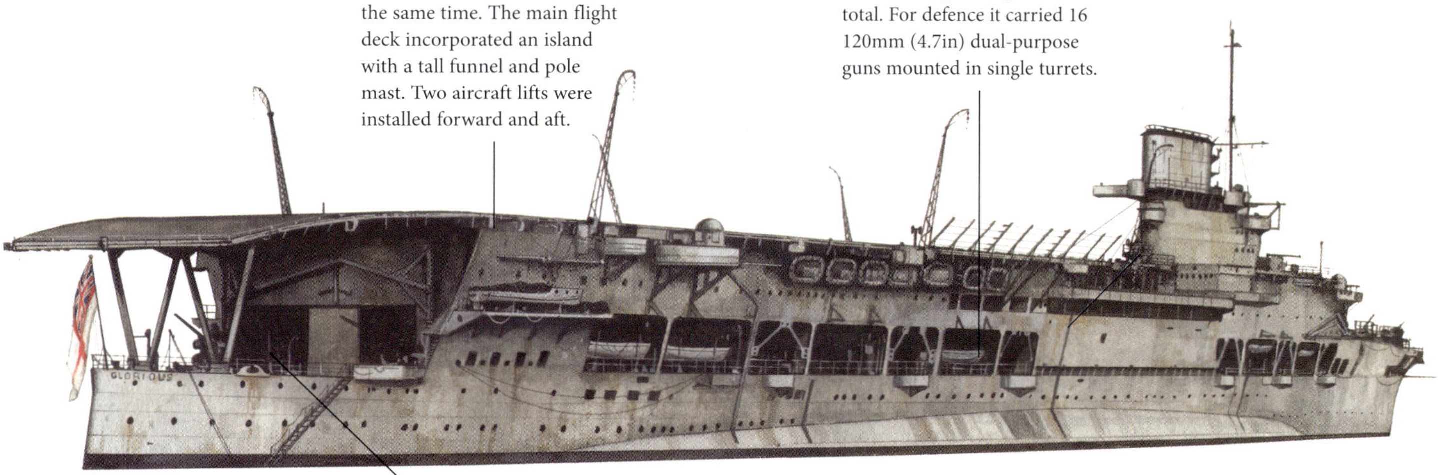

Access points
Access hatches were provided forward and aft of the conning tower. The airlock and diving compartments were located just behind the bows.

SINKING

On 8 June 1940, *Glorious* set out from Narvik for Scapa Flow, escorted only by two destroyers. It was intercepted by a German battle squadron: the battlecruisers *Scharnhorst* and *Gneisenau* and heavy cruiser *Hipper*. All British ships were sunk; in total 1474 naval and 59 RAF personnel were lost. In 1999 researchers uncovered evidence suggesting that *Glorious*'s commander, Captain Guy D'Oyly Hughes, requested to sail because he was in a hurry to get back to the UK to implement a court martial.

FLIGHT DECK
Swordfish aircraft ready for take-off from *Glorious*. Laid down on 1 May 1915 at Harland & Wolff, Belfast, *Glorious* was first commissioned as a light battlecruiser on 14 October 1916. The ship underwent carrier conversion between 1924 and 1930.

HMS *Exeter* (1931)

TYPE • *Cruiser* **COUNTRY** • *United Kingdom*

HMS *Exeter* was the sole ship of its class, a cruiser that followed the lead set by the earlier *York*, *Norfolk* and *London* classes. As such, *Exeter* was a small, lightly armoured vessel fitted with 203mm (8in) main guns.

SPECIFICATIONS

Dimensions:	Length: 175.26m (575ft); Beam: 17.68m (58ft); Draught: 6.17m (20ft 3in)
Displacement:	8390 tonnes (9245 tons) standard
Propulsion:	8 Admiralty boilers; 4-shaft Parsons geared turbines; 59,656kW (80,000shp)
Speed:	32 knots (59km/h; 37mph)
Range:	10,000nmi (19,000km; 12,000 miles) at 15 knots
Armament:	6 × 203mm (8in) guns; 4 × 102mm (4in) quick-firing guns (later 8); 2 × 2-pdr AA guns; 6 × 533mm (21in) torpedo tubes
Armour:	Box protection for ammunition storage 102–25mm (4–1in); side armour 76mm (3in)
Complement:	630

Right and below: Exeter *featured modern innovations such as shipborne aircraft, but its guns could not be elevated beyond 50 degrees, which limited their effectiveness against enemy aircraft.*

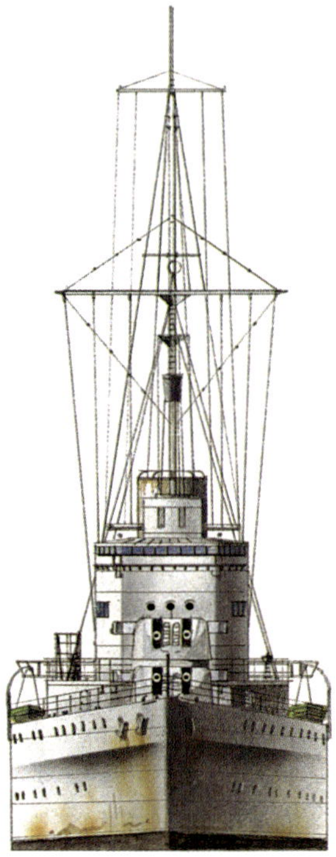

Masts
The single-pole steel masts were reconstructed in 1941 as tripods to carry heavier look-out posts and Type 286M radar equipment.

Aircraft
Two planes and catapults were first carried; by 1939 one Supermarine Walrus float plane was carried, with a single catapult.

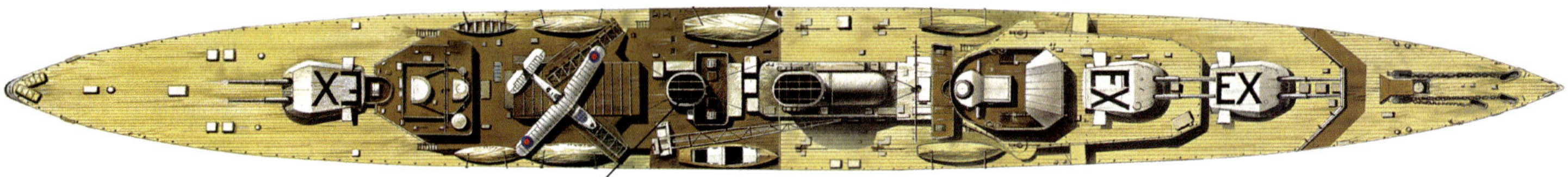

Powerplant
The ship had four Parsons geared steam turbine sets, each driving one shaft, the drive by eight Admiralty three-drum boilers.

RIVER PLATE
Together with HMS *Achilles* and HMS *Ajax*, *Exeter* was part of the British cruiser squadron commanded by Commodore (later Admiral) H.H. Harwood that was involved in the famous action against the *Graf Spee* at the River Plate in December 1939. After engaging *Graf Spee* alone, *Exeter* was hit by seven 280mm (11in) shells from the German cruiser, and fragments from other near-misses. The ship was badly damaged, with all of its own 203mm (8in) guns knocked out.

***EXETER*, 1930s**
Exeter remained unchanged throughout the 1930s until the outbreak of World War II, when it was badly damaged. The basic armament was overhauled during repairs, with the main gun turrets given a 70-degree elevation. Deployed to Far Eastern waters, the ship was sunk in action at the Battle of the Java Sea in March 1942.

Akatsuki class (1931)

TYPE • *Destroyer* **COUNTRY** • *Japan*

SPECIFICATIONS

DIMENSIONS:	Length: 118.41m (388ft 5in); Beam: 10.4m (34ft 1in); Draught: 3.2m (10ft 6in)
DISPLACEMENT:	1780 tonnes (1930 tons) normal load
PROPULSION:	4 Kampon boilers; 2 Kampon Type Ro geared turbines; 2 shafts; 37,000kW (50,000ihp)
SPEED:	38 knots (70km/h; 44mph)
RANGE:	5000nmi (9300km; 5780 miles) at 14 knots (26km/h; 16mph)
ARMAMENT:	6 × 127mm (5in) Type 3 guns; up to 28 × 25mm (1in) Type 96 AT/AA guns; up to 10 × 13.2mm (0.5in) Type 93 AA guns; 9 × 610mm (24in) torpedo tubes for Type 93 torpedoes; 36 depth charges
COMPLEMENT:	219

Generally regarded as a sub-class of the *Fubuki*-class destroyers, the four *Akatsuki*-class vessels were highly advanced warships when they emerged in the 1930s, although they wrestled with stability problems at sea.

Left: The IJN Hibiki *(1932) shown sailing off the Chinese coast in the late 1930s. In July 1947 the ship went as a prize of war to the USSR.*

MAIN GUNS
The main armament used by the class were Type 3 127mm (5in) 50-calibre naval guns, each turret featuring a pair of the weapons.

HIBIKI
The fourth of the class to be completed, *Hibiki* launched in 1932 and entered service in 1933. It was the only ship of the class to survive the war intact.

RECONSTRUCTION
The heavy armament on a small hull initially rendered the *Akatsuki* class unstable. This was rectified in a major reconstruction in the 1930s.

FATE OF THE *AKATSUKI*

The *Akatsuki* fought in both the Second Sino-Japanese War and World War II. For much of 1942, the ship had been in operation around the Solomon Islands, and by the autumn was engaged in action around Guadalcanal. On 12–13 November, it fought in the major night battle with US Navy warships, known informally as the 'The Night of the Big Guns'. *Akatsuki* was hit repeatedly by heavy naval gunfire and sank; only 18 of the crew survived.

IKAZUCHI
The IJN *Ikazuchi* (1931) is seen here off the coast of China in 1938. The ship was torpedoed and sunk by the submarine USS *Harder* in April 1944 south-southwest of Guam.

Alberto di Giussano class (1931)

TYPE • *Light cruiser* • **COUNTRY** • *Italy*

SPECIFICATIONS

DIMENSIONS:	Length: 169.29m (555ft 5in) overall; Beam: 15.49m (50ft 10in); Draught: 5.4m (17ft 9in)
DISPLACEMENT:	5130 tonnes (5656 tons) normal load
PROPULSION:	6 Yarrow boilers driving two-shaft Belluzzo geared turbines; 95,584kW (128,200hp)
SPEED:	30 knots (56km/h; 34.5mph)
RANGE:	5000nmi (9300km; 5800 miles)
ARMAMENT:	8 × 152mm (6in) guns; 6 × 102mm (4in) AA guns; 8 × 37mm (1.5in) AA guns; 8 × 13.2mm (0.5in) AA guns; 4 × 533mm (21in) torpedo tubes; 2 aircraft
ARMOUR:	Belt: 178–240mm (7–9.4in); deck: 200mm (7.9in); bulkheads: 200mm (7.9in); main turrets: 229mm (9in)
COMPLEMENT:	507

One of four ships built in response to the powerful French *Lion*-class destroyers, *Alberto di Giussano* was a large, well-armed, lightly armoured vessel. In total, four ships were built in the class.

Right: Alberto di Giussano *shown in pristine, newly completed condition. Between 1930 and 1941, Italy built five classes of destroyers, none produced in large numbers, totalling 36 new ships. These large-sized vessels were designed with close attention to contemporary French developments.*

PERFORMANCE
The class was fast – *Bartolomeo Colleoni* achieved a speed of 42 knots (77.8km/h; 46mph) during trials.

ALBERTO DI GIUSSANO
The ship was laid down in April 1928, launched in August 1930 and finally completed for service with the Regia Marina in June 1941.

ARMAMENT
As built, the class were respectably armed, with a total of 24 guns mounted (including a heavy anti-aircraft array) and four torpedo tubes.

FOUR-SHIP CLASS
The *Alberto di Giussano* class of light cruisers were part of the *Condottieri* class within the Regia Marina. There were four ships in the class: *Alberto di Giussano*, *Bartolomeo Colleoni*, *Alberico da Barbiano* and *Giovanni delle Bande Nere*. All four of the vessels were destroyed by war's end. Two of them – *Alberto di Giussano* and *Alberico da Barbiano* – were sunk in the Battle of Cape Bon off Tunisia on 13 December 1941.

GENOA, 1930
The light cruiser *Alberto di Giussano* at its launch in Genoa in 1930. Prior to World War II, the ship gained operational experience during the Spanish Civil War of 1936–39.

Leander class (1931)

TYPE • *Light cruiser* **COUNTRY** • *United Kingdom*

SPECIFICATIONS

Dimensions:	Length: 169.1m (554ft 9in); Beam: 17.1m (56ft); Draught: 5.8m (19ft 1in)
Displacement:	7390 tonnes (8146 tons) standard
Propulsion:	6 Admiralty 3-drum boilers or 4 Admiralty 3-drum boilers; 4 geared steam turbines; 4 shafts; 54,000kW (72,000shp)
Speed:	32.5 knots (60.2km/h; 37.4mph)
Range:	5730nmi (10,610km; 6590 miles) at 13 knots
Armament:	4 × twin 152mm (6in) guns; 4 × 102mm (4in) guns; 3 × quadruple 12.7mm (0.5in) AA machine guns; 2 × quadruple 533mm (21in) torpedo tubes
Armour:	Magazine 76mm (3in); deck 25mm (1in); gun turrets 25mm (1in)
Complement:	570

Named after a mythological figures, the *Leander*-class light cruisers were designed to provide heavy and fast protection for merchant shipping in contested waters. Eight vessels of the class were built, with three lost in action.

Left: Many of the Leander *class were sold or transferred to the Royal Australian Navy or Royal New Zealand Navy in the late 1930s or early 1940s, including HMS* Achilles *pictured here.*

Funnel
The truncated funnel was a feature of the five *Leander* class vessels; the three later 'Modified *Leander*' ships had two separate funnels.

Secondary armament
In addition to the main guns, the class had four high-angle QF 4 102mm (4in) Mk V guns, later replaced by twin mountings for eight guns (the later high angle QF 102mm Mk XVI naval gun).

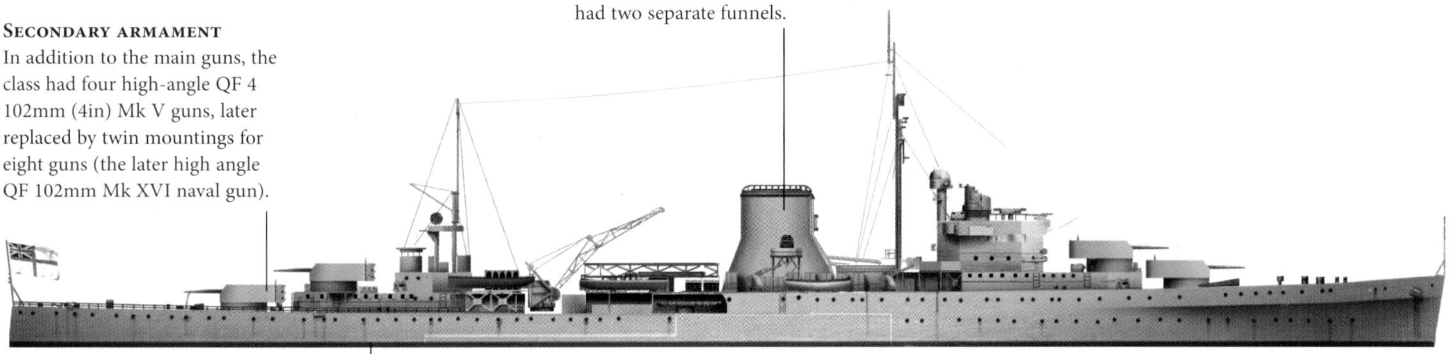

Armour
The heaviest armour applied to the ship surrounded its magazine. This armoured box had a total thickness of 76mm (3in).

RIVER PLATE
Two of the *Leander*-class warships – HMS *Achilles* and HMS *Ajax* – fought in the famous Battle of the River Plate in the South Atlantic on 13 December 1939. Along with the heavy cruiser HMS *Exeter*, they tackled the powerful German pocket battleship *Graf Spee*, both sustaining battle damage in the process. *Graf Spee* was also damaged and docked in the port of Montevideo, Uruguay, with *Ajax* and *Achilles* waiting offshore. The final showdown was avoided when the *Graf Spee* was scuttled on the orders of its captain.

HMS *ORION*
HMS *Orion* unleashes its heavy weapons in a night-firing. The principal main gun of the *Leander* class was officially termed the 50-calibre BL 6-inch gun Mark XXII. It had a maximum firing range of 23,300m (76,443ft).

Giovanni delle Bande Nere (1931)

TYPE • *Cruiser* **COUNTRY** • *Italy*

The *Giussano* class were the first cruisers built in Italy after World War I. They were part of the major 1927–28 expansion programme and were conceived as fast scout cruisers, intended to hunt and sink smaller enemy ships, such as destroyers.

SPECIFICATIONS

Dimensions:	Length: 169.29m (555ft 5in) overall; Beam: 15.49m (50ft 10in); Draught: 5.4m (17ft 9in)
Displacement:	5130 tonnes (5656 tons) standard
Propulsion:	6 Yarrow boilers; two-shaft Belluzzo geared turbines; 95,584kW (128,200hp)
Speed:	30 knots (56km/h; 34.5mph)
Range:	5000nmi (9300km; 5800 miles)
Armament:	8 × 152mm (6in) main guns; 6 × 102mm (4in) AA guns; 8 × 37mm (1.5in) AA guns; 8 × 13.2mm (0.5in) AA guns; 4 × 533mm (21in) torpedo tubes
Armour:	Belt 178–240mm (7–9.4in); deck 200mm (7.9in); bulkheads 200mm (7.9in); main turrets 229mm (9in)
Complement:	520

Right: The Giovanni delle Bande Nere *in harbour, c.1933. Italy's naval designers showed flair in developing modern fighting ships to expand the navy in the late 1930s.*

Performance
During sea trials, the ship attained a maximum speed of 38.2 knots (70.7km/h; 44mph), but regular sea speed was 30 knots (56km/h; 34.5mph).

Launch catapult
All of the ships were equipped for minelaying, and a catapult was located on the forecastle of the ship, from which it could launch its two seaplanes.

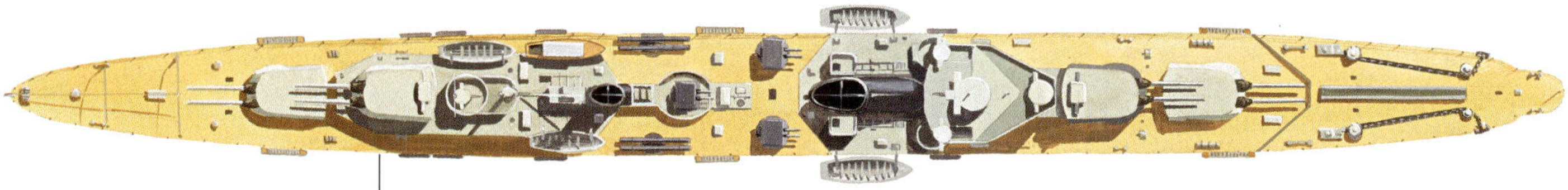

DESIGN ISSUES
Unfortunately for the *Giussano* class, too much was sacrificed for the sake of speed, and they were lightly armoured. So poor was their basic protection that a typical destroyer was capable of inflicting serious damage once within range. Built for the easy conditions of the Mediterranean, the class was rather unstable and lacking in endurance. Crews also disliked the cramped conditions inside.

Final ship in class
Giovanni delle Bande Nere was the fourth and final ship in the class. It was laid down on 31 October 1928, launched on 27 April 1930 and completed in April 1931.

AA GUNS
Here the gun crew of one of the ship's 102mm (4in) AA guns conduct a gunnery exercise. Note the style of the crew's helmets and life belts slung over the rail on the platform, as well as the apparent wicker padding used for splinter protection.

River class (1931)

TYPE • *Submarine* **COUNTRY** • *United Kingdom*

This small class of three represented the Royal Navy's final attempt to develop a fleet submarine. Given the Navy's role as a global force, a submarine with the size, speed and endurance required to operate alongside a battle fleet was seen as highly desirable.

SPECIFICATIONS

DIMENSIONS:	Length: 105m (344ft 6in); Beam: 8.61m (28ft 3in); Draught: 4.85m (15ft 11in)
DISPLACEMENT:	2199.6 tonnes (2424 tons) surfaced; 2723 tonnes (3001 tons) submerged
PROPULSION:	2 supercharged diesel engines, 3750kW (5000hp) total; 2 electric motors 950kW (1250hp); 2 screws
SPEED:	22 knots (41km/h; 25mph) surfaced; 10 knots (19km/h; 11.5mph) submerged
RANGE:	12,000nmi (22,224km; 13,809 miles) at 12 knots surfaced; 118nmi (219km; 136 miles) at 4 knots submerged
ARMAMENT:	6 × 533mm (21in) torpedo tubes; 12 torpedoes; 1 × 102mm (4in) QF gun
COMPLEMENT:	61

Left: HMS Severn *at anchor. This was the era when streamlining became an essential aspect of fast-moving machines, and the Rivers presented an elegantly rounded form of superstructure, with minimal encumbrances on the long deck and the gun, which was set on a forward extension of the tower, shielded in a trainable turret.*

CREW SPACE
Crew space was more generous than usual due to the lack of stern tubes, which allowed for an aft crew compartment above the compressed air cylinders.

ARMAMENT
The torpedo armament consisted of six 533mm (21in) bow tubes. A 119mm (4.7in) gun was originally fitted, but in keeping with submarine policy of the time was changed, after completion, to a 102mm (4in) QF gun with 120 rounds of ammunition.

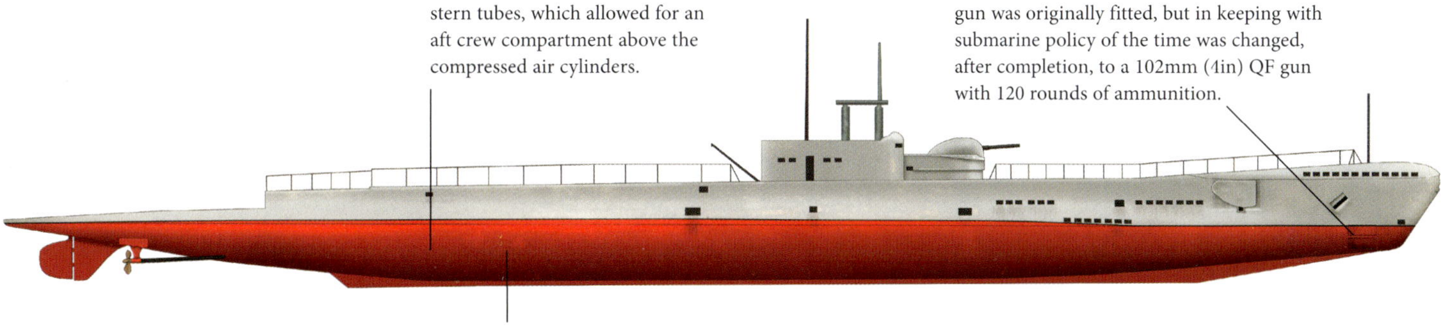

ELECTRIC MOTOR
The electric motor room was between the main and auxiliary engine rooms, and the main motors, one on each shaft, developed a total of 1864kW (250bhp) at 245rpm at a 75-minute rating and 984kW (1320bhp) at a 2-hour rating.

DRIVE SYSTEM

Apart from the U/V class, RN submarines deployed in World War II had direct drive: each diesel engine was linked by a drive shaft via a clutch to a double armature electric motor. A shaft from this, via another clutch ('tail clutch') turned the propeller shaft. By disengaging the diesel engine from the drive shaft, the boat could run on the electric motors only, drawing power from the batteries. With the tail clutch disengaged, the diesels could be used to charge the batteries.

HMS *SEVERN*
Although 20 vessels were planned in the class, only three were constructed: *Thames*, *Severn* and *Clyde*. The main differences in *Severn* from *Thames* were an increase in length of the motor room consequent on a change from the 330-volt battery grouping in *Thames* to a 220-volt grouping.

S-class (1931)

TYPE • *Submarine* **COUNTRY** • *United Kingdom*

SPECIFICATIONS (HMS *Seraph*)

Dimensions:	Length: 66.1m (216ft 7in); Beam: 7.2m (23ft 9in); Draught: 4.5m (14ft 9in)
Displacement:	879 tonnes (969 tons) surfaced; 1010 tonnes (1114 tons) submerged
Propulsion:	2 diesel engines 700kW (950bhp); 2 electric motors 435kW (650hp); 2 screws
Speed:	15 knots (28km/h; 17mph) surfaced; 10 knots (19km/h; 12mph) submerged
Range:	6000nmi (11,112km; 6905 miles) at 10 knots surfaced; 120nmi (220km; 140 miles) at 3 knots submerged
Armament:	7 x 533mm (21in) torpedo tubes; 12 torpedoes; 1 x 76mm (3in) deck gun; 1 x 20mm (0.79in) AA gun
Complement:	48

The Royal Navy's largest class of submarine, 62 S-class submarines were built in total. In design, appearance and production they formed three groups, with two sets commissioned in the 1930s, followed by the largest group of 50 between 1940 and 1945.

Below: Commissioned in November 1932, Swordfish *was among the first of the S-class, and as such was subject to many modifications througout the 1930s.*

Above: Sanguine *was the last S-class to be built and its commissioning coincided with the end of the war in Europe. It never saw active service.*

Design
The basic design was an Admiralty one, single-hulled, based in part on the WWI L-class. The original design was constrained by the terms of the 1930 naval agreement, and the designers struggled to keep the displacement weight low.

Tower
The tower, elliptical in shape, was small and low. On the first eight, the bridge was enclosed, but in war service, this was changed to an open bridge.

HMS *Storm*
Storm was commissioned in July 1943. These were the first British submarines fitted with radar for surface as well as for air search. *Storm* served in eastern waters, with a ballast tank converted to fuel-holding to allow for the lengthy patrols required.

UNDERWATER TARGET
Rumours of Germany's Type XXI 'wonder weapon' led to hurried work on HMS *Seraph* in mid-1944. *Seraph* was taken in for conversion to an underwater target training vessel. The modifications included the uprating of the main motors to 1193kW (1600shp) at full power, the fitting of coarser pitch propellers to allow the extra power to be converted to thrust and the installation of a high-capacity battery to extend endurance. To reduce drag, the hull was streamlined by fairing off all apertures and reducing the size of the hydroplanes.

HMS *Scorcher*
HMS *Scorcher* was not launched until December 1944, making its war service relatively limited. It survived the conflict and remained in service until the early 1960s.

Dekabrist and *Leninets* class (1931)

TYPE • *Submarine* COUNTRY • *Soviet Union*

SPECIFICATIONS (L-class Series III)

Dimensions:	Length: 81m (265ft 9in); Beam: 7.5m (24ft 7in); Draught: 4.8m (15ft 9in)
Displacement:	1219 tonnes (1344 US tons) surfaced; 1573 tonnes (1735 US tons) submerged
Propulsion:	2 diesels; 2 electric motors; 2 screws
Speed:	15 knots (27.8km/h; 17.3mph) surfaced; 9 knots (16.7km/h; 10.4mph) submerged
Range:	6000nmi (11,112km; 6096 miles) at 9 knots; 154nmi (285km; 177 miles) at 3 knots
Armament:	5 × 533mm (21in) torpedo tubes; 12 torpedoes; 20 mines; 1 × 100mm (3.9in) gun; 1 × 45mm (1.75in) gun
Complement:	53

Germany's invasion of the Soviet Union in June 1941 brought Russia into the Allied war effort on its western fronts (the Soviet Union did not declare war on Japan until August 1945). For the Soviet Navy, that meant operations in the Baltic, Barents and Black seas.

Above: Leninets *class* L3. *Boats* L1 *to* L6 *displayed some problems, including a slow dive speed, estimated at three minutes, which was subsequently corrected in later builds.*

Propulsion
The D-class submarines were driven by two shafts with three-bladed propellers, powered on the surface by two MAN/Kolomna diesel engines producing 809kW (1100hp) each and underwater by two PG-20 electric motors generating 392kW (525hp) each.

Dekabrist (D-class)
The D-class was the Soviet Navy's first post-World War I class. Six were built in 1927–29, double-hulled, with electric creep motors and a test depth of 90m (295ft).

Performance
During sea trials, D-class vessels reached speeds of 15.3 knots (28.3 km/h; 17.6 mph) surfaced and 8.7 knots (16.1 km/h; 10.0mph) submerged).

MINELAYER
The nature of the Baltic and Black seas made the Soviet Navy particularly concerned with mines (Russia's *Krab* of 1912 was the first submarine minelayer), and the large *Leninets* class was developed through the 1930s as a minelayer that was also an attack boat. Series I existed only as a draft design. Its prototype was the British *L55*, sunk in the Baltic in 1919, before being raised and repaired.

***L-15*, EARLY 1943**
The crew of the Soviet *Leninets*-class submarine *L-15* pose for a photograph at Greenock, Scotland, in early 1943. The submarine had an unfortunate track record of engaging German shipping with torpedoes, but reliably missing.

HMS *Porpoise* and *Grampus* class (1932)

TYPE • *Submarine* **COUNTRY** • *United Kingdom*

SPECIFICATIONS

Dimensions:	Length: 88m (288ft 9in); Beam: 9.09m (29ft 10in); Draught: 4.83m (15ft 10in)
Displacement:	1796 tonnes (1980 tons) surfaced; 2067 tonnes (2279 tons) submerged
Propulsion:	2 diesels, 1250kW (1650hp); 2 electric motors, 610kW (815hp); 2 screws
Speed:	15.5 knots (28.7km/h; 17.8mph) surfaced; 8.75kt (16.2km/h; 10mph) submerged
Range:	6300nmi (11,667km; 7251 miles) at 10.6 knots; 64nmi (118km; 73.6 mi) at 4 knots
Armament:	6 × 533mm (21in) torpedo tubes; 12 torpedoes; 50 mines; 1 × 102mm (4in) deck gun; 2 × 7.7mm (0.303in) Lewis guns
Complement:	59

The *Grampus* class was specifically designed as minelayer submarines. The immediate precursor was HMS *Porpoise* (1932), but the five subsequent *Grampus*-class boats had significant differences in design.

Left: The crew of Porpoise *cluster on its deck at Portsmouth following combat operations in the Mediterranean, January 1943.*

Propulsion
All six submarines were driven by Admiralty-type vertical four-stroke blast-injection six-cylinder diesel engines of a combined 2461kW (3300bhp), and two electric motors in tandem on each shaft, developing 1215kW (1630bhp).

Mines
Fifty Mk XVI mines were carried in a non-watertight casing above the pressure hull, set on an endless chain-and-rack system, which moved them at a controlled speed to discharge vents set in the stern.

HMS *Porpoise*
Porpoise was recognizable from the others in its class by the external fuel tanks terminating about 18m (60ft) short of the bow structure.

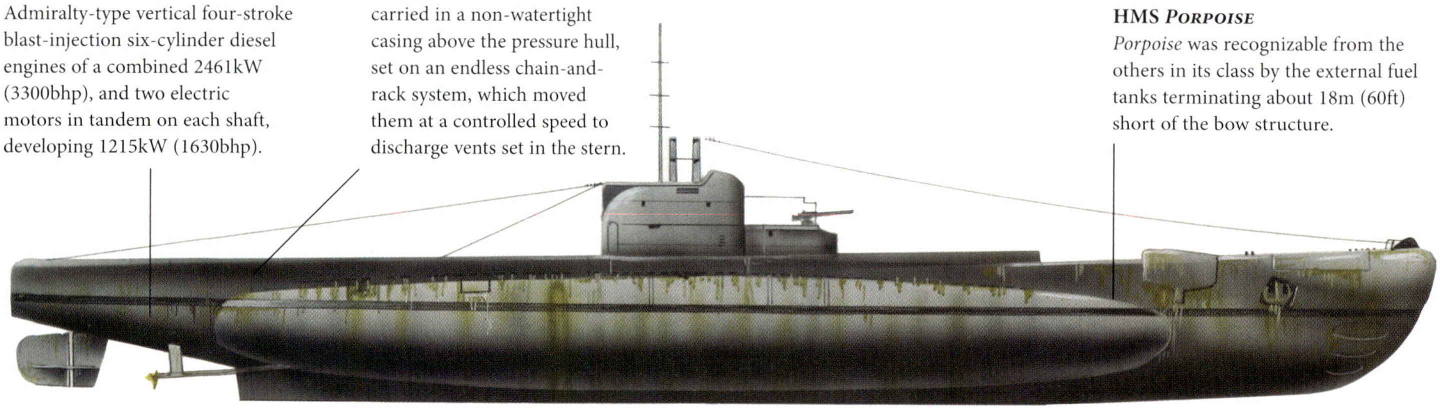

RN MINES

The RN minelayers carried specially designed Mk XVI mines, of Herz horn contact type, with a charge of 145kg (320lb) of TNT. They were of hydrostatic type, designed for the mine and its sinker to go to the seabed, where the mine would be released and rise on its cable. When the hydrostat reached the preset level (anywhere between 1.4m (5ft) and 60m (90ft), it dropped off and the mine remained moored in position.

HMS *PORPOISE* (1932)
Porpoise had welded saddle tanks and its hull form was based on the P-class. The double-hulled design of the *Grampus* boats was based on the T-class, although with some streamlining.

Krasny Kavkaz (1932)

TYPE • *Cruiser* **COUNTRY** • *Soviet Union*

SPECIFICATIONS

DIMENSIONS:	Length: 159.5m (523ft 4in); Beam: 15.7m (51ft 6in); Draught: 6.6m (21ft 8in)
DISPLACEMENT:	7560 tonnes (8333 tons) standard
PROPULSION:	10 Yarrow boilers; 4 geared steam turbines; 4 shafts; 41,000kW (55,000shp)
SPEED:	29 knots (54km/h; 33mph)
RANGE:	3500nmi (6500km; 4000mi) at 15 knots
ARMAMENT:	4 × 180mm (7.1in) guns; 8 × 100mm (3.9in) AA guns; 2 × 76mm AA guns; 4 × 45mm (1.8in) AA guns; 12 × 533mm (21in) torpedo tubes; 60–120 mines
ARMOUR:	Upper and lower armoured decks 20mm (0.79in); turrets 76mm (3in); lower armour belt 76mm (3in); upper armour belt 25mm (0.98in)
COMPLEMENT:	878

The *Krasny Kavkaz* was a Russian, later Soviet, cruiser that was laid down in October 1913, launched in 1916, but not completed to a commissioned state until 1932. The vessel thereafter served the Soviet Navy until the early 1950s.

Above: Krasny Kavkaz's *journey from an incomplete Russian cruiser to a modernized warship illustrates the challenges faced by Soviet shipbuilding across a period of intensive political and technological change.*

AA GUNS
Prior to the German invasion of the USSR in 1941, *Krasny Kavkaz* received significant anti-aircraft upgrades, replacing its original 76mm (3in) guns with more advanced 100mm (3.9in) Minizini twin mounts and additional smaller-calibre AA guns to bolster its air defence.

MAIN GUNS
Intended to carry heavier 203mm (8in) guns, *Krasny Kavkaz* was instead outfitted with 180mm (7.1in) MK-1-180 guns due to hull constraints. The ship's superstructure was heavily modified to accommodate these turrets.

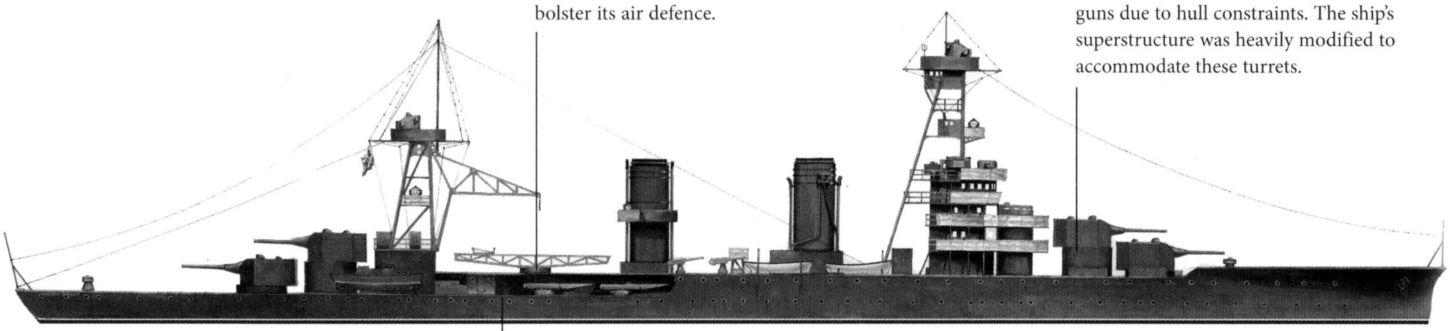

TORPEDOES AND MINES
The cruiser was equipped with four triple 533mm (21in) torpedo mounts on each side and could carry up to 120 mines. These features were added during extensive post-World War I modifications.

NAVAL SUPPORT

The Soviet Navy during World War II fought few ship versus ship engagements, as most of the conflict was fought on the land. Ships such as the *Krasny Kavkaz*, however, were often called upon to deliver offshore fire support to army operations, land troops on contested shores or perform evaucation duties. *Krasny Kavkaz* served in all these capacities, particularly around the Black Sea and Crimea.

EVOLVING DESIGN

Initially the *Krasny Kavkaz* was launched as *Admiral Lazarev*, but the ship's completion was delayed by revolution until it was completed for Soviet service in the 1930s. The vessel was modified from its original design, particularly in its armament and layout, to better suit the needs of the new Soviet Navy.

USS *Indianapolis* (1932)

TYPE • *Heavy cruiser* **COUNTRY** • *United States*

SPECIFICATIONS

Dimensions:	Length: 185.92m (610ft); Beam: 20.1m (66ft); Draught: 6.4m (21ft)
Displacement:	10,258 tonnes (11,310 tons) standard
Propulsion:	8 Yarrow boilers; 4-shaft Parsons turbines; 79,814kW (107,000hp)
Speed:	32.5 knots (60.2km/h; 37.5mph)
Range:	10,000nmi (19,000km; 12,000mi) at 15 knots
Armament:	9 × 203mm (8in) main guns; 8 × 127mm (5in) secondary guns; 40mm (1.5in) and 20mm (0.79in) AA guns; 8 × 12.7mm (0.5in) machine guns; 4 aircraft
Armour:	Belt 57mm (2.25in); deck 146–63mm (5.75–2.5in)
Complement:	917

Few ships hold such a controversial place in naval history as the USS *Indianapolis* (CA-35). It was the second and last of the Portland-class cruisers built, from an intended total of five. The subsequent ships were revised to offer better armour.

Right: USS Indianapolis *pictured off the Mare Island Navy Yard, California, on 10 July 1945. Indianapolis* served as the flagship of the US Navy's Fifth Fleet

BOILERS
The ship was powered by eight Yarrow boilers driving four-shaft Parsons turbines, giving a top speed of 32.5 knots (60.2km/h; 37.5mph).

MAIN GUNS
The cruiser had nine 203mm (8in) / 55 calibre Mark 9 guns in three triple-mount turrets, a superfiring pair fore and one pair aft.

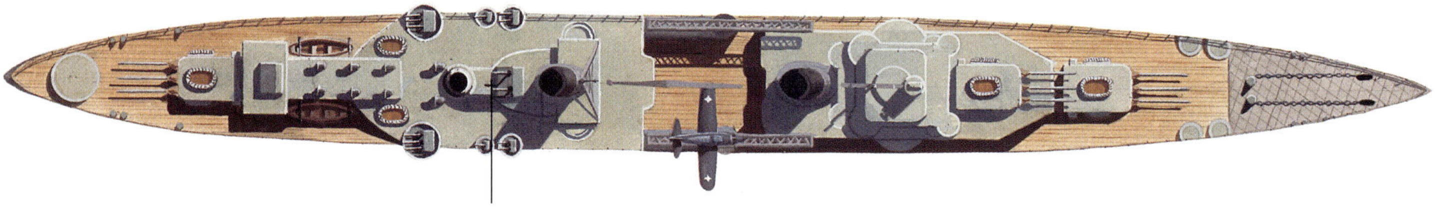

AA GUNS
The ship was fitted with anti-aircraft guns of multiple calibres, particularly automatic 40mm (1.5in) and 20mm (0.79in) cannon.

SINKING

Indianapolis was the last major US ship sunk in World War II. It carried the first atomic bomb from the United States to the island of Tinian, but on 30 July 1945, was sunk by a Japanese submarine. Of the 1196 crew on board, about 900 made it into the water. Deaths by shark attacks, drowning and exposure followed over the next four days until rescue came. Of the 900 initial survivors, only 317 were rescued.

OCEAN CROSSING

USS *Indianapolis* (CA-35) was laid down on 31 March 1930 at Camden, New Jersey, by the New York Shipbuilding Corp. It was launched on 7 November 1931 and commissioned at the Philadelphia Navy Yard on 15 November 1932, with Captain John M. Smeallie in command.

Pola (1932)

TYPE • *Cruiser* **COUNTRY** • *Italy*

SPECIFICATIONS

Dimensions:	Length: 169.8m (557ft 2in); Beam: 19.15m (62ft 10in); Draught: 6.68m (21ft 11in)
Displacement:	11,545 tonnes (12,732 tons) standard; 14,330 tonnes (15,796 tons) deep load
Propulsion:	8 Thornycroft (Yarrow, Fiume) boilers; twin-shaft Parsons geared turbines; 70,836kW (95,000hp)
Speed:	32 knots (59.3km/h; 36.8mph)
Armour:	Belt 150–120mm (5.9–4.7in); decks 70–20mm (2.75–0.78in); bulkheads 119–89mm (4.7–3.5in); main turrets 150–119mm (5.9–4.7in)
Armament:	8 × 203mm (8in) main guns; 16 × 102mm (4in) AA guns (later 14); 6 × 40mm AA guns; 8 × 37mm (1.5in)AA guns; 8 × 13.2mm (0.5in) AA guns (later 16)
Complement:	841

The Italian *Zara*-class cruisers departed from other contemporary Italian designs, opting for armour over speed. However, it proved impossible to keep the desired 203mm (8in) guns and armour within the limits of the Washington Treaty, which were therefore quietly ignored.

Left: Fiume *(1930) was another Zara-class heavy cruiser serving with the Italian Regia Marina. It just survived World War II, but in such a battered condition that it was sold for scrap in 1947.*

Bridge
Pola was different to the other ships, being fitted with an enlarged bridge structure and lacking the distinctive flutes that ran along the forecastle sideplating on the other *Zaras*.

Superstructure
To keep *Pola*'s displacement within reasonable limits, the superstructure was reduced and the flush deck of the *Trento* class (on which the design was based) was abandoned.

Weight
Some attempt was made to stay within the 10,000-ton restriction, by fitting lightweight machinery and less armour in specific areas, but the ships were still significantly overweight.

***ZARA* CLASS**
The four *Zara*-class cruisers were built in two batches, with the *Pola* coming in the second. The ships were originally classified as light cruisers, then armoured cruisers (to distinguish them from other less well-protected ships) and finally as heavy cruisers. Despite their weight the cruisers were still fast, with a maximum speed of 65km/h (35 knots; 40mph) being achieved in special trials.

***POLA* MAIN GUNS**
Officers on the deck of *Pola* stand in front of its twin-mounted 203mm (8in) main guns. During the late 1930s, the cruisers were modified to carry additional small calibre AA guns, replacing other larger guns, and more were steadily added as the war progressed.

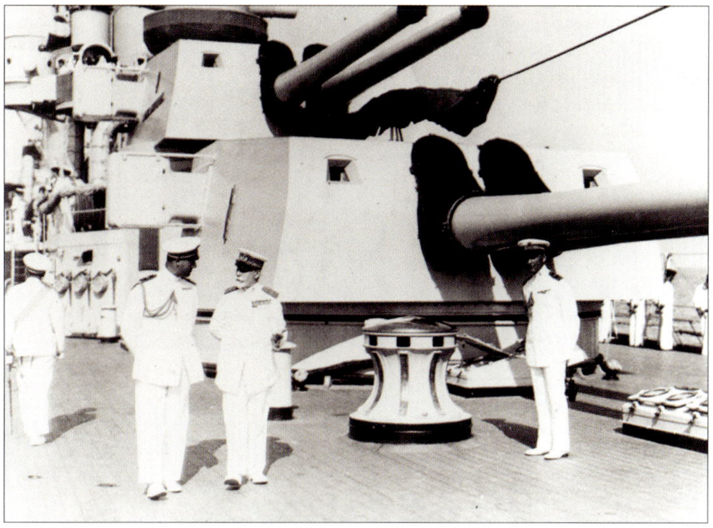

Deutschland / Lützow (1933)

TYPE • *Heavy cruiser* **COUNTRY** • *Germany*

SPECIFICATIONS

Dimensions:	Length: 186m (610ft 3in); Beam: 20.6m (67ft 7in); Draught: 7.2m (23ft 7in)
Displacement:	10,659 tonnes (11,750 tons) standard
Propulsion:	4 x 9-cylinder double-acting 2-stroke MAN diesel engines; Vulcan gearboxes; 40,268kW (54,000hp); 2 screws
Speed:	28 knots (52km/h; 32mph)
Range:	8700nmi (16,110km; 10,010 miles) at 19 knots (35km/h; 22mph)
Armament:	6 x 280mm (11in) guns; 8 x 150mm (5.9in) guns; 3 x 88mm (3.5in) guns; 8 x 500mm (19.7in) torpedo tubes
Armour:	Belt 80–60mm (3.1–2.4in); deck 45–40mm (1.8–1.6in); barbettes 150–50mm (5.9–2in); turrets 140–85mm (5.5–3.3in)
Complement:	619

In 1927, the concept of a fast *Panzerschiff* (armoured ship) began to take shape: a vessel faster than battleship, yet carrying thicker armour and heavier guns than a cruiser. It became known as the 'pocket battleship' and was the first major warship powered by diesel engines.

Left: On commissioning, Deutschland *became the flagship of the German Navy, until 1936. Its name was changed to* Lützow *on 15 November 1939.*

Torpedo tubes
Eight 500mm (19.7in) torpedo tubes were fitted in two sets of four on the upper deck, with armoured covers; later, 533mm (21in) tubes were fitted.

Refitting
In 1935, a mainmast was attached abaft the funnel, AA gun control was fitted, an aircraft catapult installed between the bridge tower and funnel, and twin-mount 88mm (3.5in) AA guns replaced the original singles.

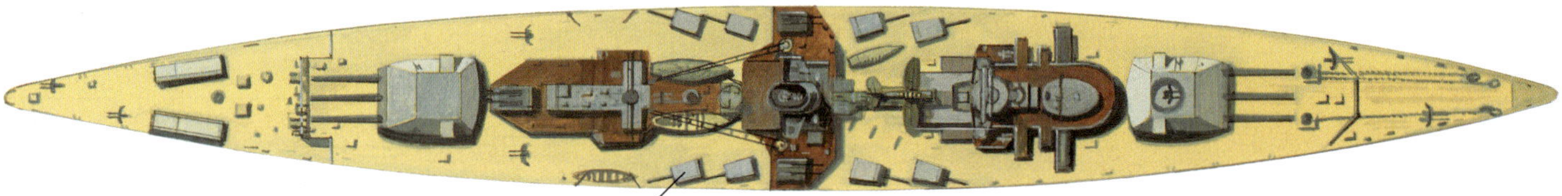

BATTLE DAMAGE
On 11 April 1940, the British submarine *Spearfish* hit *Deutschland* with a torpedo off Norway, and it was towed back to Kiel. On 13 July 1941, attempting an entry to the Atlantic, it was again badly damaged by a torpedo from an RAF Beaufort and returned to Kiel for repairs until January 1942. From October 1943, it remained at Gotenhafen (Gdynia). A bomb attack on 15 April 1945 left the ship disabled and it was scuttled by its crew on 4 May.

Side armour
Deutschland's side armour was intended to withstand fire from cruisers. It was not expected to engage heavy-gun battleships.

LIGHTWEIGHT WARSHIP
Deutschland was laid down at the Deutsche Werke, Kiel, on 5 February 1929, launched on 19 March 1931 and completed on 1 April 1933. Every effort was made to save weight, with electric welding resulting in a hull weight 15 per cent less than if riveting had been used.

Graf Spee (1933)

TYPE • *Heavy cruiser* **COUNTRY** • *Germany*

SPECIFICATIONS

DIMENSIONS:	Length: 186m (610ft 3in); Beam: 21.65m (71ft); Draught: 7.34m (24ft 1in)
DISPLACEMENT:	14,890 tonnes (16,420 tons) design; 16,280 tonnes (17,940 tons) full load
PROPULSION:	8 diesel engines; 39,720kW; 53,260shp); 2 shafts
SPEED:	28.5 knots (52.8km/h; 32.8mph)
RANGE:	16,300nmi (30,200km; 18,800 miles) at 18.69 knots
ARMAMENT:	6 × 280mm (11in) guns; 8 × 150mm (5.9in) guns; 6 × 105mm (4.1in) SK C/33 gun; 8 × 37mm (1.5in) SK C/30 guns; 10 × 20mm (0.79in) C/30s; 8 × 533mm (21in) torpedo tubes in quad mounts
ARMOUR:	Main turrets 140mm (5.5in); belt 100mm (3.9in); main deck 45–70mm (1.8–2.8in)
COMPLEMENT:	619

Strictly speaking, *Deutschland* and its companions ships *Admiral Graf Spee* and *Admiral Scheer* were *Panzerschiffe* (armoured cruisers). However, the British press derisively called them 'pocket battleships' and the name stuck.

Left: Admiral Graf Spee *underway in the English Channel in April 1939. Note the Arado Ar 196A-1 float plane on its catapult, amidships, and the freighter passing by in the background.*

MODIFICATIONS
In 1934 the port side derrick was replaced by a lattice crane. In 1935 a mainmast was attached abaft the funnel, AA gun control was fitted, an aircraft catapult installed between the bridge tower and funnel, and twin-mount 88mm (3.5in) AA guns replaced the original singles.

BRIDGE TOWER
Graf Spee and *Admiral Scheer* had angular bridge tower structures, different to *Lützow*'s. Seetakt radar, primarily for gunnery, was fitted to *Graf Spee* in 1939: the rangefinder or telemeter is prominent at the top.

GUN POSITION
The 150mm (5.9in) guns were all on the same level as the flush deck. The compact superstructure offered little room for higher gun emplacements.

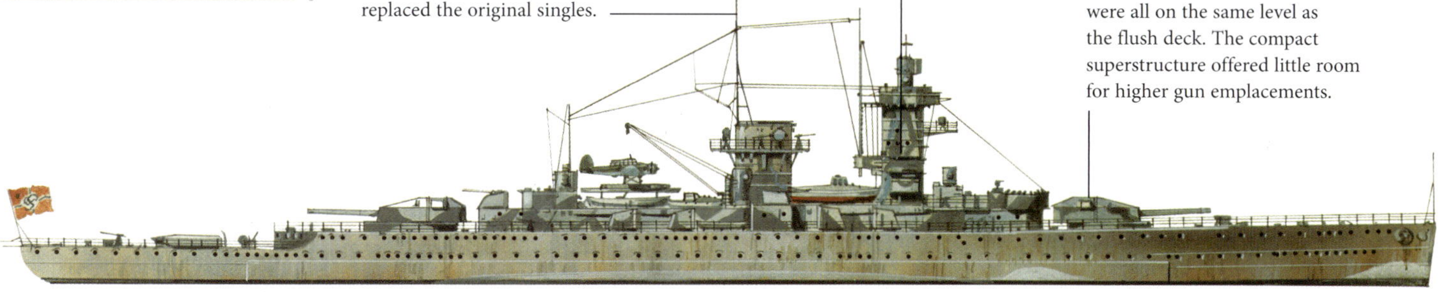

POCKET BATTLESHIP
The term 'pocket battleship', first used outside Germany, was not intended as a compliment but it became clear that *Deutschland* and its companion ships, *Admiral Scheer* (1934) and *Admiral Graf Spee* (1936), had to be taken seriously as commerce raiders. Providing a maximum speed of 28 knots (52km/h (32mph), the engines gave a wide range of 52,782km (28,500nmi) at 10 knots (18.5km/h; 11.5mph). The power came from four MAN nine-cylinder double-acting two-stroke motors, via two Vulcan gear-boxes.

RIVER PLATE
The *Graf Spee* anchored off Montevideo, Uruguay, in mid-December 1939, following the Battle of the River Plate. To avoid further battle and destruction, the ship's captain, Hans Langsdorff, famously ordered it scuttled on 17 December.

Algérie (1934)

TYPE • *Heavy cruiser* **COUNTRY** • *France*

SPECIFICATIONS

Dimensions:	Length: 186.2m (611ft); Beam: 20m (66ft); Draught: 6.15m (20ft 2in)
Displacement:	10,000 tonnes (11,023 tons) standard
Propulsion:	4 shafts; direct-geared Rateau-Bretagne SR turbines; 6 Indret boilers; 62,628kW (84,000hp)
Speed:	31 knots (57km/h; 35mph)
Range:	8700nmi (16,110km; 10,006 miles)
Armament:	8 × 203mm (8in)/55 guns; 12 × 100mm (3.9in) DP guns; 8 × 37mm (1.5in) AA guns; 16 × 13.2mm (0.5in) AA HMGs; 2 × triple 550mm (21.7in) torpedo tubes
Armour:	Belt: 120mm (4.7in); deck 80mm (3.1in); turrets 95mm (3.7in); conning tower 95mm (3.7in)
Complement:	748

***Algérie* was a fine French heavy cruiser, designed to meet the requirements of the 1922 Washington Naval Treaty. It managed to strike a good balance between the design trinity of armament, speed and armour.**

Right: Here we see Algérie *at harbour in Brest on 27 May 1937, after conducting an offshore naval review with the battleship* Dunkerque.

Armour
The warship was heavily armoured, including a 120mm (4.7in) main belt, 70mm (2.8in) transverse bulkheads and 40mm (1.6in) longitudinal bulkheads.

Improvements
In 1941 the *Algérie* was given a better AA battery (16 x 37mm and 36 x 13.2mm) and in 1942 a radar, of early French design.

Design
The *Algérie* had a distinctive flush deck, giving it a 'straight' appearance. This design helped to acheive significant weight savings.

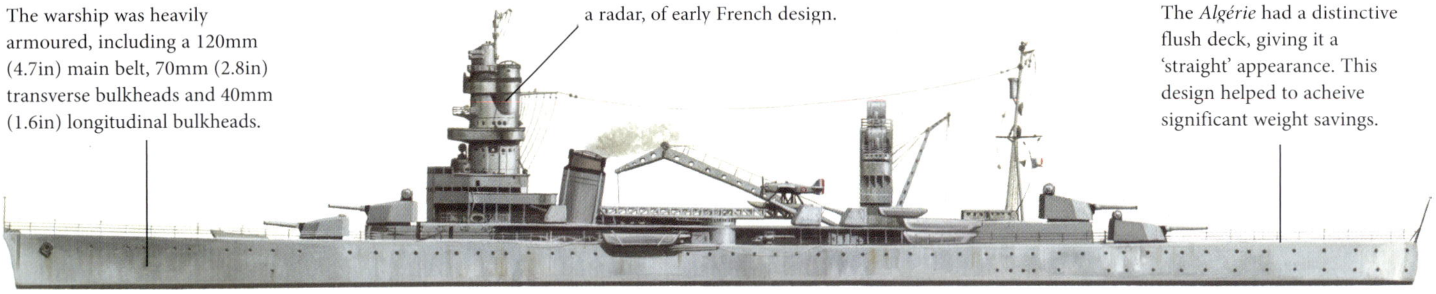

SCUTTLED
As with all French Navy vessels, *Algérie* faced an uncertain future following the defeat of France by Germany in June 1940. Like many French vessels, it remained with the navy of the Vichy fleet, based at Toulon in the south of the country, where it received significant upgrades. But on 27 November 1942, along with 77 other ships, it was scuttled by its crew to prevent it falling into German hands.

***ALGÉRIE* IN 1932**
The *Algérie* had particularly fine lines, with a sleek and slender profile that hinted at speed. At full pace, the warship could achieve a speed of 31 knots (57km/h (31km/h; 35mph).

Type IA (1934)

TYPE • *Submarine* **COUNTRY** • *Germany*

SPECIFICATIONS (*U-25* Type IA)

Dimensions:	Length: 72.39m (237ft 5in); Beam: 6.21m (20ft 4in); Draught: 4.3m (14ft 1in)
Displacement:	Surfaced 875.8 tonnes (965 tons); submerged 998.7 tonnes (1101 tons)
Propulsion:	2 1044kW (1400hp) diesel engines; 2 double 373kW (500hp) electric motors; 2 screws
Speed:	Surfaced 17.75 knots (32.8km/h; 20.4mph); submerged 8.3 knots (15.35km/h; 9.54mph)
Range:	Surfaced 6700nmi (12,395km; 7,712 miles) at 12 knots; submerged 78nmi (144km; 90 miles) at 4 knots
Armament:	6 × 533mm (21in) torpedo tubes; 14 torpedoes; 1 × 105mm (4.1in) SKC/36 gun; 1 × 20mm (0.79in) Flak 30 AA gun
Complement:	43

The two Type IA boats paved the way for the German Kriegsmarine's wartime ocean-going flotillas, particularly the Type IX submarine. Their defects, as much as their qualities, showed the way forward.

Above: U-26 sails out to open water. The Type IA had many problems. Surface balance was poor, with rolling and pitching. Rudder and propeller arrangement restricted manoeuvrability. The MAN M 8 V 40/46 eight-cylinder diesel engines were not reliable.

Tower and guns
The tower had a basic bathtub shape that would remain typical of U-boats up to Type XXI) with a narrower extension aft. A 105mm (4.1in) gun was placed forward of the tower, and a 20mm (0.79in) AA gun was mounted at bridge level on the after part.

Hull
The design was double-hulled, with the outer hull drawn down to keel level. Ballast tanks and fuel bunkers were placed between the hulls.

Electric motors
The two double Brown Boveri electric motors were rated at 373kW (500hp). Two AFA 36 MAK 40 62-cell battery sets were fitted, with a weight of 97.5 tonnes (107.4 tons).

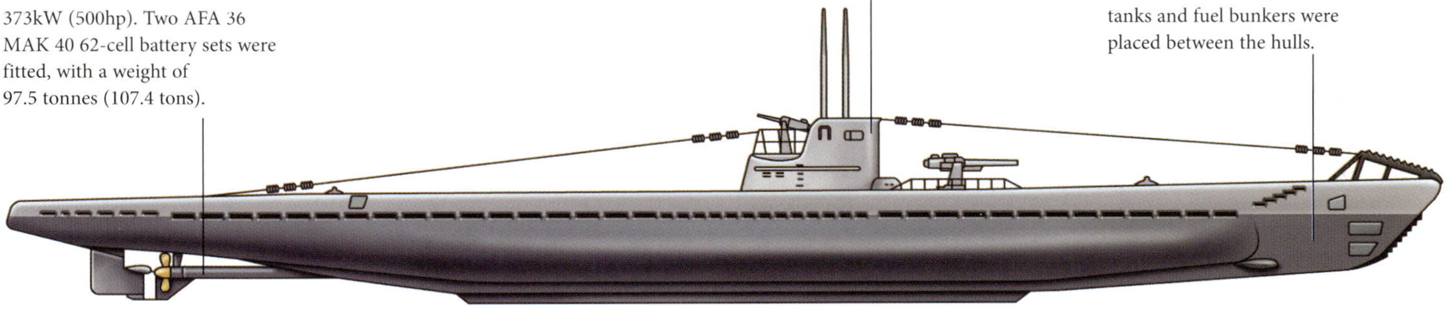

SECRET WORK

German designers working with the Dutch bureau Ingenieurskantoor voor Scheepsbouw were on the project from the mid-1920s. Semi-secret work also went on in Spain and Finland; Germany at this time was operating under the tight restrictions of the Versailles Treaty. A Spanish yard, Echevarietta in Cadiz, built a single prototype, known as *E-1*, launched in October 1930 (later sold to Turkey). Construction of Type IA began only in 1934 at the Deschimag A.G. Weser yard in Bremen.

U-25
U-25 and *U-26* proved useful as training vessels, but in 1939 they could scarcely be regarded as fit for battle. However, both saw active service and sank several merchant ships and small warships before being lost in minefields in 1940.

New Orleans class (1934)

TYPE • *Cruiser* **COUNTRY** • *United States*

The *New Orleans* class had a bruising operational life in World War II, spent mostly in the Pacific. Although the essential design was sound, three of the class were sunk in one battle alone, off Savo Island in the Pacific in August 1943.

SPECIFICATIONS

Dimensions:	Length: 179.27m (588ft 2in); Beam: 18.82m (61ft 9in); Draught: 5.92m (19ft 5in)
Displacement:	10,110 tonnes (11,146 tons) standard; 12,663 tonnes (13,961 tons) full load
Propulsion:	4 Parsons/Westinghouse geared turbines; 8 Babcock & Wilcox boilers; 4 screws; 79,800kW (107,000hp)
Speed:	32.7 knots (60.6km/h; 37.6mph)
Range:	14,000nmi (26,000km; 16,155 miles) at 10 knots
Armament:	9 x 203mm (8in)/55-calibre guns; 8 x 127mm (5in)/25-calibre guns; 8 x 12.7mm (0.5in) MGs
Armour:	Belt 76–127mm (3–5in); deck 32–57mm (1.25–2.25in); turrets 38–203mm (1.5–8in); barbettes 127mm (5in; 165mm/6.5in in CA-38); conning tower 127mm (5in)
Complement:	708

Right: The USS New Orleans *was the lead ship of the class, launched on 12 April 1943.*

Firepower
The main armament of the ships were nine 203mm (8in) guns in triple turrets. During the war, the ships were fitted with substantial AA weaponry, including Bofors and Oerlikon auto-cannon and quadruple-mount 28mm (1.1in) guns.

Upgrade
After the loss of three of the class in battle in 1942, the remaining ships were heavily upgraded, including changes to the superstructure to reduce top weight and improvements in anti-aircraft weapons and fire control.

Armour
The *New Orleans* class was heavily armoured, with belt armour that reached a depth of 130mm (5in) and turrets faced with 203mm (8in) of protection.

SAVO ISLAND

From 1942, the *New Orleans* class cruisers were heavily engaged in combat in the Pacific, fighting major surface actions against Japanese warships. A defining engagement for the class was the Battle of Savo Island (8–9 August 1942), part of the Guadalcanal campaign. In this devastating US defeat, three of the *New Orleans* class were sunk – *Astoria*, *Quincy* and *Vincennes* – as well as an Australian heavy cruiser, prompting a major investigation into US Navy tactics and technology.

USS *QUINCY*
In the background of this photograph, the USS *Quincy* (CL/CA-39) steams off Guadalcanal in August 1942. The ship would be one of the victims of the Savo Island battle, sinking after strikes from both torpedoes and shells, with 370 crew dead and 167 wounded.

‘1500-tonne’ type (1935)

TYPE • *Submarine* **COUNTRY** • *France*

SPECIFICATIONS
(*Henri Poincaré*)

Dimensions:	Length: 92.3m (302ft 10in); Beam: 8.1m (26ft 7in); Draught: 4.4m (14ft 5in)
Displacement:	1572 tonnes (1733 tons) surfaced; 2092 tonnes (2307 tons) submerged
Propulsion:	2 diesels producing 2237kW (3000hp); 2 × electric motors producing 839kW (1125hp); 2 screws
Speed:	Surfaced 17.5 knots (32.4km/h; 20.1mph); submerged 10 knots (19km/h; 12mph)
Range:	Surface 14,000nmi (25,928km; 16,111 miles) at 7 knots; submerged 90nmi (170km; 100 miles) at 7 knots
Armament:	9 × 550mm (21.45in) torpedo tubes; 2 × 400mm (15.7in) torpedo tubes; 1 × 100mm (3.9in) deck gun; 2 × 13.2mm (0.5in) machine guns
Complement:	84

From 1928 right into the war years, France’s main class of ocean-going submarines was the ‘1500-tonne’ type. The lead boat was the *Redoutable*, laid down in August 1924, launched on 1 July 1925, and commissioned on 24 February 1928.

Below: Casabianca *was the last in the class, powered by Sulzer diesels of 3206.5kW (4300hp), and Alsthom motors of 1566kW (2100hp).*

Henri Poincaré
The submarine’s length of 92.3m (302.8ft) exceeded the operational depth of 80m (262ft). The pressure hull was of riveted construction, using steel 16mm (0.62in) thick.

Torpedo tubes
The boats were equipped with 550mm (21.45in) and 400mm (15.7in) torpedo tubes, to be used according to the type of target. Four 550m tubes were placed forward, inside the hulls, a further three were mounted at deck level, aft of the tower, with two more at the stern, along with two 400mm tubes.

Machinery
The machinery consisted of two Schneider eight-cylinder double-acting diesel engines with a combined 4474kW (6000hp), and two Al23 electric motors. Fuel capacity was 320 tonnes (353 tons).

MODERNIZATION
From December 1942 the four *Redoutable*-class submarines in Free French control crossed to the United States for modernization. The smaller diameter tubes were removed and replaced by a single 533mm (21in) tube. The motors were overhauled, new batteries installed, and the pressure hulls and diving auxiliaries reinforced. The ballast tanks were altered to increase the range. Radar and underwater sound systems were installed, with up-to-date sonar capabilities.

SURFACE RAIDER
The double-hulled *Redoutable* class was an advanced version of the *Requin* boat and was heavily armed with torpedoes. Their function was that of surface raiders and defenders of the maritime supply line between France and its colonies. In World War II they did little of either before most were scuttled.

Type II (1935)

TYPE • *Submarine* **COUNTRY** • *Germany*

SPECIFICATIONS (*U-2* Type IIA)

Dimensions:	Length: 40.9m (134ft 2in); Beam: 4.1m (13ft 4in); Draught: 3.8m (12ft 5in)
Displacement:	254 tonnes (280 tons) surfaced; 303 tonnes (334 tons) submerged
Propulsion:	2 MWM 6-cylinder 261kW (350hp) diesel engines; 2 SSW 134kW (180hp) electric motors; 2 screws
Speed:	Surfaced 13 knots (24km/h; 14.9mph); submerged 7.4 knots (13.7km/h; 8.5mph)
Range:	Surfaced 2000nmi (3704km; 2302 miles) at 8 knots; submerged 35nmi (64.8km; 40.3 miles) at 4 knots
Armament:	3 × 533mm (21in) torpedo tubes; 6 torpedoes; 1 × 20mm (0.79in) Flak twin AA gun
Complement:	25

Fifty boats in four variations from IIA to IID were produced. Its design history goes back to the UF class of World War I, via the one-off Finnish-built prototype *Vesikko* (1933). The type leader was *U-1*, launched at Deutsche Werk, Kiel in 1935.

Above: The Type IIB seen here was lengthened midships by three frames compared to the Type IIA, and had an extended range and reduced diving time.

Capabilities
With only three torpedo tubes (all forward) and capacity to hold five torpedoes (six on IIC) or 18 mines, and an original range of only 1050nmi (1945km; 1208mi), the Type II was intended for short patrols.

Crew
For the 25-man crew working inside the submarine, accommodation was extremely tight: 15 bunks were provided, plus the skipper's cubicle.

***U-2* Type IIA**
Although not a large boat, *U-2* was built in a short time: laid down in February 1935 and commissioned on 25 July of the same year, at the Deutsche Werke, Kiel.

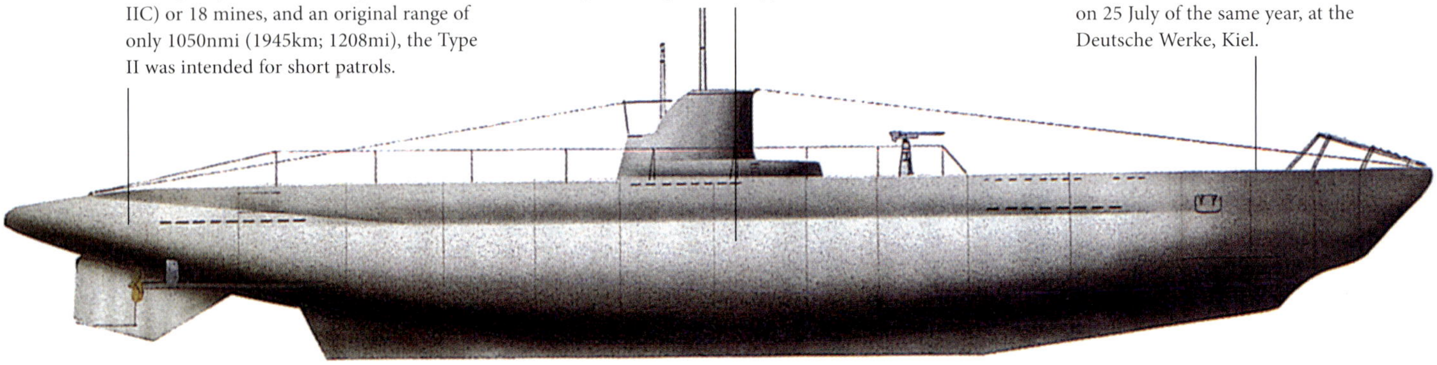

SUB-TYPES
The main differences in the various sub-types were increases in length and displacement, primarily to allow for greater fuel capacity and operating range. But the Type II remained a compact craft, nicknamed *Einbaum* (dugout) by crews, designed for coastal operations. Single-hulled, of welded construction, it was primarily intended as a training type, but the outbreak of war in September 1939 brought many into real action.

TYPE IIs AT KIEL
From left to right, *U-1*, *U-3*, *U-4*, *U-6* rest at Kiel. Known as 'dugouts' to their crews, the Type II coastal boats carried too few torpedoes to be effective in combat, and most became training vessels.

Type VIIB (1936)

SPECIFICATIONS (*U-47*)

DIMENSIONS:	Length: 66.5m (218ft); Beam: 6.2m (20ft 4in); Draught: 4.7m (15ft 5in)
DISPLACEMENT:	765 tonnes (843 tons) surfaced; 871 tonnes (960 tons) submerged
PROPULSION:	2 1722.5kW (2310hp) diesels; two 559kW (750hp) electric motors; 2 screws
SPEED:	16 knots (29.6km/h; 18.4mph) surfaced; 8 knots (14.8km/h; 9.2mph) submerged
RANGE:	4300nmi (7964km; 4949 miles) at 12 knots surfaced; 90nmi (167km; 104 miles) at 4 knots submerged
ARMAMENT:	5 × 533mm (21in) torpedo tubes; 11 torpedoes or 22 TMA/33TMB mines; 1 × 88mm (3.4in) gun; 1 × 20mm (0.79in) gun
COMPLEMENT:	44

An effective hunter-killer in the first years of the war, the Type VII U-boat eventually fell victim to intensive anti-submarine warfare both from the surface and the air.

Above: Captain Gunter Prien and the U-47 crew pose for a photograph in Wilhelmshaven on 17 October 1939, following the return from Scapa Flow and the sinking of HMS Royal Oak.

***U-47* (TYPE VIIB)**
U-47 made world headlines and won glory in Germany through its daring entry into the Scapa Flow naval anchorage in the Orkney Islands on 14 October 1939, when it sank the battleship HMS *Royal Oak* before making a successful getaway.

TWIN RUDDERS
Type VIIB had twin rudders rather than the single rudder of the VIIA boats, and its stern tube was brought inside the pressure hull.

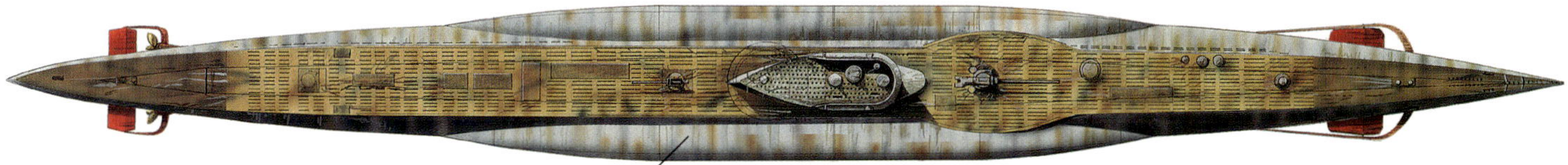

SURFACE CAMOUFLAGE
Colours of wartime U-boats varied at different times, but were almost always in shades of grey, usually in two tones: *Dunkelgrau 51* ('dark grey') above the surfaced waterline, and *Schiffsbodenfarbe III Grau* ('hull colour grey') below.

CHARACTERISTICS

The Type VIIB numbered 24 boats. The dimensions were slightly increased over the Type VIIA, enabling a longer control room and raising the displacement. The MAN diesel engines were uprated with exhaust-driven Büchigebläse supercharging to give a 1-knot (1.9km/h; 1.2mph) increase in surface speed: This was important for tactical positioning because most convoys moved at only 10–12 knots (18.5–22.2km/h; 11.5–13.8mph).

TYPE VIIB SUBMARINES
The crews of two German U-boats exchange gestures as they pass. Conditions in the U-boats were cramped and uncomfortable, ranging from stuffy tropical humidity to freezing Arctic temperatures, depending on the location of the boat.

Type VIIC (1936)

TYPE • *Submarine* **COUNTRY** • *Germany*

The Type VIIC continued the evolution of the Type VII submarine. It total the sub-class numbered 577 boats, with 17 shipyards sharing the building, making it the most prolific of the U-boat types during World War II.

SPECIFICATIONS (*U-210* Type VIIC)

Dimensions:	Length: 67.1m (220ft 2in); Beam: 6.2m (20ft 4in); Draught: 4.8m (15ft 9in)
Displacement:	773.1 tonnes (852 tons) surfaced; 878.8 tonnes (968 tons) submerged
Propulsion:	2 2087kW (2800hp) diesels; 2 559kW (750hp) electric motors; two screws
Speed:	17.2 knots (31.8km/h; 19.8mph) surfaced; 7.6 knots (14km/h; 8.7mph) submerged
Range:	6500nmi (12,038km; 7481 miles) at 12 knots; 80nmi (148km; 92 miles) at 4 knots submerged
Armament:	5 × 533mm (21in) torpedo tubes; 14 torpedoes or 26 TMA/39TMB mines; 1 × 88mm (3.4in) gun; 1 × 20mm (0.79in) gun
Complement:	44

Below: This profile of U-553 *shows adaptation of the VIIC for additional AA protection. The extended platform carries a four-barrelled Flakvierling 20mm (0.79in) gun.*

***U-210* (Type VIIC)**
U-210 was a typical Type VIIC U-boat. Almost all U-boats carried their own emblem, painted on the tower. *U-210*'s was a lobster, with claws extended in attacking mode.

Wake reduction
During 1940, anti-vibration wires were added to the tops of periscopes to help reduce the wake left by a raised periscope. All the VIICs had this feature.

Propellers
The standard U-boat propellers were three-bladed, with a diameter of 1620mm (5.26ft).

REDESIGN
With the Type VIIC, the hull was further lengthened to allow for an active S-Gerät sonar installation. The tower was slightly widened and made 30cm (11.7in) longer. Fuel capacity was increased, allowing an operating range of 8650nmi (16,020km; 9954 miles) at 12 knots. With Type VIIC/41 (88 boats from July 1943), operational depth went down to 120m (394ft).

TYPE VIIC SURRENDER, 1945
With so many boats, builders and repairers, the range of detail variations across the VIIC Type is enormous, especially in armament and tower design. Most lost their deck gun by the end of 1942 when AA defence became more vital than surface attacks on smaller vessels, replaced by a twin 20mm (0.79in) cannon mounting.

Brooklyn class (1936)

TYPE • *Light cruiser* **COUNTRY** • *United States*

SPECIFICATIONS

DIMENSIONS:	Length: 185.42m (608ft 4in); Beam: 18.77m (61ft 7in); Draught: 7.3m (24ft)
DISPLACEMENT:	9924 tonnes (10,939 tons) standard
PROPULSION:	8 steam boilers; 4 geared turbines; 4 screws; 75,000kW (100,000shp)
SPEED:	32.5 knots (60.2km/h; 37.4mph)
RANGE:	10,000nmi (18,520km; 11,510 miles)
ARMAMENT:	15 × 152mm (6in)/47-calibre guns (5 × 3); 8 × 127mm (5in)/25-calibre guns (8 × 1)
ARMOUR:	Belt 83–127mm (3.25–5in); deck 51mm (2in); burrets 32–152mm (1.25–6in); conning tower 57–127mm (2.25–5in)
COMPLEMENT:	868

In total, nine *Brooklyn*-class light cruisers were commissioned between 1937 and 1939. The 'light' designation was something of a misnomer – they had twice the displacement of the *Omaha* class they were intended to replace and carried more guns than many heavy cruisers.

Left: On 13 December 1944, the crew of the USS Nashville *(CL-43) clean up the serious damage inflicted by a Japanese kamikaze strike off Negros Island in the Philippines. A total of 133 sailors were killed and 190 wounded in the attack.*

WAR SERVICE

While many US Navy destroyers spent at least some of the war in the Pacific, USS *Brooklyn* spent its war service in more westerly waters. In early 1942 it was first deployed to the Caribbean, but from April 1942 it performed Atlantic convoy escort duties. From November 1942 to 1944, however, *Brooklyn* served principally in the Martime Trade Organization, often providing fire support for US operations around North Africa and Italy.

USS *SAVANNAH* (CL-42)

The USS *Savannah* was commissioned on 23 September 1937. It survived the war, serving in the Atlantic, Mediterranean and Caribbean, although in September 1943 it suffered serious damage from a radio-controlled Fritz X glide-bomb off Salerno.

Zuiho (1936)

SPECIFICATIONS

DIMENSIONS:	Length: 204.8m (671ft 11in); Beam: 18.2m (59ft 9in); Draught: 6.68m (21ft 11in)
DISPLACEMENT:	11,262 tonnes (12,414 tons) standard; 13,730 tonnes (15,134 tons) full load
PROPULSION:	4 Kampon boilers; 2-shaft geared turbines; 38,800kW (52,000shp)
SPEED:	28 knots (52km/h; 32mph)
RANGE:	7800nmi (14,448km; 8977 miles) at 18 knots
ARMAMENT:	8 x 127mm (5in) Type 89 dual-purpose guns; 8 x 25mm (1in) Type 96 AA guns (in twin mounts)
AIRCRAFT:	30
COMPLEMENT:	785

Zuiho **had a long combat career, starting the war with the East Indies invasions of 1942. It went on to see action in the Guadalcanal and Marianas campaigns, during December 1942 and June 1943, but was eventually sunk during the Leyte Gulf battles in October 1944.**

Left: Photographed here on 20 December 1941, Shoho *was a* Zuiho*-class aircraft carrier.* Shoho *has the unenviable distinction of being the first Japanese carrier to be sunk in World World II, at the Battle of the Coral Sea on 7 May 1942.*

FLIGHT DECK
The carrier's flight deck was 180m (590ft 6in) long and had a maximum width of 23m (75ft 6in).

CONVERSION
In the conversion from depot ship, two lifts were added to the flight deck, no island was built and the smokestack curved down to one side, in the standard Japanese fashion.

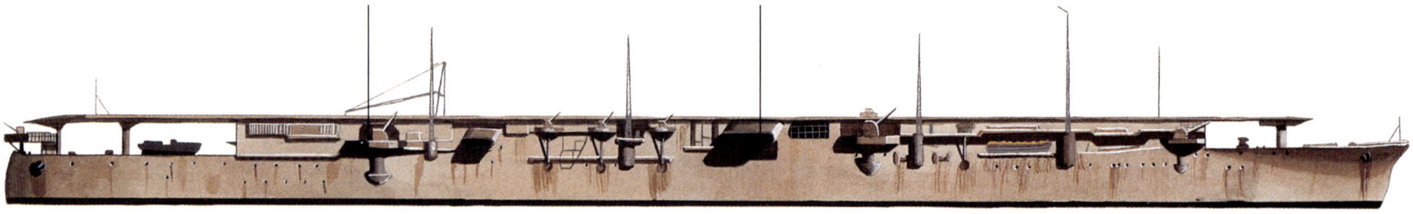

AIRCRAFT
In total, the *Zuiho* could carry 30 combat aircraft. They were stored in a single hangar that measured 120m (406ft 10in) long and 18m (59ft) wide.

DEPOT SHIP
As part of a shadow shipbuilding programme to get around Washington Treaty restrictions, Japan built two submarine depot ships, named *Takasaki* and *Tsurugisaki*, which could be converted to fleet oilers or small aircraft carriers in time of war. In the event, *Tsurugisaki* (the first to be laid down) was built as a depot ship, but *Takasagi* was modified to carrier status while still in the yard. At that point it was renamed *Zuiho*.

BATTLE OF LEYTE GULF
Zuiho maintains speed after several hits by planes from Task Force 38 during the battle off Cape Engano, 25 October 1944. Note the flight deck camouflage, buckled flight deck amidships and mast deployed horizontally, with a Japanese Naval ensign at its tip.

De Ruyter (1936)

TYPE • *Light cruiser* COUNTRY • *Netherlands*

Much of Holland's naval strength was intended to protect its interests in the Dutch East Indies, which was rich in raw materials. *De Ruyter* was one of three cruisers intended for deployment to the East Indies.

SPECIFICATIONS

DIMENSIONS:	Length: 170.9m (560ft 8in); Beam: 15.7m (51ft 6in); Draught: 5.1m (16ft 9in)
DISPLACEMENT:	6650 tonnes (7329 tons) standard
PROPULSION:	2 shafts; Parsons geared turbines; 6 Yarrow boilers; 49,216kW (66,000shp)
SPEED:	32 knots (59.3km/h; 36.8mph)
RANGE:	6800nmi (12,594km; 7819 miles) at 12 knots
ARMAMENT:	7 × 150mm (5.9in) guns; 10 × 40mm (1.6in) AA guns; 8 × Browning machine guns
ARMOUR:	Belt 50mm (2in); bulkheads 33mm (1.3in); turret faces 100mm (3.9in); deck 33mm (1.2in)
COMPLEMENT:	435

Left: De Ruyter's *twin Mk 9 150mm (5.9in) gun turrets each weighed 63.5 tonnes (70 tons), while the single Mk 10 mounting weighed 23 tonnes (25 tons). The latter was equipped to fire starshell.*

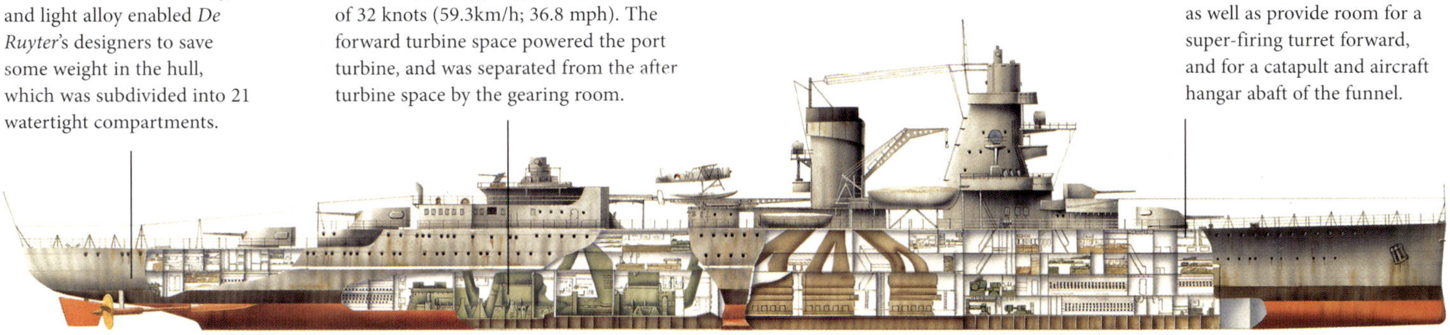

HULL
The use of electric welding and light alloy enabled *De Ruyter's* designers to save some weight in the hull, which was subdivided into 21 watertight compartments.

MACHINERY
De Ruyter was powered by two Parsons geared turbines, producing 49,216kW (66,000shp) and giving it a top speed of 32 knots (59.3km/h; 36.8 mph). The forward turbine space powered the port turbine, and was separated from the after turbine space by the gearing room.

LAYOUT
The final version was lengthened to improve the hull lines and the maximum speed, as well as provide room for a super-firing turret forward, and for a catapult and aircraft hangar abaft of the funnel.

EASTERN WATERS
The Royal Netherlands Navy maintained three cruisers on its Far East station based at Batavia (Java) and *De Ruyter* was planned as an up-to-date addition to this force. Ordered in 1932, laid down at the Wilton-Fijenoord yard, Rotterdam, on 16 September 1933, launched on 11 May 1935 and commissioned on 3 October 1936, its final specification diverged from the original plan which, due to cost limitations, had been strongly criticized as too weak for the ship's purpose.

DUTCH FLAGSHIP
Deployed to Indonesia in 1937, *De Ruyter* was flagship of the Dutch East India Squadron at Surabaja and flagship of the combined American-British-Dutch-Australian (ABDA) squadron opposing the Japanese advance on Java. The ship was torpedoed and sunk at the Battle of the Java Sea on 28 February 1942.

Duca degli Abruzzi class (1936)

TYPE • *Light cruiser* **COUNTRY** • *Italy*

The *Duca degli Abruzzi* class of light cruisers were actually one of five sub-classes of the Italian *Condottieri* class. Of the five, the *Duca degli Abruzzi* was the last to be developed, and consequently had the heaviest armour and armament.

SPECIFICATIONS

DIMENSIONS:	Length: 187m (614ft); Beam: 18.9m (62ft); Draught: 6.9m (23ft)
DISPLACEMENT:	11,350 tonnes (12,515 tons) standard
PROPULSION:	6 boilers; 2 shafts; 75,000kW (100,000hp)
SPEED:	34 knots (63km/h; 39mph)
RANGE:	4125nmi (7640km; 4747mi)
ARMAMENT:	10 × 152mm (6in) guns; 8 × 100mm (4in) guns; 8 × 37mm (1.5in) guns; 12 × 20mm (1in) AA guns; 6 × 533mm (21in) torpedo tubes; 2 × anti-submarine mortars
ARMOUR:	Outer belt 30mm (1.2in); inner belt 100mm (3.9in); main deck 40mm (1.6in); turrets 135mm (5.3in);
COMPLEMENT:	640

Above: The Luigi di Savoia Duca degli Abruzzi *photographed at Naples prior to World War II.*

FIREPOWER
Unlike the preceding *Condottieri*-class vessels, the two *Duca degli Abruzzi* cruisers received triple-gun turrets in the 'A' and 'Y' positions, rather than twin guns.

CAMOUFLAGE
Here we see the lead ship of the class as it appeared in 1939, with diagonal camouflage applied to the superstructure.

AIRCRAFT
Mounted amidships was an IMAM Ro.43 float plane. This aircraft was used for observation duties, providing over-the-horizon reconnaissance.

POSTWAR SERVICE

The lead ship of the class, *Luigi di Savoia Duca degli Abruzzi*, was commissioned in December 1937. It fought in several major engagements in 1940 and 1941, but was later interned by the Allies following the Italian armistice in 1943. After the war, however, the vessel served in the Italian Navy until 1961, its armament and search radar being upgraded to bring it into the modern age.

SHIPS IN CLASS

There were two ships in the *Duca degli Abruzzi* class, the *Luigi di Savoia Duca degli Abruzzi* (pictured here), and the lead ship, the *Giuseppe Garibaldi*.

HMS *Penelope* (1936)

TYPE • *Light cruiser* **COUNTRY** • *United Kingdom*

SPECIFICATIONS

Dimensions:	Length: 154.23m (506ft); Beam: 15.54m (51ft); Draught 5.03m (16ft 6in)
Displacement:	5270 tonnes (5806 tons) standard
Propulsion:	4 Admiralty 3-drum boilers; 4-shaft Parsons geared turbines; 47,735kW (64,000hp)
Speed:	32.3 knots (59.8km/h; 37.2mph)
Range:	5301nmi (9817km; 6100mi) at 15 knots
Armament:	6 × 152mm (6in) main guns; 8 × 102mm (4in) quick-firing guns; 6 × 533mm (21in) torpedo tubes; 1 aircraft
Armour:	Box protection to ammunition storage 25–76mm (1–3in); belt 57mm (2.25in); turrets 25mm (1in)
Complement:	500

With the *Arethusa* class, to which HMS *Penelope* belonged, the Royal Navy tried to build the smallest possible useful cruiser. It was similar in design to the cruisers of the *Perth* class, but had only three turrets instead of four.

Left: Penelope *continued to serve with the Mediterranean and Home Fleets until 1944, when it was sunk on 18 February by the* U-410. Penelope *was hit by first one and then a second torpedo and sank straight after, with the loss of 417 crew.*

Gun magazines
One unfortunate element of the design was that the magazines for the secondary 102mm (4in) guns were not near the guns themselves, which proved a serious problem in combat.

Catapults
The four *Arethusas* were initially fitted with catapults and aircraft, but these were removed during major overhauls of the class between 1940 and 1941.

Construction
HMS *Penelope* was the third of its class to be built. It was laid down on 30 May 1934, launched on 15 October 1935 and completed on 13 November 1936.

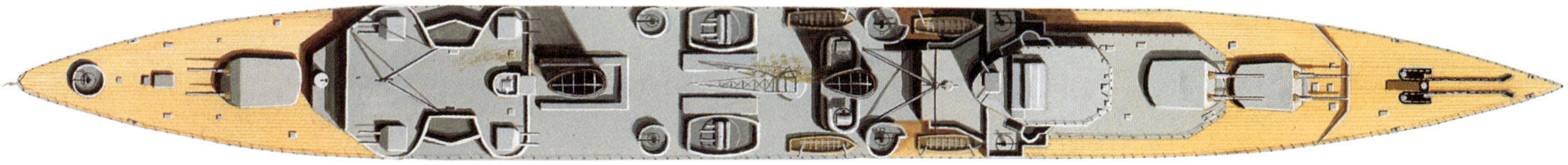

MALTA CONVOY

HMS *Penelope* distinguished itself as one of the escorts involved in the infamous Force H convoy to Malta in 1942. With the island about to succumb to Axis air attacks, a last-ditch effort was made to force supplies through the German and Italian blockade. After intense air and sea battles, only one merchant vessel made it to Malta, the US tanker *Ohio*, which was by then barely afloat. Its arrival was enough to keep Malta fighting.

HMS *AURORA*
Arethusa and three companion ships were an attempt to create the smallest possible cruiser with reasonable armament and performance. Two of the group were lost in World War II. *Aurora*, seen here, was sold to China in 1948.

Dunkerque (1937)

TYPE • *Battleship* **COUNTRY** • *France*

SPECIFICATIONS

Dimensions:	Length: 214.5m (704ft); Beam: 31.1m (102ft); Draught: 8.7m (29ft)
Displacement:	26,925 tonnes (29,677 tons) standard; 36,070 tonnes (39,756 tons) full load
Propulsion:	6 Indret boilers, 4 sets of Parsons geared turbines, 4 screws, 80,200kW (107,500shp)
Speed:	31 knots (57.5km/h; 35.7mph)
Range:	7350nmi (14,540km; 9038 miles)
Armament:	8 x 330mm (13in) guns; 16 x 130mm (5.1in) guns; 8 x 37mm (1.5in) guns; 32 x 13.2mm (0.5in) AA guns
Armour:	Belt 225mm (8.9in); main deck 115mm (4.5in); barbettes 310mm (12.2in); turrets 345–310mm (13.6–12.2in); conning tower 270mm (10.6in)
Complement:	1432

Laid down at Brest on 24 December 1932, the French battleship *Dunkerque* was not liable to the same Treaty regulations as German ships. Its larger guns and greater speed were a response to the powerful *Deutschland* class.

Left: Dunkerque*'s superstructure followed the form established with HMS* Rodney*, a tower-cum-mast rising behind the armoured conning tower, which was topped by the navigation bridge.*

Fire control
Fire control and optics, placed in three stations on the tower, were largely French-made, from OPL (Optique et Précision de Levallois-Perret).

Catapult
A 22m (72ft) compressed-air catapult and a crane were fitted on the after deck, aft of a two-level hangar, with lift, and three Loire 130 seaplanes were carried.

Side armour
The side armour covered about 60 per cent of the hull, with bow and stern unprotected. A 60mm (2.4in) layer of teak wood backed the outer plating.

FLAGSHIP
In World War II, *Dunkerque* was the proud flagship of the French Atlantic fleet at Brest before it was transferred to the Mediterranean in 1940. It was damaged in the controversial British naval attack on Mers-el-Kebir on 3 and 6 July that year. Moored facing the land, it could not return fire. Transferred to Toulon, it was scuttled by its crew in November 1944.

FORWARD GUNS
The main turrets of *Dunkerque*, built by St-Chamond, each weighed 1497 tonnes (1650 US tons). Elevating to a maximum 35 degrees, the guns had a muzzle velocity of 870m/s (2854ft/s) and a range of 41,500m (136,152ft).

Adua class (1937)

TYPE • *Submarine* COUNTRY • *Italy*

SPECIFICATIONS (*Dagabur*)

DIMENSIONS:	Length: 60.28m (197ft 9in); Beam: 6.45m (21ft 2in); Draught: 4.64m (15ft 3in)
DISPLACEMENT:	697.2 tonnes (769 tons) surfaced; 856.4 tonnes (944 tons) submerged
PROPULSION:	2 diesels of 500kW (700hp); 2 300kW (400hp) electric motors; two screws
SPEED:	14 knots (26km/h; 16mph) surfaced; 7.5 knots (13.9km/h; 8.6mph) submerged
RANGE:	3180nmi (5889km; 3659 miles) at 10.5 knots surfaced; 74nmi (137km; 85 miles) at 4 knots submerged
ARMAMENT:	6 × 533mm (21in) torpedo tubes; 12 torpedoes; 1 × 100mm (3.9in) deck gun; 2 × 13.2mm (0.5in) machine guns
COMPLEMENT:	44

The 17 *Adua* boats were constructed and commissioned rapidly, the class leader laid down in January 1936 and delivered in November of the same year. All were in service by October 1938, produced by the three yards of Monfalcone (CRDA), La Spezia (OTO) and Taranto (Tosi).

Left: Dessiè *was another of the* Adua *class. It spent its combat career in the Mediterranean and was depth-charged and sunk in November 1942.*

TOWER
The towers on most of the class were altered in 1942–43 to enable retraction of the periscopes and to provide extra AA mounts.

CABLES
Like most other classes, the majority of *Adua* class had dismountable T-form deck masts fore and aft to support radio aerial and net-avoidance cables; most sported a serrated, angled bow net-cutter.

LAYOUT
Adua submarines were short-range, single-hulled, with midships ballast tanks inside the hull, and side blisters. Minor differences in appearance reflected the different makers.

ITALIAN TORPEDOES

Italy entered the war with a stock of both 533mm (21in) and 450mm (17.7in) torpedoes. All used the wet-heater type of engine, and there was little difference in performance between the types: Italian torpedoes were recognized as superior to those of other European navies. The Type W 4x7.2 Veloce weighed 1700kg (3784lb) long, with a 270kg (595lb) explosive charge, and could run for 4000m (13,123ft) at 50 knots, or 12,000m (39,370ft) at 30 knots. SI had a similar model but with a lower speed range of 8000m (26,247ft).

ADUA DOCKED
Italian submarines tended to be underpowered compared to German or British boats and were therefore slower; otherwise they were highly effective attack boats. *Adua* was lost during an action against British ships in September 1941.

Salmon class (1937)

TYPE • *Submarine* **COUNTRY** • *United States*

SPECIFICATIONS

Dimensions:	Length: 93.9m (308ft); Beam: 7.9m (25ft 11in); Draught: 4.78m (15ft 8in)
Displacement:	1458 tonnes (1607 tons) surfaced; 2233 tonnes (2462 tons) submerged
Propulsion:	4 HOR nine-cylinder diesels of 1145kW (1535hp); 4 496kW (665hp) geared electric motors; 2 screws
Speed:	21 knots (39km/h; 24mph) surfaced; 9 knots (17km/h; 10.3mph) submerged
Range:	11,000nmi (20,372km; 12,659 miles) at 10 knots surfaced; 96nmi (178km; 110 miles) at 2 knots submerged
Armament:	8 x 533mm (21in) torpedo tubes; 24 torpedoes; 1 x 76mm (3in) 50cal deck gun; 4 machine guns
Complement:	59

The vastness of the USA's oceanic coastlines and the distance of potential aggressors led the USN to place strong emphasis on developing long-range submarines capable of keeping pace with a battle fleet on the surface. This view of the submarine's role evolved over the course of the war.

Right: Salmon*-class submarines under construction. USS* Salmon *(SS-182) was the lead boat of 16, built at Electric Boat, Groton, and the Portsmouth and Mare Island navy yards. They were partially double-hulled, and with the preceding P-class, were the firrst all-welded US submarines.*

Powerplant
The Electric Boat trio were initially powered by four HOR (Hooven-Ovens-Rentschler) nine-cylinder double-action diesels but these proved unsatisfactory in service and after Pearl Harbor were replaced by GM Winton six-cylinder engines.

Armament
Salmon boats carried eight internal torpedo tubes, four bow and four stern, with a warload of 24 torpedoes. A 76mm (3in) 50cal aft-facing deck gun was first installed, later supplemented by four 20mm (0.79in) machine guns.

Operation
For surface operation, the engines were clutched to the reduction gears and drove the propellers directly, with the generator engines providing additional voltage to the motors. For submerged operation, the diesels were declutched from the reduction gears and the motors drove the shafts with electricity supplied by the batteries.

TESTING
The class was tested to a depth of 76m (250ft). Within the basic configuration, differences between the EB and Navy builds separated *Salmon*, *Seal* (SS-183) and *Skipjack* (SS-184) from *Snapper*, *Stingray* and *Sturgeon* (SS-185-187), including tower design (though the 'sail' type of tower was reduced in height on all over the course of the war). On all vessels, the stern was designed to hold two 533mm (21in) torpedo tubes. Four spare torpedoes were initially held in wells in the casing, but this was discontinued in wartime.

USS *STINGRAY*
This forward view of the USS *Stingray* gives a direct perspective on the class's main torpedo armament. The submarine survived the war, having made 16 war patrols in the Pacific.

Type IX (1937)

Building on the lessons of Type IA, this was the Kriegsmarine's premier attack submarine. The Type IX was a great advance on Type IA – double-hulled with a streamlined outer shell and an internal pressure hull formed of 18mm (0.7in) steel.

SPECIFICATIONS

DIMENSIONS:	Length: 76.5m (251ft); Beam: 6.5m (21ft 4in); Draught: 4.7m (15ft 5in)
DISPLACEMENT:	1048.5 tonnes (1157 tons) surfaced; 1587.5 tonnes (1750 tons) submerged
PROPULSION:	2 1640.5kW (2200hp) 6-cylinder 4-stroke MAN M9V 40/46 diesels; 2 electric motors outputting 432.5kW (580hp); two screws
SPEED:	18.2 knots (33.7km/h; 21mph) surfaced; 7.7 knots (14.3km/h; 8.9mph) submerged
RANGE:	23,700nmi (43,892km; 27,273 miles) at 12 knots surfaced
ARMAMENT:	6 × 533mm (21in) torpedo tubes (4 bow, 2 stern); 22 torpedoes; 1 × 37mm (1.5in) gun; 2 × 20mm (0.79in) AA guns
COMPLEMENT:	48

Below: One of the most successful U-boats of the war, U-123 *sank 42 merchant ships totalling 219,924 GRT, one auxiliary warship of 3209 GRT and the British submarine* P615 *while serving under its three Knight's Cross-winning commanders.*

SNORKEL PIPE
The snorkel pipe angled down into a horizontal recess on the forward starboard side of the tower. Fitting of a snorkel required removal of the forward deck gun. Previously this might have been a 105mm (4.1in) SK C/32 or a 37mm (1.5in) single flak gun.

***U-156*, TYPE IXD/2**
Type IXD/2 was 11.1m (36.4ft) longer than Type IXA, with a surface displacement of 1642 tonnes (1809 tons) compared to 1048.5 tonnes (1155 tons).

EXTRA TORPEDOES
Up to 24 torpedoes were carried, including four stored in upper deck containers, below the aft flak platform.

TYPE OVERVIEW
There were eight versions of Type IX, but the first four, IXA to IXC/40, had similar vital statistics. All were powered by 1640.5kW (2200hp) MAN diesel engines, two SSW 373kW (500hp) electric motors and two 62-cell battery sets. Bunker capacity increased slightly in each revision with a consequent increase in surface range. Wartime modifications and repairs resulted in a variety of tower configurations. The foredeck gun was also removed in many cases.

SINKING SUB
U-185, a Type IXC/40, sinks in a position southwest of the Azores on 24 August 1943 after having been hit by a air strike from USN Avengers and Wildcast aircraft, flying from the escort carrier USS *Core* (CVE-13).

U-class (1937)

TYPE • *Submarine* **COUNTRY** • *United Kingdom*

SPECIFICATIONS (HMS *Undine*, Group 1)

Dimensions:	Length: 59.8m (196ft 3in); Beam: 4.8m (15ft 9in); Draught: 4.8m (15ft 9in)
Displacement:	640 tonnes (706 tons) surfaced; 741.7 tonnes (818 tons) submerged
Propulsion:	2 298kW (400hp) diesels; two 307kW (412hp) electric motors; two screws
Speed:	11.25 knots (20.8km/h; 12.9mph) surfaced; 9 knots (16.7km/h; 10.3mph) submerged
Range:	4050nmi (7500km; 4661 miles) at 10 knots surfaced; 23nmi (42.6km; 26.5 miles) at 8 knots submerged
Armament:	4 × 533mm (21in) bow torpedo tubes; 1 × 76mm (3in) deck gun; 3 × 7.7mm (0.303in) machine guns
Complement:	27

Planned in 1936, originally as unarmed training boats, the U-class was the Royal Navy's first diesel-electric submarine. In total, 49 U-class boats were commissioned into the RN. A further 21 served in other navies.

Left: RN submariners loading a torpedo onto HMS Ursula at the Malta naval base. The Mark VIII was by far the most frequently used torpedo type by the British in World War II.

Diesel-electric drive
These were the first RN submarines to have diesel-electric drive, dispensing with clutches and vibration dampers. Main power came from two Davey-Paxman RXS with welded steel frames.

Armament
The first seven of the class had two additional bow tubes, mounted externally. Total torpedo load was 10. A single deck gun was fitted close to the tower, as the design omitted an ammunition hatch: a QF 12-pdr (76mm) in the earliest boats, later a 76mm (3in) gun.

HMS *Undine*
The hull was divided into six compartments, separated by watertight bulkheads. There were four hatches to the deck, with a 'twill trunk' hatch in the torpedo and engine rooms for escape.

CLASS GROUPS
The first three U-class submarines, *Undine*, *Unity*, and *Ursula*, were modified during construction, with six internal and two external bow tubes. They were completed in 1938, with another 12 in 1940–41. A second group of 28 followed between 1940 and 1943, of similar displacement but with 1.78m (5.75ft) longer sterns, and with four internal tubes only. These also had a more streamlined bow profile.

HMS *UNITED* (P44)
HMS *United* (P44) was a Royal Navy U-class submarine built by Vickers-Armstrong at Barrow-in-Furness. Most U-class submarines were attached to the 10th flotilla at Malta, operating across the Mediterranean.

Hiryu (1937)

TYPE • *Aircraft carrier* **COUNTRY** • *Japan*

SPECIFICATIONS

Dimensions:	Length: 227.4m (746ft 1in); Beam: 22.3m (73ft 2in); Draught: 7.9m (25ft 11in)
Displacement:	17,577 tonnes (19,375 tons) standard; 22,403 tonnes (24,696 tons) full load
Propulsion:	8 Kampon Ro boilers; 4 geared turbines; 4 shafts; 114,000kW (153,000shp)
Speed:	34.3 knots (63.5km/h; 39.5mph)
Range:	10,330nmi (19,130km; 11,890 miles) at 18 knots
Armament:	12 × 127mm (5in) guns; 31 × 25mm (1in) AA guns
Aircraft:	73
Complement:	1100

The Imperial Japanese Navy was building up its carrier force in the late 1930s, with *Soryu* and *Hiryu* as fast carriers added to the fleet in 1937 and 1939. Both were to become victims of the Battle of Midway.

Left: Hiryu, *meaning Flying Dragon, was of slightly larger dimensions than* Soryu, *and was built at Yokosuka Naval Dockyard. It was laid down on 8 July 1936, launched on 16 November 1937 and commissioned on 5 July 1939.*

Island
Both *Soryu* and *Hiryu* had similar tower-type islands, although differently located, and *Hiryu*'s had an additional deck level. Both had masts rising from flight-deck level, although again of different design.

Rising sun
A large *Hinomaru* (rising sun) was painted on *Hiryu*'s flight deck, ironically providing a 'bull's-eye' target for US dive-bombers.

Hangars
Soryu and *Hiryu* retained the two-level enclosed hangar design of earlier carriers like *Kaga*. The height of the lower hangar at 4.3m (14ft) restricted the type of plane that could be carried.

COMPANION SHIPS

Hiryu and *Soryu* were companion ships, often referred to as *Soryu* class, although *Hiryu* was sufficiently different in details to the earlier ship as to form a separate, sub-class. *Soryu* (meaning 'Blue Dragon') was laid down at the Kaigun Kosho Yard at Kure on 20 November 1934, launched on 23 December 1935 and commissioned on 29 January 1937. Although designed from the start as a carrier, the ship had a conventional bow with the forward part of the flight deck extended over it on stanchions; the stern was treated in a similar way.

BATTLE OF MIDWAY
IJNS *Hiryu* makes a hard port turn to avoid US aerial bombs at the Battle of Midway shortly after 0800 hrs on 4 June 1942. US dive-bombers eventually damaged the ship so badly it was scuttled the next day.

Soryu (1937)

TYPE • *Aircraft carrier* COUNTRY • *Japan*

SPECIFICATIONS	
DIMENSIONS:	Length: 227.5m (746ft 5in); Beam: 21.3m (69ft 11in); Draught: 7.6m (24ft 11in)
DISPLACEMENT:	16,200 tonnes (17,857 tons) standard; 19,100 tonnes (21,044 tons) full load
PROPULSION:	8 water-tube boilers; 4 shafts with 4 geared steam turbine sets; 113,000kW (152,000shp)
SPEED:	34 knots (63km/h; 39mph)
RANGE:	7750nmi (14,350km; 8920 miles) at 18 knots
ARMAMENT:	6 × twin 127mm (5in) dual-purpose guns; 14 × twin 25mm (1in) AA guns
AIRCRAFT:	63 (+9 reserve)
COMPLEMENT:	1100

The Japanese carrier *Soryu* was commissioned into service on 29 December 1937. It served in operations against China and in the Pearl Harbor and early Pacific campaigns. It met its end, along with three other fleet carriers, at Midway in June 1942.

Right: Soryu *as seen in the 1930s. Prior to its loss at Midway,* Soryu *had a distinguished combat career, sinking several Allied warships and downing numerous US, British and Commonwealth aircraft.*

SPEED
Soryu's four geared steam turbine sets, which delivered a total of 113,000kW (152,000 shp), enabled the ship to reach a speed of 34 knots (63km/h; 39mph).

ARMOUR
The *Soryu* was only lightly armoured, with a waterline belt of 41mm (1.6 in) of Ducol steel over the machinery spaces and the magazines.

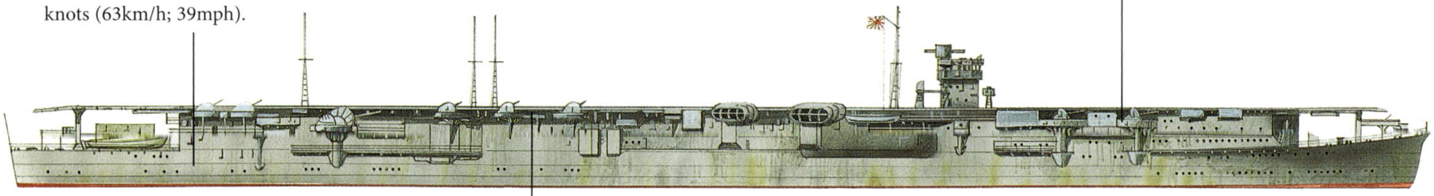

AIRCRAFT
The *Soryu*'s principal air combat component consisted of 21 Mitsubishi A6M Zero, 18 Aichi D3A and 18 Nakajima B5N aircraft. In 1942, the Zero was one of the most effective fighters in the theatre.

DISASTER AT MIDWAY

Soryu was one of four Japanese aircraft carriers deployed in a mighty Japanese fleet near Midway Island in June 1942. On 4 June, the carrier's aircraft conducted a raid on Midway Island and fought off several incoming US carrier aircraft attacks. But at 10.25am the ship was caught by a surprise attack from 14 Dauntless dive-bombers, which achieved three devastating bomb hits. Fire raged uncontrollably, and the captain eventually ordered the ship scuttled by a nearby submarine. The ship lost 711 of its crew.

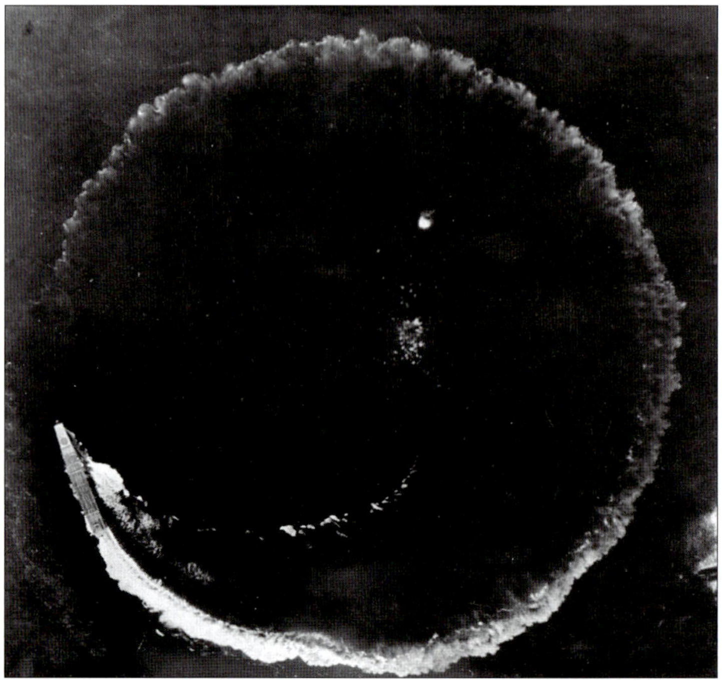

EVASIVE ACTION
The *Soryu* leaves circular wake as it evades sticks of bombs dropped by US B-17 Flying Fortress bombers at Midway on 4 June 1942. This high-altitude bombing was ineffective; what eventually caught the ship was precision US dive-bombing.

USS *Yorktown* (1937)

TYPE • *Aircraft carrier* **COUNTRY** • *United States*

SPECIFICATIONS

DIMENSIONS:	Length: 251.4m (824ft 9in); Beam: 33.4m (109ft 6in); Draught: 7.9m (25ft 11in)
DISPLACEMENT:	20,100 tonnes (22,181 tons) standard; 25,900 tonnes (28,543 tons) full load
PROPULSION:	9 Babcock & Wilcox boilers; 4 Parsons turbines; 89,000kW (120,000shp)
SPEED:	32.5 knots (60.2km/h; 37.4mph)
RANGE:	12,500nmi (23,200km; 14,400 miles) at 15 knots
ARMAMENT:	8 × 127mm (5in) 38-calibre guns; 4 × quadruple 28mm (1.1in) 75-calibre AA guns; 24 × 12.7mm (0.5in) MGs
AIRCRAFT:	90
COMPLEMENT:	2217

The first of the US Navy's large purpose-built carriers, the USS *Yorktown* (CV-5) saw fierce action in the Pacific War. Fatally crippled in the Battle of Midway, it was abandoned and subsequently torpedoed and sunk.

Left: A later Essex-class carrier, USS Yorktown *(CV-10) – seen here during its shakedown cruise – took the name of its predecessor, being commissioned in 1943 and serving through the remainder of World War II and well into the Cold War.*

FLIGHT DECK
The flight deck was formed of Douglas fir planks, 7.6cm (3in) thick and 15.2cm (6in) wide, laid over steel sheet 2.5mm (0.1in) thick. Metal aircraft tie-down strips were laid every 1.2m (4ft).

PROFILE
Yorktown in port profile (above) and in aerial plan (below), prior to 1942. As the number of carriers grew, the ship's identity was clearly designated to guide incoming pilots.

CRASH BARRIERS
Crash barriers were mounted forward of the forward elevator and at the forward end of the island.

DESIGN
Experience with USS *Ranger* (1931) convinced the Navy Board that carriers should be bigger, more powerful, faster and fitted with anti-torpedo protection through internal bulkheads and watertight compartments as well as hull armour. Island structures for better command and control were also considered essential. *Yorktown* (CV-5) was the first ship to benefit from these conclusions, although it had to be designed within the limits fixed by the Naval Treaty of 1922.

MIDWAY DAMAGE
USS *Yorktown* burns after a Japanese air attack at the Battle of Midway on 4 June 1942. The crippled carrier was subsequently torpedoed by a Japanese submarine on 6 June and it sank the next day.

Type 34 (1937)

TYPE • *Destroyer* **COUNTRY** • *Germany*

SPECIFICATIONS

Dimensions:	Length: 119m (390ft 5in); Beam: 11.3m (37ft 1in); Draught: 4.23m (13ft 11in)
Displacement:	3206 tonnes (3535 tons) full load
Propulsion:	2 shafts; 2 steam turbines; 6 water boilers; 52,199kW (70,000hp)
Speed:	38 knots (70km/h; 44mph)
Range:	1530nmi (2830km; 1760 miles) at 19 knots
Armament:	5 x 127mm (5in) guns; 2 x twin 37mm (1.5in) AA guns; 6 x 20mm (0.79in) AA guns; 2 x quadruple torpedo mounts; 60 mines
Armour:	None
Complement:	315

These four named boats (*Z1–Z4*) were laid down in 1934–35, launched in 1935 and commissioned in 1937. The design incorporated new 51,453kW (69,000hp) Wagner steam-powered turbines, which enabled the vessel to reach a maximum speed of 38 knots (70km/h; 44mph).

Right: Z12 Erich Giese *was named after a German naval officer who was killed in action during World War I.*

Leberecht Maas
In this starboard profile of the Type 1934 destroyer, *Z1 Leberecht Maas*, the ship's torpedo tubes can be seen amidships, as well as one of its lifeboats, located adjacent to the forward funnel.

Main guns
The destroyers' main armament comprised five single 12.7cm (5in) SK C/34 guns; two of these were located forward, and three were set aft.

Secondary weapons
The ships' secondary weaponry included four 3.7cm (1.46in) and six 2cm (0.79in) Flak guns, eight torpedo tubes and 60 sea mines.

EARLY LOSSES

The four ships spent most of 1938–39 undertaking training exercises. Then, during 19 February 1940, the First Destroyer Flotilla undertook a sortie into the North Sea. The flotilla comprised three Type 34s (*Z1*, *Z3* and *Z4*) and three other destroyers (*Z6*, *Z13* and *Z16*). *Z1* was attacked by a Heinkel He 111 bomber, which mistook it for a British warship. Three bomb hits caused *Z1* to break in two. When its companion ship *Z3* went to rescue the many survivors in the water, it exploded and quickly sank, probably after hitting a British mine.

Z6 Theodor Riedel
A Type 1934A-class destroyer, *Z6 Theodor Riedel* was completed in July 1937. After the war it was taken into French service in early 1946 and renamed *Kléber.*

HMS *Cossack* (1937)

TYPE • *Destroyer* **COUNTRY** • *United Kingdom*

SPECIFICATIONS

DIMENSIONS:	Length: 111.5m (365ft 5in); Beam: 11.13m (36ft 6in); Draught: 4m (13ft 1in)
DISPLACEMENT:	1696 tonnes (1870 tons) standard
PROPULSION:	2-shaft Parsons geared turbines; 33,131kW (44,000shp)
SPEED:	36 knots (66.7km/h; 41.4mph)
RANGE:	5700nmi (10,600 km; 6600 miles) at 15 knots
ARMAMENT:	4 × 2 × 120mm (4.7in) guns; 1 × quadruple 40mm (2-pdr) AA gun; 2 × quadruple 12.7mm (0.5in) machine guns; 1 × quadruple 533mm (21in) torpedo tubes; 2 × depth charge throwers; 1 × depth charge rail
COMPLEMENT:	190

Among the finest destroyers ever built for the Royal Navy, the 16 *Tribal*-class vessels, of which *Cossack* (F03) was one, were produced in response to those being built by potential enemies, rather than to fill any clearly defined role within the fleet.

Left: HMCS Haida *(1942) was a* Tribal*-class destroyer that saw service in the Royal Canadian Navy (RCN), gaining the honour of sinking more weight of enemy surface tonnage than any other ship in the RCN.*

FUNNELS
A distinctive feature of the *Tribal*-class destroyers was their two raked funnels, the rearmost of which was later reduced in size to provide a better arc of fire for the ship's AA guns.

STEM
The *Tribal*-class destroyers were quite different from previous designs, featuring a sharply raked stem. This added 3m (10ft) to the length of the forecastle deck.

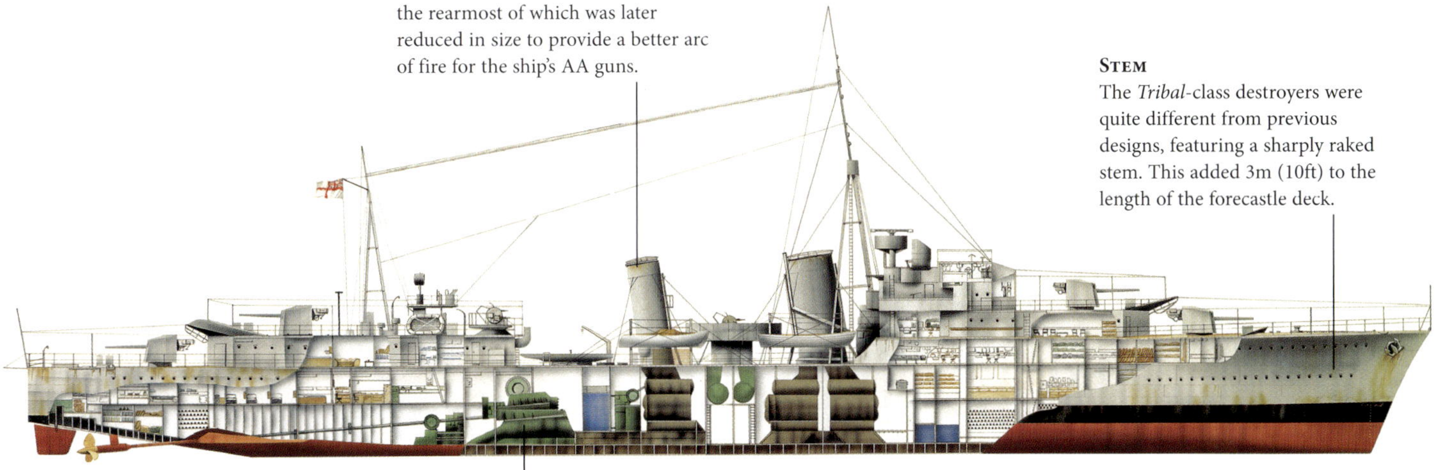

MACHINERY
Three Admiralty three-drum boilers generated steam for two Parsons geared turbines, with a maximum power output of 33,131kW (44,430shp).

WAR SERVICE

Cossack made a dramatic rescue of British prisoners from the German vessel Altmark, in Norwegian waters, in February 1940. In May 1941, it participated in the pursuit and destruction of the *Bismarck*. *Cossack* and four other destroyers made several torpedo attacks in the evening and into the next morning. No hits were scored, but they made it easier for the battleships to attack it the next morning. *Cossack* also saw action with convoys to Malta, before being sunk in the North Atlantic by *U-563* on 26 October 1941.

UNDERWAY
HMS *Cossack* gets underway on completion. During its operations in 1939–41, *Cossack* was commanded by Captain Philip Vian, who reached high rank in the Royal Navy and eventually became commander-in-chief of the Home Fleet in the years immediately after the war.

HMS *Ark Royal* (1938)

TYPE • *Aircraft carrier* **COUNTRY** • *United Kingdom*

SPECIFICATIONS

Dimensions:	Length: 208m (682ft); Beam: 28.9m (94ft 10in); Draught: 8.73m (27ft 10in)
Displacement:	22,352 tonnes (24,644 tons); 28,163 tonnes (31,042 tons) full load
Propulsion:	6 Admiralty boilers; 3 Parsons geared turbines; 76,807kW (103,000shp)
Speed:	57km/h (31 knots; 36mph)
Range:	14,100km (7600nmi; 8700 miles) at 20 knots
Armament:	16 × 110mm (4.5in) DP guns; 32 × 40mm (1.57in) 2-pdr pom-poms; 12.7mm (0.5in) AA machine guns
Aircraft:	60
Complement:	1580

The third *Ark Royal*, albeit the first aircraft carrier to bear the illustrious name, saw intensive action in the early part of World War II in the Atlantic, the Norwegian Sea and finally in the Mediterranean theatre. It was sunk in the latter by a U-boat on 13–14 November 1941.

Right: A forward profile, with a Fairey Swordfish torpedo bomber poised on the bow.

FLIGHT DECK
Ark Royal's flight deck was 20m (66ft) above the waterline. Its length exceeded that of the keel by 36m (118ft).

ISLAND
Although massive in appearance from the starboard side, the minimal space occupied by *Ark Royal's* island is clearly seen from the plan view: it is hardly wider than the ship's funnel.

AIRCRAFT
Almost from the start, naval aircraft were made with folding wings to fit lifts and maximize use of hangar space. The first naval aircraft with folding wings was the British Short Folder seaplane in 1913.

FLEET AIR ARM

After World War I the plan was for the Navy to provide the carriers while the RAF supplied the aircraft, flying crews and maintenance teams – a highly impractical decision. Despite the growing difficulties of managing naval aviation across two separate services, it was not until 1937 that the Fleet Air Arm was fully reintegrated into the Royal Navy.

FUTURE LESSONS

Ark Royal had been a model for the *Illustrious* class, and the failure of its torpedo protection caused rapid design changes in new carriers along with improved damage-control systems and practice, and the provision of diesel generators to provide electric power in the event of a mechanical failure.

Gneisenau (1938)

TYPE • *Battlecruiser* **COUNTRY** • *Germany*

SPECIFICATIONS

DIMENSIONS:	Length: 229.8m (753ft 11in); Beam: 30m (98ft 5in); Draught: 9.9m (32ft 6in)
DISPLACEMENT:	32,100 tonnes (35,400 tons) standard
PROPULSION:	3 Germania geared steam turbines, 122,040kW (163,660shp)
SPEED:	31 knots (57km/h; 36mph)
RANGE:	6200nmi (11,500km; 7100 miles)
ARMAMENT:	9 × 280mm (11in) guns; 12 × 150mm (5.9in) guns; 14 × 105mm (4.1in) guns; 16 × 37mm (1.5in) guns; 10 (later 16) × 20mm (0.79in) AA guns; 6 × 533mm (21in) torpedo tubes
ARMOUR:	Belt 350mm (13.8in); deck 50–105mm (2–4.1in); Turrets 200–360mm (7.9–14.2in); conning tower: 350mm (13.8in)
COMPLEMENT:	1669

Gneisenau ranked among mighty vessels such as _Bismarck_ and _Tirpitz_ in terms of naval firepower. This great battlecruiser was launched in December 1936 and commissioned in July 1942. However, its strength failed to protect it adequately from air power.

Right: Gneisenau *was laid down at the Deutsche Werke, Kiel, in March 1935, launched on 8 December 1936 (as pictured here) and completed on 21 May 1938.*

POWERPLANT
Diesel propulsion had been originally intended for both ships, but no diesel engines could produce the necessary power output per shaft, and 12 Wagner extra-high-pressure boilers drove the turbines.

OUTLINE
A general resemblance of outline between *Scharnhorst*, *Gneisenau*, *Bismarck* and *Tirpitz* was deliberately planned in order to confuse their identities.

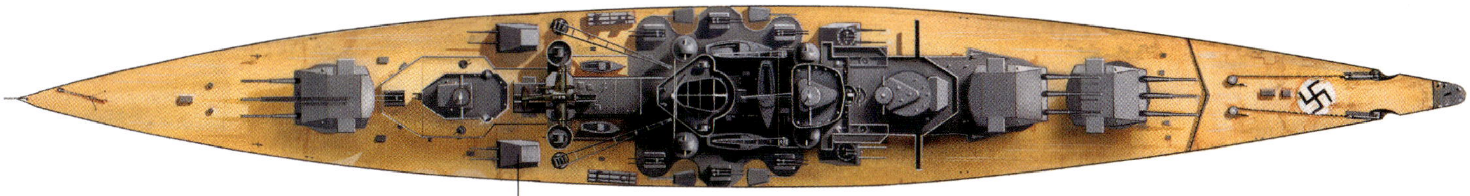

COMMERCE RAIDING

Gneisenau and its companion *Scharnhorst* had their bows lengthened in 1939, improving their seaworthiness and performance. In World War II, they attacked British commerce in the North Atlantic and also participated in the sinking of the British aircraft carrier *Glorious* on 8 June 1940. Rendered unusable by RAF bombs in 1942, *Gneisenau* was broken up between 1947 and 1951.

DECK PLAN
The 150mm (5.9in) secondary guns are clearly visible from this perspective, as are the many smaller 37mm (1.5in) and 20mm (0.79in) AA guns.

HEAVY GUNS
Gneisenau's main armament was a battery of nine 28cm (11.1in) L/54.5 guns, set in three triple gun turrets: two forward and one aft. From stern to bow the guns were known as 'Anton', 'Bruno' and 'Caesar'.

Type C (1938)

TYPE • *Submarine* COUNTRY • *Japan*

As with other navies, Japanese warship design was incremental, with successive classes building on the experience and performance of their predecessors. The Japanese Type C submarines were themselves divided into three classes, under which 12 submarines were constructed.

SPECIFICATIONS

DIMENSIONS:	Length: 109.3m (358ft 6in); Beam: 9.1m (29ft 10in); Draught:7.8m (25ft 7in)
DISPLACEMENT:	2219 tonnes (2184 tons) surfaced; 3618 tonnes (3561 tons) submerged
PROPULSION:	2 diesels of 4623kW (6200bhp); 2 745kW (1000hp) electric motors; 2 screws
SPEED:	23.6 knots (43.7km/h; 27mph) surfaced; 8 knots (14.8km/h; 9.2mph) submerged
RANGE:	14,000nmi (25,928km; 16,111 miles) surfaced; 60nmi (110km; 69 miles) submerged
ARMAMENT:	8 × 533mm (21in) torpedo tubes; 20 torpedoes; 1 × 140mm (5.5in) deck gun; 2 × Type 96 25mm (1in) AA guns
COMPLEMENT:	95

Right: Commissioned in April 1944, the Type C (C-2 sub-class) I-555 *(launched as* I-55*) was lost in July 1944 after being depth-charged by US destroyers, while taking part in the Marianas campaign.*

TYPE C-1 CLASS
The five C-1 class, double-hulled, numbered I-16, 18, 20, 22 and 24, did not carry aircraft: instead, they were fitted with cradles to hold midget submarines (A Type) and carried support equipment.

PERFORMANCE
Powerful Kampon Mk2 Model 10 diesel engines and two electric motors enabled a surface speed of 23.6 knots (43.7km/h; 27mph) and 8 knots (14.8km/h; 9.2mph) submerged.

TORPEDO TUBES
The eight torpedo tubes were all forward, in two torpedo rooms, one above the other. Twenty torpedoes were carried. A 140mm (5.5in)/50cal deck gun and two Type 96 25mm (1in) AA guns were fitted.

JAPANESE IDENTIFIERS

The designations *I*, *Ro*, and *Ha* of submarine types began as generic, denoting ocean-going, medium-range/coastal and small or midget submarines respectively, each including a range of types. Individual classes and types have their own identifiers, either a combination of letters and numbers, such as *K-6*, or simply a number, as in *I-201*. From May 1942, all submarine types were given the prefix number 1. For example, *K-6* became *K-16*.

***I-47*, KURE NAVAL YARD**
I-47, the seventh C-class submarine built for the IJN, mainly operated as a *kaiten* manned suicide attack torpedo carrier in 1944 and 1945. It survived the war but was scuttled by the US Navy in April 1946.

HMS *Eskimo* (1938)

TYPE • *Destroyer* **COUNTRY** • *United Kingdom*

SPECIFICATIONS

Dimensions:	Length: 111.5m (365ft 5in); Beam: 11.13m (36ft 6in); Draught: 4m (13ft 1in)
Displacement:	1696 tonnes (1870 tons) standard
Propulsion:	2-shaft Parsons geared turbines; 33,131kW (44,000shp)
Speed:	36 knots (66.7km/h; 41.4mph)
Range:	5700nmi (10,600 km; 6600 miles) at 15 knots
Armament:	4 x 2 x 120mm (4.7in) guns; 1 x quadruple 40mm (2-pdr) AA gun; 2 x quadruple 12.7mm (0.5in) machine guns; 1 x quadruple 533mm (21in) torpedo tubes; 2 x depth charge throwers; 1 x depth charge rail
Complement:	190

HMS *Eskimo* was the *Tribal*-class companion ship of HMS *Cossack*. Unlike many others of its class, *Eskimo* survived the war, despite seeing high levels of combat. However, its ultimate fate was to be used as a target ship in 1949.

Left: Only four of the original 16 Tribals remained afloat by the end of 1942. HMS Mohawk *(seen here) was torpedoed and sunk by the Italian destroyer* Luca Tarigo *on 16 April 1941.*

Armament
The *Tribals* carried a heavy armament of four twin 120mm (4.7in) guns and one quadruple 533mm (21in) torpedo mounting.

Fire control
The *Tribal* class introduced the Fuze Keeping Clock High Angle Fire Control Computer, which was fitted in all subsequent classes of Royal Navy destroyers during World War II.

Hull
The *Tribal*-class were visually impressive, their pleasingly balanced profile in harmony with the high freeboard hull that was introduced to improve their fighting qualities in poor weather.

WAR DAMAGE

Although *Eskimo* survived the war years, it took some particularly hard knocks along the way. During the Second Battle of Narvik in April 1940, it was hit by a torpedo fired from the German destroyer *Z2 Georg Thiele*. The explosion ripped off its entire bow. Temporary repairs were undertaken by the shipwrights of the fleet repair ship *Vindictive* at Skjelfjorden in Norway, enabling its return to the Vickers-Armstrong works at Newcastle for rebuilding.

DESTROYED BOW
This striking view shows crew members of *Eskimo* inspecting what had been the bow of the ship, blown off by a German torpedo on 12 April 1940. This resilient class was equally admired by their crews and the general public.

USS *Enterprise* (1938)

TYPE • *Aircraft carrier* **COUNTRY** • *United States*

SPECIFICATIONS

DIMENSIONS:	Length: 251.4m (824ft 9in); Beam: 34.9m (114ft 5in); Draught: 7.9m (25ft 11in)
DISPLACEMENT:	21,336 tonnes (23,519 tons) standard; 32,573 tonnes (35,906 tons) full load
PROPULSION:	9 B&W boilers; 4 Parsons geared turbines; 4 shafts; 89,484kW (120,000shp)
SPEED:	60.2km/h (32.5 knots; 37.4mph)
RANGE:	12,000nmi (23,200km; 14,400 miles) at 15 knots
ARMAMENT:	8 × 127mm (5in)/38-cal guns; 8 × twin 40mm Bofors guns; 6 × quad 40mm Bofors guns; 50 × 20mm Oerlikon guns
AIRCRAFT:	90
COMPLEMENT:	2217

The seventh US ship to carry the name, USS *Enterprise* – the 'Big E' – played a prominent part in all the great Pacific campaigns of World War II. It survived bombs and kamikaze attacks, gaining both battle scars and stars as the most decorated US ship of the conflict.

Right: USS Enterprise *(CV-6) steams ahead with the Pacific Fleet towards the Panama Canal on its way home for Navy Day Celebrations, 27 October 1945.*

ELEVATORS
To handle the large volume of aircraft, three elevators were fitted, plus two flight deck catapults and a single hangar deck catapult.

DECK TRACTORS
Deck tractors and trolleys speeded up the loading of torpedoes and bombs. This was done on the flight deck, whereas Japanese practice was to send up ready-loaded planes from the hangars.

ISLAND
Aft of the pilot house and navigation bridge, the island also held a flag officer's suite and operations room as well as the captain's sea cabin.

DEVELOPMENT

With USS *Yorktown*, *Enterprise* (CV-6) formed a class of two, built side by side at Newport News Shipbuilding & Dry Dock Co. *Enterprise* was laid down on 16 July 1934, launched on 3 October 1936 and commissioned on 12 May 1938. The *Yorktown* class followed USS *Ranger* (CV-4, 1934) in being designed as carriers from the start, and so was able to carry more aircraft and function more effectively. *Enterprise* carried as many aircraft as *Lexington* but had only half the older ship's displacement.

READY FOR LAUNCH
Torpedo Squadron Six (VT-6) TBD-1 aircraft are prepared for launching on USS *Enterprise* (CV-6) at about 0730–0740 hrs, 4 June 1942. Eleven of the 14 TBDs launched from *Enterprise* are visible. Three more TBDs and 10 F4F fighters must still be pushed into position before launching can begin.

Kirov (1938)

SPECIFICATIONS

DIMENSIONS:	Length: 191.3m (627ft 7in); Beam: 17.66m (57ft 11in); Draught: 6.15m (20ft 2in)
DISPLACEMENT:	7890 tonnes (8697 tons) standard
PROPULSION:	Steam turbines totalling 84,600kW (113,500hp)
SPEED:	66.56km/h (35.94 knots; 41mph)
RANGE:	3750nmi (6940km; 4312 miles)
ARMAMENT:	9 × 180mm (7.1in) guns; 6 × 100mm (3.9in) AA guns; 6 × 45mm (1.77in) guns; 4 × 12.7mm (0.5in) guns; 6 × 533mm (21in) torpedo tubes; 96–164 mines; 50 depth charges
ARMOUR:	Belt, deck, turrets and barbettes: 50mm (2in); conning tower: 150mm (5.9in)
COMPLEMENT:	872

The six *Kirov*-class cruisers were the first large naval vessels to be entirely built in Soviet shipyards after the Russian Revolution. The design was based on the Italian Navy's *Raimondo Montecuccoli*-class cruisers.

Left: A forward view of the bow of a Soviet Kirov*-class heavy cruiser during World War II.*
Far left: A view from the bridge of a Kirov*-class heavy cruiser, looking forward over the two forward triple 180mm (7.1in) gun turrets. Note the partially visible anti-aircraft gun atop 'B' turret.*

ARMOUR
While the conning tower had the thickest armour, the rest of the ship had a general covering of about 50mm (2in), insufficient to withstand a heavy shell impact.

CATAPULT
The ship was equipped with a Heinkel K-12 aviation catapult, which was used to deploy the ship's two Beriev Be-2 float planes.

ARMAMENT
The class suffered from an attempt to pack in too much armament on a limited displacement. Their triple 180mm (7.1in) turrets gave constant trouble.

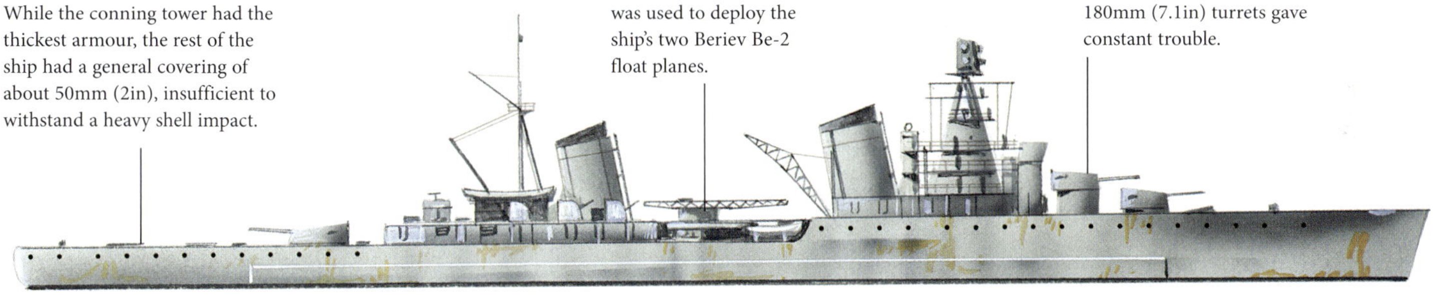

FIRST ACTION

Kirov first saw action against Finnish coastal batteries during the Winter War. It acted as the flagship for the evacuation of Tallinn in August 1941, before taking refuge in Kronstadt. It gave gunfire support during the siege of Leningrad and bombarded Finnish positions in mid-1944 during the Vyborg–Petrozavodsk offensive, but played no further part in the war. *Kirov* was reclassified as a training cruiser in 1961 and scrapped in 1974.

***KIROV* AT ANCHOR, 1936**
The first two pairs of ships (*Kirov*, *Voroshilov*, *Maxim Gorky* and *Molotov*) were deployed on shore bombardment and supply missions with the Baltic and Black Sea Fleets throughout the war, while the last pair (*Kaganovich* and *Kalinin*) were still under construction for the Pacific Fleet at the Amur Shipbuilding Plant and saw no combat during the war.

HMS *Belfast* (Town class) (1939)

TYPE • *Cruiser* **COUNTRY** • *United Kingdom*

HMS *Belfast* is the most famous of the ten Town-class light cruisers that served in World War II. Launched on 17 March 1938 and commissioned on 5 August 1939, it is today a thriving museum ship on the River Thames in London.

SPECIFICATIONS

Dimensions:	Length: 186.99m (613ft 6in); Beam: 19.3m (63ft 4in); Draught: 5.56m (18ft 3in) forward
Displacement:	11,550 tonnes (12,738 tons) full load
Propulsion:	4 Admiralty oil-fired 3-drum boilers; 4 Parsons single-reduction geared steam turbines; 60,000kW (80,000shp)
Speed:	32 knots (59km/h; 37mph)
Range:	8664nmi (16,050km; 9970mi)
Armament:	12 × 152mm (6in) Mk XXIII guns; 12 × 102mm (4in) Mk XVI dual-purpose guns; 16 × 40mm (2-pdr) AA guns; 6 × 533mm (21in) torpedo tubes (2 × 3)
Armour:	Main belt 114mm (4.5in); main turrets up to 102mm (4in); decks over magazines 76mm (3in); decks over machinery 51mm (2in); bulkheads 63.5mm (2.5in)
Complement:	781–881

Below: HMS Birmingham *was one of the first five Town class cruisers to be commissioned, in 1936.*

Aircraft
The cruisers could also transport two Supermarine Walrus amphibious biplanes. When not in use these were stored in forward hangars.

HMS *Belfast* (C35)
Belfast's main armament comprised four triple turrets mounting 12 Mk XXIII 152mm (6in) guns, augmented by a secondary armament of 12 102mm (4in) guns in six twin mounts.

Armour
The Town class had decent armour, with a main belt 114mm (4.5in) thick and 102mm (4in) over the turrets.

LONDON LANDMARK

Belfast was plunged into active service from September 1939. Severe mine damage in November 1939 put the ship out of action for two years, but following recommissioning in 1942 it performed combat duties for the remainder of the war and in the later Korean War. A much-modernized ship served the Royal Navy until decommissioning in 1963. It was saved from scrapping and since October 1971 it has served as a popular IWM museum ship, berthed prominently next to Tower Bridge in the heart of London.

KOREAN WAR SERVICE
HMS *Belfast* coming alongside USS *Bataan* (CVL-29) while off the coast of Korea on 27 May 1952. During the Korean War, *Belfast* operated as a bombardment ship, firing over 8,000 rounds to support ground forces and blockade the North Korean coast.

Z20 *Karl Galster* (1939)

TYPE • *Destroyer* **COUNTRY** • *Germany*

SPECIFICATIONS

Dimensions:	Length: 125m (410ft 1in); Beam: 11.78m (38ft 5in); Draught: 3.98m (13ft 1in)
Displacement:	1811 tonnes (1996 tons) standard; 3415 tonnes (3766 tons) full load
Propulsion:	6 Wagner boilers driving 2-shaft Wagner geared turbines, developing 52,199kW (70,000shp)
Speed:	40 knots (74km/h; 46mph)
Range:	2050nmi (3800km; 2360 miles) at 19 knots
Armament:	5 × 127mm (5in) guns; 4 × 37mm (1.45in) AA guns; 7 × 20mm (0.79in) AA guns; 8 × 533mm (21in) torpedo tubes
Armour:	None
Complement:	315

When Germany declared war, its navy had only a small force of 21 destroyers. *Karl Galster* was one of the six 1936-type destroyers laid down between 1936 and 1937. It was launched on 15 June 1938 and completed on 21 March 1939.

Above: A large-diameter signalling light was mounted above the bridge. A powerful searchlight for surface or air defence was mounted behind the aft funnel.

AA and anti-submarine
Six 37mm (1.5in) AA guns were mounted, in pairs, plus 12 20mm (0.79in) guns. Four anti-submarine depth-charge launchers were also carried.

Profile
The class is distinguished from its 1934A predecessors by flatter funnel caps, marginally wider beam and greater length (by 5m/15ft).

Machinery
The high-pressure boilers and turbines, with a substantial power output of 55,554kW (74,500shp), were more reliable than in previous classes.

SERVICE
By April 1940, the German destroyer force had been cut in half with the sinking of 10 destroyers by British forces in the Norwegian campaign. *Karl Galster* was the only one of its class to survive the war; the other five were sunk either at Narvik or in Rombaksfjord. In 1946 the *Karl Galster* was handed over to the Soviet Navy as part of Germany's war reparations to the Soviet Union. It served in the Baltic fleet as the Protshnyi into the 1950s.

IN PORT
The 1934-type destroyers were well-built and cleverly engineered but they were over-sophisticated, particularly in their powerplants, and technical troubles prevented them attaining their full design performance.

World War II

World War II (1939–45) transformed naval warfare in entirely new directions. At the beginning of the war, gun-armed battleships and cruisers were the heart of the fleets. Yet as the conflict progressed, submarines and aircraft carriers emerged as the decisive force multipliers. Huge battleships proved vulnerable, dangerously exposed to submarine-launched torpedoes and air-dropped bombs. Aircraft carriers allowed fleets to project firepower across hundreds of miles, far beyond the limited range of heavy guns, while submarines hunted in the depths of oceanic shipping lanes. In response, surface navies developed increasingly specialized anti-submarine vessels and weaponry, together with bristling arrays of anti-aircraft guns, achieving lethally effective defences against these new threats.

CARRIER OPERATION
Photographed in the Pacific in around November 1943, sailors and airmen of the *Independence*-class light aircraft carrier USS *Cowpens* (CVL-25) gather on the flight deck. The ship survived the war and served in the US Navy until the late 1950s.

Scharnhorst (1939)

TYPE • *Battleship* COUNTRY • *Germany*

SPECIFICATIONS

Dimensions:	Length: 235m (772ft); Beam: 30m (98ft 5in); Draught: 9.7m (31ft 9in)
Displacement:	29,121 tonnes (32,100 tons); 34,564 tonnes (38,100 tons) full load
Propulsion:	12 Wagner HP boilers, 3 Brown-Boveri geared turbines developing 119,312kW (160,000hp)
Speed:	31 knots (57km/h; 35.4mph)
Range:	16,298km (8800nmi; 10,127 miles) at 19 knots
Armament:	9 × 280mm (11in) guns; 12 × 150mm (5.9in) guns; 14 × 105mm (4.1in) guns; 16 × 37mm (1.5in) AA guns; 10 × 20mm (0.79in) AA guns; 6 × 533mm (21in) torpedo tubes
Armour:	Belt: 350–200mm (13.8–7.9in); bulkheads: 200–150mm (7.9–5.9in); barbettes: 350–200mm (13.8–7.9in); turrets: 350–200mm (13.8–7.9in); conning tower: 350–100mm (13.8–3.9in)
Complement:	1968

Although often described as the first German battleship to be constructed since World War I, *Scharnhorst* was laid down and commissioned after *Gneisenau*. Despite being companion ships, there were many minor differences between them.

Right: The peaked 'Atlantic' bow was added in 1939, but Scharnhorst remained a 'wet' ship, and heavy seas could put its 'A' turret out of action.

Deck plan
The 150mm (5.9in) secondary guns are clearly visible from this perspective, as are the many smaller 37mm (1.5in) and 20mm (0.79in) AA guns.

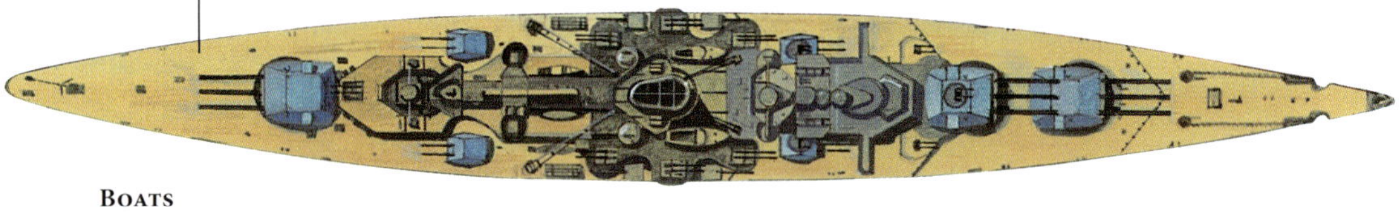

Boats
Three picket boats, two motor yawls, five cutters, one motor pinnace and one motor launch were carried.

Bow
To cope with ocean conditions, both ships underwent major bow modifications. *Gneisenau* got its clipper or 'Atlantic' stem in 1938 (modified again in 1942); *Scharnhorst* in mid-1939.

Propulsion
Diesel propulsion was originally planned for both ships, but no diesel engines could produce the necessary power output per shaft, and 12 Wagner extra-high-pressure boilers drove the turbines.

CONVOY ACTION

By the end of 1943, *Scharnhorst* was the Reich's only operational capital ship. With a force of five destroyers under Rear Admiral Bey, it was sent on 25 December to attack Convoy JW55B en route for Murmansk, and came under fire from the guns of the battleship *Duke of York*, with four escorting destroyers. A combination of torpedoes and shell fire finally destroyed *Scharnhorst*. It sank at 19.44 hrs on 26 December, in 290m (950ft) of water, with the loss of 1803 men.

IN PORT, 1939
A proposal to replace *Scharnhorst*'s main guns with 380mm (15in) guns was made but never put into action. Such armament would have produced a formidable battleship. As pictured here, note the ship's badge mounted on its bow.

USS *Sangamon* (1939)

TYPE • *Aircraft carrier* **COUNTRY** • *United States*

SPECIFICATIONS

Dimensions:	Length: 169m (553ft); Beam: 23m (75ft); Draught: 9.8m (32ft)
Displacement:	11,582 tonnes (12,768 tons) standard; 24,663 tonnes (27,194 tons) full load
Propulsion:	4 boilers, 2 steam turbines; 2 shafts; 10,067kW (13,500shp)
Speed:	18 knots (33km/h; 21mph)
Range:	23,920nmi (43,000km; 26,783 miles) at 15 knots
Armament:	2 × 127mm (5in) 38 cal guns; 2 × quad 40mm; 7 × twin 40mm 56 cal AA guns; 21 × single 20mm 70 cal AA guns
Aircraft:	32
Complement:	1080

***Sangamon* had a dual career with the US Navy, initially as a fleet oiler, then as the Navy's first escort carrier. It was first of a class of escort carriers that played an essential support role in achieving the US Navy's dominance in the Pacific:**

Right: USS Sangamon *(CVE 26) sails through Chesapeake Bay in September 1942. The first deployment of* Sangamon *was with Task Force 34 in October 1942 in the Mediterranean, supporting the Operation* Torch *landings in North Africa.*

Island
Despite its small size, the island incorporated an admiral's bridge below the captain's bridge, which was open and on the same level as the primary flight control station, one to port and one to starboard.

Side openings
The side openings show the level of the main deck, or fuel deck as the crew called it. *Sangamon* was fitted for at-sea refuelling and the pumps were housed on this deck.

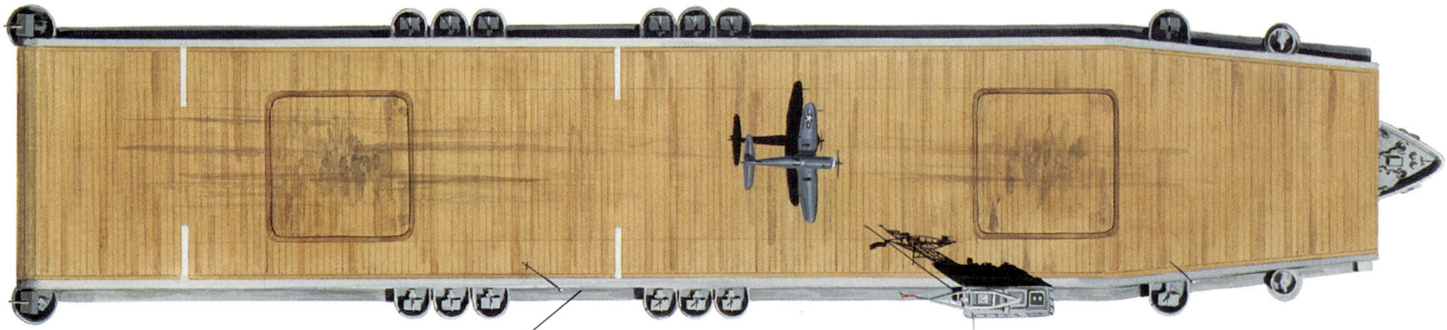

DEVELOPMENT
Laid down on 13 March 1939 for Standard Oil's fleet as *Esso Trenton,* the ship was launched on 4 November the same year. The Navy purchased it on 22 October 1940, changing the name to *Sangamon* on 12 April 1941. From December 1941 there was a search under way for ships or hulls suitable for rapid conversion to carriers. *Sangamon* was picked and commissioned as a carrier in August 1942.

Catwalks
As on most carriers, 'catwalks' for the crew were positioned all round the flight deck, partly outboard and supported by struts, to enable safe passage during take-offs and landings.

KAMIKAZE DAMAGE
In the Battle of Leyte Gulf, *Sangamon*'s aircraft were engaged in the conflict, and the carrier continued in attack rather than escort mode into 1945. On the evening of 4 May 1945 one of a flight of Japanese kamikaze planes crashed on to *Sangamon*, causing heavy damage and starting fires. There were 86 fatalities.

HMS *Kelly* (1939)

TYPE • *Destroyer* **COUNTRY** • *United States*

SPECIFICATIONS

Dimensions:	Length: 108.66m (356ft 6in); Beam: 10.87m (35ft 8in); Draught: 4.22m (13ft 10in)
Displacement:	2330 tonnes (2568 tons) standard; 2540 tonnes (2800 tons) full load
Propulsion:	2 Admiralty 3-drum boilers; 2-shaft Parsons geared turbines; 29,828kW (40,000shp)
Speed:	36 knots (67km/h; 41mph)
Range:	5500nmi (10,200km; 6338 miles) at 15 knots
Armament:	6 × 119mm (4.7in) guns; 4 × 2-pdr 'pom-pom' AA guns; 10 × 533mm (21in) torpedo tubes; 45 depth charges
Armour:	None
Complement:	183–218

HMS *Kelly* was the lead ship of the K class that was built alongside the J- and N-class destroyers laid down between 1937 and 1938. *Kelly* itself was launched on 25 October 1938 and was destined for an ill-fated war service.

Above: The J-class HMS Jupiter *(centre) conducts a flotilla exercise between* Jackal *(left) and* Vortigern *(right).*

Armour
As with many British destroyers, the *Kelly* was unarmoured and relied mainly on its speed and manoeuvrability for its survival.

Speed
The two-shaft Parsons geared turbines could take the *Kelly* to an impressive maximum speed of 36 knots (67km/h; 41mph).

Main guns
The K class featured the same 119mm (4.7in) guns as the *Tribal* destroyers, and were also armed with torpedoes and depth charges.

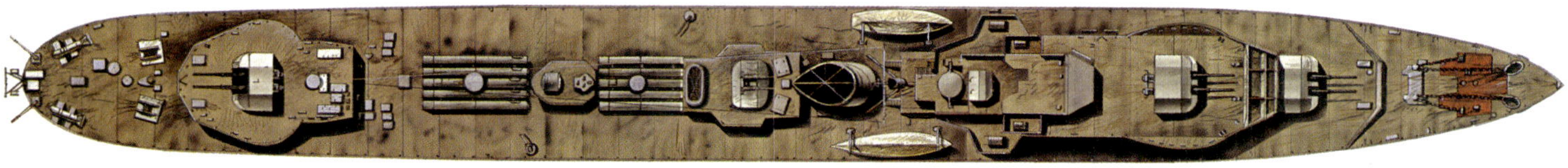

K CLASS AT WAR
In all, eight K-class destroyers were built between 1938 and 1939. Of these, four were lost to enemy action and one was destroyed by an accidental internal explosion. On 20 May 1941 off Crete, the *Kelly* and its unit came under persistent Stuka attack. While making speed and turning hard to port, the vessel suffered severe bomb damage. Still moving fast, *Kelly* rolled over and capsized, sinking in just half an hour.

TRIALS, 1939
Commissioned in August 1939, *Kelly* had a rough first year in service. It was damaged by a mine strike in British waters in December 1939 then, following lengthy repairs, collided with the destroyer HMS *Gurkha* in March 1940, requiring further time in dry dock.

Yukikaze (*Kagero* class) (1939)

TYPE • *Destroyer* **COUNTRY** • *Japan*

SPECIFICATIONS

Dimensions:	Length: 118.49m (388ft 9in); Beam: 10.79m (35ft 5in); Draught: 3.76m (12ft 4in)
Displacement:	2033 tonnes (2241 tons) standard; 2450 tonnes (2700 tons) trial
Propulsion:	3 boilers; 2-shaft geared turbines; 38,811kW (52,000hp)
Speed:	35 knots (65km/h; 40.3mph)
Range:	5000nmi (9300km; 5800 miles) at 18 knots
Armament:	6 × 127mm (5in) guns; 4 × 25mm AA guns; 8 × 609mm (24in) torpedo tubes; 16 depth charges
Armour:	None
Complement:	240

The *Kagero*-class destroyers were the most advanced of their type in service with the Japanese Navy when war broke out in Europe. Similar to the earlier *Asashio* class, they benefitted from a refined design that had been thoroughly 'wrung out'.

Above: View on the ship's deck, looking aft from the bow. The photo was taken on 26 May 1947 in Tokyo, Japan, where it was shown to representatives of the leading Allied powers.

Depth charges
The *Kagero*-class destroyers were originally equipped with 16 depth charges for anti-submarine warfare, but the number had been increased to 36 by June 1942.

Powerplant
The destroyer was powered by three boilers connected to two-shaft geared turbines, generating a total power output of 38,811kW (52,000hp).

SURVIVOR
One of the *Kagero* class, *Hamakaze*, made a small contribution to Japanese naval history by becoming the first Japanese destroyer to be fitted with radar. The smaller combat vessels of the Japanese fleet were virtually wiped out by the US Navy and the *Kagero* class was no exception, although it was virtually the only Japanese destroyer class not to have every member sunk by 1945. Of the 18 ships built, only *Yukikaze* lasted the war, one of half a dozen destroyers to survive from a total force of over 100 vessels.

Yukikaze
The sixth *Kagero*-class destroyer, *Yukikaze* was launched at the Sasebo naval yard on 24 March 1939.

POSTWAR SHIP
The *Yukikaze* as seen in Tokyo on 26 May 1947. The destroyer had been repatriating Japanese nationals from overseas, therefore was stripped of its armament. *Yukikaze* later became the Republic of China Navy's destroyer *Tan Yang*.

Katori class (1939)

TYPE • *Light cruiser* **COUNTRY** • *Japan*

SPECIFICATIONS

Dimensions:	Length: 129.77m (425ft 9in); Beam: 15.95m (52ft 4in); Draught: 5.75m (18ft 10in)
Displacement:	5985 tonnes (6598 tons) normal
Propulsion:	2-shaft geared turbines plus diesel motors; 3 Kampon boilers; 6000kW (8000shp)
Speed:	18 knots (33km/h; 21mph)
Range:	9000nmi (17,000km; 10,563 miles) at 10 knots
Armament:	4 x 140mm (5.5in)/50-cal guns (2 x 2); 2 x 127mm (5in)/40-cal AA guns (1 x 2); 4 x Type 96 AA guns (later 30); 8 x 13.2mm (0.52in) AA guns; 4 x 533mm (21in) torpedo tubes (2 x 2)
Armour:	Gun turrets: 10mm (0.4in); conning tower: 10mm (0.4in)
Complement:	315 + 275 midshipmen

The *Katori*-class cruisers were originally conceived as training vessels for the Imperial Japanese Navy from the late 1930s to the 1940s. The outbreak of the Pacific War in December 1941, however, saw them repurposed for active combat.

Left: The Katori *class were originally ordered as training ships, and therefore were designed with a a lower length-to-beam ratio than many cruisers to give better stability.*

Displacement
The *Katori* class had a normal displacement of 5985 tonnes (6598 US tons) and a full-load displacement of 6279 tonnes (6921 US tons).

AA armament
The class as-built had just four Type 96 AA guns, but this number was later increased to 30 due to the elevated threat of air attack in the Pacific theatre.

Aircraft
The ship was capable of deploying a single float plane, which was launched from a catapult positioned amidships.

KATORI
The lead ship of the class was commissioned on 20 April 1940. During the war, *Katori* was seriously damaged by US dive-bombers and torpedo-bombers in February 1942, but was repaired and returned to service. On 17–18 February 1944, off Truk, it was again damaged by a US air attack, but was subsequently finished off by heavy gunfire from the US battleships *Iowa* (BB-61) and *New Jersey* (BB-62).

BRIDGE AND GUNS
A view from the foredeck of the bridge of a *Katori*-class cruiser, with a rangefinder in the centre The guns were capable of 6–10 rounds per minute out to a range of 20,574m (67,500ft), when firing at a 35-degree angle.

Type XB (1939)

The Type XB were the Kriegsmarine's largest U-boats, used as minelayers, supply boats and transports. Type XA never went beyond the design stage; the first Type XB, U-116, was commissioned in July 1941.

SPECIFICATIONS

Dimensions:	Length: 89.8m (294ft 7in); Beam: 9.2m (30ft 2in); Draught: 4.7m (15ft 5in)
Displacement:	1791.2 tonnes (1974 tons) surfaced; 2212 tonnes (2438 tons) submerged
Propulsion:	2 diesel engines developing 1790kW (2400hp); 2 electric motors developing 410kW (550hp); 2 screws
Speed:	16.4 knots (30km/h; 19mph) surfaced; 7 knots (13km/h; 8mph) submerged
Range:	14,450nmi (26,761km; 16,632 miles) at 12 knots surfaced; 188nmi (348km; 216 miles) at 2 knots submerged
Armament:	66 × SMA mines; 1 × 105mm (4.1in) deck gun; 1 × 37mm (1.4in) AA gun; 1 × 20mm (0.79in) AA gun
Complement:	52

Left: The Type XB U-118 *comes under attack in the Atlantic by aircraft from USS* Bogue *(CVE-9) on 12 June 1943. The boat was destroyed, sharing the fate of many U-boats in the same year.*

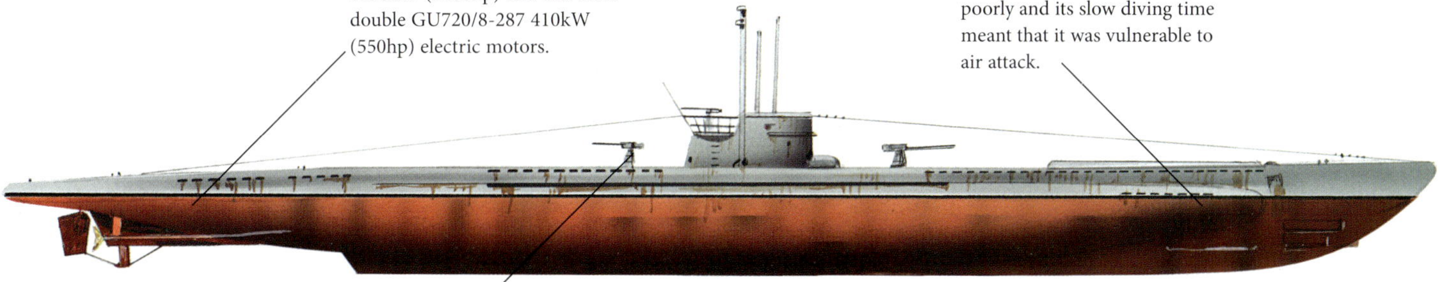

Propulsion
Propulsion was by two Germaniawerft nine-cylinder fourstroke F46a-9pu supercharged diesels each of a maximum 1790kW (2400hp) and two AEG double GU720/8-287 410kW (550hp) electric motors.

U-118
The largest submarines in service with the Kriegsmarine during World War II, the Type XB boats could carry 66 mines. However, the class handled poorly and its slow diving time meant that it was vulnerable to air attack.

Guns
Surface weaponry was one 105mm (4.1in)/45cal deck gun on the foredeck, one 37mm (1.4in) aft of the tower, and one 20mm (0.79in) AA gun mounted aft of the bridge. From 1943, the forward gun was removed and the 20mm (0.79in) gun was replaced by a twin mounting.

ADAPTED PURPOSE

The German Navy required a specialized minelayer type to plant larger mines than could be ejected from a 533mm (21in) tube. All eight of the resulting Type XB class were built by Germaniawerft in Kiel. Some of the class were adapted or converted. *U-119* was fitted as a supply ship with cradles for eight torpedoes and acted as a fuel supplier while retaining its capacity as a minelayer. *U-219* and the last of the type, *U-234*, were turned into long-range cargo carriers for communication with Japan.

DEPTH CHARGED
In the Atlantic on 7 August 1943, the Type XB submarine *U-117* (left) and Type IXC *U-66* are attacked with depth charges from TBF-1 Avengers. *U-66* escaped, but *U-117* was sunk with all crew lost.

Marconi and *Cagni* classes (1939–41)

TYPE • *Submarine* **COUNTRY** • *Italy*

SPECIFICATIONS
(Leonardo da Vinci)

DIMENSIONS:	Length: 76.5m (251ft); Beam: 6.81m (22ft 5in); Draught: 4.72m (15ft 6in)
DISPLACEMENT:	1194 tonnes (1317 tons) surfaced; 1489 tonnes (1642 tons) submerged
PROPULSION:	2 diesels of 1342.5kW (1800hp); 2 electric motors of 556kW (750hp); 2 screws
SPEED:	17.8 knots (33km/h; 20.5mph) surfaced; 8.2 knots (15.2km/h; 9.4mph) submerged
RANGE:	9124nmi (16,898km; 10,500 miles) surfaced; 95.6nmi (177km; 110 miles) submerged
ARMAMENT:	8 × 533mm (21in) torpedo tubes; 12 × torpedoes; 1 × 100mm (4in)/47-cal deck gun; 4 × 13.2mm (0.5in) machine guns
COMPLEMENT:	57

Six-strong, the *Marconis* were the Regia Marina's most successful class. A wider operating range of 10,500nmi (19,446km, 12,083 miles) was the main difference between the *Marconi*-class boats and the *Marcellos*.

Above: Ammiraglio Cagni *was built at Monfalcone to a Bernardis design and made some of the longest patrols of any World War II submarine, operating into the South Atlantic for up to 135 days.*

PROPULSION
Power came from the same CRDA diesels of a combined 2684.5kW (3600hp) as the *Marcellos*, but the two Marelli electric motors were uprated to a combined 1118.5kW (1500hp).

LEONARDO DA VINCI
This ship was fitted for a planned, but cancelled, attack on New York, with a CA-2 midget submarine in place of the deck gun. The tall fairing around the periscopes was removed to reduce the boat's silhouette.

TOWERS
The original towers, with a tall narrow periscope casing rising above the oblong base, were cut back in 1941–42, in reaction to Atlantic weather and fighting conditions.

***CAGNI* CLASS**
The *Cagni* class, the Regia Marina's largest attack submarine, was a rather strange vessel to build in 1939–41. It was designed to attack merchant shipping on the high seas, and also to operate as far away as the East African coast. For the first purpose it was fitted with 14 450mm (17.7in) torpedo tubes, eight in the bow, six at the stern and 36 torpedoes. For the second it had a fuel capacity of 183 tonnes (202 tons) and other facilities to allow for up to six months at sea.

GUGLIELMO MARCONI
The *Marconi*-class leader, *Guglielmo Marconi,* launched at Monfalcone on 27 July 1939. It disappeared in the Atlantic at the end of October 1941.

Type XIV (1940)

SPECIFICATIONS
***U-459* (Type XIV)**

DIMENSIONS:	Length: 67.10m (220ft 1in); Beam: 9.35m (30ft 8in); Draught: 6.51m (21ft 4in)
DISPLACEMENT:	1688 tonnes (1861 tons) surfaced; 1932 tonnes (2130 tons) submerged
PROPULSION:	2 × 1640.5kW (2200hp) diesels; 2 × 373kW (500hp) electric motors; two screws
SPEED:	14.9 knots (27.6km/h; 17.1mph) surfaced; 6.2 knots (11.5km/h; 7.1mph) submerged
RANGE:	12,350nmi (22,872km, 14,212 miles) at 10 knots; 55nmi (102km, 63 miles) at 4 knots
ARMAMENT:	2 × 37mm (1.4in) guns; 1 × 20mm (0.79in) AA gun
COMPLEMENT:	53

Consideration of a 'submarine' tanker went back to the 1930s. Known to German submariners as *Milchkühe* (milk cows), this was a class of 10 supply boats intended to refuel and restock U-boats on ocean patrol.

Above: U-461. *The concept of the U-boat feeder, able to avoid detection until it reached a meeting point, worked until the German radio codes were broken and their missions ceased to be secret.*

DIVING DEPTH
The Type XIV could travel at a depth of 150m (492ft) thanks to a pressure hull of 251mm (0.98in) steel.

U-459
The tower and bridge were modelled on the Type IX, although details of the aft gun platform varied in different Type XIV boats.

STORAGE
Inside the pressure hull was a double-deck layout, with cargo storage compartments where up to 20 tonnes (22 tons) of supplies could be held on pallets, ready for transfer. Fully loaded, the Type XIV could extend the patrol period of 12 Type VII U-boats for four weeks.

DESIGN
Specifications for the Type XIV were not drawn up until 1940 when the U-boat fleet was ranging far out into the Atlantic in pursuit of British convoys. Features of the Types VII and IX were incorporated in the design, but the XIV did not share their hull shape, which was of similar length but greater in beam and depth, reflecting its cargo-carrying task. It had a flat open deck to cope with the difficult business of boat-to-boat transfer.

RESUPPLY
On 21 April 1942, *U-459* resupplies multiple U-boats in the Atlantic. *U-459*, first of the 10 boats of Type XIV, was the Kriegsmarine's first purpose-built refuelling and supply submarine, a large boat intended to intensify the Atlantic war by enabling attack vessels to spend longer periods at sea.

Z30 (Type 36A) (1940)

TYPE • *Destroyer* COUNTRY • *Germany*

SPECIFICATIONS (*Z26*)

DIMENSIONS:	Length: 127m (416ft 8in); Beam: 12m (39ft 4in); Draught: 4.43m (14ft 6in)
DISPLACEMENT:	2700 tonnes (2976 tons)
PROPULSION:	Twin shafts; 2 × steam turbines; 52,199kW (70,000hp)
SPEED:	37.5 knots (69km/h; 43mph)
RANGE:	2500nmi (4600km; 2900 miles) at 19 knots
ARMAMENT:	4 × 150mm (5.9in) guns; 4 × 37mm (1.46in) SK C/30 guns; 8 × 20mm (0.79in) C/30 guns; 8 × 533mm (21in) torpedo tubes
COMPLEMENT:	320

The *Narvik* class, *Z23–30*, were officially Type 36A. Their main guns were the heaviest mounted on destroyers, but were not suitable for rapid fire, and their excessive weight reduced the ships' seakeeping qualities.

Left: German destroyers enter a Norwegian fjord, while the crew of an anti-aircraft platform keep watch for surface or aerial threats.

PROPULSION
The geared steam turbines generated 52,199kW (70,000hp), which enabled them to obtain an impressive maximum speed of 37.5 knots (69km/h; 43mph).

Z26
Z26 was one of four Type 36A destroyers that retained their original forward-located single 15cm (5.9in) guns; its two sets of quadruple torpedo tubes can be seen amidships, either side of the second funnel.

SEAKEEPING
Despite the lessons garnered from previous designs, these vessels were not regarded as good 'sea boats' when in heavy seas; the weight of the double forward turret exacerbated this problem.

GUN CONFIGURATION

As designed, this group of eight unnamed destroyers were intended to feature three single potent 15cm (5.9in) guns turrets aft with a double turret in the bow; the 15cm (5.9in) calibre was usually associated with German light cruisers rather than destroyers. All eight vessels, however, were initially built with just a single forward turret due to shortages in the twin turret. Subsequently, four vessels in the class (*Z23–25* and *Z29*) had the single fore turret replaced with a double turret.

AMERICAN DESTROYER
The destroyer *Z39* – of the Type 36A (Mob) class – survived World War II and was transferred to the US Navy as *DD-939* for conducting propulsion experiments. Here it is pictured off the Boston coastline in August 1945.

Prinz Eugen (1940)

TYPE • *Heavy cruiser* **COUNTRY** • *Germany*

Twelve 533mm (21in) torpedo tubes and three aircraft added to the capabilities of this *Admiral Hipper*-class ship, commissioned in 1940. It accompanied *Bismarck*, then *Scharnhorst* and *Gneisenau*, on their Atlantic sorties.

SPECIFICATIONS

Dimensions:	Length: 207.25m (679ft 11.5in); Beam: 21.48m (70ft 6in); Draught: 7.2m (23ft 7.5in)
Displacement:	16,974 tonnes (18,717 tons)
Propulsion:	3-shaft Deschimag geared turbines; 9 × Wagner (La Mont) boilers; 98,441kW (132,000hp)
Speed:	32.5 knots (60.2km/h; 37.4mph)
Range:	6800nmi (12,600km; 7800 miles) at 20 knots
Armament:	8 × 203mm (8in) guns; 12 × 105mm (4.1in) guns; 12 × 37mm (1.45in) AA guns; 8 × 20mm (0.79in) AA guns; 12 × 533mm (21in) torpedo tubes
Armour:	Belt 89–38mm (3.5–1.5in); deck 38–31mm (1.5–1.25in); main turrets 158–57mm (6.25–2.25in)
Complement:	1600

Right: Prinz Eugen *is seen here in a floating drydock, probably at Bremen, soon after VE Day.* Prinz Eugen *was the first of the second batch of three* Hipper-*class cruisers to be laid down. However, the other two ships –* Seydlitz *and* Lützow *– were never completed as planned.*

AIRCRAFT
The ship could carry a maximum of three float planes for reconnaissance purposes. The aircraft type was the Arado Ar 196 low-wing monoplane.

DISPLACEMENT
Standard displacement was 16,974 tonnes (18,717 US tons), but the ship's deep-load displacement was 19,042 tonnes (21,000 US tons).

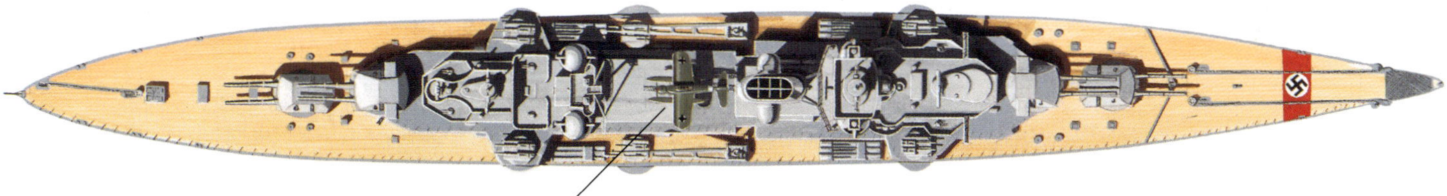

POWERPLANT
Prinz Eugen had no fewer than nine Wagner (La Mont) boilers, providing the power to three-shaft Deschimag geared turbines and generating 98,441kW (132,000hp) of energy.

EVASION

Prinz Eugen was laid down on 23 April 1936, launched on 22 August 1938 and commissioned on 1 August 1940. Together with the *Bismarck*, *Prinz Eugen* undertook the infamous 'channel dash' that took the two out into the North Atlantic. When *Bismarck* was sunk not long afterwards, *Prinz Eugen* managed to evade the British fleet and find refuge in St Nazaire.

OPERATION *CROSSROADS*

After the war, *Prinz Eugen* was taken into US Navy service as USS *Prinz Eugen*. It survived two atomic bomb blasts as a target ship for Operation *Crossroads* at Bikini Atoll in mid-1946, but finally capsized and sank at Kwajalein Atoll in December of that year.

HMS *Howe* (1940)

TYPE • *Cruiser* **COUNTRY** • *United Kingdom*

SPECIFICATIONS

Dimensions:	Length: 227.1m (744ft 11.5in); Beam: 31.4m (103ft); Draught: 9m (29ft 6in)
Displacement:	39,780 tonnes (43,313 tons) standard
Propulsion:	4 shafts; 4 sets Parsons geared turbines; 8 Admiralty 3-drum small-tube boilers; 82,000kW (110,000shp)
Speed:	28.3 knots (52.4km/h; 32.6mph)
Range:	15,600nmi (28,900km; 18,000 miles)
Armament:	10 × 356mm (14in) guns; 16 × 133mm (5.25in) guns; 48 × 40mm (1.575in) QF 2-pdr guns; 18 × 20mm (0.8in) Oerlikon AA guns
Armour:	Main belt 370mm (14.7in); lower belt 140mm (5.4in); deck 127–152mm (5–6in); main turrets 324mm (12.75in); bulkheads: 254–305mm (10–12in)
Complement:	1422

HMS *Howe* was one of the *King George V*-class battleships built between 1937 and 1940, the others being HMS *King George V* (commissioned 1940), *HMS Prince of Wales* (1941), HMS *Duke of York* (1941) and HMS *Anson* (1942).

Left: Britain's HMS Howe, *a* King George V*-class battleship in the Royal Navy, participates in trials, after its launch in February 1940. Although the class looked impressive, submarine protection proved less than satisfactory. The only member of the class lost during World War II – HMS* Prince of Wales *– was sunk by Japanese torpedoes in the Pacific.*

***George V*-class design**
Britain opted for 356mm (14in) guns but with armour capable of resisting 406mm (16in) shells.

Armour
Much effort was spent on designing and developing the ships' armour, and the five battleships of the class had extensive armour belts and internal bulkheads.

***KING GEORGE V* CLASS**
The final *King George V* design was ready by 1936. HMS *Howe* (formerly HMS *Beatty*) was laid down on 1 June 1937, launched on 9 April 1940 and completed on 29 August 1942. The other *King George V*s were involved in several major actions, but *Howe* fought well away from the front pages. Along with HMS *King George V*, HMS *Duke of York* and HMS *Anson*, *Howe* outlasted the war and remained in frontline service alongside its companions into the 1950s.

B turret
The original plan was to fit 12 main guns, but to increase the level of armour protection two guns were deleted on the forward 'B' turret.

EASTERN FLEET
Howe passes through the Suez Canal, en route to join the British Eastern Fleet in the Indian Ocean, c. 1944. *Howe* supported invasion forces in Sicily and Italy, and in 1945 was the British flagship in the Pacific. Placed in reserve in 1951, it was broken up in 1957.

HMS *Jervis Bay* (1940)

TYPE • *Armed merchant cruiser* **COUNTRY** • *United Kingdom*

SPECIFICATIONS

Dimensions:	Length: 167m (547ft 11in); Beam: 21m (68ft 11in); Draught: 10m (32ft 10in)
Displacement:	14,160 tonnes (15,609 tons)
Propulsion:	Parsons geared steam turbines; 2 shafts; 6711kW (9000shp)
Speed:	15 knots (28km/h; 17mph)
Range:	Not specified
Armament:	7 × 152mm (6in) Mk VII guns; 2 × 76mm (3in) AA guns
Armour:	None
Complement:	254

Built for the Australian emigrant trade in 1922, *Jervis Bay* was fitted out as an armed merchant cruiser with eight 152mm (6in) guns. In November 1939, its convoy was intercepted by the German battleship *Admiral Scheer*. *Jervis Bay* fought back, but was sunk with heavy loss of life.

Left: The TSS Jervis Bay *sailed under the Aberdeen & Commonwealth Line, formed in 1928 by the merging of the Australian Commonwealth Line and the Aberdeen Line.*

Capacity
In its days as a passenger liner, *Jervis Bay* could carry between 270 and c. 500 passengers, depending on the configuration.

Firepower
Like most of its fellow armed merchant cruisers, *Jervis Bay* was fitted with old Mk VII 152mm (6in) guns, which did not use high-velocity charges.

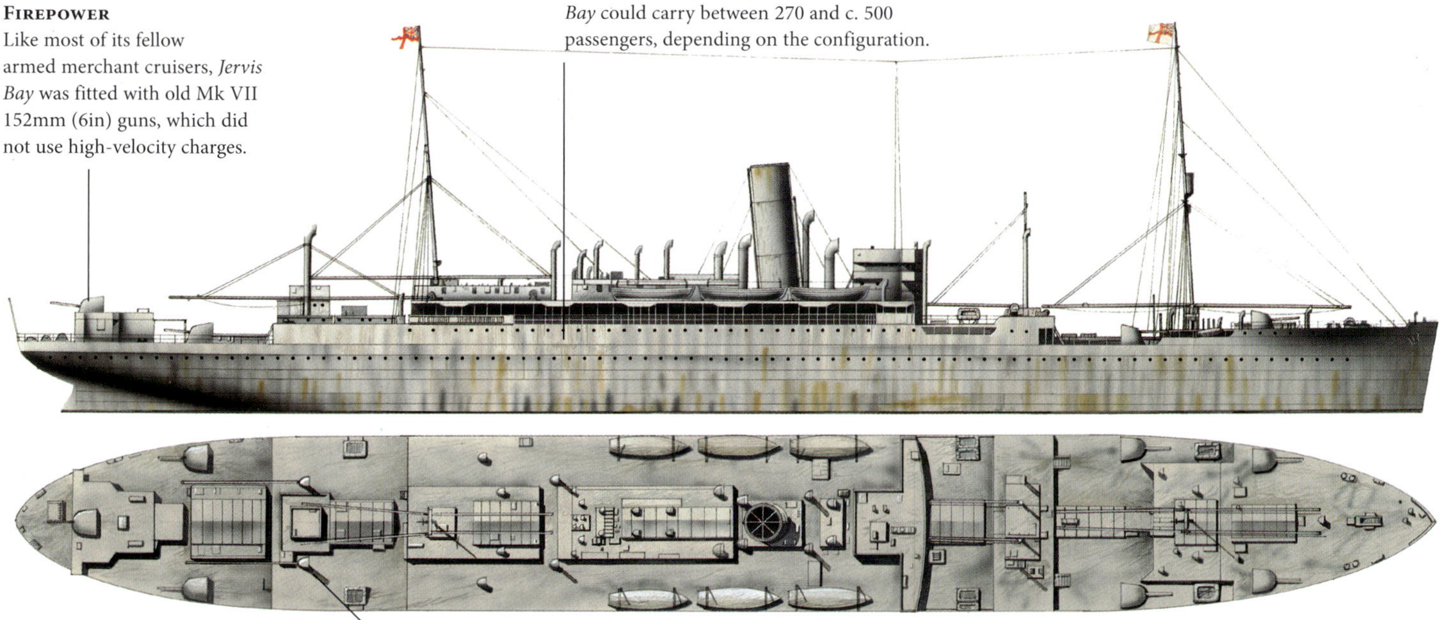

Secondary armament
In addition to the 'main' armament, the *Jervis Bay* was equipped with two 76mm (3in) anti-aircraft guns.

ARMED MERCHANTS

Atlantic convoys were tempting targets for German submarines or surface raiders, and there simply were not enough naval escorts available. To provide a degree of protection, chiefly against U-boats running on the surface, Britain adapted commercial shipping to serve as armed merchant cruisers. HMS *Jervis Bay* was one of over 50 such vessels commissioned.

***JERVIS BAY*, 1940**
Jervis Bay is here seen in Dakar in 1940. The year the war in Europe worsened, Britain became almost totally dependent on the Atlantic convoys to bring essential supplies from the United States.

Tashkent class (1940)

TYPE • *Cruiser* **COUNTRY** • *Soviet Union*

SPECIFICATIONS	
Dimensions:	Length: 139.7m (458ft 4in); Beam: 13.7m (44ft 11in); Draught: 3.7m (12ft 2in)
Displacement:	2890 tonnes (3186 tons) standard
Propulsion:	2 shafts; 2 geared steam turbines; 2 Yarrow boilers; 75,000kW (100,000shp)
Speed:	42.5 knots (78.7km/h; 48.9mph)
Range:	5030nmi (9320km; 5790 miles)
Armament:	3 × twin 130mm (5.1in) guns; 1 × twin 76mm (3in) AA guns; 6 × 37mm (1.5in) AA guns; 6 × 12.7mm (0.5in) machine guns; 3 × triple 533mm (21in) torpedo tubes; 76 mines; 24 depth charges; 2 depth-charge throwers; 1 depth-charge rack
Armour:	None
Complement:	250

A total of 11 *Tashkent*-class destroyers were planned for construction in the Soviet Union from the late 1930s. Only four went ahead and only one of those – the lead ship *Taskhkent* – was actually completed.

Above: This photograph of Tashkent *was taken in the late 1930s, likely during sea trials. The class was designed in Italy.*

Main guns
The main weapons on the *Tashkent* were 130mm (5.1in) 50-calibre B13 Pattern 1936 guns, which were standard on Soviet destroyers up to the 1950s.

Performance
Driven by two geared steam turbines, the *Tashkent* could reach a maximum speed of 42.5 knots (78.7km/h; 48.9mph).

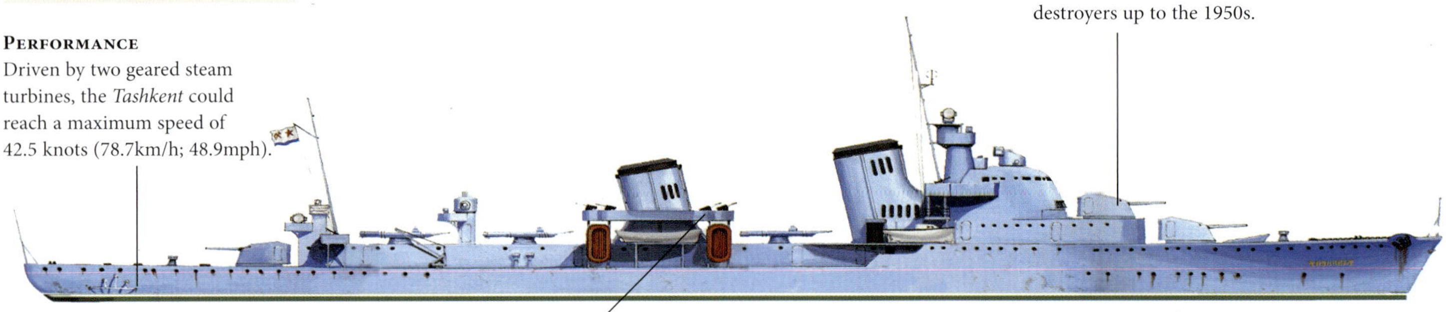

AA guns
In 1941 the vessel's 45mm (1.77in) AA guns were replaced by fully automatic 37mm (1.5in) 70-K AA guns, which gave a faster rate of fire.

***TASHKENT* AT WAR**
Tashkent was laid down in January 1937, completed in October 1939 and assigned to the Black Sea Fleet in October 1940. With the German invasion of the Soviet Union in June 1941, *Tashkent* found itself in heavy combat on the southern fringes of the Eastern Front, especially around Sevastopol. It was on the Black Sea on 2 July 1942 that the ship was bombed to destruction by German aircraft.

GUNNERY
This grainy image shows the forward guns of *Tashkent* in action. Apart from convoy escort, evacuation duties and logistics, a key role of Tashkent was to provide offshore naval gunnery support to Red Army troops.

Littorio (1940)

TYPE • *Battleship* COUNTRY • *Italy*

The *Littorio* class consisted of two batches each of two ships, laid down in 1934 and 1938 respectively. *Littorio* and its companion ship *Vittorio Veneto* were laid down on the same day and the names of both ships have been used interchangeably to identify the class.

SPECIFICATIONS

DIMENSIONS:	Length: 237.74m (780ft); Beam: 32.18m (105ft 7in); Draught: 9.6m (31ft 6in)
DISPLACEMENT:	40,724 tonnes (44,902 tons) standard
PROPULSION:	4 shaft Belluzzo geared turbines; 8 Yarrow boilers; 95,565kW (128,200hp)
SPEED:	30 knots (55.6km/h; 34.5mph)
RANGE:	4580nmi (8480km; 5269 miles) at 16 knots
ARMAMENT:	9 × 381mm (15in) guns; 12 × 152mm (6in) guns; 4 × 119mm (4.7in) guns; 12 × 89mm (3.5in) AA guns; 36 × 37mm (1.45in) or 20mm (0.79in) AA guns
ARMOUR:	Belt 280–70mm (11–2.65in); deck 162–45mm (6.37–1.75in); main turrets 350–200mm (13.78–7.87in)
COMPLEMENT:	1830–1950

Left: Littorio *was armed with high-velocity 381mm (15in) guns, which was the largest calibre that could be reliably built in Italy at the time.*

LITTORIO CLASS
The two *Littorio*-class battleships included de-capping armour plates, enhanced anti-torpedo protection, auxiliary rudders and an exceptional arc of salvo fire.

REAR TURRET
The rear main turret was mounted higher than normal to allow blast clearance for the ship's aircraft, typically a Reggiane Re 2000 fighter, which was carried unhangared.

TORPEDO PROTECTION
Vittorio Veneto and *Littorio* both incorporated Pugliese's torpedo protection system, formed of longitudinal cylinders within the ship's side, packed with tubes, intended to absorb the shock of a torpedo hit. On their inner side a longitudinal watertight bulkhead was intended to protect the machinery and magazines. It avoided the torpedo bulges that reduced the maximum speed of other battleships, but was criticized for the weakness of the inner bulkhead.

SECONDARY TURRETS
Flanking the three main turrets were the secondary 152mm (6in) guns, mounted in four triple turrets.

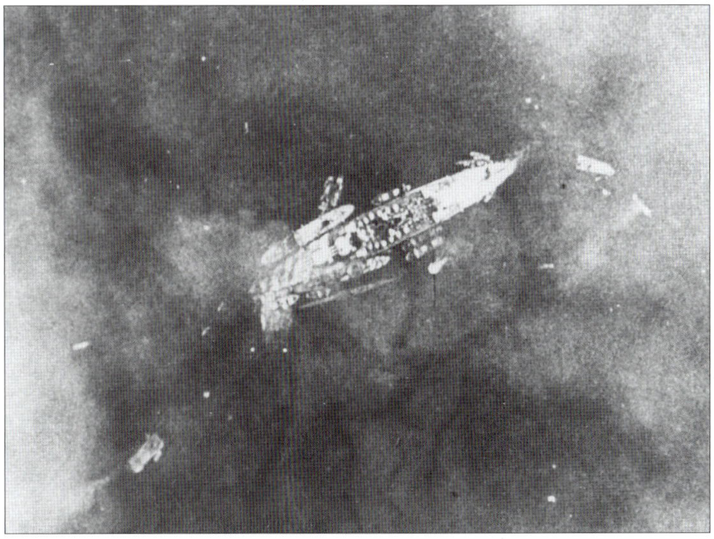

AERIAL VIEW
The *Littorio* is seen here during the famous British carrier attack on Italian warships at Taranto on 12 November 1940. The ship was hit twice by air-launched torpedoes, but survived.

Gato class (1940)

TYPE • *Submarine* **COUNTRY** • *United States*

SPECIFICATIONS
USS *Ray* (SS-271)

Dimensions:	Length: 95m (311ft 8in); Beam: 8.31m (27ft 3in); Draught: 5.2m (17ft 1in)
Displacement:	1549 tonnes (1708 tons) surfaced; 2463 tonnes (2716 tons) submerged
Propulsion:	4 diesel generator engines; combined output 4026kW (5400hp); 4 geared electric motors of 511kW (685hp) each; 2 screws
Speed:	21 knots (39km/h; 24mph) surfaced; 9 knots (16.6km/h; 10mph) submerged
Range:	11,000nmi (20,372km; 12,659 miles) at 10 knots surfaced; 95nmi (176km; 109 miles) at 5 knots submerged
Armament	(original): 10 × 533mm (21in) torpedo tubes; 24 torpedoes; 1 × 76mm (3in) 50-cal deck gun; 2 × 12.7mm (0.5in) machine guns
Complement:	60

The *Gato* class is the 'fleet submarine' that became a hunter-killer. Following on from two further prewar classes, *Tambor* and *Gar*, the *Gato* class satisfied the requirements for speed, power, range and armament.

Above: The first Gato *boat to be commissioned and enter combat, USS* Drum *(SS-228) is seen here in World War II mode. After suffering heavy depth-charging it was given a* Balao*-type conning tower at the end of 1943.*

Rate of fire
With the ammunition magazine located under the galley and crew's mess aft of the control room, a long passing chain was needed to get shells up to the gun in the forward spot, slowing down firing rates. Eventually, watertight ammunition lockers were placed in the fairwater under the forward gun platform.

USS *Ray*
This broadside view shows *Ray* in its original 1943 form. In the course of the war and afterwards, many details were changed during refits.

Operational depth
Double-hulled, except for the extremities of the submarine, built of mild steel 12mm (0.56in) thick, and with a welded outer hull, they had an operational depth of 91m (300ft).

REPURPOSING
The growing Japanese threat led the US Navy to cling to the concept of the fast fleet submarine even though the speed of battleships and carriers was starting to exceed anything a submarine could manage. When the Pearl Harbor attack changed the nature of a Pacific war beyond American expectations, the *Gato* class proved successful hunter-killers. In total, 77 were built, by Electric Boat at Groton, the US Navy yard at Kittery (Maine), Mare Island, and the Manitowoc yard in Wisconsin.

USS *GATO* (SS-212)
USS *Gato* (SS-212) photographed at the Mare Island Navy Yard, 8 August 1943. Circles mark recent alterations. The deck gun in the class was twice increased in size, first to 101mm (4in) in late 1943, then to 127mm (5in) from late 1944.

Bismarck (1940)

TYPE • *Battleship* **COUNTRY** • *Germany*

SPECIFICATIONS

DIMENSIONS:	Length: 251m (823ft 6in); Beam: 36m (118ft 1in); Draught: 9.3m (30ft 6in)
DISPLACEMENT:	44,905 tonnes (49,500 tons) standard
PROPULSION:	12 Wagner high-pressure boilers; 3 Brown-Boveri geared turbines developing 111,982kW (150,170hp)
SPEED:	29 knots (53.7km/h; 33.3mph)
RANGE:	8870nmi (16,430km; 10,210 miles) at 19 knots
ARMAMENT:	8 × 380mm (15in) guns; 12 × 150mm (5.9in) guns; 16 × 105mm (4.1in) guns; 16 × 37mm (1.5in) guns; 12 × 20mm (0.79in) AA guns
ARMOUR:	Belt 320–80mm (12.6–3.1in); bulkheads 220–45mm (8.6–1.7in); deck 120–80mm (4.7–3.1in); barbettes 340–220mm (13.4–8.6in); turrets 360–180mm (14.2–7.1in)
COMPLEMENT:	2291

***Bismarck* and *Tirpitz* represented the peak of German battleship design in terms of size, armament, speed and endurance. However, despite achieving a major success, *Bismarck*'s first combat mission was also its last.**

Right: A view of Bismarck, *as seen from* Prinz Eugen *during the early stages of Operation* Rheinübung, *the famous Atlantic commerce raid, in May 1941.*

TURRETS
A cutaway side profile of *Bismarck*. German practice was to name the main turrets A, B, C, D from front to back as Anton, Bruno, Cäsar and Dora.

RADAR EQUIPMENT
Bismarck carried FuMO radar equipment on the forward and aft rangefinders and the foretop, although blast concussion disabled it. It also had hydrophone detectors.

PROPULSION
Initial plans for propulsion – later dropped – featured turbo-electric drive, in a compact arrangement with boilers on each side, the generator in the middle and an electric motor for each shaft.

HUNTING *BISMARCK*

On 24 May 1941, the battlecruiser HMS *Hood* was sunk in the Atlantic by *Bismarck*'s fifth salvo. A huge Allied pursuit followed, and *Bismarck* was hit by Swordfish torpedo bombers from HMS *Ark Royal* on 26 May, jamming the port rudder. On 27 May, the battleships HMS *Rodney* and *King George V* opened fire. By around 10.00 hrs, all *Bismarck*'s guns were disabled and the order was given to scuttle. Hit by British torpedoes, it sank with 1977 men lost.

INSPECTION
Adolf Hitler inspects and salutes the almost completed *Bismarck* at the Gotenhafen (Gdynia) naval base on 5 May 1940. Although *Bismarck* was powerful, Hitler lost interest and belief in naval surface warfare.

Richelieu (1940)

SPECIFICATIONS

Dimensions:	Length: 247.9m (813ft 4in); Beam: 33m (108ft 3in); Draught: 9.7m (31ft 10in)
Displacement:	43,987 tonnes (48,487 tons) standard
Propulsion:	6 Indret superheated boilers; 4 Parsons geared turbines; 4 screws; 111,855kW (150,000shp)
Speed:	30 knots (56km/h; 35mph)
Range:	9850nmi (18,240km; 11,340 miles)
Armament:	8 x 380mm (15in) guns; 9 x 152mm (6in) guns; 12 x 100mm (3.9in) AA guns; 14 x 37mm (1.5in) AA guns
Armour:	Belt 343–250mm (13.5–9.8in); upper deck 170–130mm (6.7–5.1in); barbettes 405mm (16in); turrets 430–195mm (16.9–7.7in)
Complement:	1550

The *Richelieu* class of fast battleships had the honour of being the last battleships built for the French Navy. Two of the class were built between 1935 and 1955, with two planned upgraded variants never laid down.

Above: Richelieu*'s side belt went deeper than its contemporaries: 7.01m (23ft), compared to 6.09m (20ft) on* Bismarck *and 5.46m (17ft 11in) on* Iowa.

Machinery
The machinery was arranged in successive rooms, fore to aft: boiler room 1/turbines for outer screws/boiler room 2/turbines for inboard screws, with the aim of ensuring mobility even if one set were out of action.

Main guns
The eight 350mm (15in) guns were housed in two separate compartments inside each turret, with an armoured bulkead between. They could elevate to 35 degrees, with a muzzle velocity of 830m/s (2700ft/s).

Aircraft
The ship's hangar could hold two SNCAC NC.420 float planes, with a third on the ship's rear catapult.

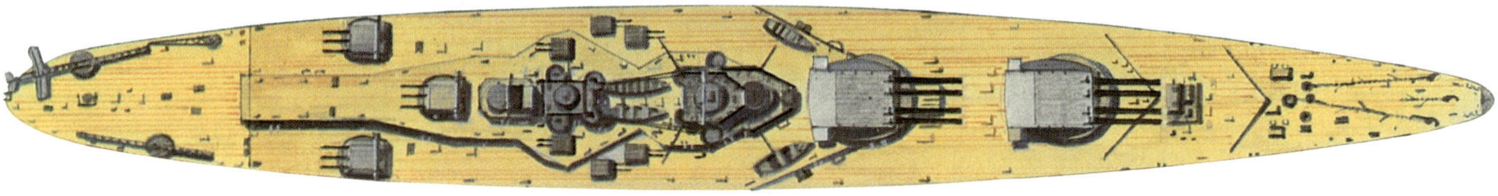

CONSTRUCTION

Richelieu was laid down at Brest in 1935. During construction, numerous changes were made to the original design. The funnel and aft tower were merged into a single structure, with the funnel angled sharply towards the stern. Despite the ever-increasing size and displacement of capital ships, designers were always looking for possible ways to save weight, if only to reuse it elsewhere, and this was one solution.

INDIAN OCEAN, 1944
The *Richelieu* cuts through heavy seas in the Indian Ocean in 1944, its quad 380mm (15in) gun turrets standing proud of the water. From 1943, *Richelieu*, as France's largest battleship, was a powerful symbol of Free France while the country was under German occupation.

Yamato (1940)

TYPE • *Battleship* **COUNTRY** • *Japan*

SPECIFICATIONS

Dimensions:	Length: 263m (862ft 9in); Beam: 36.9m (121ft 1in); Draught: 10.39m (34ft 1in)
Displacement:	61,698 tonnes (68,010 tons) standard; 65,008 tonnes (71,670 tons) full load
Propulsion:	12 Kampon high-pressure boilers; 4 Kampon turbines developing 111,855kW (150,000hp); 4 screws
Speed:	27 knots (50km/h; 31mph)
Range:	7200nmi (13,330km; 8,280 miles) at 16 knots
Armament:	9 × 460mm (18.1in) guns; 12 × 155mm (6.1in) guns; 12 × 127mm (5in) guns; 24 × 25mm (0.98in) AA guns; 4 × 13.2mm (0.52in) guns
Armour:	Belt 410mm (16in); deck 230–200mm (9.1–7.9in); barbettes 546–50mm (21.5–2in); turrets 650–193mm (19.7–11.8in)
Complement:	2500

The Japanese *Yamato* class were planned as 'super-battleships': bigger, more heavily armed and better protected than anything else afloat. Designed to enforce Japan's mastery of the Pacific, they made a minimal contribution to the country's war effort.

Forward profile
Striking aspects of the ship's appearance included the typical Japanese tall tower, more compact than on converted older ships, and a single large backwards-tilted funnel.

Above: The hull shape was carefully worked out to bear the huge stresses. The stem design was estimated to reduce water resistance by 8.2 per cent; the overall design gave a power saving of 11,797kW (15,820hp).

Turrets
The turrets were formed of two concentric armoured cylinders, the inner one of 1.5m (4ft 10in) containing a hoist for four men.

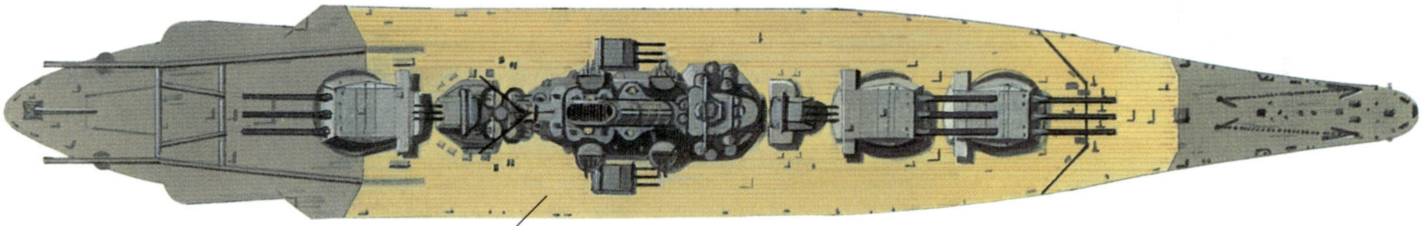

Design
The deck plan reveals the distinctive hull shape, reaching maximum beam towards the stern. The 'wings' carried Type 96 25mm (0.98in) AA guns in triple mounts.

FATE OF THE *YAMATO*
Despite their evident power, both the *Yamato* and its equally impressive companion ship *Musashi* did not survive the war. *Yamato*'s last sortie was a one-way suicide mission against US invasion forces around Okinawa. Detected before its guns could have an impact, the great ship was destroyed in a relentless attack by US carrier aircraft on 7 April 1945.

UNDER CONSTRUCTION
Here *Yamato* is seen in the late stages of construction alongside the large fitting out pontoon at the Kure Naval Base, Japan, on 20 September 1941. Note *Yamato*'s 460mm (18.1in) main battery gun turret and superfiring 155mm (6.1in) secondary battery gun turret.

Vittorio Veneto (1940)

TYPE • *Battleship* **COUNTRY** • *Italy*

SPECIFICATIONS

Dimensions:	Length: 237.8m (778ft 9in); Beam: 32.9m (107ft 9in); Draught: 9.6m (31ft 5in)
Displacement:	42,040 tonnes (46,340 tons) standard
Propulsion:	4 shafts; 8 Yarrow boilers; 4 Belluzzo geared turbines; 100,383kW (134,616shp)
Speed:	31.4 knots (58.2km/h; 36.1mph)
Armament:	9 × 381mm (15in) guns; 12 × 152mm (6in) guns; 4 × 120mm (4.7in) guns; 12 × 90mm (3.5in) AA guns; 20 × 37mm (1.5in) AA guns; 16 × 20mm (0.79in) AA guns
Armour:	Belt 350–60mm (13.8–2.4in); barbettes 350mm (13.8in); turrets 350–100mm (13.8–3.9in); deck 205–35mm (8.1–1.4in)
Complement:	1861

Classified officially as *corazzate* (literally 'armour-clads'), Italy's last battleship class consisted of four ships, but only three were completed, two of which saw active service. *Vittorio Veneto* was launched on 25 July 1937 and completed on 28 April 1940.

Above: On 8 September 1943, Italy's fleet was surrendered to the Allies. Vittorio Veneto *was interned, but returned to Italy in February 1946 and allocated to Britain as reparation. Decommissioned on 1 February 1948, it was scrapped at La Spezia between then and 1950.*

Main guns
With a maximum elevation of 35 degrees, the main guns had a range of 42,260m (138,648ft) and fired both AP (armour-piercing) shells of 885kg (1950lb) and HE shells of 774kg (1710lb). They fired a round every 45 seconds.

Design
Vittorio Veneto and *Littorio* were launched with slightly raking stems, giving a waterline length of 236m (774ft 3in). After trials, they were given straight stems down to the waterline, resulting in an increase of 1.8m (4ft 6in).

Torpedo defences
Against torpedoes it had the Pugliese system: a lateral space between the torpedo bulkhead and the inner hull bulkhead, holding a long empty cylinder, maximum diameter 3.8m (12ft 6in), suspended in oil, and intended to absorb explosive energy.

CONVOY RAIDER
Assigned to the 9th Division of the 1st Squadron, from 31 August to the end of 1940, *Vittorio Veneto* was frequently in action against British convoys to Malta. For example, it was the Italian flagship in the Battle of Cape Matapan, on 27 March 1941, when it was hit by a torpedo launched by a bomber from HMS *Formidable*. Taking on around 4000 tonnes (4409 US tons) of water, it struggled back to Taranto. The ship was torpedoed again the following August, but also survived.

VITTORIO VENETO, 1940
The lengthy construction time of the *Vittorio Veneto* reflected difficulties in the provision of materials and equipment, and much of the build quality was poor. However, the general appearance was elegant.

Type KD6 and KD7 (1940)

TYPE • *Submarine* **COUNTRY** • *Japan*

The line of KD (*Kaidai*) submarines began in 1925 with KD1. All *Kaidai*-class submarines originally had a two-digit boat name, from I–51 onwards. The KD-7 (*I-176* class) of 10 boats was the final specification.

SPECIFICATIONS

Dimensions:	Length: 105.5m (346ft 1in); Beam: 8.25m (27ft 1in); Draught: 4.6m (15ft 1in)
Displacement:	1656 tonnes (1825 tons) surfaced; 2644 tonnes (2915 tons) submerged
Propulsion:	2 diesels of 2983kW (4000hp); 2 671kW (900hp) 2 motors; 2 screws
Speed:	23.1 knots (42.8km/h; 26.6mph) surfaced; 8 knots (14.8km/h; 9.2mph) submerged
Range:	8000nmi (14,816km; 9206mi) at 16 knots; 50nmi (93km; 57.8mi) at 5 knots
Armament:	6 × 533mm (21in) torpedo tubes; 12 torpedoes; 1 × 120mm (4.7in) L/40 deck gun; 2 × 25mm (1in) Type 96 AA guns
Complement:	86

Right: The Japanese submarine I-176, a KD7 class was sunk on 16–17 May 1944 in a prolonged US depth-charge attack off the Solomon Islands.

PROPULSION
Propulsion was by two-stroke Kampon Mk 1B Model 8 diesels giving 5966kW (8000shp).

I-73
I-73 of Type KD6A, was torpedoed on the surface and sunk by USS *Gudgeon* (SS211) on 27 January 1942.

I-7
This was an expanded version of the KD-6 patrol submarine design. The intended scouting role of *I-7* and *I-8* was heavily curtailed by US use of carrier-borne attack planes, although both were refitted with extra AA guns in 1942.

TRANSPORTER
Unlike its immediate predecessors, the KD-7 class had only medium-range endurance, although it was still classified as an ocean-going attack boat. They were not well adapted to attack mode, however, despite being fast on the surface (destroyers were faster). It did not handle well when submerged, and after 1942 three were converted to transports, for which their high surface speed was well-suited.

***I-168* (KD6A CLASS)**
I-168 achieved fame for sinking two US warships during the Battle of Midway in June 1942: the damaged USS *Yorktown* aircraft carrier and the destroyer USS *Hammann*.

USS *Wasp* (1941)

TYPE • *Aircraft carrier* **COUNTRY** • *United States*

SPECIFICATIONS

Dimensions:	Length: 209.7m (687ft 10in); Beam: 24.6m (80ft 8in); Draught: 6.1m (20ft)
Displacement:	14,900 tonnes (16,425 tons) standard; 19,423 tonnes (21,410 tons) full load
Propulsion:	6 Yarrow boilers; 2 steam turbines; 2 shafts
Speed:	29.5 knots (54.6km/h; 33.9mph)
Range:	12,000nmi (22,224km; 13,810mi) at 15 knots
Armament:	8 × 127mm (5in) guns; 16 × 28mm (1.1in) AA guns; 24 × 12.7mm (0.5in) machine guns
Armour:	Belt 89mm (3.5in)
Complement:	2167

Wasp (CV-7) was the last USN carrier built under the constraint of the 1922 Washington and 1930 London naval treaties, having been ordered in 1935. It was launched on 4 April 1939 and commissioned on 25 April 1940.

Left: Anchored at Guantanamo Bay, Cuba, here we see USS Wasp *in its 'dressed ship' presentation for Navy Day on 27 October 1940.*

Design
Wasp was designed on standard lines, with a raised flight deck above the main deck, which was also the hangar deck. The island, slightly forward of midships, was of three decks.

Flight deck
The flight deck was 25m (82ft) shorter than *Yorktown*'s, with three lifts, the aft one being partly under the arrestor wires.

Deck-edge elevator
The deck-edge elevator was fitted as an experiment but was quickly recognized as beneficial in flight deck and hangar operations.

REQUIREMENTS

Intended as a replacement for the US Navy's first carrier, *Langley*, and designated CV-7, *Wasp* was laid down on 1 April 1936, launched on 4 April 1939 and commissioned on 25 April 1940. In October 1940, after further trials and training, it joined the fleet. Its dimensions and capacity were governed by the 1922 Naval Treaty. The new carrier was only about three-quarters the displacement of *Enterprise*, and it cost $20 million.

FATAL BLOWS
Wasp burns and lists after it was torpedoed by the Japanese submarine *I-19* on 15 September 1942, while operating in the southwestern Pacific in support of forces on Guadalcanal. Emergency teams were unable to control the fires, and the ship was abandoned and scuttled by the end of the day.

Type KS *RO-100* (1941)

TYPE • *Submarine* **COUNTRY** • *Japan*

SPECIFICATIONS

DIMENSIONS:	Length: 60.9m (199ft 10in); Beam: 6m (19ft 8in); Draught: 3.51m (11ft 6in)
DISPLACEMENT:	611 tonnes (674 tons) surfaced; 795 tonnes (877 tons) submerged
PROPULSION:	2 diesels developing 375kW (500hp); 2 electric motors developing 285kW (380hp); 2 screws
SPEED:	14.2 knots (26.3km/h; 16.3mph) surfaced; 8 knots (14.8km/h; 9.2mph) submerged
RANGE:	3500nmi (6482km; 4028 miles) at 12 knots surfaced; 60nmi (111km; 69 miles) at 3 knots submerged
ARMAMENT:	4 × 533mm (21in) torpedo tubes; 4 × 25mm (1in) Type 96 AA guns
COMPLEMENT:	38

The *RO-100* series, also known as *Kaisho* (small) or Type KS was an 18-strong class developed from 1940 and completed by 1944. Designed for coastal defence, with an operating depth of 75m (245ft), its size meant it carried a limited payload of eight 533mm (21in) torpedoes.

Right: An aerial photograph of a Japanese Type KS submarine. The image was taken on 16 March 1943 near Rabaul, New Britain by an Allied aircraft.

RO-100
Completed in August 1942, this was the class leader. It was based at Rabaul, New Guinea, making combat patrols and supply runs in the region until sunk by a mine near Buin, Bougainville, on 25 November 1943.

DIVING CAPABILITY
The diving limit was 74m (254ft). Despite excellent underwater manoeuvrability, the class was swept away by the ASW methods of Allied destroyers and aircraft.

RADAR
Japanese radar systems lagged behind their American and British counterparts in both sophistication and reliability. Radar was fitted on submarines only from April 1944. This was a Type 13 air search, with a 97km (60 miles) range for detection of air groups, and 48km (30 miles) for a single aircraft. It had a dipole antenna array with mat reflector. For surface search Type 22 was carried, with a range of 34km (21 miles). It had a distinctive horn-type antenna. For passive radar detection, the Type E27/3 was used.

HYDROPHONES
Japan had used hydrophones in underwater detection since 1930. The Type 93 Model 1 system had a range of 9000m (29,528ft). It used 16 receivers set in two elliptical arrays on each side of the vessel's keel.

KS CLASS
Three representatives of the KS class: (from left to right) *RO-101, 104, 102,* During the war, the need to operate further out made life difficult for the class. *RO-100*s sank 33,021 tonnes (36,399 tons) of Allied merchant shipping, but the class was severely overextended.

USS *Washington* (1941)

TYPE • *Battleship* **COUNTRY** • *United States*

SPECIFICATIONS

Dimensions:	Length: 222.1m (728ft 9in); Beam: 33m (108ft 4in); Draught: 10m (32ft 10in)
Displacement:	35,562 tonnes (39,200 tons) standard; 45,519 tonnes (50,176 tons) full load
Propulsion:	8 Babcock & Wilcox boilers; 4 General Electric steam turbines; 4 screws; 90,000kW (121,000shp)
Speed:	28 knots (52km/h; 32mph)
Range:	17,450nmi (32,320km; 20,080 miles) at 15 knots
Armament:	9 × 406mm (16in) guns; 20 × 127mm (5in) dual-purpose guns; 16 × 28mm (1.1in) AA guns; 18 × 12.7mm (0.5in) machine guns
Armour:	Belt 305mm (12in); gun turrets 406mm (16in); deck 140mm (5.5in); conning tower 406mm (16in)
Complement:	1800

Washington's plans complied with the London Treaty, but when Japan refused to ratify the agreement, 406mm (16in) guns were installed, reducing top speed by 2 knots. With _South Dakota_, it sank the Japanese battlecruiser _Kirishima_ at Guadalcanal in November 1942.

Left: This view of Washington, *taken in September 1945, shows six of its nine 406mm (16in) Mark 6 guns. As well as naval surface engagements, the guns proved useful for long-range offshore fire support of land forces.*

Catapults
Washington carried three Vought OS2U Kingfisher float planes. These were launched by a pair of aircraft catapults on the fantail.

Advanced upgrades
During its service career the ship received three Mark 3 fire-control radar sets to control the main guns, four Mark 4 radars for the secondary armament, a CXAM air-search radar, and an SG surface-search radar.

AA guns
The ship's AA armament was constantly improved over the course of the war, until it was bristling with air defence firepower. By August 1943, it had 60 40mm (1.6 in) Bofors guns alone, set in quad mounts.

WAR SERVICE
Given its commission date – 15 May 1941 – the USS *Washington* was virtually destined to define itself through combat service. Following the US entry into war in December 1941, *Washington* was initially deployed to reinforce the British Home Fleet and conduct Atlantic operations, but from August 1942 was deployed to the Pacific. There it found itself in frequent combat, supporting US campaigns from Guadalcanal to Okinawa. The ship survived the war and stayed on the US inventory until 1960.

PUGET SOUND NAVY YARD, 1944
USS *Washington*, as seen on 26 April 1944, following repair of massive damage to its bow after it collided with the battleship *Indiana* on 1 February 1944. Six sailors aboard *Washington* were killed in the accident, four on *Indiana*.

HMS *Prince of Wales* (1941)

TYPE • *Battleship* **COUNTRY** • *United Kingdom*

SPECIFICATIONS

Dimensions:	Length 227m (745ft 1in); Beam: 31.4m (103ft 2in); Draught: 10.5m (34ft 4in)
Displacement:	38,610 tonnes (42,560 tons) standard
Propulsion:	8 Admiralty 3-drum boilers; 4 Parsons geared steam turbines; 4 screws; 82,000kW (110,000shp)
Speed:	28.3 knots (52km/h; 32mph)
Range:	15,600nmi (28,900km; 17,957 miles) at 15 knots
Armament:	10 × 356mm (14in) guns; 16 × 133mm (5.25in) dual-purpose guns; 32 × 40mm (1.5in) AA guns; 80 × UP projectors
Armour:	Belt 381–114mm (15–4.5in); bulkheads 381mm (15in); deck 152–127mm (6–5in); barbettes 330–280mm (13–11in); turrets 330–152mm (13–6in)
Complement:	1612

All five ships of the *King George V* class were laid down in 1937, with *Prince of Wales* beginning construction at Cammell Laird's yard in Birkenhead in January. The second ship of the class, it had a brutally short career, becoming the first battleship to be sunk by aircraft on the open sea.

Design
The design was balanced and handsome, with two widely spaced funnels, the fore-funnel marginally taller than the aft one, between two tall tripod masts, and the now standard tower structure rising directly behind the forward turrets, with a high-set enclosed navigation bridge.

Armour
Protection was on the 'all or nothing' principle, and armour amounted to 40 per cent of design displacement, based on the realistic awareness that it would have to face 406mm (16in) guns in combat.

Above: Aircraft were withdrawn from surviving King George V*-class ships during the course of the war.*

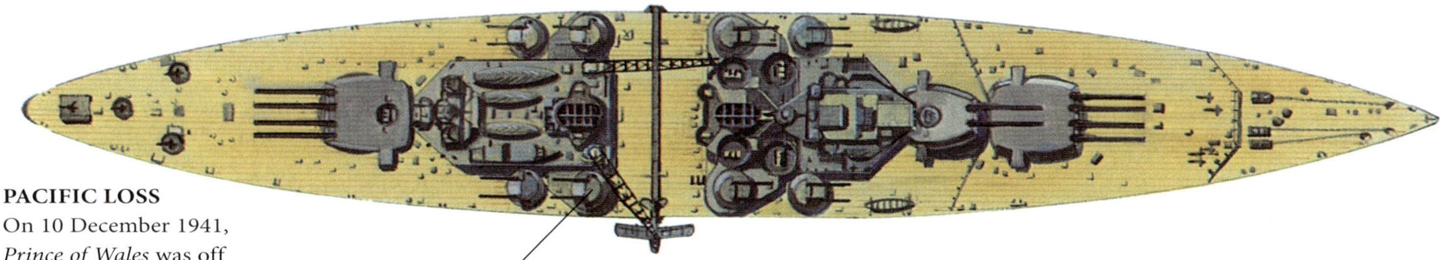

PACIFIC LOSS
On 10 December 1941, *Prince of Wales* was off the east Malayan coast as flagship of Force 'Z' with the battlecruiser *Repulse* and four destroyers when they were attacked by waves of Japanese naval aircraft. The ships opened intensive AA fire, increased speed and began zig-zagging, but control of *Prince of Wales*'s port propellers was lost and its manoeuvrability was restricted. After many bomb and torpedo hits, both battleships were sunk. The escorting destroyers rescued 1285 men from *Prince of Wales*.

Turrets
The *King George V*-class were the first British battleships to have quadruple main turrets; they carried 12 356mm (14in) guns in three quadruple turrets.

DEFENSIVE FIREPOWER
Prince of Wales carried far less AA defence than major ships would do later in the war, although it was probably regarded at the time as well-equipped to deal with air attack, through its High Angle Control System, which was linked to long-range air-search radar.

Tirpitz (1941)

TYPE • *Battleship* **COUNTRY** • *Germany*

SPECIFICATIONS

Dimensions:	Length: 251m (823ft 6in); Beam: 36m (118ft 1in); Draught: 9.3m (30ft 6in)
Displacement:	38,283 tonnes (42,200 tons) standard
Propulsion:	12 Wagner superheated boilers; 3 Brown, Boveri geared steam turbines; 3 screws; 102,906kW (138,000hp)
Speed:	30 knots (56km/h; 35mph)
Range:	8870nmi (16,430km; 10,210 miles) at 19 knots
Armament:	8 × 380mm (15in) guns; 12 × 150mm (5.9in) guns; 16 × 105mm (4.1in) guns; 16 × 37mm (1.5in) AA guns; 12 × 20mm (0.79in) AA guns
Armour:	Belt 320–80mm (12.6–3.1in); deck 120–80mm (4.3–1.7in); barbettes 340–220mm (13.4–8.6in); turrets 360–180mm (14.2–7in)
Complement:	2065

As *Bismarck*'s companion ship, *Tirpitz* shared similar qualities of power and resistance. Although its active service was limited, its mere presence tied up substantial British resources until its ultimate destruction.

Above: The open fore part of the superstructure, hardly higher than 'B' turret, was rimmed out to deflect blast from gunfire.

Paint scheme
Tirpitz is depicted here in 'Paint Scheme K', which it carried between March and July 1944. The camouflage style was frequently changed to confuse aerial observers.

Funnel
At first black, the funnel cowling was painted silver or light grey from 1942. The catwalk from the funnel platform to the tower was added in 1942.

DESTRUCTION
On 12 November 1944, 32 RAF Lancaster bombers converged on *Tirpitz* from 09.35 hrs, when it was at anchor in Norway. The ship's main guns opened up on them, but there was no aerial opposition from the Luftwaffe. The Lancasters scored at least two direct hits and four near misses with 'Tallboy' 5080kg (11,199lb) bombs carrying around 3000kg (6614lb) of high explosive. The most destructive hit was between the aircraft catapult and funnel, which blew a massive hole in the side armour.

Anti-torpedo protection
Void spaces of 5.4–3m (17ft 8in–9ft 10in) between the outer hull and inner longitudinal bulkheads of *Wotan weich* (Ww) armour provided anti-torpedo protection.

***TIRPITZ* IN NORWAY**
Tirpitz spent much of its wartime service in Norwegian waters. Although the ship did not have a 'glorious' career. Its influence on the Arctic naval war was important.

Ariete (1941)

TYPE • *Torpedo boat* COUNTRY • *Italy*

SPECIFICATIONS

DIMENSIONS:	Length: 83.5m (273ft 11in); Beam: 8.62m (28ft 3in); Draught: 3.15m (10ft 4in)
DISPLACEMENT:	757 tonnes (833 tons) standard; 1118 tonnes (1232 tons) full load
PROPULSION:	2 boilers; 2-shaft geared steam turbines; 16,400kW (22,000hp)
SPEED:	31.5 knots (58.3km/h; 36.2mph)
ARMAMENT:	2 × 100mm (4in) dual-purpose guns; 10 × 20mm (0.79in) AA guns; 6 × 450mm (18in) torpedo tubes (2 triple mounts); 20 mines
ARMOUR:	None
COMPLEMENT:	158

Built at Trieste, *Ariete* was the lead ship of Italy's final class of World War II destroyers, and the only one of the class to be commissioned before the surrender of Italy in 1943. After the war, it was transferred to Yugoslavia, and served as *Durmitor* until 1963.

Above: Seen here off the coast of Yugoslavia is the ex-German TA47 Balestra, *an* Ariete*-class torpedo boat. In Yugoslav Navy service from 1949, it was known as* Ucka.

POWERPLANT
The ship was propelled by two-shaft geared steam turbines, developing a total of 16,400kW (22,000hp).

CAMOUFLAGE
Ariete is shown here in a fractured camouflage pattern, ideal for breaking up the ship's profile against a grey seascape.

TORPEDOES
The *Ariete* class put a heavier focus on torpedo armament than the previous *Spica* class of torpedo boats, with six tubes mounted on the centreline.

FAILED PRODUCTION

The *Ariete* class had a particularly troubled production history. In total, 42 of the vessels were planned for development, but in the early 1940s only 16 were ultimately laid for construction, owing to supply problems caused by the war. Of these 16, only one was actually completed, the rest taken into the hands of the Germans and completed for Kriegsmarine service.

ARIETE CLASS
Dragone was constructed by the famous Ansaldo shipbuilders in Genoa, laid down in July 1942 and completed in August 1943.

HMS *Anchusa* (*Flower* class) (1941)

TYPE • *Corvette* COUNTRY • *United Kingdom*

SPECIFICATIONS

Dimensions:	Length: 62m (205ft); Beam: 10.11m (33ft 2in); Draught: 4.8m (15ft 9in)
Displacement:	1190 tonnes (1,312 tons) standard; 1410 tonnes (1554 tons) full load
Propulsion:	2 boilers; 1 vertical triple-expansion steam engine generating 2051kW (2750hp); 1 shaft
Speed:	16.5 knots (30.6km/h; 19mph)
Armament:	1 × 102mm (4in) BL Mk IX gun; 40 depth charges; 72 Hedgehog rockets
Armour:	None
Complement:	85–109

The bulk of convoy escort duties were handled by corvettes from the Royal Navy and the Royal Canadian Navy. Most of these were *Flower*-class or modified *Flower*-class ships, which played one of the most important yet unsung roles of the entire war at sea.

Left: Anchusa rests at anchor in Liverpool on 21 January 1943. The ship would survive the war.

Hedgehog
The chief weapon was the Hedgehog anti-submarine mortar. Hedgehog consisted of high-explosive depth-charges fired in groups, which when fired straddled the target U-boat and detonated on impact.

Propulsion
The ship was powered by two boilers driving one VTE engine generating 2051kW (2750hp).

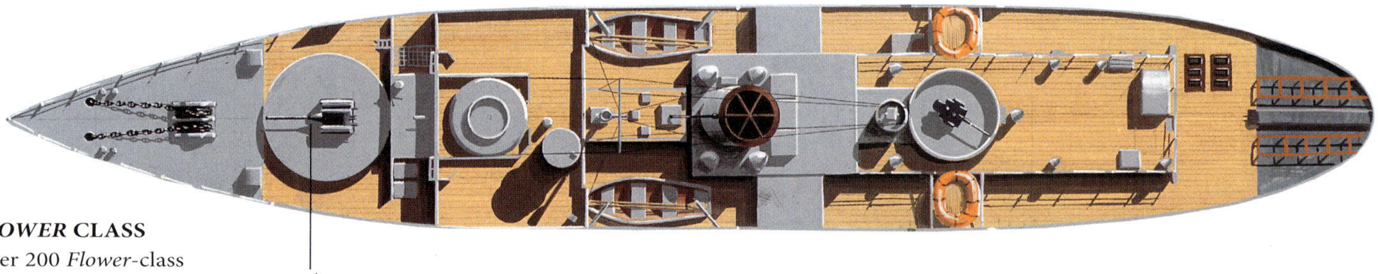

***Flower* Class**
Over 200 *Flower*-class ships were built. The class was based on a prewar whaler design drawn up by Smith's Docks, which was lengthened and fitted out with military equipment. They were built in 30 different ship yards in the UK and Canada, with the highest number being produced by Belfast's Harland & Wolff. HMS *Anchusa* was one of the 34 *Flower*-class ships built in Northern Ireland, launched between 1940 and 1942.

Armament
By the end of the war the weapons fit on most of the *Flowers* had changed at least once, although most ships kept the 102mm (4in) main gun.

HMS *Polyanthus*
HMS *Polyanthus* was launched in November 1940, but sunk in a U-boat attack in September 1943. A total of 34 *Flower*-class ships were lost during the course of the war. The role of the corvette was to escort merchant shipping in convoy and, if possible, to find and attack enemy submarines.

Fairmile Type C (1941)

TYPE • *Motor gun boat* **COUNTRY** • *United Kingdom*

SPECIFICATIONS

DIMENSIONS:	Length: 35.66m (117ft); Beam: 5.3m (17ft 5in); Draught: 1.73m (5ft 8in)
DISPLACEMENT:	69 tonnes (76 tons) standard; 75 tonnes (83 tons) deep load
PROPULSION:	3 Hall-Scott petrol engines; 3 shafts; 2013kW (2700hp)
SPEED:	27 knots (50km/h; 31mph)
RANGE:	500nmi (930km; 580 miles)
ARMAMENT:	2 × 40mm (2-pounder) AA guns; 4 × 12.7mm (0.5in) machine guns; 4 × 7.6mm (0.3in) machine guns (later 6 × 20mm (0.79in) and 4 × 7.6mm (0.3in) guns)
ARMOUR:	None
COMPLEMENT:	16

The Fairmile series of motor gun boats (MGBs) were developed from the earlier Type A and Type B motor launches/minelayers, which led to the Type C and larger Type D MGBs. The first Type C was launched in 1941 and, of the 23 built, five were lost during the war.

Right: MTBs were built in large numbers by a range of shipyards such as Vosper, Whites, Camper & Nicholson, British Power Boats and Thornycroft, and to a range of different designs.

PERFORMANCE
The Type C could achieve speeds of 50km/h (27 knots; 31mph), and its shallow draught meant it could sail close to the shoreline.

ARMAMENT
The Type C had only light firepower, but plenty of it for its size, the heaviest weapons being two 2-pounder automatic AA guns, which could also be used against surface targets.

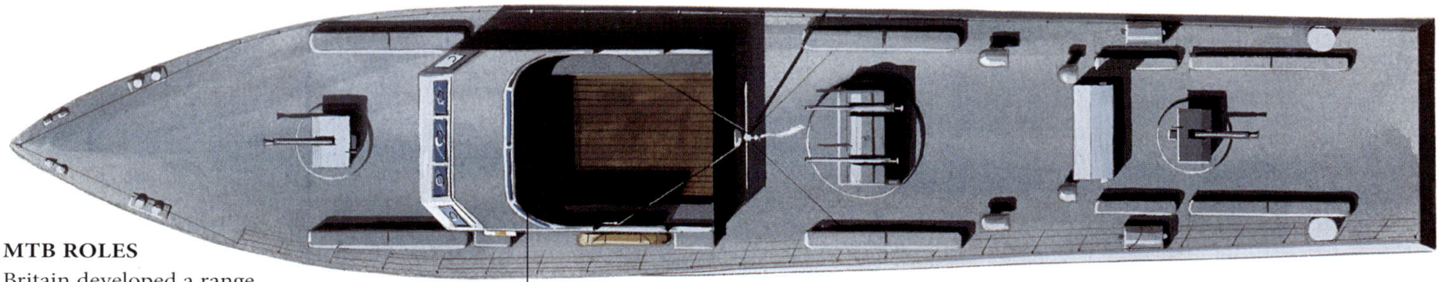

MTB ROLES
Britain developed a range of motor torpedo boats (MTBs), motor gun boats (MGBs) and motor anti-submarine boats (MA/SBs) to serve with the coastal forces. All eventually became grouped under the MTB classification. Their role was to harry enemy shipping and guard against intruders. They were also tasked with specialist roles such as inserting commando teams, minelaying/minesweeping and providing fast targets for other units.

COMPLEMENT
The vessel had an overall length of 35.66m (117ft) and was operated by a crew of two officers and 14 enlisted men.

TYPE C MOTOR GUN BOATS
Fairmile Type C MGBs off Dover (MGB Q328 + MGB 330). MGBs came in various formats, some heavily armed with deck-mounted weapons, although the movement of such light ships in heavier seas made accurate gunfire difficult.

CAM and MAC Ships (1941)

TYPE • *Merchant aircraft carrier* **COUNTRY** • *United Kingdom*

SPECIFICATIONS
MAC tanker MV *Ancylus*

Dimensions:	Length: 142m (465ft); Beam: 18m (59ft); Draught: 8.4m (27ft 6in)
Displacement:	8145 tonnes (8979 tons) standard; 16,256 tonnes (17,917 tons) full load
Propulsion:	1 diesel engine; 1 shaft; 2610kW (3500bhp)
Speed:	13 knots (24km/h; 15mph)
Range:	10,000nmi (18,520km; 11,510 miles)
Armament:	1 × 100mm (4in) gun; 8 × 20mm (0.787in) AA guns
Armour:	None
Aircraft:	3

Mobilized into action as convoy protection vessels, the CAM (catapult aircraft merchant) and MAC (merchant aircraft carrier) ships played an important role during the Battle of the Atlantic, where warfare was conducted on three levels: sea surface, undersea and in the air.

Left: MAC ships were a more highly developed form of the CAM ships. As converted bulk carriers, mostly built as grain ships and oil tankers, they were both cargo carriers and capable of launching and landing assault planes.

Crew
CAM ships had a larger crew than normal merchantmen and carried additional rescue boats in the form of Carley floats.

Island
The island, like the rest of the structure, was made as much as possible of prefabricated parts.

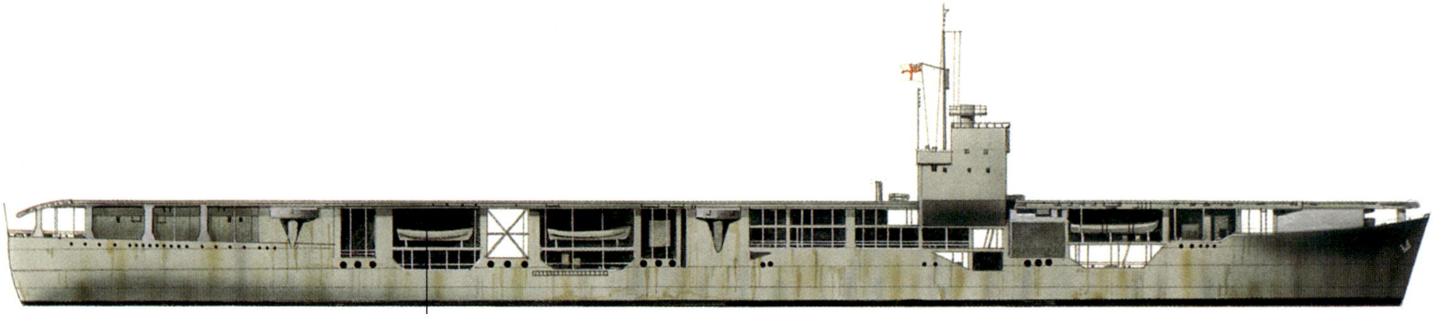

CAM SHIPS

In 1941 the Royal Navy had no carriers to spare for Atlantic patrols. Instead, the idea of the CAM (catapult aircraft merchant ship) was developed. Ships were already available: what was needed was a system mounted on the forecastle to launch a fighter plane defence against air attack. In total, 35 merchant ships were fitted with bow catapults that used a cordite-powered rocket to drive a launch trolley.

Aircraft
The 19 'Empire'-named MAC ships flew Swordfish Mk I and Mk II torpedo bombers whose biplane construction allowed take-off at lower speeds than Sea Hurricanes.

ROCKET CATAPULT
A Sea Hurricane IA (V6756) aboard CAM ship *Empire Tide.* The rocket catapult was 22.8m (75ft) long, sited on the port side of the ship's forecastle and extending to the foremast.

HMS *Audacity* (1941)

TYPE • *Light aircraft carrier* **COUNTRY** • *United Kingdom*

SPECIFICATIONS

DIMENSIONS:	Length: 142.4m (467ft 3in); Beam: 17.15m (56ft 3in); Draught: 8.4m (27ft 6in)
DISPLACEMENT:	10,230 tonnes (10,068 tons) standard; 12,192 tonnes (12,000 tons) full load
PROPULSION:	One 7-cylinder MAN diesel engine; 1 shaft; 3900kW (5200bhp)
SPEED:	15 knots (28km/h; 17mph)
RANGE:	Not known
ARMAMENT:	1 × 102mm (4in) gun; 4 × 40mm (2-pdr) AA guns; 4 × 20mm (0.787in) AA guns
AIRCRAFT:	6
COMPLEMENT:	480

There was an element of audacity in the decision to convert the German merchant ship *Hannover*, seized in the Atlantic in March 1940, into Britiain's first escort carrier. Its brief career as a warship clearly showed the value of such vessels.

Above: HMS Nairana *was first of a class of three escort carriers built in the UK on merchant ship hulls. Entering service on 26 November 1943, it carried up to 20 aircraft and escorted Arctic and Atlantic convoys.*

HMS AUDACITY
A minimal island unit, placed forward, was added to *Audacity* some time after completion to provide better shelter for flight control and navigation.

AIRCRAFT
A Fairey Swordfish is shown here on deck. *Audacity* was the first British carrier to operate Grumman Martlet Mk II fighters. The F4F Wildcat was renamed Martlet by its British users, but from March 1944 the Royal Navy also used the Wildcat name.

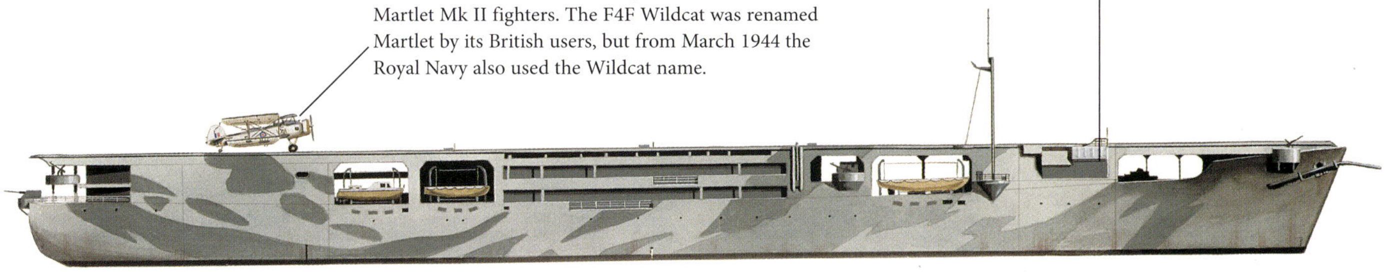

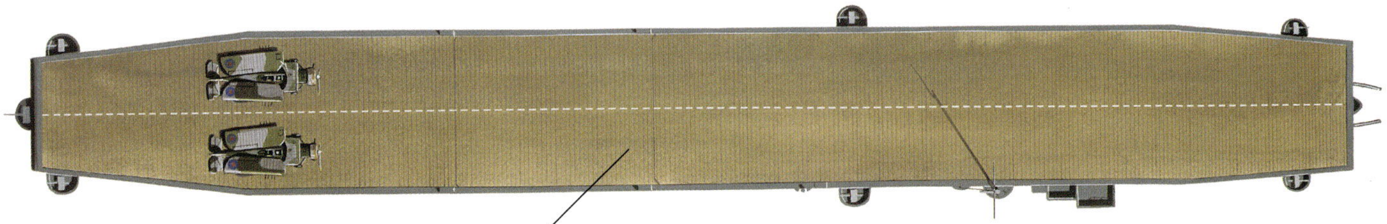

DECK CONSTRUCTION
The flight deck was wood-surfaced, unlike standard RN carriers, in the interest of rapid construction, saving of steel and reducing top weight. In action it was painted in camouflage colours.

RN CARRIERS
Between August 1942 and June 1943, 18 purpose-built escort carriers were manufactured in the USA and supplied both to the US Navy and the Royal Navy. The RN had eight, supplied on Lend-Lease terms to Britain. Based on a cargo ship hull design adapted to form a light carrier with hangar deck and lifts, they could hold 24 aircraft. A further 24 were built in 1943–44, almost all of which went to the Royal Navy.

***AUDACITY*'S FATE**
Audacity was sunk by a German U-boat on 21 December 1941, while performing escort duties off the coast of Portugal. It foundered at 22.10 hrs, with the loss of 73 crew members, most of them drowned.

Shokaku (1941)

TYPE • *Aircraft carrier* **COUNTRY** • *Japan*

SPECIFICATIONS

DIMENSIONS:	Length: 257.5m (844ft 10in); Beam: 29m (95ft 2in); Draught: 9.3m (30ft 7in)
DISPLACEMENT:	26,086 tonnes (28,758 tons) standard; 32,619 tonnes (35,944 tons) full load
PROPULSION:	8 boilers; 4 Kampon geared steam turbines; 4 shafts; 120,000kW (160,000shp)
SPEED:	34.5 knots (64km/h; 39.7mph)
RANGE:	9700nmi (18,000km; 11,200 miles) at 18 knots
ARMAMENT:	(original) 8 × Type 89 127mm (5in) 40-cal twin HA guns; 12 × Type 96 triple 25mm machine guns
AIRCRAFT:	84
COMPLEMENT:	1660

The large carriers *Shokaku* and *Zuikaku* were key elements in Japan's Pacific war plan. With long-range and high-endurance capacity, and able to hold 84 aircraft, they were viewed as battle-winners for the Imperial Navy.

Above: Shokaku *in 1941. The long, low profiles of* Shokaku *and* Zuikaku *were not intentionally planned to minimize radar detection, but they certainly helped. In daylight combat, however, they presented a substantial target area.*

WINDBREAK
A recessable windbreak, raised when aircraft were on the flight deck, was set in the flight deck just forward of the island.

BOATS
In total, 12 boats were carried on the stern, with four-tonne folding cranes to unship and retrieve them.

RADIO MASTS
The radio masts along the aft part of the flight deck were lowered when landing and take-offs were in progress.

VAPOUR EXPLOSIONS
Aviation fuel tanks in Japanese carriers were not independently installed and protected by water-filled spaces but integral with the hull and far more prone to shocks and leakage. Following the loss of *Kaga* at Midway, concrete protection was applied to the fuel tanks on *Shokaku* and *Zuikaku*, although this was of limited value against leaking vapour. Historians have pointed out that *Shokaku*, *Taiyo* and *Hiyo* all suffered similar vapour-induced explosions after torpedo attacks.

PEARL HARBOR ATTACK
A Japanese Navy Type 97 Carrier Attack Plane ('Kate') takes off from the aircraft carrier *Shokaku*, en route to attack Pearl Harbor, during the morning of 7 December 1941.

HMS *Indomitable* (1941)

TYPE • *Aircraft carrier* **COUNTRY** • *United Kingdom*

SPECIFICATIONS

DIMENSIONS:	Length: 230m (754ft 7in); Beam: 29.2m (95ft 10in); Draught: 8.8m (29ft)
DISPLACEMENT:	23,368 tonnes (25,763 tons) standard; 30,206 tonnes (33,297 tons) full load
PROPULSION:	6 boilers; Parsons geared steam turbines; 3 shafts; 83,000kW (111,000shp)
SPEED:	30.5 knots (56km/h; 35mph)
RANGE:	11,000nmi (20,000km; 13,000 miles) at 14 knots
ARMAMENT:	16 × 114mm (4.5in) guns; 48 × 40mm (2-pdr) and 10 × 20mm AA guns
AIRCRAFT:	48
COMPLEMENT:	1600

Commissioned in 1941, *Indomitable* was a ruggedly built and well-equipped fleet carrier. It was the fourth member of a class that gained a distinguished service record through World War II and endured into the 1950s.

Right: HMS Indomitable *towers over local fishing junks in Victoria Bay, Hong Kong, September 1945.*

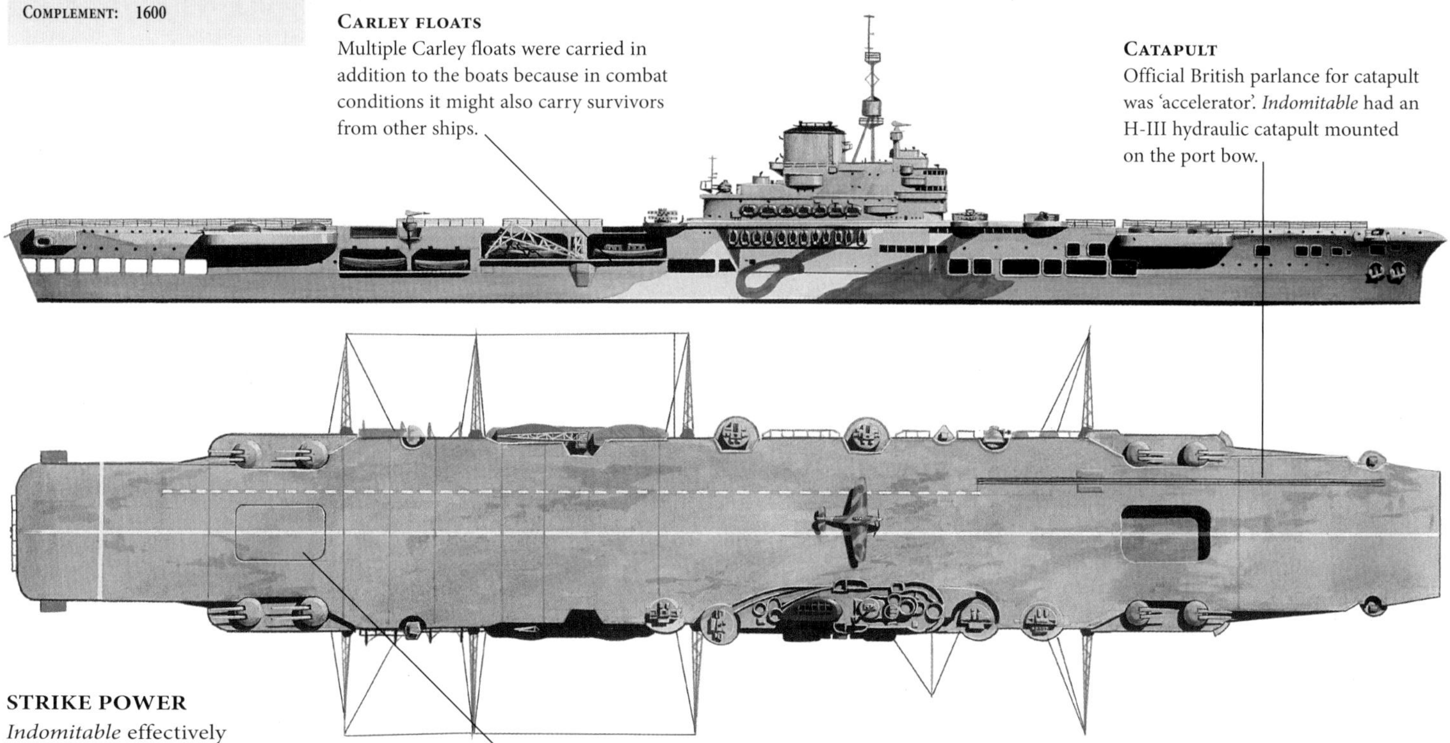

CARLEY FLOATS
Multiple Carley floats were carried in addition to the boats because in combat conditions it might also carry survivors from other ships.

CATAPULT
Official British parlance for catapult was 'accelerator'. *Indomitable* had an H-III hydraulic catapult mounted on the port bow.

HANGARS
The 4.3m (14ft) height of the lower hangar was not enough to accommodate later planes such as the F4U Corsair.

STRIKE POWER
Indomitable effectively represented an interim design between the *Illustrious* and *Implacable* classes. *Implacable* (in service from 1944 to 1955) was slightly larger and faster, also with two hangar decks, but with hangar-side armour of 51mm (2in). It had more powerful engines, with eight boilers and four shafts, and made a faster speed. Most importantly, it could carry up to 81 aircraft, giving it considerably greater strike power.

TAKE-OFF VIEW
HMS *Indomitable* pictured from one of its departing aircraft in 1940. The carrier remained in service into the 1950s, until an explosion on board on 3 February 1953, caused by a leak of aviation fuel and resulting in nine deaths, led to it being struck from the Navy list in 1955. It was scrapped in October that year.

CB class (1941)

TYPE • *Midget submarine* COUNTRY • *Italy*

SPECIFICATIONS (*CB-12*)

Dimensions:	Length: 15m (49ft 5in); Beam: 3m (9ft 10in); Draught: 2.05m (6ft 9in)
Displacement:	36 tonnes (40 tons) surfaced; 45 tonnes (50 tons) submerged
Propulsion:	1 diesel engine 67kW (90hp); 1 electric motor 37kW (50hp); 1 screw
Speed:	7.5 knots (13.9km/h; 8.6mph) surfaced; 7 knots (13km/h; 8.1mph) submerged
Range:	1216nmi (2252km; 1400 miles) at 5 knots surfaced; 43nmi (80km; 50 miles) at 3 knots submerged
Armament:	2 × 457mm (18in) torpedo tubes; 2 torpedoes or 2 mines
Complement:	4

Manned torpedoes and midget submarines were important tactical weapons in Italy's naval strategy, both in the Mediterranean and Black seas. A total of 72 were planned for construction, but only 21 were laid down.

Above: A CB-class midget submarine docked at Taranto, Italy, following the surrender of the Italian fleet to the Allies in November 1943. The torpedo tubes can clearly be seen fixed to the side of the boat.

Equipment
The boats were fitted with a gyrocompass, magnetic compass, radio gear and periscope. A hydrophone set was mounted in a frame at the fore-end of the keel.

CB-12
The torpedoes were mounted in external collars. Maximum diving depth was 55m (180ft), and they were equipped to carry 10 man-days of provisions.

Propulsion
The boat was driven and powered by an Isotta-Fraschini 67kW (90hp) diesel engine and a 37kW (50hp) Brown Boveri electric motor. Batteries were placed below the navigation room.

PROFILE
At the start of hostilities, Italy gained experience from the CA class of midget submarines. CB was a much-improved development, 21 in number, intended for penetration of anchorages. Built by the Caproni company in Milan, they had a considerable surface profile compared to the British X-craft, with a fairing built on top of the pressure hull supporting a breakwater and small observation tower, which would certainly have made surface movement easier.

CB-12
The CBs were built far inland, in Milan by Caproni Taliedo, but could be moved on special rail trucks. They had similar dimensions to the British X-boats, and, although still slow-speed, were slightly faster.

Akizuki (1942)

TYPE • *Destroyer* **COUNTRY** • *Japan*

SPECIFICATIONS

DIMENSIONS:	Length: 134.2m (440ft 3in); Beam: 11.6m (38ft 1in); Draught: 4.15m (13ft)
DISPLACEMENT:	3759 tonnes (3700 tons) full load
PROPULSION:	3 Kampon boilers; 2 Kampon geared steam turbines; 2 shafts; 38,000kW (52,000shp)
SPEED:	33 knots (61km/h; 38mph)
RANGE:	8300nmi (15,370km; 9550 miles) at 18 knots
ARMAMENT:	(1942) 8 × 100mm (4in) guns; 4 × Type 96 25mm (0.98in) AA guns; 4 × 610mm (24in) torpedo tubes; 8 × Type 93 torpedoes; 56 × Type 95 depth charges
ARMOUR:	None
COMPLEMENT:	300

Japan's final class of wartime destroyers carried a newly developed dual-purpose gun for anti-aircraft and anti-surface roles, together with long-range torpedoes. Although nearly 40 vessels were planned, wartime production constraints limited completion to 12 ships.

Left: Pictured here in Kure on 16 October 1945, shortly after war's end, the Yoitzki *(1944), was an* Akizuki*-class destroyer whose late commission in January 1945 meant that it did not see significant wartime combat service.*

DEPTH CHARGES
The *Akizuki* class carried two Type 94 depth-charge projectors and 54 depth charges, increased to 72 depth charges for some of the later vessels in the class.

PROPULSION
Propulsion came from two Kampon geared steam turbines, each driving one propeller shaft with power from three Kampon water-tube boilers. The turbines could deliver a designed speed of 33 knots (61km/h; 38mph).

TYPE 98 GUNS
The principal firepower of the *Akizuki* class was provided by the 10cm/65 Type 98 naval gun. A chief strength was its rate of fire: a good gun crew could achieve 22rpm.

BATTLE DAMAGE
After its completion on 11 June 1942, *Akizuki* was sent soon after to support Japanese operations around the Solomon Islands. It was seriously damaged by a bomb hit on 25 October then, following repairs, it was torpedoed by a US submarine and had to be beached. More, and massive, repairs followed, but the ill-fated ship was permamently sunk on 25 October, in the Battle off Cape Engaño.

SEA TRIALS
Akizuki conducts sea trials in Miryazu Bay in May 1942. The ship was constructed at the Maizuru Naval Arsenal, being launched on 2 July 1941 and completed and commissioned on 11 June 1942.

Jun'yo (1942)

TYPE • *Aircraft carrier* **COUNTRY** • *Japan*

SPECIFICATIONS

DIMENSIONS:	Length: 219.3m (719ft 7in); Beam: 26.7m (87ft 7in); Draught: 8.15m (26ft 9in)
DISPLACEMENT:	24,526 tonnes (27,045 tons) standard; 28,753 tonnes (31,698 tons) full load
PROPULSION:	6 Mitsubishi boilers; Mitsubishi geared steam turbines; 2 shafts; 41,906kW (56,250shp)
SPEED:	26 knots (48km/h; 30mph)
RANGE:	10,000nmi (18,500km; 11,500 miles)
ARMAMENT:	6 × twin 127mm (5in) Type 89 dual-purpose guns; 8 × triple 25mm (1in) Type 96 AA guns
AIRCRAFT:	53
COMPLEMENT:	1224

Jun'yo **was laid down as a passenger liner and underwent conversion to a carrier to boost Japan's considerable naval air power. It was one of two Japanese carriers to survive action in World War II, a remarkable feat given Japanese naval losses.**

Left: Junyo*'s aft flight deck.* Jun'yo *carried 48 planes ready to fly, plus five in kit form: 12 Mitsubishi Type Zero A6M fighters, 18 Aichi Type 99 D3A bombers and 18 Nakajima Type 97 B5N attack planes, known to Allied Intelligence as 'Zeke', 'Val' and 'Kate' types respectively.*

GAS DISPERSION
Jun'yo's funnel was angled out to starboard at 25 degrees to help in dispersing hot gases away from the flight deck.

ROCKETS
In September 1944, six racks of 30-tube launchers for 127mm (5in) phosphorus rockets were fitted, three on each side, on stern sponsons. These rockets were armed with multiple incendiary shrapnel charges on a time fuse.

PROTECTION
In March 1944 additional concrete protection was applied around the aviation fuel tanks in an attempt to contain the leakage and build-up of oil vapour that had been the bane of other Japanese carriers.

MODIFICATIONS

After Midway in June 1942, as one of only four large carriers remaining in service, *Jun'yo* was redesignated as a fleet carrier. In the light of combat experience its aircraft wing was altered to 21 A6M Zero fighters, 12 D3A dive bombers and nine B5N torpedo bombers. During a quick refit at Kure in the second half of July 1942, Type 21 air search radar was fitted on the top of the island, along with two Type 94 high-angle fire control directors.

JUN'YO
The ship, planned as *Kashiwara Maru* for NYK Lines, was laid down at Mitsubishi Heavy Industries Shipyard in Nagasaki on 20 March 1939. It was one of a number of ships, including its twin carrier *Hiyo*, built as a 'shadow fleet' that could rapidly be converted into warships.

USS *Bogue* (1942)

TYPE • *Aircraft carrier* **COUNTRY** • *United States*

SPECIFICATIONS

DIMENSIONS:	Length: 151.08m (495ft 8in); Beam: 25m (82ft); Draught: 7.9m (26ft)
DISPLACEMENT:	8524 tonnes (9400 tons) standard
PROPULSION:	2 Foster Wheeler boilers; 1 Allis-Chalmers steam turbine; 1 shaft; 6300kW (8,500shp)
SPEED:	18 knots (33km/h; 21mph)
RANGE:	26,300nmi (48,700km; 30,300 miles)
ARMAMENT:	2 × 127mm (5in)/38-cal dual-purpose guns; 8 × twin 40mm (1.57in) Bofors AA guns; 20 × 20mm (0.79in) Oerlikon AA cannons
AIRCRAFT:	19–24
COMPLEMENT:	890

Commissioned on 26 September 1942, USS *Bogue* (CVE-9) was the lead ship of the *Bogue* class of escort carrier, known in the British Navy as the *Attacker* class. The vessel survived service in the Atlantic and Pacific and remained on reserve until 1959.

Above: Bogue *was originally classified as AVG-9, but this was changed to ACV-9 in August 1942, CVE-9 in July 1943 and CVHE-9 in 1955.*

AIRCRAFT
The *Bogue* class could carry up to 24 aircraft, typically a mixture of Grumman F4F Wildcat fighters and Grumman TBF Avenger torpedo bombers.

***ATTACKER* CLASS**
The Royal Navy had eight *Attacker*-class carriers supplied on Lend-Lease terms to Britain.

ACCOMMODATION
While all *Bogue* carriers went through modification before commissioning in the British fleet, RN sailors were pleasantly surprised by the standard of crew accommodation on a US warship, with bunks and cabins instead of mess-decks and hammocks.

ESCORT CARRIERS
Escort carriers were a distinct category of aircraft carrier during World War II. Smaller and lighter than the large fleet carriers – they were about half the length and a third the displacement – they therefore carried far fewer aircraft, about 24–36. However, because they were built on modified merchant ship hulls, they were fast to produce and served in significant numbers: 122 escort carriers were produced during World War II.

***BOGUE*-CLASS CARRIER**
Bogue-class carriers served in both the Atlantic and Pacific theatres. In the Atlantic, the ships' aircraft destroyed or damaged multiple German U-boats, and they repeated their successes against Japanese submarines in the Pacific.

HMS *Jamaica* (1942)

TYPE • *Cruiser* **COUNTRY** • *United Kingdom*

SPECIFICATIONS

Dimensions:	Length: 169.32m (555ft 6in); Beam: 19m (62ft); Draught: 5.03m (16ft 6in)
Displacement:	8670 tonnes (9560 tons) standard
Propulsion:	4 Admiralty 3-drum boilers; 4 geared steam turbines; 4 shafts; 54,100kW (72,500shp)
Speed:	31.5 knots (58.3km/h; 36.2mph)
Range:	10,100nmi (18,700km; 11,600 miles)
Armament:	9 x 152mm (6in) guns; 8 x 102mm (4in) guns; 8 x 40mm (2-pdr) pom-pom guns; 2 x triple 533mm (21in) torpedo tubes
Armour:	Belt 89–83mm (3.5–3.25in); bulkheads 51–38mm (2–1.5in); turrets 51–25mm (2–1in);
Complement:	730; 920 wartime

The *Fiji*-class cruisers to which HMS *Jamaica* belonged were considered the best available to the Royal Navy in the early years of the war. All 11 ships were named after British possessions or overseas territories.

Left: HMS Jamaica *(C44), as seen in 1940. The ship had a distinguished postwar career, supporting the Inchon Landings in September 1950 during the Korean War and participating in Operation* Musketeer*, the Anglo-French invasion of Egypt during the Suez Crisis of 1956.*

Firepower
The *Fiji*-class vessels were mostly armed with 12 152mm (6in) guns, although some had a reduced armament and carried only nine.

Displacement
Following later modifications to the class, the full load displacement rose to 11,270 tonnes (12,423 US tons).

Armour
Jamaica's thickest concentration of armour was over the engine and boiler rooms, with a depth of 89mm (3.5in).

SURFACE CLASHES

HMS *Jamaica* is best known for its clashes with larger German warships. While serving as part of Convoy 51B to Murmansk, *Jamaica* encountered the German raiders *Lützow* and *Admiral Hipper*. The *Jamaica* scored direct hits on the *Hipper*, causing damage to its boiler room before contact was broken. Later, *Jamaica* participated in the sinking of the *Scharnhorst* on 26 December 1943.

AT BERMUDA, 1941
HMS *Jamaica* was the sixth ship of the class to be laid down, on 28 April 1939. It was launched on 16 November 1940 and completed on 29 June 1942. By then several of its companion ships had seen considerable action, and HMS *Fiji* had been sunk.

USS *Indiana* (1942)

TYPE • *Battleship* **COUNTRY** • *United States*

SPECIFICATIONS

Dimensions:	Length: 210m (680ft); Beam: 32.9m (107ft 8in); Draught: 8.9m (29ft 3in)
Displacement:	34,446 tonnes (37,970 tons) standard
Propulsion:	8 Foster Wheeler boilers; 4 Westinghouse geared steam turbines; 4 shafts; 96,941kW (130,000hp)
Speed:	27 knots (50km/h; 31mph)
Range:	15,000nmi (27,750km; 17,245 miles) at 15 knots
Armament:	9 × 406mm (16in) guns; 20 × 127mm (5in) guns; 24 × 40mm (1.6in) AA guns; 50 × 20mm (0.79in) AA guns
Armour:	Belt 310–22mm (12.2–0.87in); bulkheads 279mm (11in); barbettes 439–287mm (17.3–11.3in); turrets 457–184mm (18–7.25in); deck 152–146mm (6–5.75in)
Complement:	1793

Indiana was the second of the four _South Dakota_-class ships. Its 127mm (5in) guns were on two levels amidships, and the single funnel was faired into the rear of the bridge. In World War II, it was deployed mostly for carrier protection and shore bombardment.

Right: USS Indiana *(BB-58) off Norfolk, Virginia, September 1942. Decommissioned in 1947, it was scrapped in 1963.*

Aircraft
A Vought OS 2 U-1 Kingfisher seen on the catapult. Two of these aircraft were carried on board the ship.

AA guns
A 40mm (1.6in) four-barrelled anti-aircraft gun was placed on 'B' turret in 1944 to enhance the volume of forward AA fire.

Cranes
The heavy boat cranes by the aft superstructure were carried only on the USS *Indiana* and the USS *Massachusetts*.

SOUTH DAKOTA CLASS

These were the first American ships to be given inclined internal side armour, reaching from the armoured deck to the inner bottom and 310mm (12.2in) thick, tapering to 25mm (1in). This layout gave the ships a long indented inward-angled stretch of the central hull, just below the flush deck, reminiscent of the old casemate structure. Torpedo bulges were not fitted to this class, but a splinter protection deck placed 0.8cm (2ft 7in) below the main armour deck was a new feature.

SHORE BOMBARDMENT
The USS *Indiana* pounds Kamaishi, Honshu, Japan, with its forward 406mm (16in) guns on 14 July 1945. This was the last salvo of the bombardment; note a projectile is just visible in the air at left.

Balao class (1942)

TYPE • *Submarine* **COUNTRY** • *United States*

SPECIFICATIONS

Dimensions:	Length: 95m (311ft 9in); Beam: 8.3m (27ft 3in); Draught: 5.1m (16ft 9in)
Displacement:	1550 tonnes (1708 tons) surfaced; 2440 tonnes (2689 tons) submerged
Propulsion:	4 F-M diesel generator engines of 4026kW (5400hp), 2 511kW (685hp) Elliott geared electric motors, 2 screws
Speed:	21 knots (39km/h; 24mph) surfaced; 9 knots (16.6km/h; 10mph) submerged
Range:	11,000nmi (20,372km; 12,659 miles) surfaced; 95nmi (176km; 109 miles) submerged
Armament:	10 × 533mm (21in) torpedo tubes; 24 torpedoes; 1 × 127mm (5in) 25-cal deck gun; 1 × 40mm (1.5in) Bofors AA gun; 1 × twin 20mm (0.79in) Oerlikon AA gun
Complement:	80

The *Balao* was the largest class of submarine in the US Navy. The design was so similar to the later *Gato* class that USS *Balao* (SS-285) and others of its class were being laid down even as the last of the *Gato* boats were nearing completion.

Above: The complex rig above the shears reflects the variety of sensory and communications equipment carried by 1943.

USS *Entemedor*
The USS *Entemedor* (SS-340) was a *Balao*-class submarine that with a postwar GUPPY IIA modernization served the US Navy until 1972.

USS *Piper* (SS-409)
The two tapering and robust periscopes were the main visual definer between the 121 *Balaos* and the *Gatos*, the latter with thinner scopes and supporting ribs.

Armament
Some *Balao*-class submarines acquired a second Mk 17 127mm (5in) deck gun from late 1943, as torpedo-worthy targets were becoming scarce.

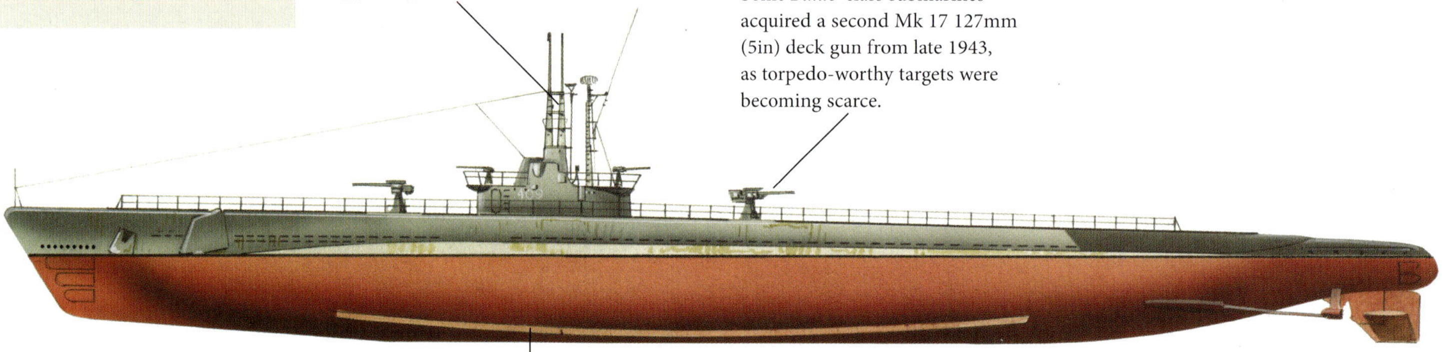

Watertight trunk
A useful feature of all the class was a watertight trunk from the deck to the control room, allowing gun crews to reach position without going through the packed conning tower.

DIVE DEPTH
The US Navy was aware that Japanese depth charges were effective to 90m (295ft) and that the *Gato* class's operating depth was down to only 91.4m (300ft). Greater depth ability was needed. By changing the pressure hull cover to high tensile steel (HTS) 20mm (0.79in) thick, a maximum diving depth of 198m (650ft) could be achieved. In the interest of safety, the test depth of the *Balao* design was finally fixed at 123m (400ft). The *Gato* plans were modified to incorporate the thicker pressure hull.

RETURN TO PORT
A *Balao*-class submarine returns to port after a successful war patrol in early 1945. The location is probably Guam. Blindfolded Japanese prisoners of war are being escorted along the deck. Note the deck gun at right and battle flag hung from the forward 20mm gun.

X class (1942)

TYPE • *Midget submarine* **COUNTRY** • *United Kingdom*

SPECIFICATIONS (*X-5*)

Dimensions:	Length: 15.6m (51ft 3in); Beam: 1.75m (5ft 9in); Draught: 1.6m (5ft 4in)
Displacement:	27.4 tonnes (30.2 tons) surfaced; 30.48 tonnes (33.6 tons) submerged
Propulsion:	1 diesel engine, 31.3kW (42hp); 1 electric motor, 22.3kW (30hp); 1 screw
Speed:	6.5 knots (12km/h; 7.5mph) surfaced; 5 knots (9.2km/h; 5.1mph) submerged
Range:	1320nmi (2444km; 1519 miles) at 4 knots surfaced; 80nmi (148km; 92 miles) at 2 knots submerged
Armament:	2 × 2000kg (4400lb) detachable charges
Complement:	4

The Italians, Japanese and British all made extensive use of small two- to four-man submarines, for covert penetration of protected anchorages and bases and for coastal reconnaissance operations.

Left: The X-craft were towed, usually by a submarine, to the starting point (across the North Sea in X-5's case) while manned by a 'passage crew' who then handed over to the mission crew.

Hydroplane
A single hydroplane was fitted at the stern, across the rudder which was divided in two to allow this. An air induction mast for the diesel engine was folded against the hull for submerged running.

Hull
The hull, formed of welded 0.703kg (1.5lb) 'S' type steel, was formed of three sections, bolted together. It had a 'wet and dry' airlock hatch for a diver in addition to a surface hatch.

Warheads
Bolted on each side were its mine-type warloads: cased amatol explosive charges each of two tons, to be released from inside and deposited beneath a target vessel.

RN SUBS
The Royal Navy's versions of the midget submarine were known as X-craft and were developed under tight security from Varley Marine mini-subs built for river work. The first operational type was *X-3* (1942). *X-5* was built by Vickers at Barrow in the winter of 1942–43. It had all the key features of a full-size submarine incorporated in a hull 15.5m (51ft) long, with a loaded surface displacement of around 37 tonnes (41 tons).

PERISCOPE VIEW
Sub-Lt K.C.J. Robinson uses the periscope of an X-craft while training in Rothesay Bay, Scotland. The confined environment is apparent. Habitability was minimal, even allowing for the expectation that missions would be relatively short.

River class (1942)

TYPE • *Frigate* **COUNTRY** • *United Kingdom*

The *River* class were a major part of the Royal Navy's surface fleet during World War II, with a total of 151 of the frigates (although initially classified as corvettes) built and launched between 1941 and 1944.

Right: Six River*-class frigates were transferred to France in 1944, including F712* La Découverte *(ex-HMS* Windrush*) pictured here.*

SPECIFICATIONS

Dimensions:	Length: 91.8m (301ft 3in); Beam: 11.1m (36ft 6in); Draught: 2.7m (9ft)
Displacement:	1392 tonnes (1535 tons) standard
Propulsion:	2 Admiralty 3-drum boilers; reciprocating vertical triple-expansion engines, 4100kW (5500hp); 2 shafts;
Speed:	20 knots (37km/h; 23mph); 20.5 knots (38km/h; 23.6mph) for turbine ships
Range:	7200nmi (13,300km; 8300 miles)
Armament:	2 × QF 102mm (4in) guns; up to 10 × QF 20mm (0.79in) AA guns; 1 × Hedgehog 24-spigot A/S projector; 8 depth charge throwers; 2 rails; up to 150 depth charges
Armour:	None
Complement:	107

Machinery
The *River* class achieved an improved performance over the *Flower* class by doubling the size of the powerplant and using more efficient water-tube boilers.

Forecastle
The *River*-class ships had a distinctive raised forecastle that was fitted with a single 102mm (4in) gun.

Hull
To speed up production, the *River* class used merchant-shipping models in the hull design.

INTERNATIONAL
The *River*-class frigates served extensively during World War II, and not only with the Royal Navy (RN). Other users of this class included the Royal Australian Navy, the Royal Canadian Navy (the next most prolific user after the RN), the Free French Naval Forces, the Royal Netherlands Navy and the South African Navy. Two were even used by the US Navy: USS *Asheville* and USS *Natchez*.

ATLANTIC CONVOY
HMS *Spey* (K246) was a *River*-class frigate that served in the Atlantic, Mediterranean and Indian Ocean during the war, principally providing convoy-escort duties, as seen here.

USS *Johnston* (1943)

TYPE • *Destroyer* **COUNTRY** • *United States*

In October 2019, the wreck of the USS *Johnston* (DD-557) was discovered more than 6100m (20,000ft) below the surface in the Philippine Sea. The ship had been sunk following the thunderous naval engagement now known as the Battle off Samar.

SPECIFICATIONS

Dimensions:	Length: 114.7m (376ft 6in); Beam: 12.1m (39ft 8in); Draught: 5.4m (17ft 9in)
Displacement:	1859 tonnes (2050 tons) standard
Propulsion:	2 General Electric steam turbines;45,000kW (60,000shp); 2 screws
Speed:	38 knots (70.4km/h; 44mph)
Range:	6500nmi (12,000km; 7456 miles) at 15 knots
Armament:	5 × 127mm (5in) guns; 10 × 40mm (1.57in) AA guns; 7 × 20mm (0.79in) AA guns; 10 × 533mm (21in) torpedo tubes; 6 depth charge projectors and tracks
Armour:	None
Complement:	208

Above: USS Johnston *is pictured here in the United States (off Seattle or Tacoma, Washington) on 27 October 1943.*

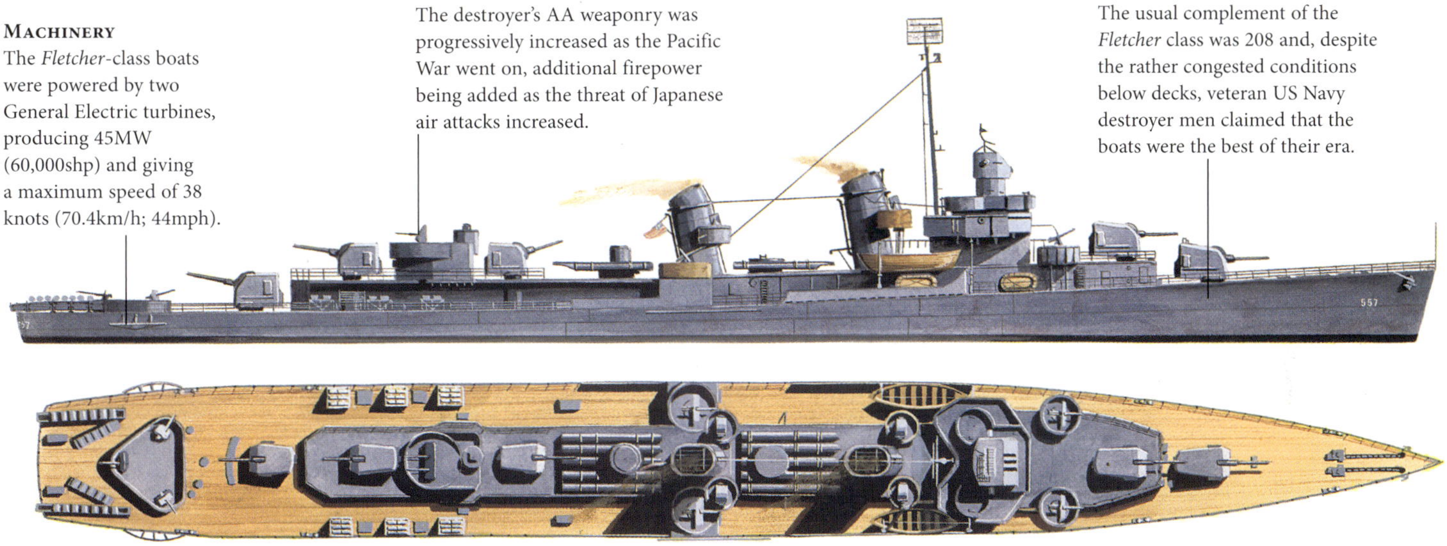

Machinery
The *Fletcher*-class boats were powered by two General Electric turbines, producing 45MW (60,000shp) and giving a maximum speed of 38 knots (70.4km/h; 44mph).

AA armament
The destroyer's AA weaponry was progressively increased as the Pacific War went on, additional firepower being added as the threat of Japanese air attacks increased.

Complement
The usual complement of the *Fletcher* class was 208 and, despite the rather congested conditions below decks, veteran US Navy destroyer men claimed that the boats were the best of their era.

BATTLE OFF SAMAR

The Battle off Samar was part of the Battle of Leyte Gulf, a vast clash of Japanese and US surface vessels and carrier aircraft off the coast of the Philippines in October 1944. At one point in the battle, *Johnston* made a valiant torpedo charge against a superior Japanese force that included the battleship *Yamato* and three other battleships. The valiant destroyer was eventually sunk by heavy shellfire, with 186 lives lost.

LAUNCH
USS *Johnston* slips down the building ways at the Seattle-Tacoma Shipbuilding Corporation shipyard, Seattle, Washington, 25 March 1943.

Matsu class (1943)

TYPE • *Destroyer* COUNTRY • *Japan*

The *Matsu* class of destroyers were built from 1943 onwards, by which time the war in the Pacific had irreversibly turned against the Japanese. Rapid losses in naval shipping compelled the Japanese to design the class for rapid construction.

SPECIFICATIONS

DIMENSIONS:	Length: 100m (328ft 1in) overall; Beam: 9.35m (30ft 8in); Draught: 3.30m (10ft 10in)
DISPLACEMENT:	1260 tonnes (1389 tons) standard
PROPULSION:	2 Kampon water-tube boilers; 2 Kampon impulse geared steam turbines, 14,000kW (19,000shp); 2 shafts
SPEED:	27.8 knots (51.5km/h; 32mph)
RANGE:	3500nmi (6500km; 4039 miles)
ARMAMENT:	3 × 127mm (5in) AA guns; 24 × 25mm (1in) AA guns; 4 × 610mm (24in) torpedo tubes; 4 × Type 93 torpedoes; 4 depth-charge throwers; 36 × depth charges
ARMOUR:	None
COMPLEMENT:	211

Left: Momi (5489) was sunk on 5 January 1945 by US Navy torpedo bombers off Manila, with all hands lost.

MAIN GUN
The main gun on the forward deck was a 127mm (5in) Type 89 dual-purpose gun that could engage both surface and air targets.

PERFORMANCE
Despite being underpowered compared to many other Japanese destroyer types, the Matsu class could still achieve 27.8 knots (51.5km/h; 32mph).

FUNNELS
The two widely spaced funnels were a distinguishing feature of the class. They also made the ships look bigger than they actually were.

CONSTRUCTION

The *Matsu* class of destroyers, also known as the D-class, were designed from the outset for rapid construction and fitting out. Compared to previous destroyer types, economies were made in every area, including dimensions, armament and armour, and a powerplant producing only about a third of the output of a fleet destroyer. Nevertheless, the vessels were serviceable and effective, and they were further streamlined for production in the modular *Tachibana* sub-class.

DESTROYER *MAKI*
The *Matsu*-class destroyer *Maki* was a late-war arrival, only being completed in August 1944 and damaged shortly thereafter during the Battle off Cape Engaño. It survived the war and was surrendered to the Allies in September 1945.

USS *The Sullivans* (*Fletcher* class) (1943)

TYPE • *Destroyer* **COUNTRY** • *United States*

SPECIFICATIONS

Dimensions:	Length: 114.7m (376ft 6in); Beam: 12.1m (39ft 8in); Draught: 5.4m (17ft 9in)
Displacement:	1859 tonnes (2050 tons) standard
Propulsion:	2 General Electric steam turbines; 45,000kW (60,000shp); 2 screws;)
Speed:	38 knots (70.4km/h; 44mph)
Range:	6500nmi (12,000km; 7456 miles) at 15 knots
Armament:	5 × 127mm (5in) guns; 10 × 40mm (1.57in) AA guns; 7 × 20mm (0.79in) AA guns; 10 × 533mm (21in) torpedo tubes; 6 depth charge projectors and tracks
Armour:	None
Complement:	208

The *Fletcher*-class destroyer *The Sullivans* was named in honour of the five Sullivan brothers, who lost their lives when *Juneau* was sunk by a Japanese submarine off Guadalcanal in November 1942. It was the greatest combat loss suffered by an American family in World War II.

Left: USS The Sullivans *(DD-537) underway on 29 October 1962. Postwar, the vessel eventually became a museum ship.*

Superstructure
The US Navy ordered the *Fletcher* class to be built with a simplified superstructure to hasten production. A new bridge design also increased the overhead view available to the captain.

Machinery
The *Fletcher*-class boats were powered by two General Electric turbines, producing 45MW (60,000shp) and giving a maximum speed of 38 knots (70.4km/h; 44mph).

Depth-charge racks
The *Fletcher*-class ships were fitted with two racks, each holding eight 270kg (600lb) depth charges, and adjacent to them were two storage racks with five depth charges each.

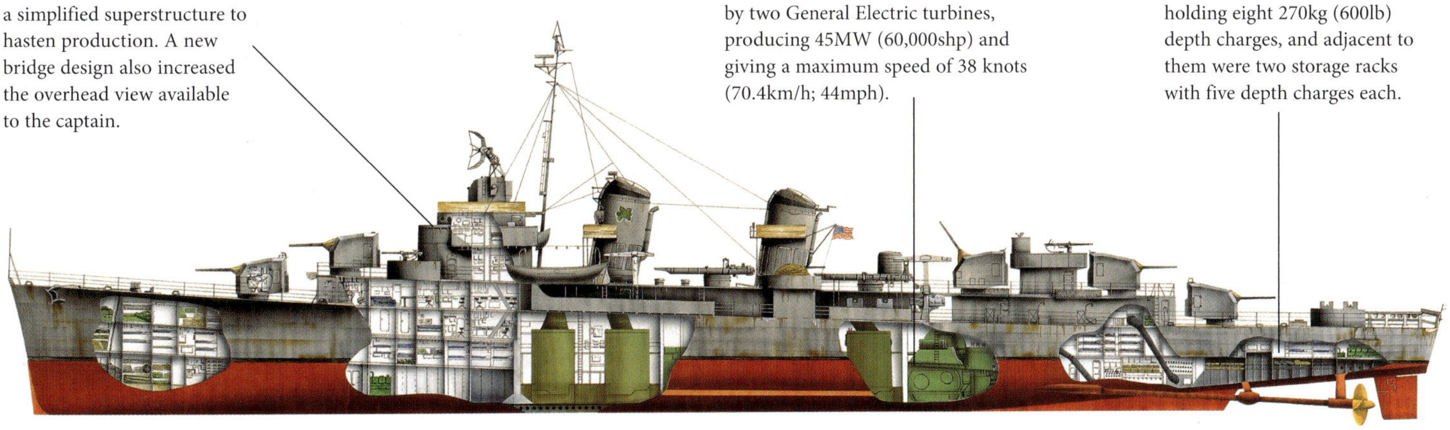

***FLETCHER* CLASS**
A total of 175 *Fletcher*-class destroyers were completed between 1941 and 1945. Although vessels in the class were generally rushed out to the Pacific on completion, those built on the Atlantic seaboard saw some service there. The *Fletcher* itself served on convoy-protection duty in the western Atlantic before its deployment to the Pacific in time for the naval actions off Guadalcanal. All ships of this class were named after US Navy personnel or civilians who had rendered outstanding service.

***THE SULLIVANS* (DD-537)**
Although the extensive *Benson* class achieved the aim of putting the US destroyer-building industry on to a war footing, the design had limitations for a Pacific war, both in terms of endurance and in its curtailed weapons fit. The *Fletcher* class remedied these deficiencies.

USS *Wasp* (1943)

TYPE • *Aircraft carrier* **COUNTRY** • *United States*

SPECIFICATIONS

Dimensions:	Length: 250m (820ft); Beam: 28m (93ft); Draught: 10.41m (34ft 2in)
Displacement:	27,534 tonnes (30,365 tons) standard; 36,962 tonnes (40,748 tons) full load
Propulsion:	8 boilers; 4 Westinghouse geared steam turbines, 110,000kW (150,000shp); 4 shafts;
Speed:	33 knots (61km/h; 38mph)
Range:	15,440nmi (28,564km; 17,756 miles) at 15 knots
Armament:	12 × 127mm (5in) guns (4 twin, 4 single); 8 × quad 40mm (1.5in) AA guns; 46 × 20mm (0.79in) AA guns
Aircraft:	110
Complement:	2600

The tenth ship of the *Essex* class, USS *Wasp* (CV-18) enjoyed an almost 30-year, postwar career serving as an attack carrier, an anti-submarine warfare (ASW) carrier and in support of the manned NASA space programme.

Above: USS Essex *(CV-9) lead ship of the largest-ever class of aircraft carriers, with 24 launched between 1942 and 1945.*

Catapults
Wasp was one of the *Essex*-class ships to carry athwartships catapults on the hangar deck.

USS *Wasp*
En echelon (alternately located) engine and boiler rooms ensured that if one set of boilers and turbines was flooded or put out of action, the other could still function.

Designation
The ship's number was originally painted in reverse on the forward edge of the flight deck.

AVIATION

A complement of up to 90 planes could be carried in four squadrons, plus a fifth in reserve. Careful internal planning enabled the carrying of large stocks of spare parts, as well as the tools and machinery for rapid repair, all located on hangar deck level. *Wasp* and the other *Essex*-class ships could put more planes into the air faster than any other carriers of World War II.

USS *WASP*, PACIFIC, AUGUST 1945
Wasp first went into active service on 10 January 1944 in the Atlantic, serving off the East Coast of the USA until early March 1944, when it was deployed to the Pacific.

USS *Iowa* (1943)

TYPE • *Battleship* **COUNTRY** • *United States*

SPECIFICATIONS

DIMENSIONS:	Length: 270.4m (887ft 2in); Beam: 33.5m (108ft 3in); Draught: 11.5m (38ft)
DISPLACEMENT:	52,834 tonnes (58,268 tons) standard; 58,372 tonnes (64,360 tons) full load
PROPULSION:	8 Babcock & Wilcox boilers; 4 General Electric steam turbines; 4 screws; 158,088kW (212,000shp)
SPEED:	33 knots (61.2km/h; 38.25mph)
RANGE:	12,937nmi (23,960km; 14,889 miles) at 12 knots
ARMAMENT:	9 × 406mm (16in) guns (3 turrets); 20 × 127mm (5in) guns; 60 × 40mm (1.6in) four-barrelled AA guns
ARMOUR:	Belt 310mm (12.2in); barbettes 440–287mm (17.3–11.3in); turret faces 500mm (19.7in); deck 152mm (6in)
COMPLEMENT:	1921

The *Iowa* class of fast battleships, America's last and largest, began in 1936 and had heavier power and armour protection than the previous *South Dakota* class. A carrier escort in World War II, *Iowa* was reactivated in 1951–58 and in 1984–90 for bombardment of shore targets.

Left: The main firepower was nine Mk 7 406mm (16in) guns in three triple turrets. 'Special charge' shells were supplied, to increase muzzle velocity and penetrative effect.

AIRCRAFT
Four reconnaissance float planes were carried, launched from two stern-mounted catapults. There was no hangar accommodation and the catapults were removed in 1950 to allow for a helicopter landing pad.

RADAR
Radar was installed from the start, although numerous additions were made. In 1943, SK and SRa aerials were fitted on the foremast and in 1945, type SK-2 was added, with type SC-2 on the aft mast.

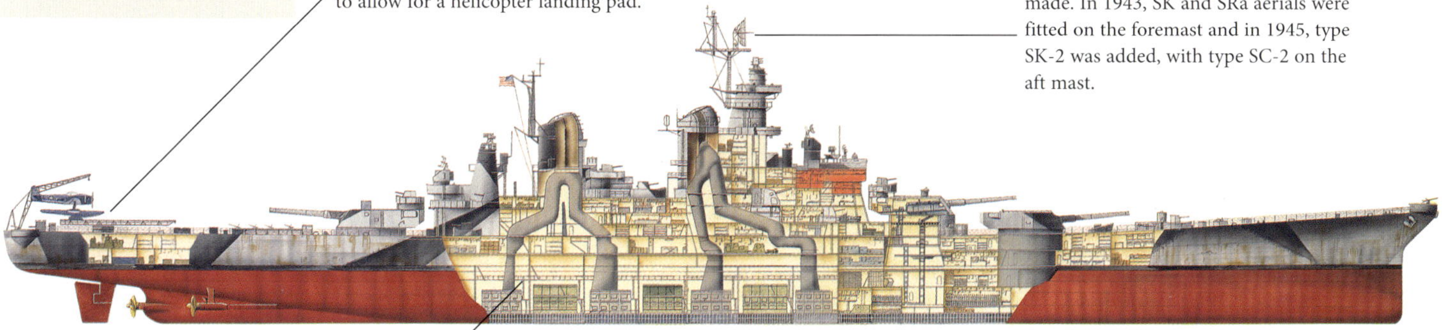

BOILERS
Eight high-pressure boilers powered four turbine sets, with the machinery arranged in alternate fire room/turbine room spaces.

DESIGN

Iowa (BB-61) was built at the New York Navy Yard and commissioned on 22 February 1943. Longer than Indiana by 64m (210ft) but less than a metre wider (the Panama Canal locks had to be borne in mind), it had a long, narrow forecastle, a distinctive bulbed bow and a broad cruiser-type stern. The long superstructure was packed with guns, aerials and searchlight platforms, with the tower partially enclosing the forward funnel. The masts were mere appendages to the tower and aft funnel (the mainmast was heightened in 1945 and replaced by a light tripod in 1948).

BROADSIDE
USS *Iowa* (BB-61) fires a full broadside of nine 16/50 and six 5/38 guns during a target exercise near Vieques Island, Puerto Rico, 1 July 1984, the shockwaves of the blasts rippling out across the water.

Landing Ship Tank, Mk III (1943)

TYPE • *Landing craft* **COUNTRY** • *United Kingdom/Canada*

SPECIFICATIONS

DIMENSIONS:	Length: 105.8m (347ft 6in) o/a; Beam: 16.84m (55ft 3in); Draught: 3.78m (11ft 6in)
DISPLACEMENT:	2300 tonnes (2535 tons) standard; 4980 tonnes (5488 tons) full load
PROPULSION:	Reciprocating vertical triple-expansion steam engine; 2 shafts; 4100kW (5500hp)
SPEED:	13.5 knots (25km/h; 16mph)
ARMAMENT:	4 × 20mm Bofors AA guns; 6 × 20mm Oerlikon AA guns
ARMOUR:	None
COMPLEMENT:	118–190

The standard US tank landing craft (LST), which has become an almost timeless design, grew from a British requirement for a tank carrier capable of transporting new tanks from US factories directly to Europe.

Left: A Canadian-built LST(3), HMS Striker *L3516.* Alongside the tanks, the LSTs could carry up to 200 troops (depending on the class), enabling them to land infantry and armour in a single pass.

ARMAMENT
By the end of the war, most US LSTs were fitted with seven 40mm (1.5in) and 12 20mm (0.79in) AA guns, but there were many exceptions to this rule.

LST MK III
All LSTs built/designated by the Royal Navy were known as LST(3)s. The LST(3) was a compromise between the sea-going qualities of the first British-designed tank landing ships

TANK TRANSPORTATION
The ships could carry 15 40-ton tanks or 20 25-ton tanks, plus 14 3-ton trucks and up to 170 troops.

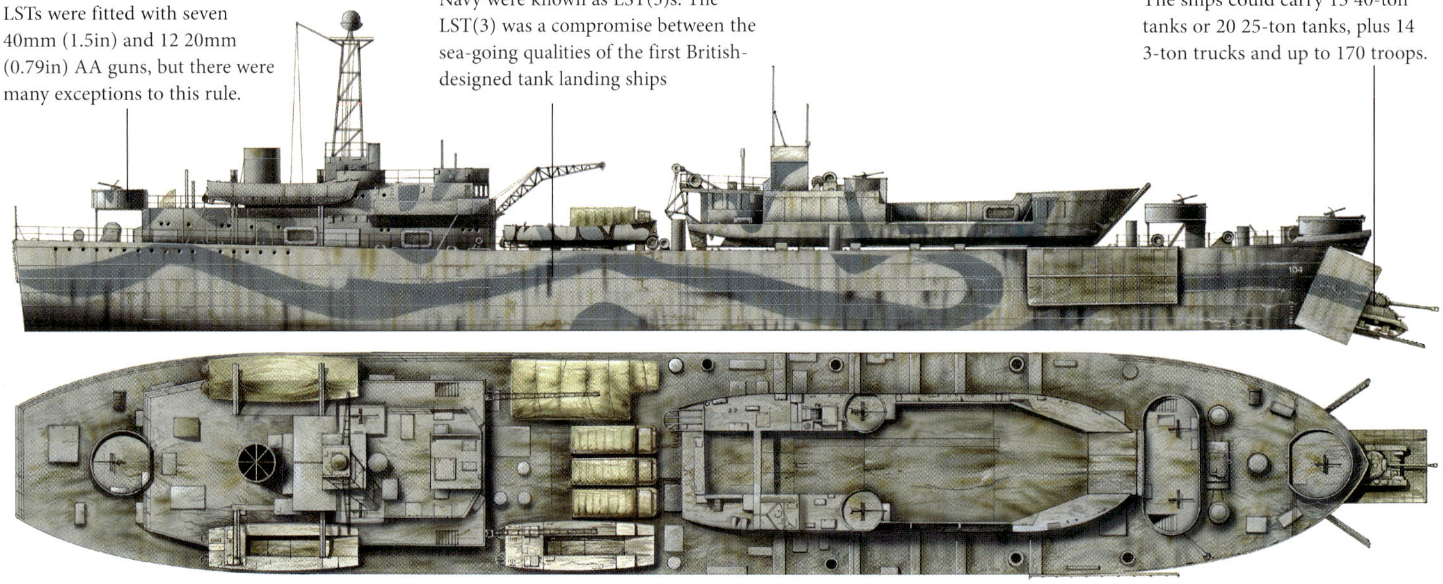

LST(3)
To remedy a deficit in numbers of LSTs in the later years of the war, the UK developed another class, the LST Mk III or LST(3). These had roughly the same dimensions as the US LSTs, but were much heavier (due to their diesel engines and riveted construction) and therefore more powerful. The LST(3)s were also launched between 1944 and 1945. A total of 62 were built in the UK and Canada. More were ordered for after the war, but were cancelled on its conclusion.

LST(3) HMS *NARVIK*, 1948
The LST(3)s proved to be remarkably versatile and after the war most remained in service even into the 1970s.

USS *Independence* (1943)

TYPE • *Aircraft carrier* **COUNTRY** • *United States*

SPECIFICATIONS

Dimensions:	Length: 190m (623ft); Beam: 21.8m (71ft 6in); Draught: 7.4m (24ft 4in)
Displacement:	10,833 tonnes (11,942 tons) standard; 14,987 tonnes (16,520 tons) full load
Propulsion:	4 boilers; General Electric steam turbines; 4 shafts; 75,470kW (100,000shp)
Speed:	31 knots (57km/h; 35.6mph)
Range:	13,000nmi (24,000km; 14,963 miles) at 15 knots
Armament:	24 x 40mm AA guns; 16 x 20mm AA guns
Aircraft:	34
Complement:	1569

As the lead ship of its class of light carriers, USS *Independence* (CV-22) fought with fast carrier battle groups during World War II and operated as the first night carrier, but was ultimately chosen as a target for postwar atom bomb tests.

Left: This stern view of Independence *taken on 30 April 1943 shows the unusual tumble-home hull form of the class.*

Antennas
By mid-1944 the antennas mounted on the mast included a YE radio beacon, on top, with an SG surface search radar just below.

Armour
Independence had no side armour, apart from a 51mm (2in) section protecting the forward magazine. The armoured deck had 51mm (2in) of plate and the bulkheads ranged between 63mm and 127mm (2.5–5in).

Flight deck
The flight deck, narrowed towards the bow, was 186m (610ft) long. It did not extend over the finely tapered bow, and a 127mm (5in) 38-cal gun was mounted right on the bows (this was removed early on).

ATOM BOMB TESTS

In 1946 *Independence* was selected as one of the test vessels to be used in the Operation *Crossroads* atomic bomb tests of Bikini Atoll. The hull survived two explosions without sinking. Despite being highly radioactive, it was towed first to Kwajalein, then to Pearl Harbor and finally to San Francisco for examination. It was ultimately scuttled in the Pacific on 29 January 1951.

CALIFORNIA, 1943
Independence in San Francisco Bay, California, on 15 July 1943, the day its hull number was changed from CV-22 to CVL-22 (indicating a light aircraft carrier). Here it has nine SBD scout bombers parked amidships and aft and nine TBM torpedo planes parked amidships and forward.

USS *Princeton* (1943)

TYPE • *Aircraft carrier* **COUNTRY** • *United States*

SPECIFICATIONS

Dimensions:	Length: 189.7m (622ft 6in); Beam: 21.8m (71ft 6in); Draught: 7.9m (26ft)
Displacement:	11,176 tonnes (12,320 tons) standard
Propulsion:	General Electric geared steam turbines; 2 shafts; 74,570kW (100,000shp)
Speed:	31.6 knots (58.5km/h; 36.4mph)
Range:	12,500nmi (23,125km; 14,375 miles)
Armament:	24 x 40mm (1.5in) AA guns; 22 x 20mm (0.79in) AA guns
Aircraft:	34
Complement:	1569

The fourth US Navy ship to bear the name, the USS *Princeton* (CV-23) was originally intended to be a light cruiser, but the need for more carriers saw it converted to an *Independence*-class aircraft carrier.

Left: This side view of USS Princeton *(CVL-23) clearly shows the ship's distinctive funnel arrangement.* Princeton *was scuttled after sustaining severe damage from air attack at the Battle of Leyte Gulf in October 1944.*

Island
The small island was almost completely clear of the flight deck, which because of the relatively narrow cruiser-form hull needed all possible space.

Aircraft wing
The planned aircraft wing was formed of 30 planes – 12 fighters, nine dive-bombers and nine torpedo bombers – although the *Independence* class normally carried around 28.

Measure 21 camouflage
Princeton's paint scheme went through different camouflage types; here its original dark grey. Measure 21, all-over sea-blue, was frequently used on the large surface areas of carriers.

DEVELOPMENT

The future *Princeton* had been ordered as a light cruiser in July 1940. To be named *Tallahassee*, it was laid down at the New York Shipbuilding Corporation yard, Camden, New Jersey, on 2 June 1941. On 6 February 1942 the change of plan was made, by which time the hull was complete up to the main deck level. As a carrier it was launched on 18 October 1942 and commissioned at the Philadelphia Navy Yard on 25 February 1943 as USS *Princeton* (CV-23). In July 1943 the designation was modified to CVL to indicate its 'light' status.

USS *PRINCETON*, CV-37
The *Princeton* name was carried forward into the postwar years by the *Essex*-class USS *Princeton* (CV-37), which was commissioned in November 1945 and served in the Korean and Vietnam wars.

USS *Lexington* (1943)

TYPE • *Aircraft carrier* **COUNTRY** • *United States*

SPECIFICATIONS

Dimensions:	Length: 250m (820ft); Beam: 28m (93ft); Draught: 10.41m (34ft 2in)
Displacement:	27,534 tonnes (30,365 tons) standard; 36,962 tonnes (40,748 tons) full load
Propulsion:	8 boilers; 4 Westinghouse geared steam turbines; 4 shafts; 110,000kW (150,000shp)
Speed:	33 knots (61km/h; 38mph)
Range:	15,440nmi (28,564km; 17,756 miles) at 15 knots
Armament:	12 × 127mm (5in)/38-cal guns (4 twin, 4 single); 8 × quad 40mm (1.5in)/56-cal AA guns; 46 × 20mm/78-cal AA guns
Aircraft:	110
Complement:	2600

Commissioned in 1943 and reported sunk on four occasions, the *Essex*-class USS *Lexington* (CV-16) enjoyed an illustrious career spanning almost 50 years. The ship remains intact and is preserved as a naval monument, serving as an evocative memory of its era.

Left: USS Lexington *(CV-16) under way on 12 November 1943.* Lexington*'s war record was second to none even in the annals of US carriers. It spent a total of 21 months on combat duty, shot down 372 enemy aircraft and destroyed a further 475 on the ground.*

Paint scheme
Lexington usually sported a blue camouflage paint scheme (Measure 33) and this, together with four successive Japanese claims that it had been sunk, gained it the nickname of 'The Blue Ghost'.

Aviation fuel
The ship's fuel bunkers held 5,955,000 litres (1,500,000 gals) plus 1,747,000 litres (440,000 gals) of aviation fuel.

Internal arrangements
The machinery rooms were arranged en echelon (alternately located) for maximum survivability. The three hangar bays had 3600 sq m (40,000 sq ft) of space.

CLASS DESIGN

The plan of USS *Yorktown* (CV-5) was taken as a basis for the *Essexes*, but despite the urgency of construction, much experience, thought and new technology went into the design. It was finally signed off on 20 February 1940, although the possibility of adaptation to new aircraft types and technology had been incorporated, and modification would go on throughout the war. *Lexington* was built to the specifications and standards of the *Essex* class as a fleet carrier capable of fielding 90 aircraft.

LAUNCH
USS *Lexington* (CV-16) slides down the shipways during its launching at the Bethlehem Steel Company shipyard, Quincy, Massachusetts, on 26 September 1942.

USS *Intrepid* (1943)

TYPE • *Aircraft carrier* **COUNTRY** • *United States*

SPECIFICATIONS

Dimensions:	Length: 250m (820ft); Beam: 28m (93ft); Draught: 10.41m (34ft 2in)
Displacement:	27,534 tonnes (30,365 tons) standard; 36,962 tonnes (40,744 tons) full load
Propulsion:	8 boilers; 4 Westinghouse geared steam turbines; 4 shafts; 110,000kW (150,000shp)
Speed:	33 knots (61km/h; 38mph)
Range:	15,440nmi (28,564km; 17,756 miles)
Armament:	12 × 127mm (5in)/38-cal guns (4 twin, 4 single); 8 × quad 40mm (1.5in)/56-cal AA guns; 46 × 20mm/78-cal AA guns
Aircraft:	110
Complement:	2600

Despite being the most frequently hit US carrier in World War II, *Intrepid* (CV-11) served in the US Navy for more than 30 years on missions ranging from Pacific battles to spacecraft recovery. It remains one of the four surviving *Essex*-class carriers.

Left: Forward and aft profiles of USS Intrepid. *In design and appearance,* Intrepid *followed the basic* Essex *pattern. The strength deck, providing the ship with rigidity, was the hangar deck.*

Radio antennas
On completion there were four lattice masts holding radio antennas, and these could be hinged horizontally outwards during flight operations.

Aircraft elevators
Like other *Essex*-class carriers, *Intrepid* had three aircraft elevators, originally wooden-planked like the flight deck, two on the centreline and one on the port side of the flight deck.

Aircraft
The *Intrepid* carried 110 aircraft of various types, including Hellcats and Helldivers. The flight deck area was 262.8m x 32.9m (862ft 2in x 107ft 11in).

TWO-OCEAN NAVY

On 10 July 1940, President Franklin D. Roosevelt signed the 'Two-Ocean Navy' Act, affirming the USA's aim to have a powerful naval presence in both the Atlantic and Pacific zones, and authorizing funding for three more fleet carriers of the *Essex* class: *Yorktown*, *Intrepid* and *Hornet*. As American involvement in the war grew more likely, a further seven fleet carriers were approved in the same month.

PACIFIC OPERATIONS
USS *Intrepid* (CV-11) on operations in the Philippine Sea in November 1944. Note the F6F Hellcat fighter parked on an outrigger forward of the island.

Unryu (1943)

TYPE • *Aircraft carrier* **COUNTRY** • *Japan*

SPECIFICATIONS

Dimensions:	Length: 227.35m (745ft 11in); Beam: 22m (72ft 2in); Draught: 8.73m (28ft 8in)
Displacement:	20,450 tonnes (22,545 tons) standard
Propulsion:	8 Kampon water-tube boilers; 4 geared steam turbine sets; 4 shafts; 113,000kW (152,000shp)
Speed:	34 knots (63km/h; 39mph)
Range:	8000nmi (15,000km; 9200 miles)
Armament:	6 × twin 127mm (5in) dual-purpose guns; 16 × triple and 3 × single 25mm (0.98in) AA guns; 6 × 28-tube 120mm (4.7in) AA rocket launchers
Armour:	Belt 46–140mm (1.8–5.5in); deck 25–56mm (0.98–2.2in)
Aircraft:	48 (up to 65)
Complement:	1595

A strike carrier, for use against convoys, *Unryu* was one of only three to be completed out of a planned class of 17. With the island sponsoned out, it had a wide flight deck, served by two lifts. *Unryu* was sunk by the US submarine *Redfish* in December 1944, before it saw any action.

Above: Unryu *was sunk on its maiden sea voyage, at a time when US submarines were inflicting record losses on both merchant and naval vessels.*

Aircraft
The standard aircraft carried aboard the *Unryu* were the Mitsubishi A6M fighter, Aichi D3A dive-bomber and Nakajima B5N torpedo-bomber

Island
Although the *Unryu* class was based on the carrier *Hiryu*, the island was placed on the starboard side instead of the port side.

Powerplant
The ship had eight water-tube boilers linked to Kampon geared turbines, in total generating 113,000kW (152,000shp) of propulsive energy.

TORPEDOED

Unryu arrived late in the war, only commissioned in August 1944. Its service history lasted a mere five months. On 19 December 1944, while heading to the island of Luzon to reinforce Japanese operations, *Unryu* was torpedoed by the submarine USS *Redfish*, causing major explosions on board. A total of 1238 Japanese sailors and aviators went down with the ship.

***AMAGI*, MARCH 1946**
The carrier *Amagi* capsized after bomb strikes from American carrier aircraft while moored at Kure Naval Base on 29 July 1945. In total three aircraft carriers of the *Unryu* class were completed – *Unryu*, *Amagi* and *Katsuragi* – while three others of the class were unfinished by war's end.

Type XXI (1943)

SPECIFICATIONS

Dimensions: Length: 76.7m (251ft 10in); Beam: 8m (26ft 3in); Draught: 6.32m (20ft 9in)

Displacement: 1621 tonnes (1787 tons) surfaced; 1819 tonnes (2006 tons) submerged

Propulsion: 2 1640kW (2200hp) diesels; 2 1864kW (500hp) electric motors; 2 241kW (323hp) creep motors, 2 screws

Speed: 15.6 knots (28.9km/h; 18mph) surfaced; 17.2 knots (31.9km/h; 19.8mph) submerged

Range: 11,150nmi (20,650km; 12,831 miles) surfaced; 285nmi (528km; 328 miles) submerged

Armament: 6 × 533mm (21in) torpedo tubes; 2 × twin 20mm (0.79in) Flak mountings

Complement: 57

German scientists and naval architects, led by Dr Helmuth Walter, were developing a 'true' submarine capable of remaining submerged for days and that would be virtually undetectable. However, such a design was still a good way off. The Type XXI was planned to fill the gap.

Left: A Type XXI sails out of port. The amount of technical equipment packed into the hull meant that only 47 berths were available for a crew of 57. Hammock hooks were provided.

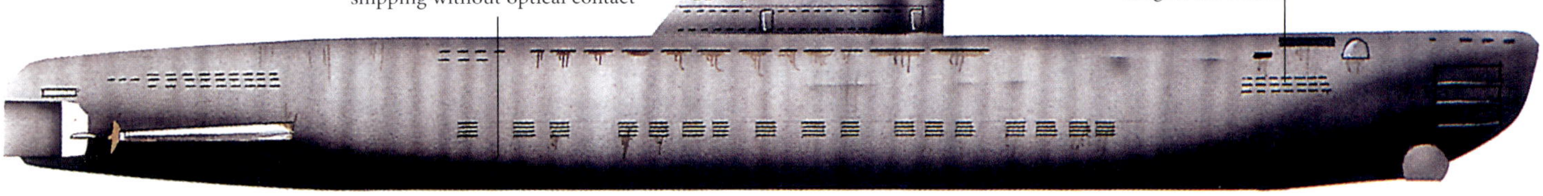

Passive/active sonar
An improved passive sonar system was fitted beneath the keel, and a new active system enabled detection and attack of enemy shipping without optical contact

U-2051
The Type XXI's smooth lines made existing submarines look antique: everything – guns, aerials, bollards – was retractable (including the bow diving planes) to ensure minimum drag in the water.

EVASION
Although the Type XXI attacked at slow speed, with the listening gear unaffected by its own propeller noise, it used high speed to evade escort hunts. Its manoeuvrability (despite a large submerged turning circle) posed difficulties for a single escort trying to hold contact; several escorts might be needed to hem it in. Depth-charge attacks would be wholly ineffective if the U-boat got any warning of the attack.

Engines
Walter's ultimate aim was for an air-independent main engine, but he had to compromise with two MAN M6V409/46KBB supercharged six-cylinder diesels delivering 2900kW (3900hp).

UNDER CONSTRUCTION
These incomplete Type XXI submarines were photographed at Bremen at war's end in 1945. The speed of design and production for the Type XXI meant that many modifications had to be made, which often required retrofitting.

Type XXIII (1944)

TYPE • *Submarine* **COUNTRY** • *Germany*

SPECIFICATIONS

DIMENSIONS:	Length: 34.68m (113ft 9in); Beam: 3.02m (9ft 11in); Draught: 3.66m (12ft)
DISPLACEMENT:	237.7 tonnes (262 tons) surfaced; 262 tonnes (289 tons) submerged
PROPULSION:	1 diesel engine, 469.8kW (630hp); 1 electric motor, 426.5kW (572hp); 1 creep motor, 26kW (35hp); 1 screw
SPEED:	9.7 knots (18km/h; 11mph) surfaced; 12.5 knots (23km/h; 14.4mph) submerged
RANGE:	2600nmi (4815km; 2992 miles) surfaced; 194nmi (359km; 223 miles) submerged
ARMAMENT:	2 × bow torpedo tubes
COMPLEMENT:	14

This urgently needed coastal patrol submarine shared many characteristics of the Type XXI. *U-2321*, the class leader, was laid down on 10 March 1944 and commissioned on 12 June that year, indicating its high priority.

Left: U-2361, *a Type XXIII U-boat seen here in Stranraer, Scotland, after VE Day in 1945, was commissioned in February 1945. It was scuttled in November 1945 as part of Operation* Deadlight, *the victorious Allies' effort to dispose of the U-boat fleet.*

MACHINERY
The machinery had to be as compact as possible: a single MWM RS-348 469.8kW (630hp) diesel and an AEG GU 4463/8 426.5kW (572hp) electric motor.

SAIL
The sail was teardrop-shaped, tapering to a point, as was the stern. A single periscope, contained inside the tower casing with the snorkel, was fitted at the back of the tiny open bridge.

NO GUNS
The Type XXIII carried no guns and presumably relied on its nine-second crash-diving ability to avoid aerial attack – it could go down to 180m (590ft). The operational Type XXIIIs were appreciated by crews for their manoeuvrability both on the surface and submerged. While the surface speed was moderate, the submerged speed of 12.5 knots (22km/h; 14.4mph) could get them away from depth-charge attacks, assuming the boat was detected.

HULL
The streamlined hull formation was designed for submerged speed and silence. In cross-section, the forward part of the hull was of expanded figure-of-eight form, with the tubes above and the battery compartment in a narrower lower section.

S171 *HECHT*
S171 *Hecht* was formerly the Type XXIII *U-2367*, which sank on 5 May 1945 near Schleimünde after a collision with another U-boat, but was raised and recommissioned into the Federal German Navy in 1956.

Taiho (1944)

TYPE • *Aircraft carrier* **COUNTRY** • *Japan*

Marking an advance in Japanese carrier construction and considered unsinkable, *Taiho* was a powerful equipped and formidably armed ship. However, its fighting life lasted only three months as the Japanese lost grip of the Pacific waters.

SPECIFICATIONS

Dimensions:	Length: 260.6m (855ft); Beam: 27.4m (89ft 11in); Draught: 9.6m (31ft 6in)
Displacement:	29,769 tonnes (32,812 tons) standard; 33,243 tonnes (36,644 tons) full load
Propulsion:	8 Kampon boilers; 4 Kampon geared steam turbines; 4 shafts; 120,000kW (160,000shp)
Speed:	33.3 knots (61.7km/h; 38.3mph)
Range:	10,000nmi (19,000km; 12,000 miles) at 18 knots
Armament:	6 × twin 100mm (3.9in)/65-cal Type 98 guns; 17 × triple 25mm (1in) AA cannon
Aircraft:	84
Complement:	1751

Left: Taiho *took almost two years to fit out after launching, an indication of the difficulties that Japanese industry had in sustaining the huge war effort.*

Armament
The 100mm (3.9in) guns were controlled by two Type 89 Mod 1 directors, with 4.5m (14ft 9in) Type 94 rangefinders, one mounted forward of the island, the other amidships on the port side below the flight deck.

Radar
Two sets of Type 21 Mod 3 radars, one in front of the bridge, the other abaft the funnel, gave 360 degrees coverage, with a maximum 150km (92 miles) in the air and 20km (12.5 miles) on the surface.

AVIATION ADVANTAGES
Japanese aircraft carriers never employed catapult launchers, so aircraft had to rely on their own speed and windspeed to launch. They were helped by the planes' relatively light weight, which also gave the Japanese planes a somewhat greater range, especially in reconnaissance and long-range strikes. On the negative side, it meant a lesser degree of protection for the aircrew and greater liability in the airframe to break up under attack.

Arrestor wires
Taiho mounted 14 arrestor wires, six forward and eight aft, with electrically operated deceleration gear. Two crash barriers were placed aft of the island and a third to forward.

FLAGSHIP
Taiho was Vice-Admiral Jisaburo Ozawa's flagship in the Battle of the Philippine Sea. The carrier was torpedoed and sunk by a US submarine on 19 June 1944.

Shinano (1944)

TYPE • *Aircraft carrier* **COUNTRY** • *Japan*

The *Shinano* was originally intended to be another mighty *Yamato*-class battleship, but changing wartime requirements led to its conversion into the largest carrier yet built. However, it was sunk even before entering operational service.

SPECIFICATIONS

Dimensions:	Length: 265.8m (872ft 1in); Beam: 36.3m (119ft 1in); Draught: 10.3m (33ft 10in)
Displacement:	65,800 tonnes (72,547 tons) standard; 73,000 tonnes (80,469 tons) full load
Propulsion:	12 Kampon boilers; 4 geared steam turbines; 4 shafts; 110,000kW (150,000shp)
Speed:	27 knots (50km/h; 31mph)
Range:	10,000nmi (19,000km; 12,000 miles)
Armament:	8 × twin 127mm (5in) Type 89 guns; 35 × triple 25mm Type 96 guns; 12 × 28-barrel 127mm rocket launchers
Aircraft:	139
Complement:	2400

Left: A sketch of the Shinano *by Shizuo Fukui. The huge* Shinano *was viewed as a floating naval base, not unlike the present-day strategic concept of the aircraft carrier.*

Hangar area
The hangar area, 163.4m by 33.8m (536ft by 111ft), had unarmoured sides that were mostly open-sided – the enclosed hangar sides of earlier carriers had contributed to their destruction – but with fire curtains to close off sections.

Island
The island was placed slightly forward of midships and sponsoned well out to give the maximum flight deck and plane parking area. The funnel was integrated with the island.

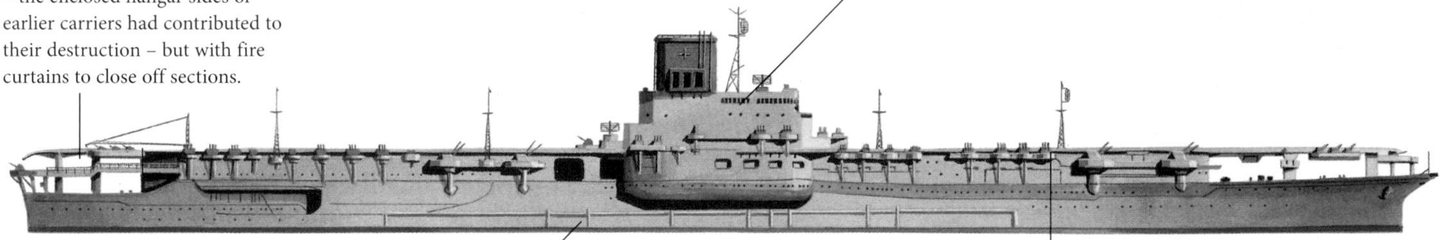

Armour
The *Yamato* class had 400mm (15.7in) belt armour, reduced on *Shinano* to 205mm (8in), and lateral anti-torpedo bulges were fitted, backed up internally by armoured and watertight bulkheads, making a double hull below the waterline and a triple bottom.

Flight deck
The armoured flight deck was 256m (839ft 11in) long and 40m (131ft 3in) wide. Like other Japanese carriers, multiple arrestor wires were fitted, 10 towards the stern and five towards the bow, for forward landings.

ROCKET PLANES
Shinano was transporting 30 Ohka MXY-7 rocket-powered attack aircraft when it was sunk. These were kamikaze craft, essentially human-guided flying bombs, with a 1200kg (2650lb) high-explosive bomb forming the nose of the craft. Normally transported beneath Mitsubishi G4M2e 'Betty' bombers, they would be released and glide towards the target vessel. The pilot would then fire the solid fuel rocket motors and crash-dive his plane on to the target.

ON TRIALS, 11 NOVEMBER 1944
Shinano was sunk on 29 November 1944 off the coast of Japan, hit by four torpedoes from the US Navy submarine *Archerfish*. A total of 1435 lives were lost.

Tench class (1944)

TYPE • *Submarine* **COUNTRY** • *United States*

SPECIFICATIONS	
Dimensions:	Length: 95m (311ft 8in); Beam: 8.33m (27ft 3in); Draught: 5.18m (17ft)
Displacement:	1800 tonnes (1984 tons) surfaced; 2455 tonnes (2706 tons) submerged
Propulsion:	4 diesel generator engines, 4026kW (5400hp); 4 electric motors, 2044kW (2740hp); 2 screws
Speed:	21 knots (39km/h; 24mph) surfaced; 9 knots (16.6km/h; 10mph) submerged
Range:	11,000nmi (20,372km; 12,659 miles) surfaced; 95nmi (176km; 109 miles) submerged
Armament:	10 × 533mm (21in) torpedo tubes; 28 × torpedoes; 1 × 127mm (5in) deck gun; 1 × 40mm (1.5in) AA gun; 1 × twin 20mm (0.79in) AA gun
Complement:	81

The US Navy's final submarine class of the war, the *Tench*-class served in modernized form for another 30 years. Ten boats were completed in time for war service. Externally, they were almost identical to the *Balao*-class, although their fairwater was shorter.

Above: USS Tigrone *(SSR-419), September 1963. The* Tench *class had a low superstructure and hull profile, as well as streamlined lines that gave it a good surface speed.*

Loop antenna
This was mounted between the periscopes, although later *Tench*-class boats had it mounted on a bracket aft of the SJ mast.

Ballast tanks
Re-planning of the ballast tanks also allowed space for an additional four torpedoes compared with the *Balao* and *Gato* boats.

Deck armament
As designed, the *Tench* boats carried a single 127mm (5in) 25-cal gun aft of the tower and two 40mm (1.5in) AA guns on the fairwater platforms. Pintle mounts for portable .50-cal machine guns were mounted on the bridge and alongside the fairwater.

ACCOMMODATION
From stem to stern, the compartments consisted of a forward torpedo room with escape trunk; officers' quarters (above the forward battery); control room, with pump room, stores and access hatch to conning tower above; crew's mess and crew's quarters, with aft battery below; forward engine room; aft engine room; machine room with manoeuvring room above; aft torpedo room. Crew accommodation was the best of all wartime submarines.

USS *PICKEREL* (SS-524)
On the surface, forward-diving planes were stowed ready-tilted so that when extended they would immediately force the bow down. Diving to periscope depth could be achieved in 30–40 seconds.

Type I-400 (1944)

TYPE • *Submarine* **COUNTRY** • *Japan*

SPECIFICATIONS

Dimensions:	Length: 122m (400ft); Beam: 12m (39ft 5in); Draught: 7m (23ft)
Displacement:	5307 tonnes (5850 tons) surfaced; 6760 tonnes (7454 tons) submerged
Propulsion:	4 diesel engines, 1680kW (2250hp) each; 2 electric motors, 1600kW (2100hp) each; 2 screws
Speed:	18.7 knots (34.6km/h; 21.5mph) surfaced; 6.5 knots (12km/h; 7.5mph) submerged
Range:	37,000nmi (68,524km; 42,579 miles) at 14 knots surfaced; 60nmi (111km; 69 miles) at 3 knots submerged
Armament:	8 × 533mm (21in) torpedo tubes; 12 × torpedoes; 1 × Type 11 140mm (5.5in) gun; 3 × triple and 1 × single Type 96 25mm AA guns
Complement:	157 including aircrew

In 1942, under Japan's Fifth Fleet Replenishment Programme, a total of 18 super-submarines, designated *Sen-Toku* (special class), were included. Work commenced immediately, and *I-400* was laid down at Kure on 18 January 1943.

Left: The two Type I-400s, *seen here after the Japanese surrender. The type struck awe into American observers in 1945, but the Japanese war effort would have been better served by more conventional submarines in greater numbers.*

Deck layout
The upper section with the bridge deck was canted out on the port side, with the pressurized conning tower alongside. The AA placings were to starboard and port of the centreline.

Float plane
This *I-400* mounts a Aichi M6A1 'Seiran' float plane. The Seirans were launched from a 26m (85ft) Type 4 No. 2 Model 10 compressed-air catapult, rising at a shallow angle from the hangar door.

Galley
The galley was in the starboard hull, fitted with giant steam kettles for preparing rice for the 157-strong crew.

LAYOUT
Only two *I-400*s were commissioned into service, in early 1945. Special sea-air missions were intended, including an attempt to block the Panama Canal (the USN's most vital artery). A single pressure hull could not have provided sufficient space, and the unique design was based on a pair of parallel linked cylindrical pressure hulls inside an outer hull. The tower was above the port hull, with the control centre below.

RESUPPLY
Crew of the *I-400* pass supplies down the line as they restock the submarine. The *I-400* could carry enough fuel to cruise for 37,000nmi (68,524km, 42,579 miles) at 14 knots, and maintain a patrol for 90 days (capacities never fully tested: the vessel had no air conditioning and no flushing toilets).

Landing Ship Medium (Rocket) (1944)

TYPE • *Landing craft* **COUNTRY** • *United States*

SPECIFICATIONS

Dimensions:	Length: 62.03m (203ft 6in); Beam: 10m (34ft); Draught: 1.63m (5ft 4in) light
Displacement:	1194 tonnes (1316 tons) full load
Propulsion:	2 General Motors Cleveland diesel engines; 2 screws; 2088kW (2800shp)
Speed:	13.2 knots (24.4km/h; 15.2mph)
Range:	5000nmi (9300km; 5800 miles)
Armament:	1 × 127mm (5in) gun; 2 × 40mm (1.5in) Bofors guns; 3 × 20mm (0.79in) Oerlikon cannon; 75 × 4-rail Mark 36 automatic rocket launchers; 30 × 6-rail Mark 30 launchers (removed April 1945)
Armour:	None
Complement:	81

The Landing Ship Medium (Rocket) was a curious vessel designed to bombard the shorelines of landing beaches with ripples of '3-inch' rockets. The ship was used by both the Americans and the British.

Right: USS LCS-6 *photographed from* LSM(R)-509 *in Lock 41 on the Ohio River, October 1945. Landing Craft, Support (LCS) vessels could be adapted for a range of purposes, and were armed with three guns and 10 rocket launchers.*

Rocket launchers
The ship was armed with 75 four-rail Mark 36 automatic rocket launchers on the topside rocket deck and 30 six-rail Mark 30 launchers mounted along the gunwales.

Main gun
In addition to its rockets, the ship was fitted with a single 127mm (5in) naval gun.

Trajectory
The rocket racks were set to send the unguided rockets on a high arcing trajectory, the intention being to drop the rockets down behind enemy forward defences.

ROCKET BARRAGE

The 3-inch rockets fired from the LSM(R) and larger LCT(R) vessels each carried a 27kg (60lb) warhead and could reach ranges of up to about 1600m (5250ft). Individually, the weapons were quite inaccurate, but when fired en masse the area effects compensated for that lack of precision. In the Pacific where the LSM(R)s served, however, sometimes barrages did little more than erase sand and coral or churn the water.

USS *BLACKSTONE RIVER*
USS *Blackstone River* (LSMR-403) was laid down on 6 January 1945 at Charleston, SC, launched on 26 January 1945 and commissioned on 25 April 1945. It did not see combat in World War II, but it did during the Korean War (1950–53).

USS *Missouri* (1944)

TYPE • *Battleship* **COUNTRY** • *United States*

SPECIFICATIONS

DIMENSIONS:	Length: 270.4m (887ft 3in); Beam: 33m (108ft 2in); Draught: 11.5m (37ft 9in)
DISPLACEMENT:	52,834 tonnes (58,268 tons) standard; 58,372 tonnes (64,360 tons) full load
PROPULSION:	8 Babcock & Wilcox boilers; 4 General Electric steam turbines; 4 screws; 158,088kW (212,000shp)
SPEED:	33 knots (61.2km/h; 38mph)
RANGE:	15,000nmi (28,000km; 17,400 miles) at 15 knots
ARMAMENT:	9 × 406mm (16in) guns; 20 × 127mm (5in) guns; 60 × 40mm (1.6in) four-barrelled AA guns
ARMOUR:	Belt: 310mm (12.2in); barbettes: 440–287mm (17.3–11.3in); turret faces: 500mm (19.7in); deck 152mm (6in)
COMPLEMENT:	1921

The USS *Missouri* (BB-63) was the last of the *Iowa* class to be completed, when commissioned on 11 June 1944 at the New York Navy Yard. In profile and general detail, the ships were very similar.

Right: An unwelcome feature of the widening hull design was that the ship's 'shoulders' pushed out waves that broke against the sea waves and caused heavy spray to break over the bridge.

HULL DESIGN
One unwelcome feature of the widening hull design was that the ship's 'shoulders' pushed out waves that broke against the sea waves and caused heavy spray to break over the bridge.

MAIN GUNS
As with the other ships in the class, the main armament was nine Mk 7 406mm (16in) guns in three turrets. Each gun could be aimed and fired independently.

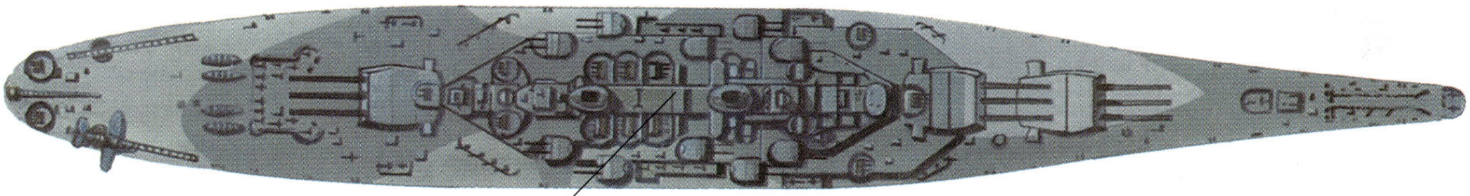

ENGINES
The engines were reliable: *Missouri* is recorded as steaming continuously for 58 days with no mechanical problems.

POSTWAR SERVICE
Missouri is famous as the site of Japan's formal surrender on 31 August 1945. However, between periods in reserve, it was put back in service in 1950–55 and 1984–91, including during the Korean and Gulf wars. In the 1950s a nuclear-capped shell was made specifically for the 406mm (16in) guns. Weighing 862kg (1900lb), its estimated yield was between 15 and 20 kilotons of TNT. Information on whether these shells were actually carried on the ships remains classified.

STARBOARD SALVO
Missouri fires a salvo from the forward 16/50 gun turret, during it shakedown period, circa August 1944. *Missouri* was finally decommissioned on 1 March 1992. As a museum ship, the battleship today stands moored to a pier at Naval Station Pearl Harbor, Hawaii.

USS *Guam* (1944)

TYPE • *Battleship* **COUNTRY** • *United States*

SPECIFICATIONS

DIMENSIONS:	Length: 264.4m (808ft); Beam: 27.8m (91ft 1in); Draught: 9.7m (31ft 10in)
DISPLACEMENT:	30,257 tonnes (33,364 tons) standard; 34,803 tonnes (38,360 tons) full load
PROPULSION:	8 Babcock & Wilcox boilers; 4 General Electric geared steam turbines; 4 screws; 114,092kW (153,000shp)
SPEED:	33 knots (61.1km/h; 38mph)
RANGE:	12,000nmi (22,094km; 13,700 miles) at 15 knots
ARMAMENT:	9 × 305mm (12in) guns; 12 × 127mm (5in) AA guns; 56 × 40mm (1.6in) guns; 34 × 20mm (0.8in) AA guns
ARMOUR:	Belt 229–127mm (9–5in); turrets 325mm (12.8in); deck 102mm (4in)
COMPLEMENT:	1517

The last battlecruisers to be built, the two-ship *Alaska* class was intended as a counter to the German Navy's *Scharnhorst* and *Gneisenau*. By the time *Guam* (CB-2) was launched in 1944, however, such a role was no longer required.

Left: The flag is raised over Guam *for the first time during its commissioning ceremony at the Philadelphia Navy Yard, 17 September 1944. The photograph comes from the ship's wartime cruise book,* U.S.S. Guam: Her Story, *1944–1945.*

ELEVATION
The 127mm (5in) guns could elevate to 85 degrees for AA fire (their number was increased to 68 in 1945) and the 20mm to 90 degrees.

ARMOUR
The *Alaska* class was armoured only against 305mm (12in) shells. Total thickness of the three armoured decks over the magazines was 169mm (6.65in).

BOILERS
Steam was raised by eight Babcock & Wilcox express boilers. Fitted with superheaters of the same type as the *Iowa*-class ships, they worked at a maximum pressure of 44.6kg/sq cm (634psi).

FIREPOWER
Nine Mk 8 305mm (12in) guns were carried. The three-gun turret, with each gun elevating independently up to 40 degrees, was preferred for its compactness. Shell weight was 57kg (1140lb) and the muzzle velocity of 762m/s (2500ft/s) gave a maximum range of 35,271m (115,711ft). Magazine capacity was 1500 rounds. The Mk 8 was generally regarded as an excellent gun, capable of penetrating deck armour 121mm (4.75in) thick.

OFF TRINIDAD, 1944
USS *Guam* photographed off Trinidad on 13 November 1944 during shakedown training. The ship entered service in early 1945 and operated with *Alaska* in supporting carrier raids on the Japanese mainland and patrols in the China Sea.

T class (1944)

TYPE • *Submarine* **COUNTRY** • *United Kingdom*

SPECIFICATIONS (HMS *Thule*)

Dimensions:	Length: 84.28m (276ft 6in); Beam: 7.77m (25ft 6in); Draught: 4.45m (14ft 7in)
Displacement:	1310 tonnes (1445 tons) surfaced; 1585 tonnes (1748 tons) submerged
Propulsion:	2 diesel engines, 1900kW (2500hp); 2 electric motors, 1080kW (1450hp); 2 screws
Speed:	15.5 knots (28.7km/h; 17.8mph) surfaced; 9 knots (17km/h; 10mph) submerged
Range:	4500nmi (8334km; 5178 miles) at 11 knots surfaced; 121nmi (224km; 139 miles) at 2.5 knots submerged
Armament:	11 × 533mm (21in) torpedo tubes (6 internal forward, 5 external aft-facing); 17 torpedoes; 1 × QF 102mm (4in) deck gun; 1 × 20mm (0.79in) AA gun; 2 × 7.7mm (0.303in) AA guns
Complement:	61

This was another large class (53 boats) with boats launched in three phases. The original brief was for an ocean-going submarine of around 1200–1500 tonnes surfaced, with an endurance capability of 42 days and a large forward salvo of torpedoes.

Left: HMS Torbay *(N79) was launched in April 1940 and went on to have a highly successful wartime career, sinking 17 merchant vessels and five warships.*

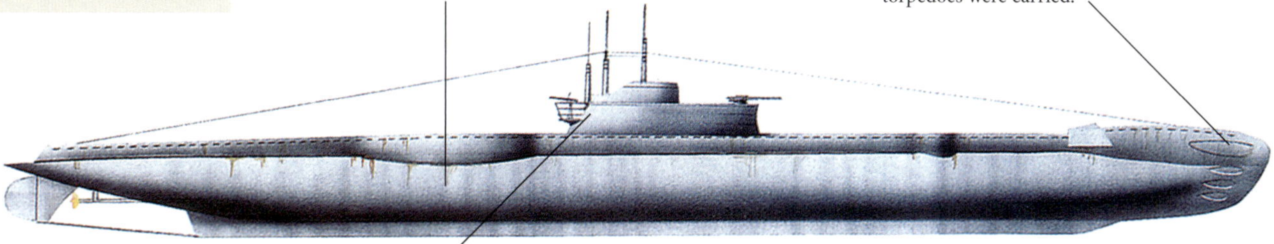

Engines
Diesel power came from several manufacturers. Most had Vickers four-stroke engines. Boats built by Cammell Laird had Sulzer two-stroke engines, and the Royal Dockyards used Admiralty-type. Four from Scotts had MAN engines: these were replaced in wartime.

HMS *Thule* (Group III)
The T-class was formidably well-armed. Six internal forward tubes were supplemented by four forward-facing external tubes, two in the bow and two at the base of the tower. Sixteen torpedoes were carried.

Equipment
From 1941, in addition to Type 129 ASDIC, fitted below the bow, the T-class were fitted with a deck-mounted DF detecting coil, Type 267 Seaguard and Type 291 Airguard radar. ASDIC Type 129 was fitted below and aft of the forward tubes, its dome protected by an obstruction rod.

SUB-GROUPS

Group I of the class consisted of 22 boats of riveted construction, single-hulled with saddle tanks. Diving depth was 91m (303ft). Group II and III boats had welded hulls, formed in prefabricated sections, which were combined on the launching berth. In order to serve in the Far East and Pacific, a number of Group II and III boats had their operating range increased to 9559nmi (17,703km, 11,000mi) at 10 knots (18.5km/h; 11.5mph).

HMS *TRIBUNE*
HMS *Tribune* (N76), a British T-class submarine, ploughs through heavy seas in the North Atlantic, c. 1943. It survived the war and was scrapped in November 1947.

Type I-201 (1945)

A late arrival into the Pacific War, the powerful Type 201 was Japan's belated hope to stem the US seaborne onslaught. Using the 1938 design as a basis, planning was rapid and the first of the new *Sen-Taka* class, *I-201*, was laid down at Kure Naval Arsenal in March 1944.

SPECIFICATIONS

DIMENSIONS:	Length: 79m (259ft); Beam: 5.8m (19ft); Draught: 7m (23ft)
DISPLACEMENT:	1312 tonnes (1446 tons) surfaced; 1470 tonnes (1621 tons) submerged
PROPULSION:	2 diesel engines, 1025kW (1375hp); 4 electric motors, 932kW (1250hp); 2 screws
SPEED:	15.75 knots (29.2km/h; 18mph) surfaced; 19 knots (35km/h; 21.8mph) submerged
RANGE:	15,000nmi (27,780km; 17,262 miles) at 6 knots surfaced; 135nmi (250km; 155.4 miles) at 3 knots submerged
ARMAMENT:	4 x 533mm (21in) torpedo tubes; 10 torpedoes; 2 x Type 96 25mm (1in) AA guns
COMPLEMENT:	50

Left: I-201 *survived the war and was handed over to the US Navy, who sank it as a target ship in 1946.*

WEAPONRY
As with previous IJN classes, the four torpedo tubes were all in the bow, with six reloads. The two deck-mounted 25mm Type 96 AA guns, forward and aft of the tower, were retractable to reduce underwater resistance.

I-201
I-201 was completed on 2 February 1945. An anechoic rubberized coating was applied to reduce noise. The bow planes were retractable.

HULL CONSTRUCTION
Electric welding was used as much as possible in the hull construction, with the complete hull assembled from prefabricated sections transported to the launch slip.

ATTACK SUBMARINE

As the American war effort in the Pacific intensified through 1943, the IJN command had to accept that its existing submarines were simply incapable of defending themselves against the range of ASW detections and weaponry being deployed against them. Although Japan had built a small experimental high-speed submarine in 1938, this had not been followed up and the boat had been scrapped in 1940. It was not until October 1943 that the formal request was made for an attack submarine that would have high submerged speed and silent running.

***I-201* CLASS, KURE**
An aerial photograph captures an *I-201* class submarine off Kure in 1945. As this aerial view shows, the beam of the Type I-201 was unusually narrow for the boat's length, with retractable forward dive planes in order to reduce underwater resistance, although low-set broad horizontal stern fins increased it to 9.2m (30ft).

USS *Midway* (1945)

TYPE • *Aircraft carrier* **COUNTRY** • *United States*

SPECIFICATIONS

Dimensions:	Length: 295m (968ft); Beam: 34.4m (113ft); Draught: 10.7m (35ft)
Displacement:	45,720 tonnes (50,400 tons) standard; 60,960 tonnes (67,200 tons) full load
Propulsion:	12 boilers; 4 steam turbines; 4 shafts; 158,000kW (212,000shp)
Speed:	33 knots (61km/h; 38mph)
Range:	12,000nmi (22,200km; 13,800 miles)
Armament:	18 × 127mm (5in)/54-cal Mk 16 guns; 84 × 40mm Bofors guns; 68 × 20mm Oerlikon AA cannon
Aircraft:	145
Complement:	3443

A wartime design but not commissioned until after the conflict had ended, USS *Midway* (CVA-41) was the first of a new generation of USN carriers, and underwent major changes through a working life of almost half a century.

Left: An early photograph of USS Midway. *From November 1957 onwards,* Midway *served in the Pacific and Indian oceans. In 1965 it was deployed on active combat duties in the Vietnam war zone, flying air strikes into North Vietnam, shooting down three MiG jets but also losing 11 of its own planes.*

Gun positions
All guns were mounted below the flight deck, but as they were mounted on barbettes, and the flight deck was 15m (50ft) above the surface, they commanded a good angle of fire.

Flight deck
Midway class's armoured flight deck was not part of the hull structure, as in British Navy carriers, but part of the ship's superstructure, mounted above the hull. Subsequent US classes followed the British format.

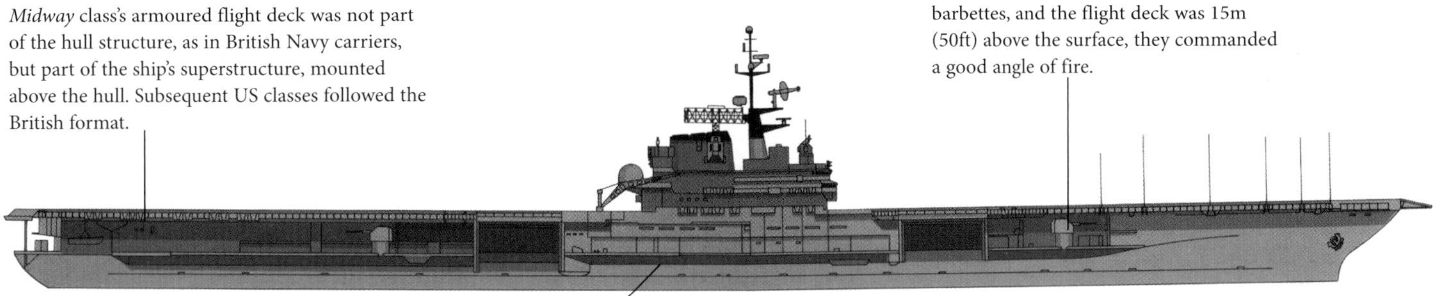

Size
Midway was the largest warship in the world until 1955. An overall width of 31.45m (136ft) prevented it from using the Panama Canal. Atlantic–Pacific transits had to be made via Cape Horn.

REQUIREMENTS

Laid down at Newport News Shipbuilding Co., Virginia, on 27 October 1943, launched on 20 March 1945, *Midway* (CVB-41) was commissioned on 10 September 1945, a month after Japan's surrender, at a construction cost of $86.5 million. Battle conditions of the Pacific War determined the ship's characteristics: more heavily armoured than any of its predecessors, with heavy AA armament, it was the first US carrier class to have an armoured flight deck of 89mm (3.5in). At the same time, it had to have a large air wing, and the ability to make high speed.

NEW SHIP
USS *Midway* (CVA-41) leaves its builders. It would subsequently undergo several major postwar upgrades and modifications, radically changing its appearance and capabilities.

Cold War

On both sea and land, the Cold War marked the ascendancy of the guided missile. As the last battleships faded into history – some lingered on in service into the 1990s – the era of the naval gun drew to a close. Although many cruisers, destroyers, frigates and other vessels retained small numbers of auto-loading guns, the weight of armament shifted to a new generation of guided missiles, including surface-to-air, anti-ship and land attack types. Submarine-launched ballistic missiles (SLBMs) gave submarines the ability to strike globally with nuclear weapons without ever surfacing. The Cold War also saw the widespread introduction of nuclear powerplants, particularly for aircraft carriers and submarines, providing endurance limited only by food and water supplies. Combined with increasing computerization of ship systems, these developments meant that even relatively small warships often had over-the-horizon striking power.

SLAVA/MOSKVA
Lead ship of Project 1164 Atlant class, *Slava* was a Soviet navy missile cruiser, renamed *Moskva* in 1995 as part of the Russian navy. It was heavily armed with 16 P-1000 Vulcan anti-ship missiles, S-300F Fort anti-aircraft launchers and AK-630 30mm guns.

HMS *Vanguard* (1946)

TYPE • *Battleship* **COUNTRY** • *United Kingdom*

SPECIFICATIONS

DIMENSIONS:	Length: 249m (814ft 6in); Beam: 32.8m (107ft 7in); Draught: 9.3m (30ft 6in)
DISPLACEMENT:	41,871 tonnes (46,155 tons) standard; 46,665 tonnes (51,440 tons) full load
PROPULSION:	8 Admiralty 3-drum superheated boilers; 4 Parsons single-reduction geared steam turbines; 4 screws; 96,941kW (130,000hp)
SPEED:	29.75 knots (55.1km/h; 34.2mph)
RANGE:	9000nmi (16,670km; 10,360 miles) at 20 knots
ARMAMENT:	8 x 381mm (15in) Mk I guns; 16 x 133mm (5.25in) guns; 73 x 40mm (1.6in) Bofors AA guns
ARMOUR:	Belt 355–330mm (14–13in); bulkheads 305–101mm (12–4in); deck 152–133mm (6–5in); barbettes 330–279mm (13–11in); turrets 330–152mm (13–6in)
COMPLEMENT:	1500

Britain's last and most powerful battleship was completed too late to see action in World War II. Althoughit was a potent symbol of naval power, the changing post-imperial demands left the Royal Navy left with little practical role for it.

Right: HMS Vanguard *underway at sea, photographed from the island of a nearby British aircraft carrier.* Vanguard *was the last classic battleship to be built; the atomic era had already begun by the time it was commissioned.*

RADAR
Several types of radar were fitted: AR262 (AA fire control), SW268 (surface warning), SR274 (main gun fire control), AR275 (secondary gun fire control), AW277 (height-finding), TI293 (target indication) and GW960 (air warning).

POWERPLANT
The powerplant consisted of four units, each of two boilers with a separate set of turbines, able to operate independently of the others.

SUPERSTRUCTURE
The superstructure is built straight up behind 'B' turret, a substantial edifice that contained the admiral's quarters and command post as well as the ship's own bridges.

CONSTRUCTION

Vanguard was a one-off, laid down at John Brown's yard in Clydebank on 2 October 1941, launched on 30 November 1944 and completed on 25 April 1946. Much of the *Lion* and *King George V* class design work was incorporated, with modifications. A straight bow and level forecastle, as in *King George V*, was replaced by a handsome cutaway bow with considerable flare, which helped to make *Vanguard* an exceptionally good sea boat. It was also the only 20th-century British battleship with a flat transom stern.

SPEED RUN
Heavy seas break over the bow of HMS *Vanguard* during NATO exercise Operation *Mainbrace*, 19 September 1952. At the time, *Vanguard* was making about 30 knots (55.6km/h; 34.5mph) during a speed run.

HMS *Eagle* (1951)

TYPE • *Aircraft carrier* **COUNTRY** • *United Kingdom*

The Royal Navy's largest carrier yet, *Eagle* played a major role in British and NATO naval operations until the decision to drop fleet carriers was taken. Laid down at Harland & Wolff's yard in Belfast on 24 October 1942, it was not commissioned until 5 October 1951.

SPECIFICATIONS

DIMENSIONS:	Length: 247.4m (811ft 10in); Beam: 34.4m (112ft 9in); Draught: 10.13m (33ft 3in)
DISPLACEMENT:	37,400 tonnes (41,232 tons) standard; 47,000 tonnes (51,809 tons) full load
PROPULSION:	8 boilers; 4 geared steam turbines; 4 shafts; 113,000kW (152,000shp)
SPEED:	31 knots (57km/h; 36mph)
RANGE:	7000nmi (13,000km; 8050 miles) at 18 knots
ARMAMENT:	16 × 114mm (4.5in) guns; 61 × 40mm AA guns
AIRCRAFT:	60
COMPLEMENT:	2500

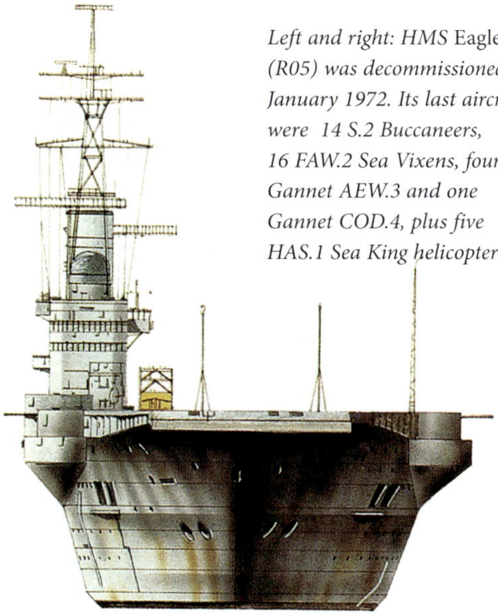

Left and right: HMS Eagle *(R05) was decommissioned in January 1972. Its last aircraft were 14 S.2 Buccaneers, 16 FAW.2 Sea Vixens, four Gannet AEW.3 and one Gannet COD.4, plus five HAS.1 Sea King helicopters.*

SEA VENOM
The de Havilland Sea Venom was an all-weather twin-boom turbo-jet interceptor, introduced in 1951 and operational until 1970.

FLIGHT DECK
Configuration prior to the fitting of the 8.5 degree angled deck and mirror-landing system.

HANGAR
In the established Royal Navy design, the hangar was an all-round armoured box structure, beneath a flight deck 25–100mm (1–4in) thick, and with 25mm (1in) sides and deck.

SUEZ OPERATIONS
Eagle was the only operational RN carrier in the Mediterranean during the Suez conflict of 1956, carrying 17 Sea Venom FAW 21, 24 Sea Hawk FGA-6, nine Westland Wyvern S-4 and four Douglas Skyraider AEW-1 planes, and two Westland Whirlwind HAR-3 helicopters. The carrier aircraft established air supremacy over Egypt, and in what was otherwise a fruitless campaign confirmed the value of offshore carriers.

MODERNIZATION
HMS *Eagle* was completed in 1951, but underwent a major modernization between 1959 and 1964 at Devonport. This photograph was taken after that modernization, and shows it berthed at Portsmouth's South Railway Jetty.

Project 615 *Quebec* class (1952)

TYPE • *Submarine* **COUNTRY** • *Soviet Union*

This was the only air-independent propulsion (AIP) engined submarine to be in series production before the 1990s, but technical problems limited its usefulness. Theoretically capable of remaining 14 days submerged, the class suffered numerous accidents and the design was abandoned after 1957.

SPECIFICATIONS

DIMENSIONS:	Length: 56m (183ft 9in); Beam: 5.1m (16ft 5in); Draught: 3.8m (12ft 6in)
DISPLACEMENT:	460.4 tonnes (508 tons) surfaced; 548.6 tonnes (605 tons) submerged
PROPULSION:	2 diesel engines; 1 AIP diesel engine, 670kW (900hp); 1 creep motor; 3 shafts
SPEED:	18 knots (33km/h; 20.5mph) surfaced; 16 knots (30km/h; 18.6mph) submerged
RANGE:	2750nmi (5090km; 3163 miles) at 9 knots surfaced
ARMAMENT:	4 × 533mm (21in) torpedo tubes; 8 torpedoes
COMPLEMENT:	30

Right: The Soviet Navy's development of nuclear submarines overtook the Quebecs, *although they were modernized in the later 1950s under Project M615.*

PROJECT 615

Project 615, initiated in the late 1940s, was intended to provide the Soviet fleet with the most up-to-date coastal attack submarine. Before World War II, the Soviets had already been working on a closed-cycle engine that did not need to draw air from above the surface, and the experimental craft *M-401* had been launched in 1941 and tested in the secluded waters of the Caspian Sea. Liquid oxygen mixed with purified exhaust gases powered the underwater motor.

***M-296*, 2016**

Codenamed *Quebec* by NATO, 30 of the class were built between 1952 and 1957, out of an original 100 planned. Although some notable performances were achieved, the fuel caused frequent problems because of its tendency to catch fire – two boats were lost to accidents.

USS *Nautilus* (1954)

TYPE • *Submarine* **COUNTRY** • *United States*

SPECIFICATIONS

Dimensions:	Length: 97m (323ft 6in); Beam: 8.4m (27ft 6in); Draught: 6.6m (21ft 9in)
Displacement:	3589 tonnes (3956 tons) surfaced; 4168 tonnes (4595 tons) submerged
Propulsion:	1 S2W pressurized water-cooled nuclear reactor; geared steam turbines; 2 screws; 10,000kW (13,400hp)
Speed:	22 knots (40.7km/h; 25.3mph) surfaced; 25 knots (46.2km/h; 28.7mph) submerged
Range:	Unlimited
Armament:	6 × 533mm (21in) torpedo tubes
Complement:	105

To some experts, *Nautilus* (SSN-571) is the first true submarine – not a surface ship that could submerge for limited periods, but a vessel capable of staying underwater for lengthy periods without needing to draw any sustenance from the surface.

Left: The launch of USS Nautilus *(SSN-571) at the Electric Boat Company, Groton, Connecticut in January 1954 signalled the beginning of the age of nuclear-powered carriers.*

Steel plate
The hull is formed of HY-42 steel plate, of 290 MPa (2704 ton/sq ft) yield strength. The improved HY-75 steel used on later boats was not available until the mid-1950s.

Nuclear reactor
The reactor compartment, with the reactor, primary coolant circuit and steam generators, was not accessible when the reactor was operating.

Sonar array
Nautilus had a BQR-4 passive sonar system with 48 vertical stave hydrophones mounted in the forefoot of the bow. But the boat's own noise precluded effective use of sonar at speeds above 7 knots (12.9km/h; 8mph).

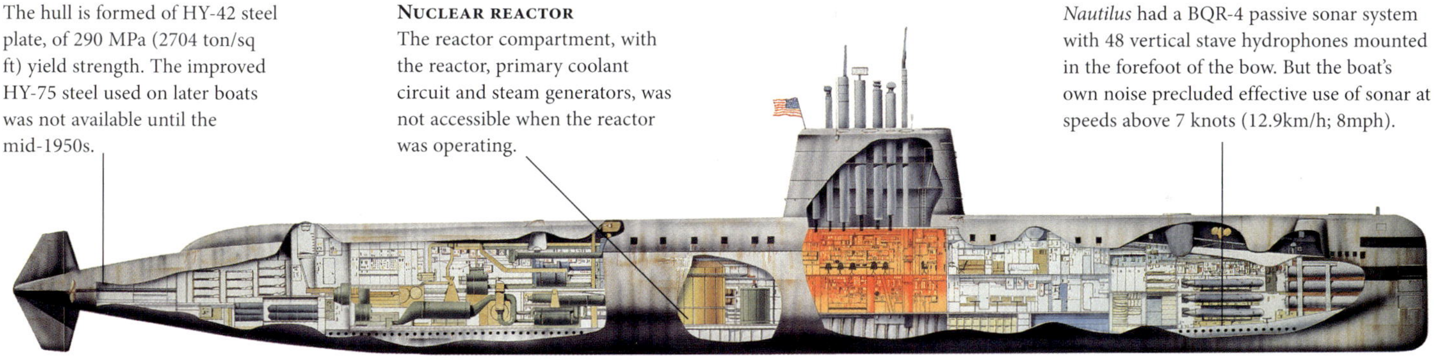

NUCLEAR POWERED

In July 1951, Congress authorized construction of the world's first nuclear-powered submarine. *Nautilus* was fitted with the S2W reactor built by the Westinghouse Corporation and was launched on 21 January 1954. It was commissioned on 30 September of that year as SSN-571. Apart from its nuclear drive, *Nautilus* was modelled on the *Tang* class of diesel-electric drive submarines, although with greater dimensions to allow for the atomic reactor.

SEA TRIALS
Nautilus conducts sea trials in 1954. Four years later, on 3 August 1958, *Nautilus* became the first vessel to pass over the North Pole, using an inertial navigation system developed by North American Aviation. This Arctic mission was no mere 'flag-waving' display, but a major technological challenge that took three attempts to achieve.

USS *Forrestal* (1955)

TYPE • *Aircraft carrier* **COUNTRY** • *United States*

The USS *Forrestal* (CVA-59, later CV-59) introduced the concept and strategic power-projection of the supercarrier and was the lead ship of its class. It was also the last conventionally powered US carrier.

SPECIFICATIONS

DIMENSIONS:	Length: 326.14m (1070ft); Beam: 39.6m (130ft); Draught: 11.3m (37ft)
DISPLACEMENT:	60,610 tonnes (66,820 tons) standard; 82,402 tonnes (90,832 tons) full load
PROPULSION:	8 Babcock & Wilcox boilers; 4 Westinghouse geared steam turbines; 4 shafts; 190,000kW (280,000shp)
SPEED:	33 knots (61km/h; 38mph)
RANGE:	7995nmi (14,806km; 9200 miles)
ARMAMENT:	8 × 127mm (5in)/54-cal Mk 42 guns
AIRCRAFT:	85
COMPLEMENT:	5540

Above and right: USS Forrestal *(CVA-59), taken by W.F. Radcliff in 1955, showing the Navy's first supercarrier shortly after commissioning, featuring its early configuration before major deployments and the 1967 flight deck fire.*

RADAR ANTENNAS
In the original design there was no island, although the hull design was the same. Radar antennas were to be mounted on retractable poles.

RUDDERS
Originally *Forrestal* had three rudders. The centre rudder was found to be unnecessary and was welded as an extension of the keel.

JET AIRCRAFT
Forrestal in 1979. The first carrier specifically designed to embark and service jet aircraft, it accomplished 21 successful operational deployments.

LETHAL ACCIDENT
In June 1967, *Forrestal* was deployed to Vietnam. On 29 July, the ship was on Yankee Station off the Vietnamese coast, launching combat flights. A Zuni rocket, accidentally fired on deck from a F-4 Phantom, struck an armed and loaded A-4 Skyhawk. A resultant series of explosions and fires spread through the lower decks. With the help of sprays from nearby ships, the flight deck fire was put out quickly, but damage crews fought inside the hull for 24 hours to extinguish the flames. The incident cost the lives of 134 men.

DECK DEVASTATION
Forrestal's flight deck is littered with debris from burning aircraft as crewmen and personnel from USS *Rupertus* (DD-851) fight fires after the rocket accident aboard the carrier in the Gulf of Tonkin, 1967.

Project 633 *Romeo* class (1957)

TYPE • *Submarine* **COUNTRY** • *Soviet Union*

SPECIFICATIONS

Dimensions:	Length: 77m (252ft 8in); Beam: 6.7m (22ft); Draught: 4.9m (16ft 1in)
Displacement:	1351 tonnes (1489 tons) surfaced; 1727 tonnes (1903 tons) submerged
Propulsion:	2 diesel engines delivering 2.94MW (4000shp); 2 electric motors; 2 shafts
Speed:	16 knots (29.6km/h; 18.4mph) surfaced; 13 knots (24km/h; 14.9mph) submerged
Range:	16,000nmi (29,632km; 18,414 miles) at 10 knots surfaced
Armament:	8 × 533mm (21in) torpedo tubes
Complement:	60

In the second half of the 1950s, the Soviet Union developed an advance on its *Whiskey* classes of submarine. The new programme was termed Project 633 and produced a diesel-electric attack class given the name *Romeo* by NATO.

Left: A Soviet Romeo*-class submarine underway in April 1979. Now obsolete, the* Romeos *survive perhaps only in the North Korean fleet, although other navies may have mothballed versions.*

PROPULSION
On the surface, the submarine's two diesel engines could deliver a maximum speed of 16 knots (29.6km/h; 18.4mph), or 13 knots (24km/h; 14.9mph) when submerged.

TORPEDOES
The *Romeos* had eight torpedo tubes, six mounted in the bows and two in the stern. It could also carry 28 mines as an alternative to its 14 533mm (21in) torpedoes.

SENSORS
The main sensor array included the Arktika-M sonar system for target tracking and the MG-15 passive sonar system. There was also a retractable surface search radar in the tower.

CHINESE VARIANTS

Apart from two British submarines acquired in the immediate postwar period, China's first submarine classes originated in the Soviet Union. In 1956, China began construction of 21 boats identical to the *Whiskey* class. The Soviet Union then developed Project 633. After 20 Soviet boats had been built, the designs were passed to China, and production began in 1962 at the Jiangnan yard in Shanghai. In China it was known as Type 033 and the first was launched in December 1965.

MISSILE LAUNCH
An upgrade of the *Romeo* class began in the mid-1980s. The main retrofitted improvement in this class was the ability to launch wire-guided torpedoes and anti-ship missiles (ASM) while submerged.

USS *Skate* (1957)

TYPE • *Submarine* **COUNTRY** • *United States*

The progress of nuclear technology and the performance of the first nuclear submarines convinced the US Navy to press on with nuclear propulsion. An order was placed in July 1955 for USS *Skate*, lead ship of a class of four, and it was commissioned on 23 December 1957.

SPECIFICATIONS

DIMENSIONS:	Length: 81.5m (267ft 7in); Beam: 7.6m (25ft); Draught: 6.4m (21ft)
DISPLACEMENT:	2611 tonnes (2879 tons) surfaced; 2907 tonnes (3205 tons) submerged
PROPULSION:	1 pressurized water-cooled nuclear reactor (S3W); geared steam turbines; 2 shafts; approx. 4900kW (6600hp)
SPEED:	20 knots (37km/h; 23mph) surfaced; 25 knots (46.3km/h; 28.8mph) submerged
RANGE:	Unlimited
ARMAMENT:	6 × 533mm (21in) torpedo tubes
COMPLEMENT:	95

Right: The US Navy's eighth nuclear submarine, USS Skate *(SSN-578), emulated* Nautilus*'s sub-Arctic transit, but actually surfaced very close to the North Pole, on 11 August 1958. There the crew scattered the ashes of the renowned arctic explorer Sir Hubert Wilkens.*

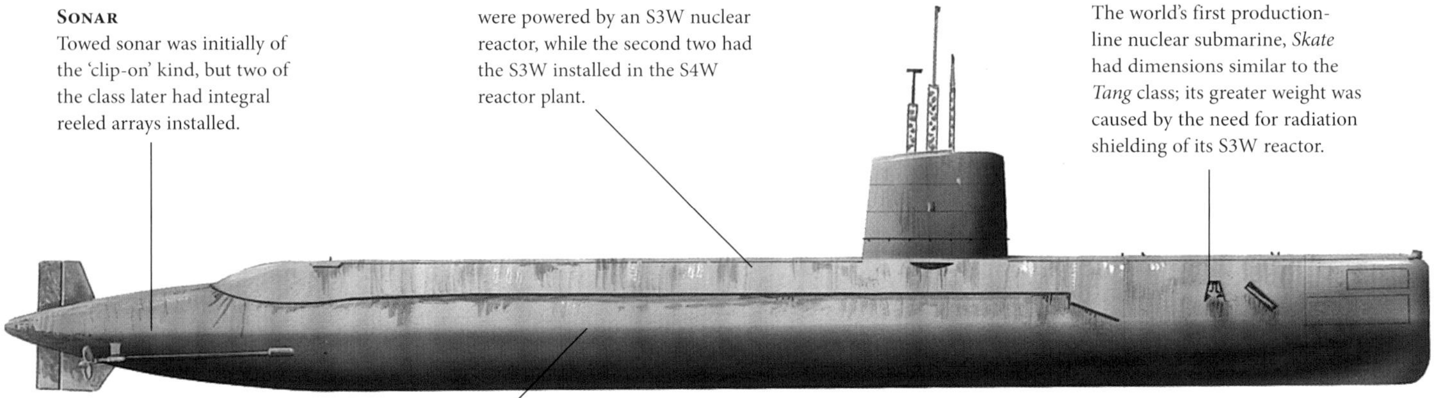

SONAR
Towed sonar was initially of the 'clip-on' kind, but two of the class later had integral reeled arrays installed.

REACTOR
The first two *Skate*-class boats were powered by an S3W nuclear reactor, while the second two had the S3W installed in the S4W reactor plant.

DIMENSIONS
The world's first production-line nuclear submarine, *Skate* had dimensions similar to the *Tang* class; its greater weight was caused by the need for radiation shielding of its S3W reactor.

PURPOSE
Up to 1962, submarines played a key strategic role as radar picket boats, equipped with air search radar to give early warning of missile or aircraft attacks aimed at land bases or at fleet groups.

DEPLOYMENT

The design of the *Skate* class was based on that of the *Tang* class and they were among the smallest nuclear-powered attack submarines. At the time, however, they were considered large boats, and were certainly effective in operation. *Skate* went on sub-Arctic patrols and was the first boat to surface at the North Pole (17 March 1959). Most of its almost 30-year career was spent with Atlantic Fleet, based at New London, Connecticut.

USS *SEADRAGON*
USS *Seadragon* (SSN-584) was the fourth in the series of *Skate*-class submarines. Built at Portsmouth Naval Shipyard in Kittery, Maine, it was commissioned in 1959 and served all the way through to 1984.

Project 641 *Foxtrot* class (1957)

TYPE • *Submarine* **COUNTRY** • *Soviet Union*

The Project 641 boats were the Soviet Navy's largest conventionally powered submarines and the last to be constructed with old-style hulls, although the sail was of streamlined design. Incorporating many novel features, they also served with other navies.

SPECIFICATIONS

DIMENSIONS:	Length: 91.5m (300ft 2in); Beam: 8m (26ft 3in); Draught: 6.1m (20ft)
DISPLACEMENT:	1983 tonnes (2187 tons) surfaced; 2515 tonnes (2773 tons) submerged
PROPULSION:	3 Kolomna 2D42M diesel engine, 2000kW (2700hp); 2 1010kW (1350hp) and 1 2000kW (2700hp) electric motors; 3 screws
SPEED:	16 knots (29.6km/h; 18.4mph) surfaced; 15 knots (27.7km/h; 17.2mph) submerged
RANGE:	20,000nmi (37,040km; 23,015 miles) at 8 knots surfaced
ARMAMENT:	10 × 533mm (21in) torpedo tubes; 22 torpedoes
COMPLEMENT:	80

BOW VIEW
The bow antenna and the starboard-positioned UHF mast gave the class a distinctive forward appearance.

DESIGN
Introduced on the *Foxtrots*, this construction was typical of later Russian classes. Snorkel intake and exhaust were located at the rear. Additional sonar arrays were incorporated in the sides of the sail.

ANTENNA
Underwater telephone antenna MG-18 and passive sonar antenna.

HULL CONSTRUCTION
The outer hull was formed from 9.5mm (0.37in) steel. The pressure hull was formed from 22.2mm (0.87in) AK-25 high tensile steel with a yield strength of 590MPa (5501 tons/sq ft).

STRATEGIC SHIFT
Project 641 was intended to provide a more up-to-date patrol/attack boat than the *Whiskey* class of 1949 and the subsequent 1952 *Zulu* class. The *Foxtrot*, its hull built of newly developed AK-25 high-tensile steel, had a greater underwater speed and went deeper than the *Zulu*. These classes marked a shift in Soviet strategic thinking. Now, with improved communications and other technology, the focus was on large attack submarines with long range and endurance.

BLUE-WATER OPERATIONS
Foxtrot operated at a depth of 250m (820ft), could dive to more than 300m (984ft) and was capable of reaching 16.8 knots (31.1km/h; 19.3mph). Its hydroplanes enabled the *Foxtrot* to dive and surface quite horizontally. Angled at more than 30 degrees, however, it lost control.

Duperré class (1957)

TYPE • *Destroyer* **COUNTRY** • *France*

SPECIFICATIONS

DIMENSIONS:	Length: 128.6m (421ft 11in); Beam: 12.7m (41ft 8in); Draught: 5.4m (17ft 9in)
DISPLACEMENT:	2794 tonnes (3081 tons) standard; 3800 tonnes (4189 tons) full load
PROPULSION:	4 boilers; 2 geared steam turbines; 2 shafts; 46,979kW (63,000shp)
SPEED:	34 knots (63km/h; 39mph)
RANGE:	5000nmi (9300km; 5800 miles) at 18 knots
ARMAMENT:	6 × 127mm (5in) guns (3 × twin); 6 × 57mm (2.2in) guns (3 × twin); 4 × 20mm (0.79in) guns; 6 × Bofors 375mm ASW rocket launcher (1 × 6); 12 × 550mm (21.7in) torpedo tubes (4 × 3)
COMPLEMENT:	347

Also known as the T53 class, the *Duperré* destroyers formed France's second generation of postwar destroyers, improving on the earlier T47 class. They fulfilled the classic destroyer role, providing air defence and anti-submarine duties.

Right: Duperré *(D633) was the lead ship of six vessels in the class, commissioned between October 1957 and July 1962.*

SENSORS
The class carried an advanced suite of sensors, including DRBV31 surface sentry and navigation radar, DRBC30 fire control radars, aDRBI10B altimetry radar, DRBV22A combined sentry radar and DUBV 1B sonar.

MAIN GUNS
The standard main gun of French destroyers and cruisers during the 1950s was the 127mm (5in)/54 Model 1948, a dual-purpose gun able to engage air and surface targets.

PROPULSION
The class had four boilers, delivering 46,979kW (63,000hp) to geared turbines that pushed the maximum speed to 34 knots (63km/h; 39mph).

MODIFICATIONS

In 1972–73, two ships of the class – *Duperré* (D633)and *Forbin* (D635) underwent significant modifications, reflecting evolving Cold War technologies and priorities. Both ships received helipads on the afterdeck, which meant losing rear guns. *Duperré* was substantially rearmed with a Model 1968 100mm (4in) main gun, four Exocet MM38 missiles, two quadruple L5 torpedo launchers, and two 20mm (0.79in) AA cannon.

***DUPERRÉ* (D633)**
This rear view of *Duperré* shows the vessel after the alterations of the early 1970s, as shown by the presence of the helipad. This feature enabled the deployment of anti-submarine helicopters.

Tribal class (1958)

TYPE • *Frigate* **COUNTRY** • *United Kingdom*

SPECIFICATIONS

Dimensions:	Length: 109.73m (360ft); Beam: 12.88m (42ft 3in); Draught: 4.04m (13ft 3in);
Displacement:	2337 tonnes (2576 tons) standard; 2743 tonnes (3024 tons) full load
Propulsion:	COSAG system: 1 steam turbine, 9300kW (12,500shp); 1 Metrovick G-6 gas turbine, 5600kW (7500shp); 1 shaft
Speed:	27 knots (50km/h; 31mph)
Range:	4500nmi (8300km; 5200 miles) at 12 knots
Armament:	2 × single QF 113mm (4.5in) guns; 2 × single 40mm (1.57in) Bofors guns; 2 × four-rail GWS-20 Seacat missile systems; 2 × single 20mm (0.79in) guns; 1 × Mark 10 Limbo ASW mortar
Complement:	253

The Type 81, or *Tribal* class, was a seven-ship class of frigates developed for the Royal Navy in the 1950s and 1960s. They were initially ordered as sloops, but were reclassified as general-purpose frigates during the 1960s.

Left: A Royal Navy Westland Wasp helicopter lands on the flight deck of HMS Nubian *(F131). The* Tribal *class were the first British frigates designed from the outset to operate helicopters.*

Bofors guns
Until the 1970s, 40mm (1.57in) Bofors guns stood in for the intended GWS-21 Seacat anti-aircraft missile system, the exception being *Zulu* (F124).

Helicopter flight deck
The *Tribal* class had a dedicated helicopter hangar built into the superstructure, just forward of the flight deck. The ship deployed a single Westland Wasp, armed with anti-submarine homing torpedoes.

Propulsion
The COSAG system was expensive and could be complex. The ship had only a single shaft, and top speed was 27 knots (50km/h; 31mph)

COSAG SYSTEM

The *Tribal* class utilized the Combined Steam and Gas (COSAG) propulsion system, mixing gas turbine and steam turbine powerplants to drive the ship's shafts, either together or independently. The advantages of COSAG were that the gas turbine had a quick start-up time and delivered powerful acceleration, while the steam turbine offered better cruising efficiency and reliability.

HMS *ASHANTI*
HMS *Ashanti* (F117) was the first of the *Tribal*-class frigates to enter service, being commissioned on 23 September 1961. After a long and lively service career, it was sunk by anti-ship missiles as a target ship in 1988.

USS *Long Beach* (1959)

TYPE • *Missile cruiser* **COUNTRY** • *United States*

SPECIFICATIONS

Dimensions:	Length: 219.8m (721ft 3in); Beam: 22.3m (73ft 4in); Draught: 7.2m (23ft 9in)
Displacement:	14,061 tonnes (15,500 tons) standard
Propulsion:	2 C1W nuclear reactors; 2 General Electric steam turbines; 2 shafts; 60,000kW (80,000shp)
Speed:	30 knots (55.6km/h; 34.5mph)
Range:	Unlimited
Armament:	8 × Harpoon SSM; 2 × Terrier SAM systems (later Standard ER SAMs); 2 × 127mm (5in) guns; 2 × 20mm (0.79in) Phalanx CIWS; 1 × ASROC launcher; 6 × 324mm (12.75in) torpedo tubes
Complement:	1107

Completed in September 1961, *Long Beach* (CGN-9) was the largest non-carrier surface warship built by the United States since 1945. Off Vietnam in 1968, it shot down two MiG fighters in the first successful ship-to-air missile engagement in history.

Right: USS Long Beach *(CGN-9) underway off Oahu, Hawaii, 9 May 1973. Its deactivation ceremony took place on 2 July 1994 at Norfolk Naval Station.* Long Beach *was decommissioned on 1 May 1995, more than 33 years after it had entered service.*

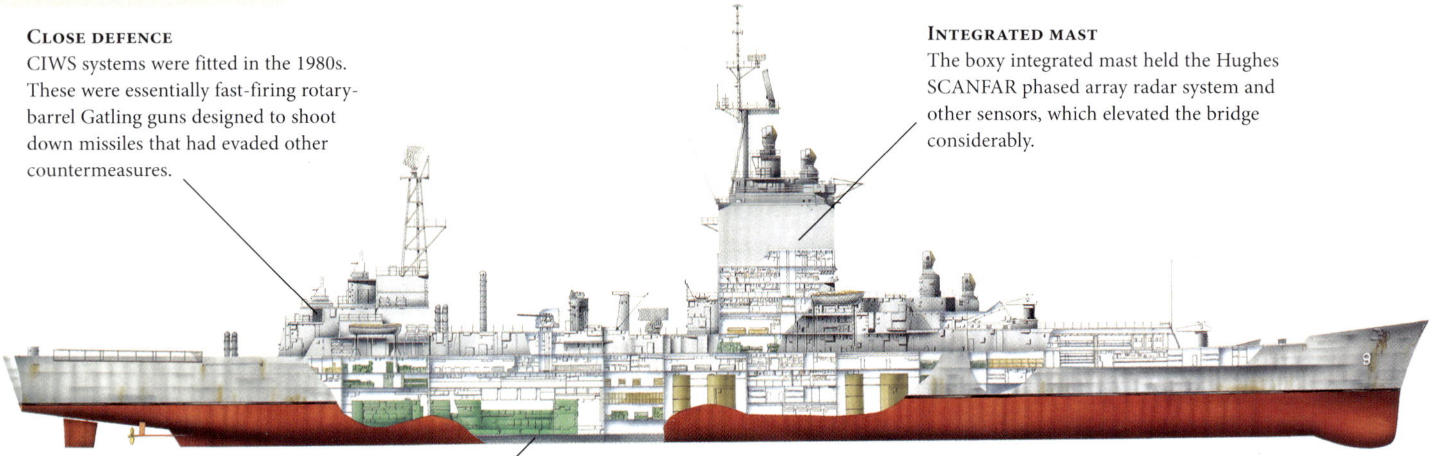

Close defence
CIWS systems were fitted in the 1980s. These were essentially fast-firing rotary-barrel Gatling guns designed to shoot down missiles that had evaded other countermeasures.

Integrated mast
The boxy integrated mast held the Hughes SCANFAR phased array radar system and other sensors, which elevated the bridge considerably.

Hull
Long Beach was the last US cruiser to be built with a traditional cruiser-type hull that was long and relatively narrow.

TALOS
A Talos missile fired by *Long Beach* in 1968 downed a North Vietnamese jet 112.6km (70 miles) away: it was the first ship-to-air missile kill in naval history. The Bendix RIM-8 Talos was one of the first generations of US Navy SAM systems. The RIM-8A had a range of 92km (57 miles), while the RIM-8J extended to 241km (150 miles). *Long Beach* carried Mark 12 Guided Missile Launching Systems, with a 52-round magazine below the main deck.

FITTING OUT
USS *Long Beach* is fitted out at the Bethlehem Steel Co. Shipyard, Quincy, Massachusetts, 5 July 1960. Note the barracks craft (APL) on the opposite side of the slip.

HMS *Hermes*/INS *Viraat* (1959)

TYPE • *Aircraft carrier* **COUNTRY** • *United Kingdom/India*

In service with the Royal Navy from 1959 to 1984, HMS *Hermes* achieved fame at home as flagship of the British naval force in the Falklands War of 1982. It went on to have a second career with the Indian Navy until 2016.

SPECIFICATIONS (original)

DIMENSIONS:	Length: 225.2m (738ft 10in); Beam: 27.4m (90ft); Draught: 8.5m (27ft 11in)
DISPLACEMENT:	23,000 tonnes (25,353 tons) standard; 28,000 tonnes (30,865 tons) full load
PROPULSION:	4 Admiralty 3-drum boilers; 2 Parsons single-reduction geared steam turbines; 2 shafts; 57,000kW (76,000shp)
SPEED:	28 knots (52km/h; 32mph)
RANGE:	7000nmi (13,000km; 8050 miles) at 18 knots
ARMAMENT:	10 × 40mm Bofors AA guns
AIRCRAFT:	30
COMPLEMENT:	2100

Left: From 1959 to 1961 Hermes *flew de Havilland Sea Vixen fighters of 890 Squadron and Supermarine Scimitars of 803 Squadron, seven Blackburn Buccaneer bombers (as seen here), five turboprop Fairey Gannet ASW planes and six Westland Sea King helicopters.*

SEA HARRIER
These served with the Fleet Air Arm from 1980 to 2006. Sea Harriers shot down 20 Argentinian aircraft in the Falklands conflict.

SKI JUMP
A refit between 20 March 1980 and 12 May 1981 gave the ship a 12 degree ski jump on the port foredeck to enable short take-off Sea Harrier jets to be carried.

CONTAMINATION MEASURES
Hermes was the first British carrier to be designed so that internal accommodation, including the boiler and turbine rooms, could be sealed off and controlled remotely in the event of radioactive contamination.

BACKGROUND

In 1943, among its other plans, the Royal Navy was proposing to build eight carriers to be classed as light fleet carriers. However, none were completed before the end of hostilities, when four were immediately cancelled. Four others remained on the stocks for several years. First to be launched was HMS *Centaur* in 1954, which gave its name to the class. *Hermes* was the final one, originally destined to be HMS *Elephant* but renamed in honour of its predecessor *Hermes*, sunk in the Pacific in 1942.

SPEED TRIAL
HMS *Hermes* makes a hard starboard turn while undergoing a speed trial on 18 November 1959. It was commissioned on 25 November 1959, leaving immediately for a shakedown cruise in the Mediterranean.

Project 659 *Echo* class (1960)

TYPE • *Submarine* **COUNTRY** • *Soviet Union*

SPECIFICATIONS (*Echo I*)

Dimensions:	Length: 115m (377ft 2in); Beam: 9m (29ft 6in); Draught: 7.5m (24ft 6in)
Displacement:	4486 tonnes (4943 tons) surfaced; 5852 tonnes (6453 tons) submerged
Propulsion:	2 VM-A pressurized water nuclear reactors; each 70,000hp (52MW), powering two steam turbines delivering a total of 30,000hp (8MW); auxiliary diesel engine; 2 screws
Speed:	20 knots (37km/h; 23mph) surfaced; 23 knots (42.5km/h; 26.4mph) submerged
Range:	Unlimited
Armament:	4 × 533mm (21in) torpedo tubes and 4 × 406mm (16in) torpedo tubes (2 bow, 2 stern); 8 × P-6 (SS-N-3) cruise missiles
Complement:	90

The *Echo* class originated as Soviet Russia's first five guided-missile (SSGN) submarines, designated Project 659 by the Russian Navy and *Echo I* by NATO. The two *Echo* sub-classes evolved in terms of armament and role over 35 years of fleet service.

Left: An aerial port beam view of a Soviet Echo II *class nuclear-powered cruise missile submarine underway in 1982.*

Cruise missile
In total, 14 boats were modified from 1970 to carry the P-500 Bazalt (SS-N-12 Sandbox) anti-ship cruise missile and three were further upgraded with the P-1000 Vulkan (GRAU 3M70).

Armament
The *Echo* submarines were formidably armed. The *Echo II* had eight P-6 cruise missiles, four 533mm (21in) bow torpedo tubes and two 400mm (16in) stern torpedo tubes.

Echo II
The 29 *Echo II* boats, Project 675, were built at Severodvinsk and Komsomolsk between 1962 and 1967 for anti-ship warfare, with aircraft carriers particularly in mind. At 4m (13ft) longer than *Echo I*, they were fitted with four launchers for eight P-6 (SS-N-3a Shaddock A) anti-ship cruise missiles, which as with *Echo I* could only be fired from a surfaced position. In order to effect guidance of a fired missile, the boat had to remain surfaced until mid-course correction and final target selection had been made.

Echo I
The profile shows an *Echo I* submarine with two launchers in the open position. The P-6 was the first practical anti-ship missile to be deployed on a submarine. Between 1977 and 1983, 10 *Echo IIs* were fitted with the Kasatka-B satellite-guided target system and the P-500 Bazalt missile.

***ECHO II* CLASS (NO 714)**
The *Echo* class were one of the first generation of Soviet nuclear submarines, and at least four had serious internal accidents. All of *Echo I* were decommissioned by 1989, and the *Echo II* boats were decommissioned between 1989 and 1995.

Clemenceau (1960)

TYPE • *Aircraft carrier* **COUNTRY** • *France*

SPECIFICATIONS	
Dimensions:	Length: 265m (869ft); Beam: 51.2m (168ft); Draught: 8.6m (28ft)
Displacement:	22,710 tonnes (25,047 tons) standard; 32,000 tonnes (35,274 tons) full load
Propulsion:	6 Indret boilers; 4 steam turbines; 2 shafts; 94,000kW (126,000shp)
Speed:	32 knots (59km/h; 37mph)
Range:	7500nmi (13,875km; 8625 miles) at 18 knots
Armament:	8 × 100mm (4in)/55-cal Mod 53 guns
Aircraft:	38
Complement:	1338

A strong symbol of French power, *Clemenceau* undertook many operational deployments in European and Pacific waters, mostly in support of French actions and policies, but also as part of international peacekeeping operations.

Left: French carrier Clemenceau *(R98) underway in 1981.* Clemenceau *and its companion ship* Foch *were planned when France still had overseas imperial colonies.*

Steam catapults
The two Mitchell-Brown BS-5 52m (170ft) steam catapults could put a 12–15 tonne (13–16.5 ton) aircraft into the air at 150 knots (277.8km/h; 172.6mph).

Flight deck
The landing area on *Clemenceau* was 165m (543ft) by 29.5m (97ft), and was set at 8.5 degrees to the ship's axis.

Armament
On commission, the armament was eight 100mm (4in) turret-mounted guns, and five 127mm (5in) machine guns. Four of the turrets were replaced around 1997 by two SACP Crotale EDIR air defence missile systems.

Powerplant
Six Indret boilers supplying steam to four turbines, with two shafts and two four-bladed propellers, drove the ship. Electric power was supplied via three turbo-alternators and six MGO V12 SACM diesel alternators.

REFIT AND OPS

Between 13 July 1977 and 22 November 1978, a major overhaul saw the renewal of *Clemenceau*'s crew accommodation, installation of a new control and command centre, adaptation to carry the Super-Etendard aircraft and provision of capacity to carry nuclear weapons. The Lebanese civil war crisis of 1982–84 brought it to the eastern Mediterranean; it also operated in the Gulf of Oman during the Iran–Iraq war of 1987–88, in Operation *Prometheus*, and was in the Red Sea and Arabian Sea at the end of 1990 in defence of Kuwait against the Iraqi invasion.

TOULON HARBOUR
Clemenceau arrives in Toulon Harbour in 1993. A last refit period was from 2 October 1994 to 28 June 1995. The carrier's final cruise was made from 16 July 1997, and it was disarmed in October of the same year. Against much opposition, *Clemenceau* was sold for scrapping to a Spanish yard on 14 April 2003.

County class (1960)

TYPE • *Destroyer* **COUNTRY** • *United Kingdom*

SPECIFICATIONS

Dimensions:	Length: 160m (520ft); Beam: 16m (54ft); Draught: 6.4m (21ft)
Displacement:	6200 tonnes (6834 tons) standard
Propulsion:	COSAG; 2 Babcock & Wilcox boilers; geared steam turbines; 2 shafts; 22,370kW (30,000shp); 4 Metrovick G6 gas turbines, 22,370kW (30,000shp)
Speed:	30 knots (56km/h; 35mph)
Range:	3500nmi (6500km; 4000 miles)
Armament:	2 × twin 113mm (4.5in) Mark N6 gun turrets; 2 × 20mm (0.79in) Oerlikon cannon; 1 × Seaslug GWS.1 or GWS.2 SAM system (24 missiles); 2 × Seacat GWS-22 SAM launchers; 2 × triple shipborne torpedo tubes
Complement:	471

The *County*-class destroyers had the capabilities of guided-missile cruisers; the lesser 'destroyer' classification basically helped the Royal Navy get approval for their construction. Their capabilities were centred on modern SSM and SAM systems.

Left: HMS Glamorgan *was built by Vickers-Armstrongs in Newcastle upon Tyne and commissioned on 14 October 1966.*

HMS *Kent*
HMS *Kent* was one of four 'Batch 1' *County*-class destroyers; the remaining 'Batch 2' vessels had one of their 114mm (4.5in) gun mounts replaced with a quad Exocet missile launcher.

Main guns
The Batch 1 *County* class had a pair of QF Mk V 114mm (4.5in) guns. Each gun had a magazine of 225 rounds; rate of fire was 24 rounds per minute when power-loaded.

Seaslug
The Seaslug Mk 1 SAM system was a beam-riding weapon, guided by information from the Type 901 fire-control radar. 'Batch 2' vessels received the Mk 2 upgrade.

FALKLANDS WAR
Two of the *County*-class destroyers served with distinction during the Falklands War of 1982. HMS *Antrim* assisted in the recapture of South Georgia, and later provided fire support for the main landings on East Falkland, surviving an Argentine bomb strike. HMS *Glamorgan* also gave fire support – firing more than 1000 shells – but it was hit by an Exocet missile on 12 June and heavily damaged, with 14 crew killed.

HMS *NORFOLK*
In total, eight *County*-class destroyers were constructed between 1959 and 1967, forming a critical element of the Royal Navy's postwar surface fleet.

USS *Enterprise* (1961)

TYPE • *Aircraft carrier* **COUNTRY** • *United States*

The world's first nuclear-powered aircraft carrier, the USS *Enterprise* (CVAN-65) became the US Navy's third-longest serving surface ship. In commission from 1961 to 2012, it made a total of 24 operational deployments around the world.

SPECIFICATIONS

Dimensions:	Length: 342m (1123ft); Beam: 40.5m (132ft 9in); Draught: 12m (39ft)
Displacement:	84,626 tonnes (93,283 tons) standard
Propulsion:	8 Westinghouse A2W nuclear reactors; 4 steam turbines; 4 shafts; 208,600kW (280,000shp)
Speed:	33.6 knots (62.2km/h; 38.7mph)
Range:	Unlimited
Armament:	2 × Sea Sparrow launchers; 2 × RIM-116 RAM launchers; 2 × 20mm Phalanx CIWS
Aircraft:	90
Complement:	5828

Left: Enterprise *underway during the early years of its career. On trial,* Enterprise *reached a speed in excess of 40 knots (74km/h; 46mph).*

Propulsion
Eight Westinghouse A2W nuclear reactors were installed, with two reactors providing power to each of four Westinghouse geared steam turbines to turn the four shafts. Uniquely, the ship had four rudders.

Bridle-catcher ramps
Enterprise was the last USN operational carrier to have bridle-catcher ramps at the catapult ends, which extended over the bow.

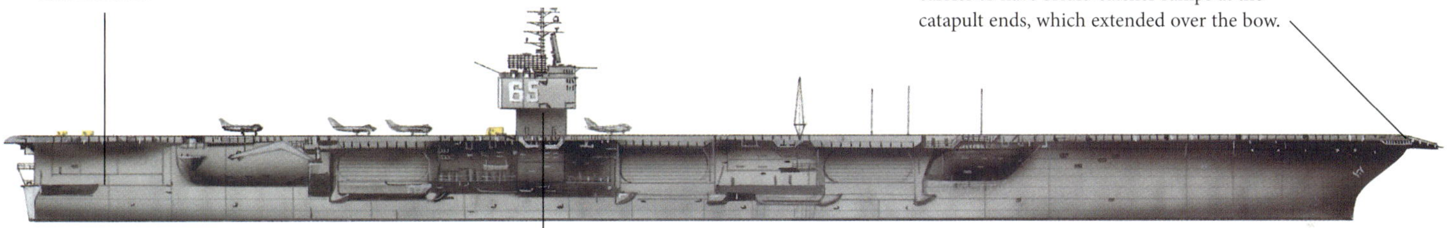

Island
Enterprise's vast surface area allowed for a square-shaped island rather than the rectangular structure of previous carriers, its upper section built out above a narrower base

ADVANCED TECH

Enterprise was intended to mark a milestone in electronic technology as well as in its propulsion system. The SCANFAR radar system employed two radars, AN/SPS-32 and AN/SPS-33, giving combined search and targeting. Although a pioneer example of a phased array, the most advanced for the time, it had reliability and functional problems and used a huge amount of electric power. It was not fitted to any subsequent carriers and was replaced in 1980.

PEARL HARBOR, 1989
Its deck crowded with aircraft, USS *Enterprise* (CVN-65) leaves Pearl Harbor in February 1989. The ship underwent a further major overhaul at Newport News from April 1990 to 27 September 1994.

Project 58 *Kynda* class (1962)

TYPE • *Missile cruiser* **COUNTRY** • *Soviet Union*

SPECIFICATIONS

Dimensions:	Length: 141.9m (465ft 7in); Beam: 15.8m (51ft 10in); Draught: 5.3m (17ft 5in)
Displacement:	4400 tonnes (4850 tons) standard; 5500 tonnes (6063 tons) full load
Propulsion:	4 boilers; 2 steam turbines; 2 shafts; 75,000kW (100,000shp)
Speed:	34 knots (63km/h; 39mph)
Range:	7000nmi (13,000km; 8100 miles) at 14.5 knots
Armament:	2 × quad SS-N-3b ASM launchers; 1 × twin SA-N-1 Goa SAM launcher; 2 × 76mm (3in) AK-726 guns; 4 × 30mm (1.2in) AK-630 guns; 2 × RBU-6000 ASW rocket launchers; 2 × triple 533mm(21in) torpedo tubes
Complement:	390

Also known as the *Grozny* class, the Project 58 *Kynda* class represented a new generation of Soviet anti-surface 'rocket cruisers', although they were originally designated as destroyers. Four ships in total were built – *Grozny*, *Admiral Fokin*, *Admiral Golovko* and *Varyag*.

Right: The Shaddock missile launchers dominate the foredeck of Grozny *as shown here. Note the bristling array of radar and sensors.*

Propulsion
The geared steam turbines could take the ship up to a maximum speed of 34 knots (63km/h (39mph), although cruising speeds were far slower.

Superstructure
The class were the first in the Soviet Navy to introduce pyramid-type masts to support the ship's extensiveradar, communications and surveillance equipment.

Missile launchers
The *Kynda* class had two quadruple SS-N-3B Shaddock missile launchers, one at each end of the ship, and a total of 16 missiles.

ANTI-SHIP MISSILE
The main weapon of the *Kydna* class was the P-5 Pityorka (NATO: SS-N-3B 'Shaddock') anti-ship missile. Designed by the Chelomey design bureau, the missile had true over-the-horizon capability, with an operational range of up to 450km (280 miles), depending on the generation of missile. The missile was guided to its target either by on-ship radar or via aerial data link.

***KYNDA* CLASS**
Photographed in the 1990s (the decade in which all the class were decommissioned), this *Kynda*-class vessel shows the aft helicopter platform, although the ship had no embarked aircraft.

HMS *Dreadnought* (1962)

TYPE • *Submarine* **COUNTRY** • *United Kingdom*

SPECIFICATIONS

DIMENSIONS:	Length: 81m (265ft 9in); Beam: 9.8m (32ft 3in); Draught: 8m (26ft 3in)
DISPLACEMENT:	3556 tonnes (3920 tons) surfaced; 4064 tonnes (4481 tons) submerged
PROPULSION:	1 Westinghouse S5W reactor, 2 geared steam turbines, 1 shaft, 11,000kW (15,000shp)
SPEED:	20 knots (37km/h; 23mph) surfaced; 30 knots (55.6km/h; 34.5mph) submerged
RANGE:	Unlimited
ARMAMENT:	6 x 533mm (21in) torpedo tubes
COMPLEMENT:	88

Britain had experimented with hydrogen-peroxide drive, but by 1956 it was clear that any first-class naval power needed to have nuclear submarines on a par with those being built by the United States and the Soviets.

Left: HMS Dreadnought *(S101) in Singapore in 1973. The ship was decommissioned in 1980, after long service.*

ARMAMENT
Dreadnought had six bow tubes for 533mm (21in) torpedoes. Twenty-four rounds were carried, including the wire-guided Mark 23 Grog.

HULL DESIGN
The hull design of *Dreadnought* resembled that of the *Skipjack* class, although navigational and combat systems were largely based on British technology.

REACTOR
HMS *Dreadnought* was the first ship, launched on 21 October 1960 and powered by a US S5W pressurized water reactor, designed for the US *Skipjack* class (the aft machinery area was known as 'the American sector').

SERVICE

Commissioned on 17 April 1963, *Dreadnought* was inevitably used as a training and testing boat, although it was a fully effective SSN (fast attack), whose high underwater speed of 30 knots (5km/h; 34.5mph) made it suitable for use as an escort to fast carrier groups. In 1980 it was withdrawn following damage to the machinery, which was not considered repairable, and the hulk remains at Rosyth naval dockyard. *Dreadnought*, built rapidly with US help, remained a one-off.

NIGHT PATROL
HMS *Dreadnought* (S101) slips out of harbour on a moonlit night. Obtaining a new US reactor design was a considerable achievement for the British, although future Royal Navy nuclear boats would have British reactors. A major refit and refuelling was carried out in 1970.

USS *Permit/Thresher* Class (1962)

TYPE • *Submarine* **COUNTRY** • *United States*

SPECIFICATIONS

Dimensions:	Length 84.9m (278ft 6in); beam 9.6m (31ft 6in); draught 8.8m (28ft 10in)
Displacement:	3810 tonnes (4199 tons) surfaced; 4380 tonnes (4828 tons) submerged
Propulsion:	1 S5W pressurised water reactor; 2 steam turbines delivering 11,000kW (15,000shp); single shaft
Speed:	18 knots (33.3km/h; 20.7mph) surfaced; 27 knots (50km/h; 31mph) submerged
Range:	Unlimited
Armament:	4 × 533mm (21in) torpedo tubes; 12–18 × Mark 37 torpedoes (later Mark 48); 4–6 × UUM-44 SUBROC anti-submarine missiles; 4 × UGM-84 Harpoon anti-ship missiles
Complement:	134

The first of the SSNs in the US Navy with a deep-diving capability, advanced sonars mounted in the optimum bow position, amidships angled torpedo tubes with the SUBROC ASW missile, and a high degree of machinery-quieting, the *Thresher* class formed an important part of the US subsurface attack capability for 20 years.

Left: USS Thresher *(SSN-593, the photograph taken while the submarine was underway on 30 April 1961. After its loss, the boats were renamed the Permit class. The loss of Thresher prompted the introduction of the Deep Sea Rescue Vessel (DSRV).*

Noise control
Beginning with the *Thresher* class, successive SSN designs demonstrated a willingness to prioritize quietness over other operational capabilities such as speed and diving performance.

SUBROC
The *Permit* class of submarine introduced the SUBROC anti-submarine missile, which was a nuclear depth charge propelled by a solid-fuel rocket motor.

USS *Permit*
The class was the first to carry the spherical sonar array that could track fast submarines in all three dimensions. These boats could detect the standard Mk 37 ASW torpedo, and outrun it.

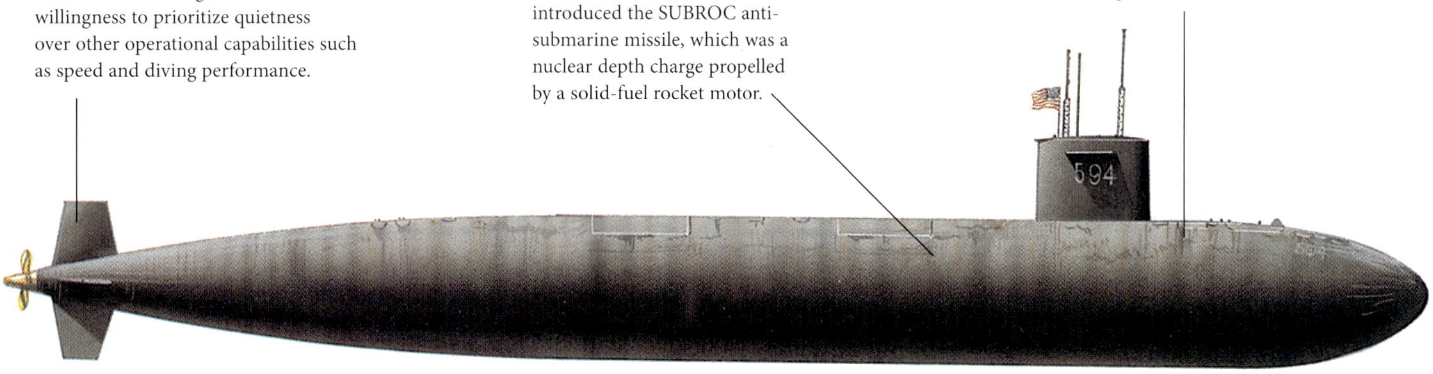

LOST SUBMARINE
The lead boat of the class, USS *Thresher*, was lost in a deep-diving accident with all 129 crew on board, off New England on 10 April 1963, midway through the building period 1960–66. The loss of *Thresher* was a major shock, as this was considered the best nuclear submarine yet. The class was renamed after *Permit*, the second ship. As a result of the enquiry following the loss of Thresher, the last three of the class were modified during construction with SUBSAFE features.

USS *THRESHER*
USS *Thresher* enters the water for the first time, during launching ceremonies at the Portsmouth Naval Shipyard, Kittery, Maine, 9 July 1960.

Suffren (1965)

TYPE • *Missile destroyer* **COUNTRY** • *France*

SPECIFICATIONS

Dimensions:	Length: 157.6m (517ft 1in); Beam: 15.54m (51ft); Draught: 7.4m (24ft 3in)
Displacement:	5335 tonnes (5880 tons) standard; 6780 tonnes (7475 tons) full load
Propulsion:	4 boilers; 4 geared steam turbines; 2 shafts; 54,063kW (72,500shp)
Speed:	34 knots (63km/h; 39mph)
Range:	5100nmi (9400km; 5900 miles) at 18 knots
Armament:	48 × Masurca surface-to-air missiles; 4 × MM38 Exocet anti-ship missiles; 1 × Malafon anti-submarine missile system; 10 × L5 anti-submarine torpedoes; 2 × single 100mm (3.9in) turrets; 4 × single 20mm cannon; 4 × 12.7mm machine guns
Complement:	355

***Suffren* and *Duquesne* were the first French frigates designed to carry surface-to-air missiles, serving as escort ships for the new generation of French aircraft carriers. *Duquesne* also received four Exocet missile launchers.**

Left: This view of Suffren *(D602) gives a good impression of the sheer size of the radome that dominated the central part of the superstructure.*

Masurca
The main air defence system of the ship was the Masurca, a semi-active radar homing weapon with an operational range of 55km (34 miles) and a flight ceiling of 30,000m (98,000ft).

Radome
The bulky radome housed the DRBI-23 3D radar system, which provided target data for the SAM systems.

Malafron
The Malafron ASW system entered service in 1966. It was a glider-delivery weapon, deploying an L4 homing torpedo.

CARRIER DEFENCE

The *Suffren* class were designed primarily to provide anti-submarine air-defence protection for the French *Clemenceau* class of aircraft carriers. Initially, the construction of five ships in the class was planned, but ultimately only two were built and commissioned – *Suffren* (D602) and *Duquesne* (D603), both of the ships named after French admirals. Refits extended the ship's service life through 1990s, and both were replaced in the early 2000s.

***SUFFREN* (D602)**
Suffren cuts through the seas at pace in the mid-1960s, likely during sea trials. It served with both the French Atlantic and Mediterranean fleets.

HMS *Fearless* (1965)

TYPE • *Helicopter carrier* **COUNTRY** • *United Kingdom*

Designed as a highly specialized assault ship capable of operating landing craft and helicopters in any part of the world, *Fearless* played a central part in the Falklands War and later served as a training ship.

SPECIFICATIONS

Dimensions:	Length: 160m (520ft); Beam: 24m (80ft); Draught: 6.4m (21ft)
Displacement:	12,802 tonnes (14,112 tons) standard; 15,138 tonnes (16,687 tons) full load
Propulsion:	2 Babcock & Wilcox Y24A boilers; 2 English Electric steam turbines; 2 shafts; 8203kW (11,000shp)
Speed:	21 knots (39km/h; 24mph)
Range:	10,000nmi (18,500km; 11,500 miles)
Armament:	2 × BMARC GAM B01 20mm single mounts; 2 × 20mm Vulcan Phalanx CIWS
Aircraft:	5
Complement:	580

Right: Perhaps the most remarkable aspect of Fearless*'s full-frontal view was the apparent absence of weaponry, although self-defence armament was mounted on the superstructure.*

Capacity
By the Falklands War, the capacity of the Landing Craft Utility (LCU) ships had been derated to 61 tonnes (67 tons), although this stipulation was ignored in the combat zone.

Crane
The mobile gantry crane from Houlder and SCX Special Projects could lift 4.5 tonnes (5 tons).

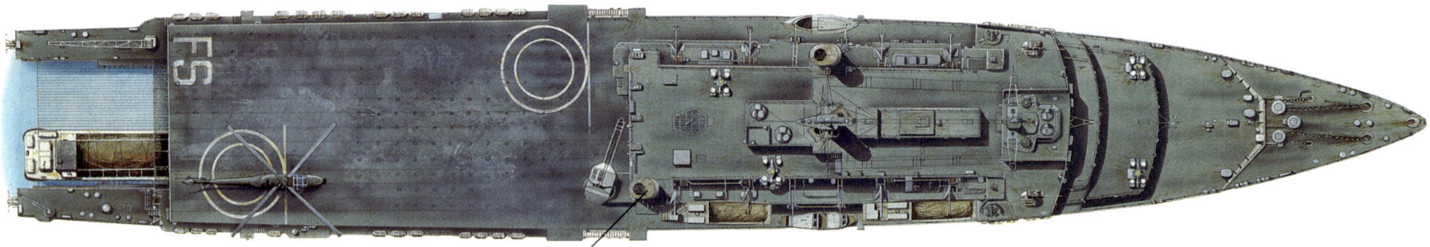

FALKLANDS WAR
In the Falklands War, *Fearless* had an enhanced satellite communications system to enable its use as command ship for the amphibious task force. On this task it spent 100 consecutive days at sea. On 27 May its 20mm (0.79in) Bofors guns downed an Argentinian US-built Douglas A-4B attack plane. In the campaign its flight deck was used at times by Harrier VTOL jets. One of its LCU boats, *Foxtrot 4*, was sunk by a bomb, killing six Marines, on 8 June.

Funnel ducts
Ducts to the funnels were located to port and starboard to maximize interior space – a system developed on car ferries.

HELICOPTER LOAD
A Wessex helicopter hovers over the deck of HMS *Fearless* during the ship's working-up trials in the Solent, March 1966. *Fearless* had no permanently assigned aircraft but could operate up to four Sea King HC4 assault or multi-role helicopters from two take-off positions on the flight deck.

Sturgeon class (1966)

TYPE • *Submarine* **COUNTRY** • *United States*

This class of nuclear-powered SSNs formed the core of the US Navy's attack and espionage submarine fleet in the later stages of the Cold War, carrying a range of tactical missiles, special mission gear and detection equipment.

SPECIFICATIONS

DIMENSIONS:	Length: 89m (292ft 3in); Beam: 9.7m (31ft 8in); Draught: 8.8m (28ft 10in)
DISPLACEMENT:	3698 tonnes (4077 tons) surfaced; 4714 tonnes (5196 tons) submerged
PROPULSION:	1 S5W pressurized water nuclear reactor; steam turbines; single screw
SPEED:	15 knots (27.7km/h; 17.2mph) surfaced; 25 knots (46.2km/h; 28.7mph) submerged
RANGE:	Unlimited
ARMAMENT:	4 × 533mm (21in) torpedo tubes; Harpoon and Tomahawk cruise missiles; torpedoes or mines
COMPLEMENT:	109

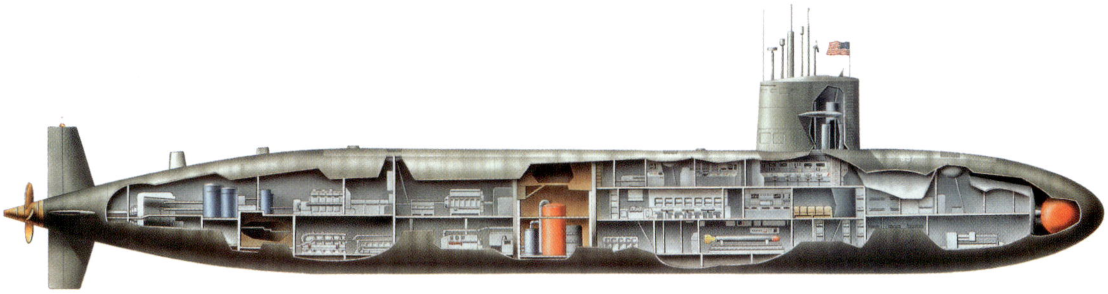

Above: The five-compartment arrangement of the Permit-*class hull was retained in the design of* Sturgeon: *bow compartment with diesel generator, operations compartment with torpedo room beneath, reactor compartment, auxiliary machinery room no. 2 and engine room.*

ARMAMENT
From 1965 until around 1980 some of the class were equipped with a SUBROC (UUM-44A) nuclear-tipped rocket-propelled ballistic missile, launched from a 533mm (21in) tube, with a range of 48km (30 miles).

SONAR ARRAY
The *Sturgeon* class first towed the TB-16 Fat Line sonar array. Later the TB-23 Thin Line array, with greater sensor capacity, was used.

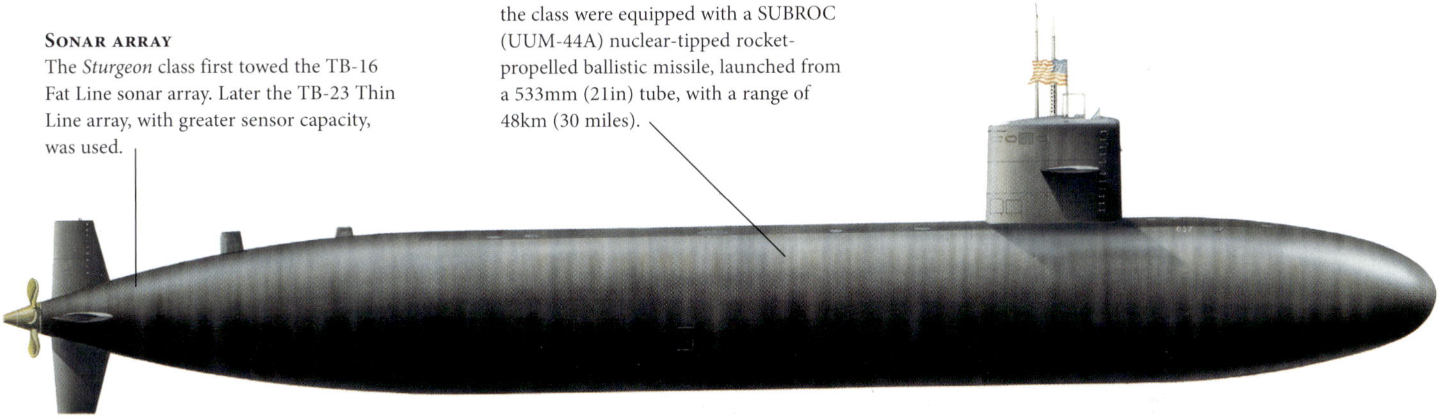

ATTACK BOAT
The *Sturgeon* class were equipped to carry the Harpoon missile, the Tomahawk cruise missile and a mixed load of torpedoes, including Mk 14 Mod 6, Mk 37 and Mk 37TS, Mk 48 and Mk 48 ADCAP. Bow positioning of the BQQ-2 active and BQQ-7 passive sonar sphere required the Mk 63 torpedo tubes that were positioned aft of the array to be canted some 15 degrees outwards. They could also carry Mk 57 deep-water mines and Mk 60 CAPTOR mines.

MODIFICATIONS
Beginning with SSN-678 *Archerfish*, 10 units of the class were lengthened by a 3m (10ft) plug in order to experiment with new propulsion systems and increase mission capabilities.

USS *BILLFISH* (SSN-676)
Between 1982 and 1991 six boats were modified to carry the SEAL Dry Deck Shelter (DDS), a 30-tonne (33-ton) launch hangar that attached to the weapon-shipping hatch and could be used to service undersea or surface operations.

HMS *Resolution* (1966)

TYPE • *Submarine* **COUNTRY** • *United Kingdom*

SPECIFICATIONS

Dimensions:	Length: 129.5m (425ft); Beam: 10.1m (33ft); Draught: 9.1m (30ft)
Displacement:	7620 tonnes (8402 tons) surfaced; 8535 tonnes (9410 tons) submerged
Propulsion:	1 Vickers/Rolls-Royce PWR1 pressurized water nuclear reactor; steam turbines; single screw; 20,500kW (27,500hp)
Speed:	20 knots (37km/h; 23mph) surfaced; 25 knots (46.2km/h; 28.7mph) submerged
Range:	Unlimited
Armament:	16 × Polaris A3TK SLBMs; 6 × 533mm (21in) torpedo tubes
Complement:	154

The *Resolution* class were the Royal Navy's first ballistic missile submarine. HMS *Resolution* fired the RN's first nuclear-capable Polaris missile on 15 February 1968, assuming the nuclear deterrent role previously held by the Royal Air Force.

Right: A view inside HMS Resolution *shows the Polaris missile access hatches.*

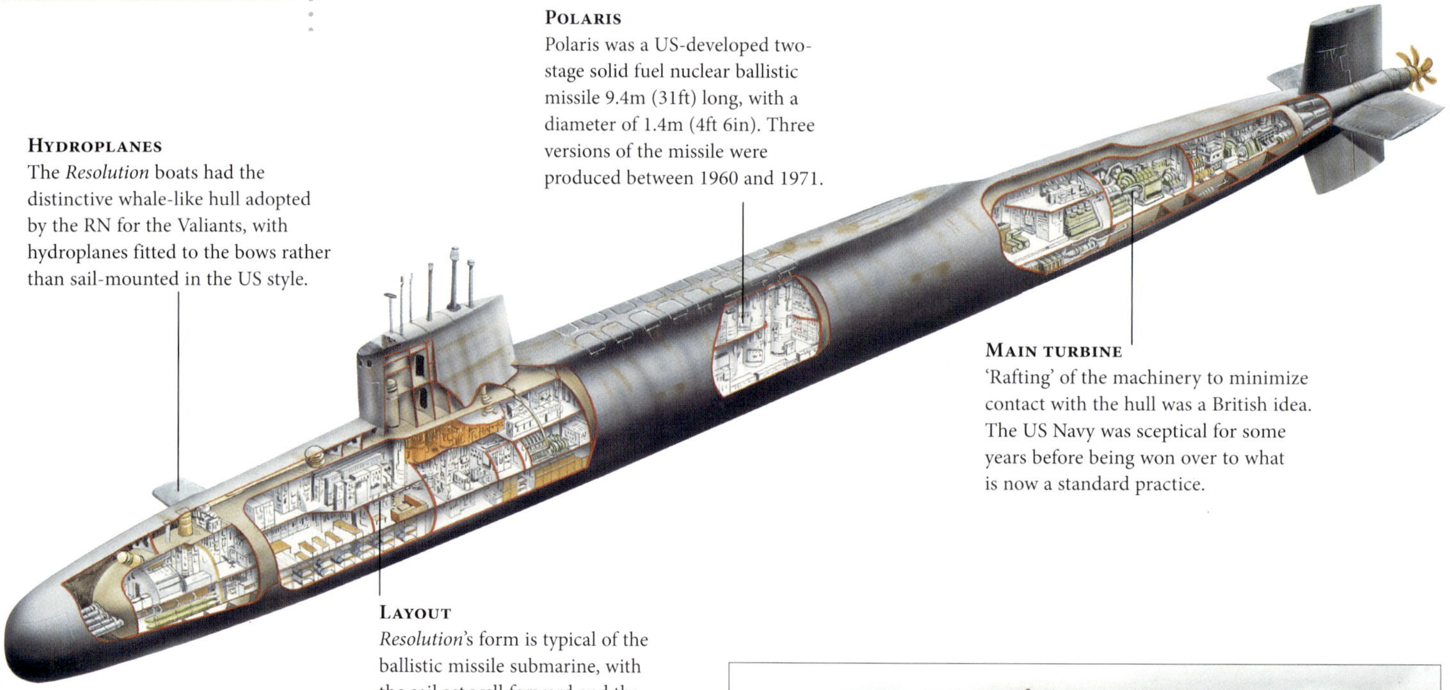

Polaris
Polaris was a US-developed two-stage solid fuel nuclear ballistic missile 9.4m (31ft) long, with a diameter of 1.4m (4ft 6in). Three versions of the missile were produced between 1960 and 1971.

Hydroplanes
The *Resolution* boats had the distinctive whale-like hull adopted by the RN for the Valiants, with hydroplanes fitted to the bows rather than sail-mounted in the US style.

Main turbine
'Rafting' of the machinery to minimize contact with the hull was a British idea. The US Navy was sceptical for some years before being won over to what is now a standard practice.

Layout
Resolution's form is typical of the ballistic missile submarine, with the sail set well forward and the missile compartments in the widest section.

NUCLEAR ROLE

Four nuclear-powered submarines were ordered, with HMS *Resolution* as class leader. The designated missile was the American Polaris. The bow and stern sections were constructed separately, and the American-designed missile compartment inserted between them. The design has been called a lengthened modification of the British *Valiant* class of first-generation SSNs.

CLASS LEADER

Laid down at the Vickers-Armstrong yard at Barrow-in-Furness on 26 February 1964 and launched on 15 September 1966, HMS *Resolution* was commissioned on 2 October 1967. Its cost was recorded as £40,240,000.

Dédalo (1967)

TYPE • *Aircraft carrier* **COUNTRY** • *Spain*

SPECIFICATIONS

DIMENSIONS:	Length: 189.7m (622ft 6in); Beam: 33.27m (109ft 2in) (overall); Draught: 7.9m (26ft)
DISPLACEMENT:	13,000 tonnes (12,125 tons) standard
PROPULSION:	4 Babcock & Wilcox boilers; 4 General Electric steam turbines; 4 shafts; 75,000kW (100,000shp)
SPEED:	32 knots (59km/h; 37mph)
ARMAMENT:	26 × 40mm (1.6in) Bofors guns
AIRCRAFT:	8 × AV-8S; 8 × SH-3D; up to 33 fixed-wing aircraft and helicopters
COMPLEMENT:	1112

The Spanish Navy's first fixed-wing carrier was *Dédalo*, which was ex-USS *Cabot* (CVL-28), an *Independence*-class carrier acquired on loan from the USA in August 1967 and formally purchased in 1972.

Left: A Hawker Siddeley AV-8S Matador flies over Dédalo *(R01) in the Mediterranean, 1 June 1988. From this perspective, the ship's initial World War II design is evident.*

HELICOPTERS
In addition to its fixed-wing air component, the carrier also operated four Sea King anti-submarine warfare helicopters and four AB 212ASW Twin Hueys.

FLIGHT DECK
The carrier's flight deck measured 168 × 22m (551 × 72ft) and had to be metal sheathed in places to project the original wooden deck from the downblast of Matador vertical take-offs.

POWERPLANT
The ship's four General Electric steam turbines outputted to four shafts, giving the vessel a top speed of 32 knots (59km/h; 37mph).

AIR WING

Used originally for ASW helicopter duties, *Dédalo* was converted to carry Harrier jump jets in 1972 but was never really suitable for jet aircraft, and was retired in 1989 after *Principe de Asturias* came into service. For the original air wing on *Dédalo*, eight AV-8B (VA.2) Harrier II V/STOL multi-purpose attack planes were ordered (known as the 'Matador' in Spanish service, with a further four in 1980).

DÉDALO
The *Dédalo* went into Spanish service in 1967 and was not struck from the fleet until 1989. Efforts to turn it into a museum ship failed, and it was scraped in 2002.

Moskva class (1967)

TYPE • *Helicopter carrier* **COUNTRY** • *Soviet Union*

SPECIFICATIONS

Dimensions:	Length: 196.6m (645ft); Beam: 35m (115ft); Draught: 7.6m (25ft)
Displacement:	14,800 tonnes (16,319 tons) standard
Propulsion:	4 pressure-fired boilers; 2 steam turbines; 2 shafts; 75,000kW (100,000shp)
Speed:	31 knots (57km/h; 35.4mph)
Range:	14,000nmi (25,928km; 16,110 miles) at 12 knots
Armament:	1 × twin SUW-N-1 launcher; 2 × twin SA-N-3 missile launchers
Aircraft:	18 helicopters
Complement:	850 including air wing

Moskva was the first 'helicopter cruiser' built for the Russian Navy. It was completed in 1967 to counteract the threat from the US nuclear-powered missile submarines that began to enter service in 1960.

Right: A side view of the Moskva. *The vessel could carry an impressive total of 18 naval helicopters.*

Radar mast assembly
Moskva had an extensive radar and surveillance suite aboard, including the Top Sail (air warning) system and the 'Head Net A' 2D air surveillance and surface search radar.

Flight deck
The flight deck measures 86m (282ft) by 34m (111ft 6in) and has four mesh-covered landing areas; a fifth landing area is located in the centre of the ship.

Flight deck
Primary air defence came from eight twin SA-N-3 'Goblet' surface-to-air missile (SAM) launchers fed from a magazine of 48 missiles.

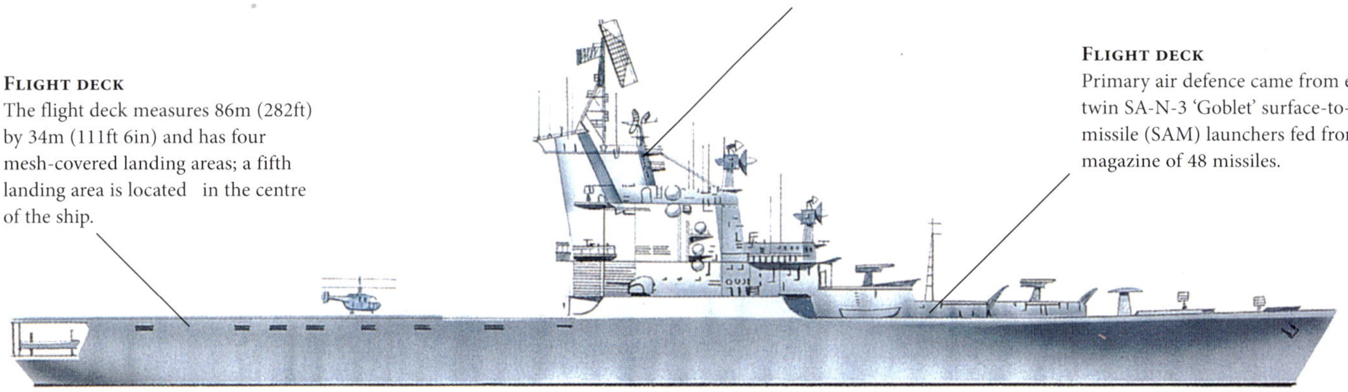

STRATEGIC PURPOSE
The *Moskva* and its sister ship *Leningrad* were designed primarily to hunt and destroy NATO strategic missile submarines in the North Sea, North Atlantic, Baltic and Indian Ocean. They would also act as command vessels for larger anti-submarine hunter-killer groups. In reality, however, their systems were soon outdated; both ships were decommissioned in the 1990s.

AT ANCHOR
The *Moskva*, seen here at anchor in the Mediterranean Sea in September 1968. The central block dominated the vessel and housed the major weapons systems.

Enrico Toti (1967)

TYPE • *Submarine* COUNTRY • *Italy*

SPECIFICATIONS

DIMENSIONS:	Length: 46.2m (151ft 8in); Beam: 4.7m (15ft 5in); Draught: 4m (13ft)
DISPLACEMENT:	532 tonnes (587 tons) surfaced; 591 tonnes (652 tons) submerged
PROPULSION:	2 Fiat MB 820 N/I diesel engines; 1 electric motor; 1600kW (2200hp); single screw
SPEED:	14 knots (25.8km/h; 16mph) surfaced; 15 knots (27.8km/h; 17.3mph) submerged
RANGE:	3000nmi (5556km; 3452 miles) at 5 knots surfaced
ARMAMENT:	4 x 533mm (21in) torpedo tubes
COMPLEMENT:	26

After training experience with ex-US submarines, Italy completed its first home-built submarines in 1968. These formed the four-strong *Enrico Toti* class, small vessels that were intended as interceptors and attack boats.

Left: The submarine Enrico Toti (S 506) *sails along the picturesque Ligurian coast of Italy in October 1970. As well as coastal patrols, the submarine participated in NATO exercises.*

PROPULSION
Two Fiat MB820 diesels generated 1640kW (2220hp) and drove the vessel to a maximum speed of 14 knots (25.8km/h; 16mph) surfaced and 15 knots (27.8km/h; 17.3mph) submerged.

TORPEDOES
The four torpedo tubes could fire the Whitehead Motofides A184 wire-guided torpedo, intended both for anti-ship and anti-submarine engagements, with a range of some 25km (15.5 miles) and an active/passive acousti homing head.

COMPACT DESIGN
Being designed primarily for short-range coastal work, the *Enrico Toti* class had only a small crew – four officers and 22 men.

COASTAL PATROL
Built by Fincantieri in Monfalcone, the *Toti* class can be compared to the French *Aréthuse* and the German Type 205 classes in size and functions. *Enrico Toti* itself served in the Mediterranean Sea, achieving 220,480km (137,000 miles) on patrol duty around the coasts of Italy. All were decommissioned in 1991–93. *Enrico Toti* is preserved as a museum boat at Milan.

***LAZARRO MOCENIGO* (S 514)**
The *Lazarro Moncenigo* (S 514) was the fourth ship of the *Enrico Toti* class. The bulbous projection at the bow of the submarines housed the sonar array.

Project 671 *Shchuka* class (1967)

TYPE • *Submarine* **COUNTRY** • *Soviet Union*

SPECIFICATIONS
(*Victor III*)

Dimensions:	Length: 107m (351ft); Beam: 10.8m (35ft 5in); Draught: 7.2m (23ft 6in)
Displacement:	5283 tonnes (5824 tons) surfaced; 7366 tonnes (8119 tons) submerged
Propulsion:	2 OK-300 VM-4 pressurized water nuclear reactors; steam turbines delivering 23,000kW (31,000shp); single screw
Speed:	18 knots (33.3km/h; 20.7mph) surfaced; 30 knots (55.5km/h; 34.5mph) submerged
Range:	Unlimited
Armament:	2 × 650mm (25.3in) and 6 × 533mm (21in) bow torpedo tubes; RPK-7 anti-submarine missiles, S-10 Granat cruise missiles, up to 24 weapons; up to 36 mines as an alternative load
Complement:	115

To the Soviet Navy, it was Project 671, known as *Shchuka* ('Pike'). To NATO, it was designated the *Victor* class. Built between 1967 and 1992, the final *Victor III* submarines were contemporary with the US *Sturgeon* class, significantly faster and with comparable stealth characteristics.

Left: A Victor-class Russian nuclear submarine seen here on exercise in the North Atlantic. The Victor II version was enlarged to provide additional weapons capabilities and improved fire-control system. The new generation of 650mm (25.3in) heavy torpedoes were longer than earlier models, and required power assistance for handling in the torpedo room.

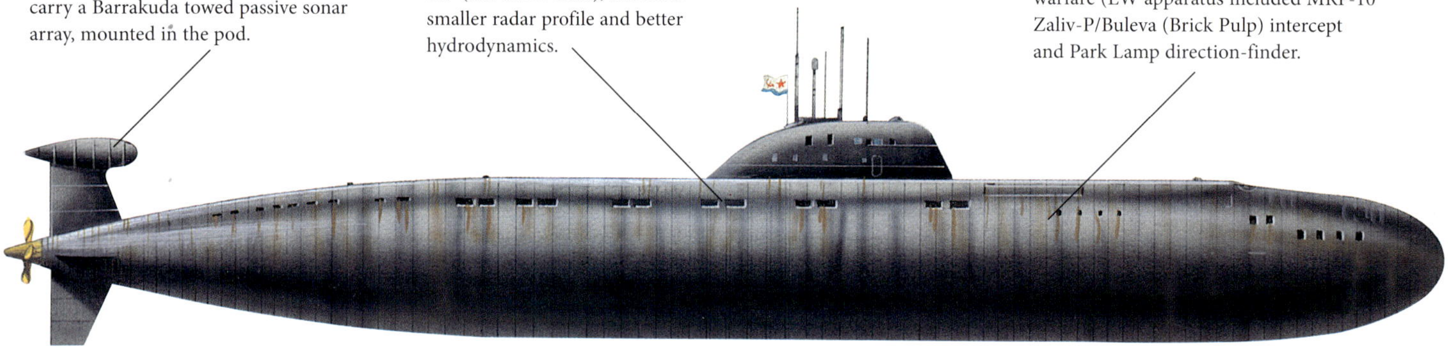

SUB-HUNTER

The Project 671 brief, given to the Malakhit Central Design Bureau (TsKB 143), was for a new class to engage hostile ballistic missile submarines and anti-submarine task forces, and protect friendly vessels and convoys from attacks. A total of 48 were built in three phases. *Victor I* featured an advanced teardrop hull design for high underwater speeds, used also in *Victor II* and *III*. A total of 15 units were built.

VICTOR III

A Soviet *Victor III* class submarine disabled and adrift off the coast of South Carolina, 2 November 1983. A total of 26 *Victor III* class boats were built between 1976 and 1985. In the new design, the hull was lengthened by nearly 6m (20ft) forward of the sail to make room for rafting and sound insulation of the turbine machinery.

Leander class (1967)

TYPE • *Frigate* COUNTRY • *United Kingdom*

SPECIFICATIONS (Batch 1)

DIMENSIONS:	Length: 113.4m (372ft); Beam: 12.5m (41ft); Draught: 4.5m (15ft)
DISPLACEMENT:	2388 tonnes (2633 tons) standard; 2906 tonnes (3203 tons) full load (broad-beam:
PROPULSION:	2 Babcock & Wilcox oil-fired boilers; geared steam turbines; 2 shafts; 22,370kW (30,000shp)
SPEED:	27 knots (50km/h; 31mph)
RANGE:	4000nmi (7400km; 4600 miles) at 15 knots
ARMAMENT:	1 × Ikara ASM launcher; 2 × Seacat SAM launchers; 2 × single 40mm (1.6in) guns; 2 × triple 324mm (12.75in) STWS-1 torpedo tubes for Mk 46 and Stingray ASW torpedoes; 1 × Limbo ASW mortar
COMPLEMENT:	260

The *Leander*-class frigate was one of the most successful surface warfare types of the Cold War era. Not only were 26 of these general-purpose frigates built for the Royal Navy, but they were also built for or sold to numerous other navies around the world.

Right: HMS Leander *(F109) was the lead ship of the class. The absence of forward guns indicates that it has been through the 'Ikara conversion', replacing the guns with the GWS40 Ikara anti-submarine warfare missile plus an additional Seacat SAM launcher.*

HELICOPTER
The Initial and Batch 1 *Leander* class carried a single Westland Wasp ASW helicopter. Batches 2 and 3 took the Westland Lynx HAS.Mk 2.

HMS *LEANDER*
The presence of the forward gun turret indicates that the ship is in the 'Initial' configuration, before the Ikara conversion that took place between 1970 and 1973.

PROPULSION
The class were powered by two Babcock & Wilcox oil-fired boilers feeding to geared steam turbines, delivering 22,370kW (30,000hp) to two shafts.

BATCHES

The Royal Navy *Leander*-class frigates were built as or upgraded to three specific batches, modifications of the 'Initial' design. The variations mostly focused on changing armament. Batch 1, for example, had twin 114mm (4.5in) guns, but in Batches 2 and 3 these had been replaced by Exocet missiles. Batch 3 had the more capable Seawolf SAM system, as compared to the preceding Seacat. The Batch 3 ships also had a wider beam, to improve stability and internal space.

HMS *SCYLLA* (F71)
HMS *Scylla* (F71) was a *Leander*-class frigate built at Devonport Royal Dockyard, being launched in 1968 and commissioned in 1970. It was one of the Batch 3 ships and served the Royal Navy until 1990.

Sjöörmen class (1968)

TYPE • *Submarine* COUNTRY • *Sweden*

With the *Sjöörmen* class, Sweden once again claimed a place among the leaders of diesel-electric submarine design. Five boats formed the class, all commissioned between July 1968 and September 1969.

SPECIFICATIONS

DIMENSIONS:	Length: 51m (167ft 3in); Beam: 6.1m (20ft); Draught: 5.8m (19ft)
DISPLACEMENT:	1143 tonnes (1260 tons) surfaced; 1422 tonnes (1567 tons) submerged
PROPULSION:	Diesel-electric drive; 2 Pielstick diesel engines; 1 ASEA electric motor; single shaft with five-bladed propeller; 1600kW (2200bhp)
SPEED:	15 knots (27.8km/h; 17.3mph) surfaced; 20 knots (37km/h; 23mph) submerged
RANGE:	Not available
ARMAMENT:	4 × 533mm (21in) and 2 × 400mm (15.75in) torpedo tubes
COMPLEMENT:	18

Right: HSwMS Sjöormen *(seen here) – its name meaning 'sea serpent – was the lead ship of the class.*

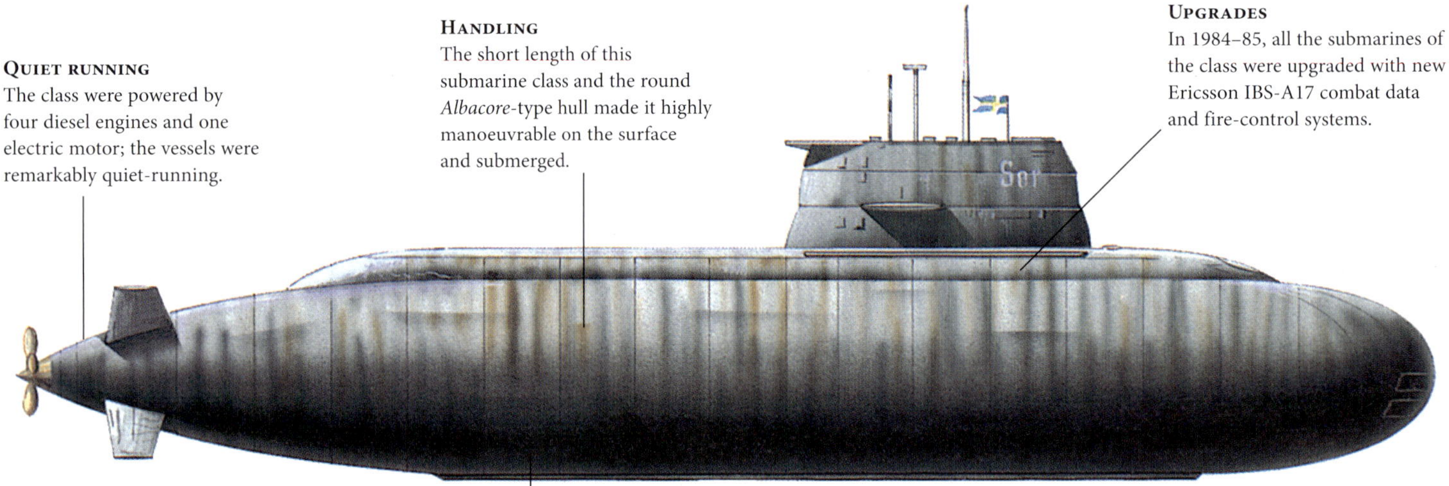

QUIET RUNNING
The class were powered by four diesel engines and one electric motor; the vessels were remarkably quiet-running.

HANDLING
The short length of this submarine class and the round *Albacore*-type hull made it highly manoeuvrable on the surface and submerged.

UPGRADES
In 1984–85, all the submarines of the class were upgraded with new Ericsson IBS-A17 combat data and fire-control systems.

ANECHOIC TILES
Anechoic tiles damped the sound and the submerged speed of 20 knots (37km/h; 23mph) was fast for a conventional submarine of the time.

SINGAPORE SERVICE

The *Sjöörmen* class incorporated numerous features developed by the Kockums shipbuilding form and the Swedish Navy, including a form of the X-rudder first seen on USS *Albacore*. In Swedish service until the early 1990s, they were modernized and refitted for tropical waters and became Singapore's *Challenger* class, four for service plus one for spare parts.

SJÖHUNDEN
This photograph shows the Swedish Navy's HSwMS *Sjöhunden*, the fourth of five *Sjöörmen*-class submarines. Anti-submarine action was the prime purpose of the class.

Veinticinco de Mayo (1969)

TYPE • *Aircraft carrier* COUNTRY • *Argentina*

SPECIFICATIONS	
Dimensions:	Length: 192m (585ft 2in); Beam: 24.4m (74ft 4in); Draught: 7.6m (23ft 2in)
Displacement:	16,146 tonnes (17,797 tons) standard; 20,214 tonnes (22,285 tons) full load
Propulsion:	4 Admiralty 3-drum boilers; Parsons geared steam turbines; 2 shafts; 29,828kW (40,000shp)
Speed:	25 knots (46.25km/h; 28.75mph)
Range:	12,000nmi (22,200km; 13,800 miles)
Armament:	12 × 40mm AA guns
Aircraft:	37
Complement:	1300

Commissioned into the Argentinian Navy in 1969, *Veinticinco de Mayo* already had a long career behind it, having served with the British and Dutch navies. The carrier's story began in World War II, when it was ordered from the Cammell Laird yard in Birkenhead, England.

Left: Veinticinco de Mayo *pictured at the time of the Falklands War between Argentina and the United Kingdom, April–June 1982.*

Propulsion
As designed, the ships in the class were driven by Parsons geared turbines, supplied with steam by Admiralty three-drum boilers in two separate engine rooms, the starboard room set forward of the port one.

Falklands action
The ship supported troop landings on the Falklands on 1 April 1982, but operational difficulties, including its low speed, then 24 knots (44.4km/h; 27.6mph) at maximum, made it impossible to launch jet strikes from the flight deck and its air wing operated from land bases.

Modernization
From 1955 to 1958 the carrier was docked for extensive modernization. The island was rebuilt, the funnel lengthened and its cap tilted back, the masts replaced and fitted with extensive LW-01 and LW-02 radar antennas, a steam catapult was fitted, and the flight deck was reinforced and extended to port by an 8 degree angled flight deck.

INTERNATIONAL SHIP
The ship was launched on 30 December 1943 and commissioned on 17 January 1945 as HMS *Venerable*. In 1948 it was sold to the Royal Netherlands Navy and recommissioned as *Karel Doorman*. From 1955 to 1958 it was docked for extensive modernization. The ship was sold to Argentina in October 1968 and refitted for the Argentinian Navy as *Veinticinco de Mayo*. It was recommissioned on 12 March 1969 and entered service on 22 August.

***VEINTICINCO DE MAYO* (R81)**
As *Veinticinco de Mayo*, Argentina's only operative carrier from 1970, it shipped first Grumman F9F Panther jets, then A-4Q Skyhawks, Grumman S-2 twin-propeller ASW planes, and Sikorsky SH34 helicopters. Its home port was Puerto Belgrano Naval Base, near Buenos Aires.

USS *California* (1970)

TYPE • *Missile cruiser* **COUNTRY** • *United States*

SPECIFICATIONS

Dimensions:	Length: 181.6m (596ft); Beam: 18.5m (61ft); Draught: 10m (33ft)
Displacement:	10,814 tonnes (11,923 tons) standard; 11,733 tonnes (12,933 tons) full load
Propulsion:	2 General Electric nuclear reactors; steam turbines
Speed:	32 knots (59.3km/h; 36.8mph)
Range:	Unlimited
Armament:	2 × Mk 141 Harpoon missile launchers; 2 × 127mm (5in) Mk 45 guns; 2 × 20mm (0.79in) Phalanx CIWS; 1 × ASROC missile launcher; 2 × Mk 13 missile launchers; Mark 46 torpedoes for anti-submarine use
Complement:	584

Originally planned as a five-ship class of nuclear-powered guided missile frigates, the *California* class were reduced to only two vessels. These ships were commissioned as frigates and later redesignated as cruisers.

Ship role
California's principal task was to act in the 'goalkeeper' role as escort to an aircraft carrier, her missiles eliminating any air threats that leaked through the carrier task group's outer defence zone.

Superstructure
Vital parts of the superstructure were protected by Kevlar armour, based on a type of strong and heat-resistant synthetic fibre.

Sensors
Sensors included an SPS-48A 3D air search radar, 2D air search radar, and an array of surface search radars and fire control radar systems. These enabled the vessel to engage targets up to 130km (80 miles) distant.

Gun armament
California's main gun armament comprised two 127mm (5in) Mk 45 guns. The gun mount featured an automatic loader with a capacity of 20 rounds.

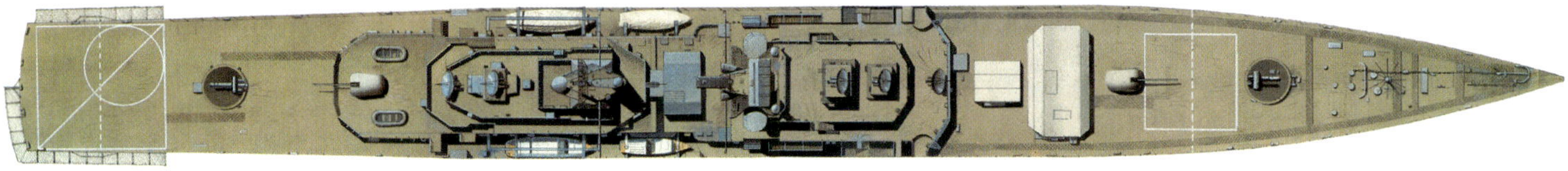

CGNS
Two nuclear-powered *California*-class cruisers were built for the US Navy after plans for a pair of guided missile destroyers fell through. They differed markedly from their cancelled predecessors, incorporating new systems, most notably their nuclear powerplants. In 1975 the *Californias* were redesignated as nuclear-powered cruisers (CGNs), and their advanced systems enhanced the Standard SAM's capabilities against high-performance aircraft.

STANDARD SAM
California's main weapon, the semi-active radar homing Standard SAM, began development in the early 1960s as a replacement for the Tartar and Terrier systems. Here one is launched from *California*'s foredeck.

Type 21 *Amazon* class (1970)

TYPE • *Frigate* **COUNTRY** • *United Kingdom*

SPECIFICATIONS

Dimensions:	Length: 117m (384ft) overall; Beam: 12.7m (41ft 10in); Draught: 5.8m (19ft)
Displacement:	2794 tonnes (3081 tons) standard; 3302 tonnes (3640 tons) full load
Propulsion:	COGOG; 2 Tyne cruise gas turbines, 6300kW (8500shp); 2 Olympus boost gas turbines, 37,000kW (50,000shp); 2 shafts
Speed:	32 knots (59km/h; 37mph)
Range:	1200nmi (2200km; 1400 miles) at 30 knots
Armament:	(Initial) 1 × 114mm (4.5in) Vickers Mark 8 gun; 2 × 20mm (0.79in) Oerlikon guns (later 4); 1 × 4-rail GWS-24 Sea Cat launcher; 2 × 8-barrel Knebworth Corvus countermeasures launchers
Complement:	177

Battered veterans of the Falklands War, the Type 21 *Amazon*-class destroyers were a private design by Vosper Thorneycroft, with their development and early service marked by political controversy.

Left: HMS Amazon *(F169) takes on fuel at sea.* Amazon *was decommissioned and sold to Pakistan on 30 September 1993. Renamed PNS* Babur, *it served until 2014.*

Sensors
The sensor package included Type 992Q surface/air search and target indication radar, Type 978 navigation radar, two Type 912 fire-control radars, a Type 184M medium-range search sonar and a Type 162M bottom-search sonar.

Armament
Initial armament consisted of a single 114mm (4.5in) Mark 8 naval gun and a four-round launcher for the Seacat SAM aft, plus two 20mm (0.79in) cannon. Later the ships received the Exocet anti-ship missile.

Propulsion
The ship was powered by two Rolls-Royce Olympus TM3B gas turbines and two Rolls-Royce Tynes in a combined gas or gas (COGOG) arrangement.

DESIGN ISSUES

In service, significant design issues were discovered in the Type 21 vessels, eight of which were commissioned between May 1974 and April 1978. There was an error in weight and vertical centre of gravity, and the hulls exhibited cracks from use in heavy seas, as demonstrated during the class's deployment to the Falklands in 1982. It also proved more difficult than ideal to upgrade the ships with newer generations of weapons and sensors.

HMS *ARROW* (F173)
HMS *Arrow* (F173) underway, c. 1982. During the Falklands War, two Type 21 frigates were lost to enemy action – HMS *Ardent* and HMS *Antelope* – the class proving vulnerable to modern air attack.

Krivak class (1970)

TYPE • *Frigate* **COUNTRY** • *Soviet Union*

SPECIFICATIONS (Project 1135)

Dimensions:	Length: 123m (404ft); Beam: 14.2m (47ft); Draught: 4.5m (15ft)
Displacement:	2810 tonnes (3098 tons) standard; 3200 tonnes (3527 tons) full load
Propulsion:	COGAG; 4 gas turbines; 2 shafts; 36,000kW (48,000shp)
Speed:	32 knots (59km/h; 37mph)
Range:	4000nmi (7408km; 4600 miles) at 14 knots
Armament:	4 × URPK-3 Metel (SS-N-14 Silex) ASMs (1 × quad launcher); 4 × ZIF-122 4K33 launchers (2 × twin) with 40 × 4K33 Osa-M (SA-N-4 Gecko) SAMs; 4 × 76mm (3in) AK-726 guns (2 × twin); 2 × RBU-6000 Smerch-2 ASW rocket launchers; 8 × 533mm (21in) torpedo tubes; 18 mines
Complement:	197

The *Krivak* class is a prolific series of Soviet and Russian frigates and patrol ships, first emerging in the 1970s at the height of the Cold War, with highly modernized derivatives remaining in active service to this day.

Right: Poryvistyy *was a* Krivak I*-class vessel commissioned in December 1981. This photograph was taken by US Navy personnel in March 1987.*

Sensors
Sensors mounted on the upper superstructure included the MR-755 Fregat-M/Half Plate air/surface search radar and Drakon/Eye Bowl SSM targeting system.

Bow missiles
Atop the foredeck sat the boxy SS-N-14 anti-submarine missile quad launcher; a blast screen extends out to the front of the launcher unit.

Propulsion
The *Krivak* ships used a combined gas turbine and gas turbine (COGAG) propulsion system; in this case four gas turbines are connected to two propeller shafts.

SUB-CLASSES

Including all sub-classes and derivatives, the *Krivak* class has produced 52 vessels, with three more under construction in 2026. The principal Cold War sub-classes were the *Krivak I, II, III* and *IV*, the latter involving the modernization of existing ships to meet the requirements of coast guard duties. The Indian Navy also took six new-build *Krivak III* frigates as the *Talwar* class. Six vessels based on the *Talwar* class were ordered for Russian service in the 2000s as Project 11356R/M.

DOBLESTNYY
The *Doblestnyy* was a Project 1135 *Krivak*-class vessel. At the rear of the ship note the housing for the variable-depth sonar system, its protective door in the closed position.

Type 209 (1971)

TYPE • *Submarine* COUNTRY • *Germany*

SPECIFICATIONS (Type 209/1200)

Dimensions:	Length: 56m (183ft 9in); Beam: 6.2m (20ft 4in); Draught: 5.5m (18ft 1in)
Displacement:	1122 tonnes (1236 tons) surfaced; 1249 tonnes (1377 tons) submerged
Propulsion:	Diesel-electric drive; 4 diesel engines; 1 electric motor; single shaft; approx. 3700kW (5000shp)
Speed:	10 knots (18.5km/h; 11.5mph) surfaced; 22 knots (40.6km/h; 25.2mph) submerged
Range:	2400nmi (4447km; 2763 miles) at 8 knots surfaced
Armament:	8 × 533mm (21in) torpedo tubes
Complement:	35

The German-designed Type 209 is one of the most widely used Cold War submarine types. Although the Type 209 has never been used by the German Navy, between 1971 and 2008, 61 were commissioned for 13 other navies.

Left: The Jang Bogo *(SS-061) is a South Korean Type 209/1200, the lead ship of its class. The Type 209/1200s used by Greece and South Korea and the Type 209/1400s used by Turkey are also armed with Sub-Harpoon missiles. Those used by South Korea can be armed with 28 mines in place of torpedoes and Harpoon missiles.*

Hull configuration
The design is single hulled, with a pressure hull diameter of 6.8m (22.3ft) on all sub-types. The entire submarine from bow to stern can be viewed while standing at the periscope.

Weaponry
CSU-83 active/passive sonar is fitted as well as DUUX 2 intercept sonar. They carry 14 A-184 SST (Special Surface Target) torpedoes.

***Angamos* (Peruvian Navy)**
The Type 209 submarines are designed for coastal operations rather than oceanic patrols. The two 209/1100 boats were upgraded in Peru in 2008. The Type 209/1200 *Angamos* was formerly named *Casma*.

PERUVIAN BOATS

The Type 209 is closely based on the IKL Type 206 design. Single-hulled, with a test depth of 500m (1640ft), it is a versatile craft that lends itself to many tasks and can carry a variety of different armaments. In 1975, the Peruvian Navy acquired two Type 209/1100 submarines, and between 1980 and 1983 a further four of Type 209/1200.

TYPE 209/1300
The S-32 *Caribe* seen here was one of two Type 209/1300 submarines produced for Venezuela and designated as the *Sabalo* class, after the lead ship S-31 *Sabalo*.

Type 42 *Sheffield* class (1971)

TYPE • *Destroyer* **COUNTRY** • *United Kingdom*

SPECIFICATIONS (Batch 1 and 2)

Dimensions:	Length: 125m (410ft); Beam: 14.3m (47ft); Draught: 4.2m (14ft)
Displacement:	3600 tonnes (3968 tons) standard; 4350 tonnes (4795 tons) full load
Propulsion:	COGOG, 2 × Rolls-Royce Olympus TM3B gas turbines, 37,000kW (50,000shp); 2 × Rolls-Royce Tyne RM1C cruise gas turbines, 3980kW (5340shp); 2 shafts
Speed:	30 knots (56km/h; 35mph)
Range:	4200nmi (7800km; 4800 miles) at 13.8 knots
Armament:	1 × twin GWS-30 Sea Dart missile launcher; 1 × 114mm (4.5in) Mark 8 gun; 2 × 20mm (0.79in) Phalanx CIWS; 2 × 20mm Oerlikon/ BMARC cannon; 2 × STWS II triple ASW torpedo tubes
Complement:	274

The primary purpose of the Type 42 *Sheffield*-class destroyer was fleet air defence, but budget constraints shaped the design and limited its effectiveness in that role, as was arguably demonstrated during the Falklands War in 1982.

Left: HMS Manchester *was one of the Batch 3 Type 42 vessels, which incorporated lessons learned from the Falklands War, and featured a lengthened hull and increased displacement.*

Propulsion
The Type 42 was powered by two Rolls-Royce Olympus and two Tyne gas turbines, which offered fast acceleration but were heavy on fuel consumption.

Sea Dart
The Sea Dart SAM launcher on the foredeck was capable of engaging targets out to at least 56km (35 miles). During the Falklands War, Sea Darts had seven confirmed kills.

Helicopter
The Type 42 carried a Westland Lynx HAS / HMA armed with four anti-ship missiles and two anti-submarine torpedoes.

FALKLANDS LOSSES
The Type 42 was produced in three steadily improving 'batches'. In total, five Type 42 destroyers saw service in the Falkands War, four from Batch 1 and one (*Exeter*) from Batch 2. The conflict was hard on the class. Two vessels were sunk: HMS *Sheffield* by an Exocet missile and HMS *Coventry* by multiple bomb hits. HMS *Glasgow* was also struck by a bomb, although this passed through without exploding.

HMS *GLOUCESTER*
HMS *Gloucester* was a Batch 3 Type 42. It served in the Persian Gulf War in 1991, where it shot down an incoming Iraqi Silkworm (targeting the USS *Missouri*) with Sea Dart missiles – the first validated missile-versus-missile strike of its kind.

Swiftsure class (1973)

TYPE • *Submarine* COUNTRY • *United Kingdom*

SPECIFICATIONS

DIMENSIONS:	Length: 82.9m (272ft); Beam: 9.8m (32ft 2in); Draught: 8.5m (28ft)
DISPLACEMENT:	4471 tonnes (4929 tons) surfaced; 4979 tonnes (5487 tons) submerged
PROPULSION:	1 Rolls-Royce Mk 1 pressurized water nuclear reactor; 2 GEC steam turbines; single screw; 11,175kW (15,000hp)
SPEED:	20 knots (37km/h; 23mph) surfaced; 30 knots (55.5km/h; 34.5mph) submerged
RANGE:	Unlimited
ARMAMENT:	5 × 533mm (21in) torpedo tubes; Tomahawk land-attack cruise missiles; Sub-Harpoon anti-ship missiles
COMPLEMENT:	116

HMS *Swiftsure* was the lead ship of a six-strong British hunter-killer class, which showed significant design changes from its predecessors. The class took part in several campaigns, including the Falklands War of 1982 and the NATO strikes against Serbia in 1999.

Above: Spartan *and* Splendid *(seen here) took part in the 1982 Falklands War, conducting reconnaissance missions and enforcing the maritime exclusion zone around the islands.*

PAIRED TORPEDO TUBES
Mounted to port and starboard of the sonar sphere, with a fifth beneath it. Compared with the *Valiant* class, the armament was reduced by one tube and seven torpedoes. Each tube could be reloaded in 15 seconds.

SKEW-BLADE PROPELLER
Swiftsure was the only submarine of its class to have a propeller rather than a propulsor. It had a large skew-blade propeller, perhaps intended for comparative testing of mobility and speed.

BALLAST TANKS
During surface travel, the stern ballast tanks had to remain flooded to maintain trim, lowering the boat in the water and leaving the bridge – already lower in height than that of the preceding *Valiant* class – more exposed to spray and waves.

DEVELOPMENT

From its first generation of nuclear submarines in the 1960s, the British Royal Navy's submarine arm was based on two types, the SSN hunter-killer, intended to be fast and tactically-armed; and the SSBN missile-carrier, intended for long-range sub-surface cruising and armed with ballistic missiles. The *Swiftsure* class formed the second generation of SSNs. Laid down on 6 June 1969 at the Vickers Yard at Barrow-in-Furness, HMS *Swiftsure* was launched on 7 September 1971 and commissioned on 17 April 1973.

HMS *SUPERB* (S109)
HMS *Superb* was commissioned on 13 November 1976 and served for more than four decades with the Royal Navy, finally being decommissioned in September 2008.

Delta class (1973)

TYPE • *Submarine* **COUNTRY** • *Soviet Union*

SPECIFICATIONS (*Delta III*)

DIMENSIONS:	Length: 160m (524ft 11in); Beam: 12m (39ft 4in); Draught: 8.7m (28ft 6in)
DISPLACEMENT:	10,791 tonnes (11,897 tons) surfaced; 13,463 tonnes (14,846 tons) submerged
PROPULSION:	OK-700A reactor system comprising 2 × VM-4S pressurized water reactors; steam turbines delivering 44,700kW (59,900shp); 2 screws
SPEED:	14 knots (26km/h; 16mph) surfaced; 24 knots (44.4km/h; 27.6mph) submerged
RANGE:	Unlimited
ARMAMENT:	16 × RSM-50 R-29R Vysota SLBMs; 4 × 533mm (21in) bow torpedo tubes; 16 torpedoes
COMPLEMENT:	130

After the shock of the *Alfa* class, the Soviet Navy unveiled what was perhaps arguably an even greater threat to NATO security: Project 667B *Murena* (Eel), known to NATO as the *Delta 1* class. This large SSBN was a major advancement over American submarines.

Above: The Delta III *was fitted with the Almaz-BDR battle-management system for firing deep-water torpedoes. It had the Tobo-M-1 and then the Tobol-M-2 inertial navigation systems and a hydroacoustic navigational system codenamed by NATO as* Bumblebee*, using hydroacoustic buoys to determine position.*

ENDURANCE
In nuclear-powered submarines, endurance is theoretically unlimited. However, practically it is constrained by mundanities such as crew food supply and mechanical repairs.

DELTA IV (PROJECT 667 BDRM)
Similar to the *Delta IIIs*, the *Delta IVs* are quieter in operation, with more effective hydroacoustic coating and new five-bladed propellers.

MISSILE LAUNCH
The submarine's ballistic missiles can be launched individually or even as a single salvo at a submerged depth of 55m (180ft) while cruising at a speed of 6–7 knots (11–13km/h; 6.9–8.1mph).

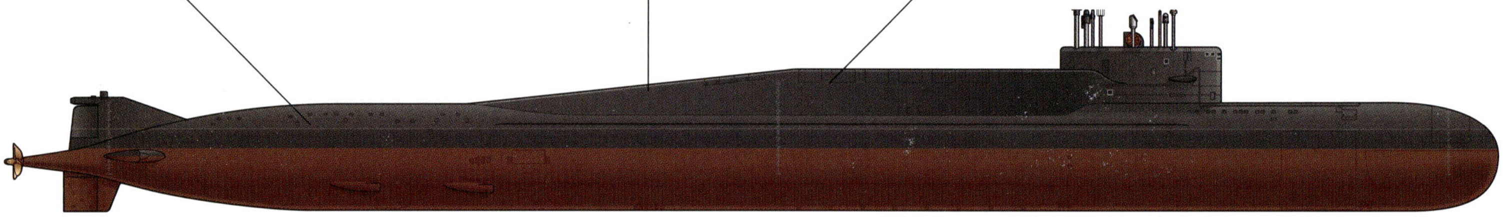

DELTA III
From 1972, Soviet Russia developed the Project 667 BDR *Kalmar* class, codenamed *Delta III* by NATO. Fourteen were built between 1976 and 1982, double-hulled, with a large, broad turtle-back missile silo set behind the sail. The 16 missiles were R-29R (SS-N-18, Stingray), the first multiple-warhead rockets to be carried on Soviet submarines. *Delta III* boats were assigned both to the Northern Fleet from the early 1990s and the Pacific Fleet based on the Kamchatka Peninsula.

ADVANCED HUNTER
The *Delta* class reverted to the Soviet two-reactor arrangement (pressurized water-cooled system), with twin propellers. Capable of operating under the Arctic ice cap, it was equipped with the Tobol-B navigation system and the Cyclone-B satellite navigation system.

Tromp class (1973)

TYPE • *Destroyer* COUNTRY • *Netherlands*

SPECIFICATIONS

DIMENSIONS:	Length: 133.2m (437ft); Beam: 14.8m (48ft 7in); Draught: 6.6m (21ft 8in)
DISPLACEMENT:	3724 tonnes (4107 tons) standard; 4377 tonnes (4827 tons) full load
PROPULSION:	COGOG, 2 shafts; Rolls-Royce Olympus gas turbine, 32,811kW (44,000hp); Rolls-Royce Tyne gas turbine, 6115kW (8200hp)
SPEED:	28 knots (52km/h; 32mph)
RANGE:	5000nmi (9300km; 5780 miles) at 18 knots
ARMAMENT:	1 × Standard SAM launcher; 1 × octuple Sea Sparrow SAM launcher; 8 × RGM-84 Harpoon SSM; 2 × 120mm (4.7in) Bofors guns (twin turret); triple ASW torpedo tubes; 1 × Goalkeeper CIWS
COMPLEMENT:	306

Replacing two cruisers in service with the Royal Netherlands navy, HNLMS *Tromp* (F801) and *De Ruyter* (F806) were among the largest and most capable of frigates afloat. Weapons fitted included Harpoon, Standard and Sea Sparrow missiles.

Above: This forward view of Tromp *(F801) shows the voluminous polyester radome at the top, which housed a 3D radar.*

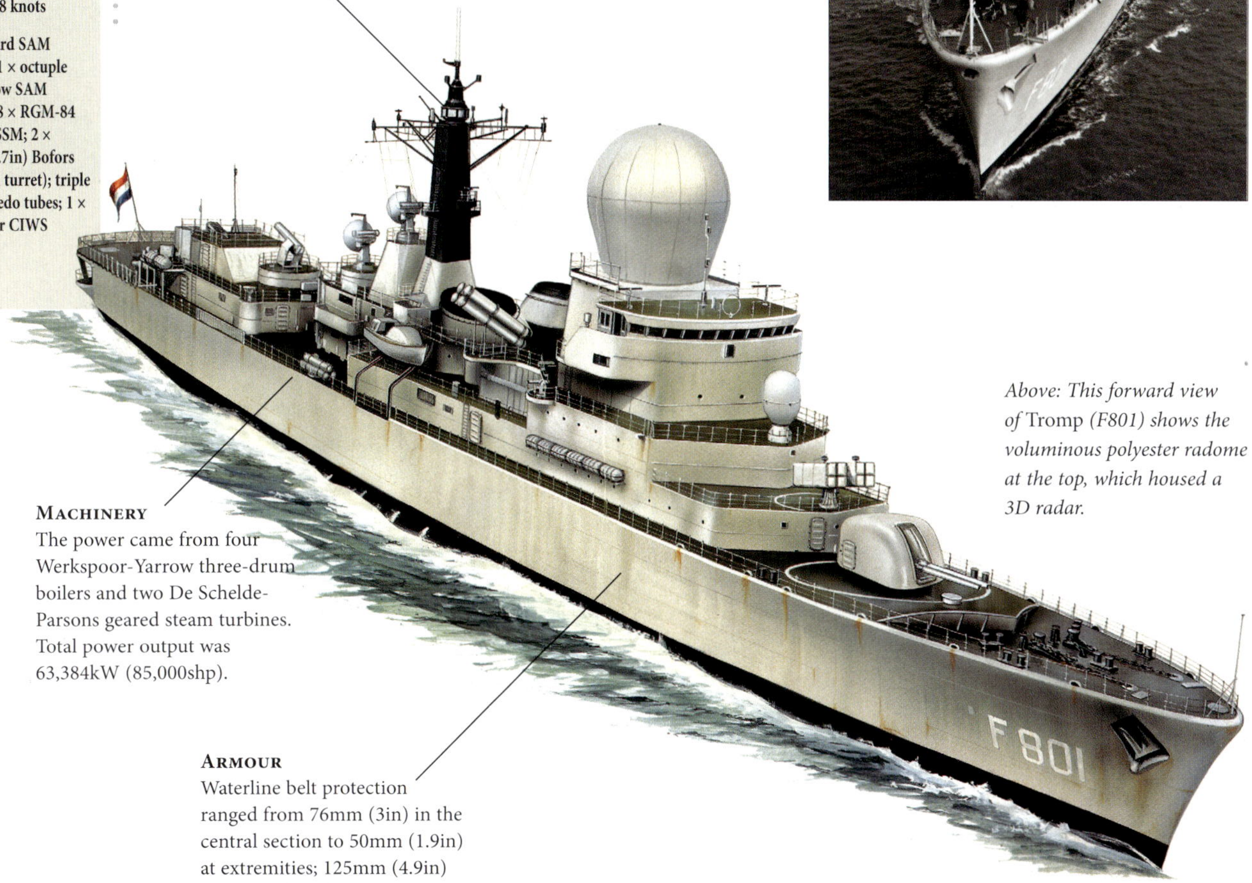

TROMP
A major upgrade in 1985–88 brought new search and fire-control radar systems, decoy launchers and other countermeasures systems, and data links.

MACHINERY
The power came from four Werkspoor-Yarrow three-drum boilers and two De Schelde-Parsons geared steam turbines. Total power output was 63,384kW (85,000shp).

ARMOUR
Waterline belt protection ranged from 76mm (3in) in the central section to 50mm (1.9in) at extremities; 125mm (4.9in) to 50mm (1.9in) on turrets.

FLAGSHIPS

Tromp and companion ship *De Ruyter* acted as flagships to two long-range NATO task groups, for operating in the Eastern Atlantic. An octuple Sea Sparrow launcher with 60 reloads provided short-range anti-aircraft and anti-missile defence, and a later refit added a Goalkeeper point defence gun system. A single Lynx helicopter was carried.

DE RUYTER
In a picture-perfect moment, *De Ruyter* (F806) sails under Tower Bridge in London in December 1976. The ship served the Royal Netherlands Navy until it was decommissioned in October 2001.

Type 053 *Jiangdong* class (1974)

SPECIFICATIONS (Type 053K)

Dimensions:	Length: 103.2m (338ft 7in); Beam: 10.2m (33ft 6in); Draught: 3.1m (10ft 2in)
Displacement:	1700 tonnes (1874 tons) standard; 2000 tonnes (2205 tons) full load
Propulsion:	2 SEMT Pielstick 12PA6 diesel engines; 12,000kW (16,000shp); 2 shafts
Speed:	26 knots (48km/h; 30mph)
Range:	7400km (4000nmi; 4600 miles) at 15 knots
Armament:	2 × twin 100mm (3.9in) guns; 4 × twin 37mm (1.5in) AA guns; 2 × HQ-61 twin SAM launchers; 2 × BMB-2 depth charge projectors; 2 × depth charge racks; 2 × RBU-1200 ASW rocket launchers
Complement:	185

In service by 1974, the Type 053 is a prolific class of Chinese frigates, with 35 vessels completed for the Chinese People's Liberation Army Navy Surface Force and a further four built for the Royal Thai Navy.

Right: Here we see the Type 053H1G vessel Beihai *(558), which until 2010 was called the* Zigong*. It was completed and commissioned in 1993.*

Sensors and processors
The ship included Type 381 air search and Square Tie surface search radars, plus the Sun Visor 8, Fog Lamp and Type 341 fire-control systems.

SAM system
The *Yingtan* was equipped with the HQ-61 SAM system fore and aft. It was a semi-active radar homing type with a range of around 10km (6 miles).

***Yingtan* (531)**
The *Yingtan* was the first ship in the Type 053 class. It was in service for 20 years and saw some combat in 1988 in clashes between Chinese and Vietnamese warships in the Union Banks region in the South China Sea.

TYPE 053 FAMILY
The Type 053 class is not so much a single specification of ships, but a highly varied family of sub-classes, commissioned over a period of 20 years and frequently upgraded and refitted. The original sub-class was the Type 053K (*Jiangdong*), to which belonged the *Yingtan*, primarily tasked with air-defence roles. This was followed by the most extensive sub-class, the Type 053H, with 14 vessels, given an improved surface-warfare capability. There were four further sub-classes, terminating in the Type 053H1G.

TYPE 053H
The key difference of the Type 053H from the Type 053K was the fitting of four SY-1 anti-ship missiles in two twin-missile box launchers (here visible amidships), plus other armament changes.

USS *Virginia* (1974)

TYPE • *Missile cruiser* **COUNTRY** • *United States*

SPECIFICATIONS

DIMENSIONS:	Length: 178m (585ft); Beam: 19m (63ft); Draught: 9.4m (31ft)
DISPLACEMENT:	11,176 tonnes (12,325 tons) standard; 13,106 tonnes (14,450 tons) full load
PROPULSION:	2 General Electric D2G nuclear reactors; geared steam turbines; 44,742kW (60,000shp); 2 screws
SPEED:	30 knots (56km/h; 35mph)
RANGE:	Unlimited
ARMAMENT:	2 × Mk 26 launchers for Standard SAM and ASROC; 2 × Mk 141 Harpoon launchers; 2 × Armoured Box Launchers for Tomahawk missiles; 2 × triple 324mm (12.75in) torpedo tube mounts for Mk 46 torpedoes; 2 × 127mm (5in) guns; 2 × 20mm (0.79in) Phalanx CIWS; 4 × machine guns
COMPLEMENT:	500

Bristling with anti-ship and surface-to-air missiles, USS *Virginia* (CGN-38) was the lead ship in a class of four nuclear-powered guided-missile cruisers. Originally planned as part of the *Spruance* class, they evolved into a class of their own.

Left: USS Virginia *presents an impressive sight underway during the NATO Exercise* Distant Drum *in May 1983.* Virginia *undertook combat deployments around the Lebanon in 1983 and in the Persian Gulf during Operation* Desert Storm *in 1991.*

TOWED DECOYS
The class carried the AN/SLQ-25 Nixie towed torpedo decoy system, which had a TB-14A decoy device to draw guided torpedoes away from the ship.

GUN ARMAMENT
Despite its extensive missile weaponry, the class had two Mk 45 127mm (5in)/54 caliber rapid-fire guns, used against surface warships, aircraft and shore targets.

POWERPLANT
The nuclear powerplant consisted of two General Electric D2G nuclear reactors, driving two shafts with 45,000kW (60,000shp).

HARPOON MISSILE

A key element of the *Virginia*'s armament was the Harpoon all-weather, over-the-horizon, anti-ship missile one of the most successful missile types of the postwar era. The sea-skimming cruise missile flies at Mach 071 and is guided to its target by active radar terminal homing. In its original Block 1 version, it had a range of more than 124km (77 miles) and carried a devastating 221kg (448lb) warhead.

USS *ARKANSAS*
The USS *Arkansas* (CGN-41) was the last of the four ships in the *Virginia* class. In this dramatic photograph, an explosive charge is detonated off the ship as part of a hull shock test on 17 March 1982.

F67 Type *Tourville* class (1974)

TYPE • *Destroyer* **COUNTRY** • *France*

SPECIFICATIONS

Dimensions:	Length: 152.8m (501ft 4in); Beam: 15.3m (50ft 2in); Draught: 5.7m (18ft 8in)
Displacement:	4580 tonnes (5050 tons) standard; 5745 tonnes (6334 tons) full load
Propulsion:	4 water-tube boilers; 2 steam turbines; 2 shafts; 40,566kW (54,400shp)
Speed:	32 knots (59km/h; 37mph)
Range:	5000nmi (9300km; 5800 miles) at 18 knots
Armament:	1 × Malafon ASW missile launcher (13 missiles); 6 × MM38 Exocet anti-ship missiles; 2 × L5 ASW torpedo launchers; 2–3 × 100mm (3.9in) guns; 2 × 20mm (0.79in) autocannon; 4 × 12.7mm (0.5in) machine guns
Complement:	282

The three *Tourville*-class guided-missile destroyers were the first French warships of destroyer size purpose-built to operate two anti-submarine helicopters. Their missile armament gave them over-the-horizon attack capabilities.

Right: D610 Tourville *was the lead ship of the class. The double hangar seen here at the stern accommodated a pair of Lynx ASW helicopters.*

De Grasse
The third and final ship in the class, *De Grasse* (D612) was launched in November 1974 and commissioned in October 1977.

Crotale launcher
De Grasse had the aft gun replaced by a Crotale short-range all-weather SAM system, designed to intercept a range of incoming aerial threats.

Performance
The ships of the class had a maximum speed of 32 knots (59km/h; 37mph) at 18 knots (33km/h; 21mph) and a range of 5000nmi (9300km; 5800 miles).

RATIONALE

The development of the *Tourville* class, and similar guided-missile surface warships, was as a French counter to the expanding threat from Soviet nuclear-powered and ballistic missile submarines operating in the North Atlantic, the North Sea and the Norwegian Sea. The ships also helped free France from what was seen as excessive reliance upon NATO forces; France withdrew from NATO's integrated military command structure in 1966.

Duguay-Trouin
Duguay-Trouin (D611) was the second of the *Tourville* class, commissioned on 17 September 1975. It was scrapped in 2020, 10 years after its decommissioning. From 1999 it served as a breakwater at the École Navale in Lanvéoc-Poulmic.

Spruance class (1975)

TYPE • *Destroyer* **COUNTRY** • *United States*

SPECIFICATIONS

DIMENSIONS:	Length: 172m (563ft); Beam: 17m (55ft); Draught: 8.8m (29ft)
DISPLACEMENT:	8170 tonnes (9010 tons) full load
PROPULSION:	4 General Electric LM2500 gas turbines; 60,000kW (80,000shp); 2 shafts
SPEED:	32.5 knots (60.2km/h; 37.4mph)
RANGE:	6000nmi (11,000km; 6900 miles) at 20 knots
ARMAMENT:	2 × 127mm (5in) guns; 2 × Phalanx CIWS; 1 × 8-cell Sea Sparrow Mk 29 launcher; 2 × quad Harpoon missile canisters; 2 × triple 324mm (12.75in) torpedo tubes; 1 × Mk 41 VLS launcher for Tomahawk missiles
COMPLEMENT:	334

Larger than many former cruisers, the *Spruance* class were designed to offer a stable weapon-launcher capable of operating in difficult sea conditions. Its successful hull design was used, with modifications, on two other classes of US warship.

Left: This official US Navy photograph shows the USS Spruance *(DD-963) in February 1975, during its builder's trials in the Gulf of Mexico.*

USS *SPRUANCE* (DD-963)
Spruance was commissioned on 20 September 1975. It was the first gas-turbine powered ship in the US fleet.

FLIGHT DECK
The *Spruance*-class ships were bigger than previous US destroyers, enabling them to mount a helicopter flight deck. An internal hangar held up to two medium-lift helicopters.

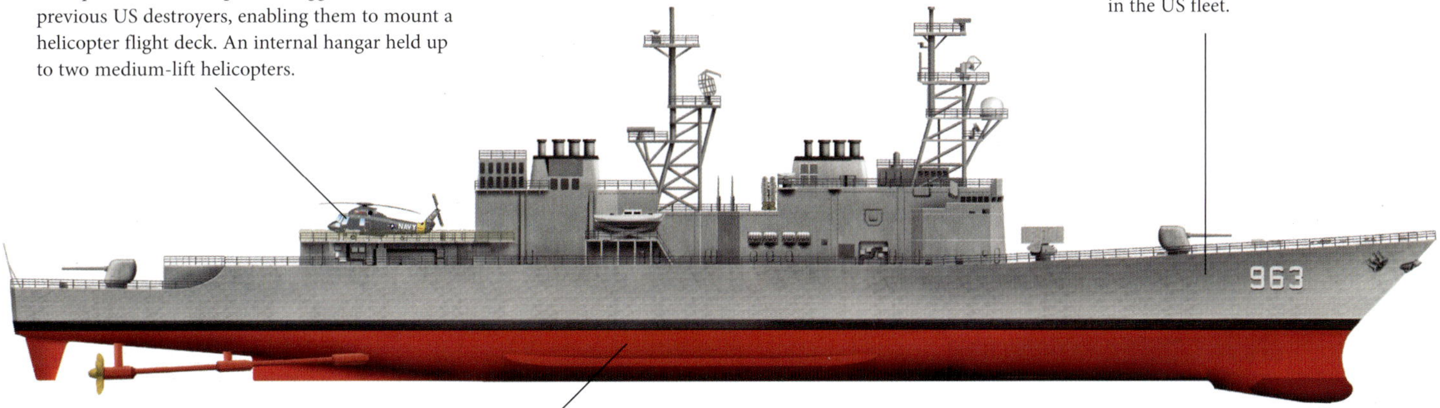

PROPULSION
The ship was powered by four General Electric LM2500 gas turbines, generating 60,000kW (80,000shp delivered to two shafts.

BEST IN CLASS

For nearly three decades, the *Spruance* class were one of the most formidable surface warship types on the planet. Thirty-one vessels were completed in total, and although there was initial controversy about the class specifications, regular upgrades and refits kept their capabilities at the cutting edge. The fitting of Tomahawk cruise missile systems in the 1980s gave the ships strategic influence in regional theatres.

USS *ARTHUR W. RADFORD*
USS *Arthur W. Radford* (DD-968) photographed in the Mediterranean Sea in late November 2002, equipped with the eye-catching experimental Advanced Enclosed Mast/Sensor, designed to reduce the ship's radar cross-section.

Kiev (1975)

SPECIFICATIONS

DIMENSIONS:	Length: 273.1m (896ft); Beam: 31m (102ft); Draught: 8.95m (29ft 5in)
DISPLACEMENT:	31,018 tonnes (34,196 tons); 42,032 tonnes (46,332 tons) full load
PROPULSION:	4 geared steam turbines; 4 shafts; 100,000kW (140,000shp)
SPEED:	32 knots (59km/h; 37mph)
RANGE:	13,500nmi (25,000km; 15,500 miles) at 18 knots
ARMAMENT:	2 × 12-barrel RBU-6000 ASW launchers; 1 × twin SUW-N-1 launcher for FRAS-1 missiles; 2 × twin SA-N-3 and 2 × twin SA-N-4 SAM launchers; 4 × twin SS-N-12 SSM launchers; 2 × twin 76mm cannon; 8 × ADMG-630 30mm six-barrel Gatling CIWS; 10 × 533mm (21in) torpedo tubes
AIRCRAFT:	32
COMPLEMENT:	2042

Described as a 'heavy aviation cruiser', *Kiev* was the Soviet Navy's answer to the requirement of multi-role warships in the 1970s, and Russia's first carrier for fixed-wing aircraft. The Russian Navy had introduced helicopter carriers with the *Moskva*-class in the 1960s.

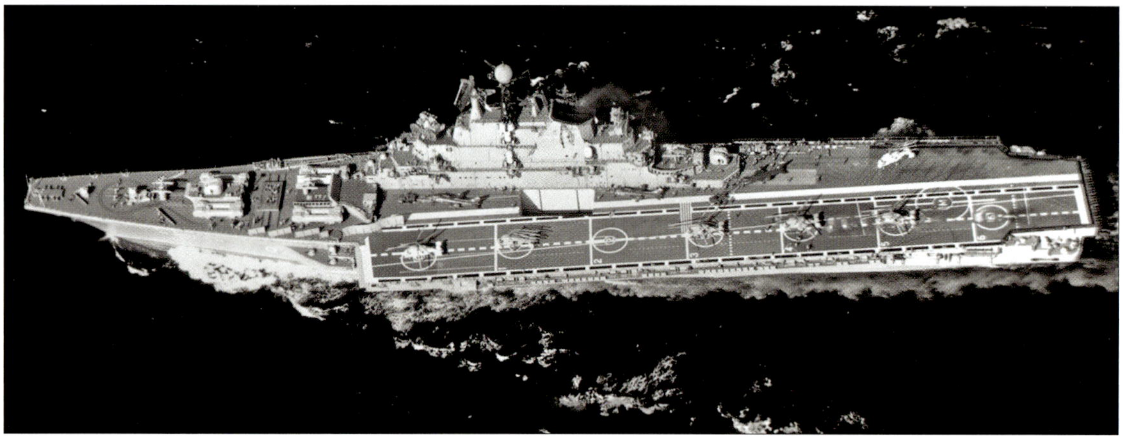

Above: The Minsk*, seen here in an impressive aerial photograph, was the second ship of the* Kiev *class.*

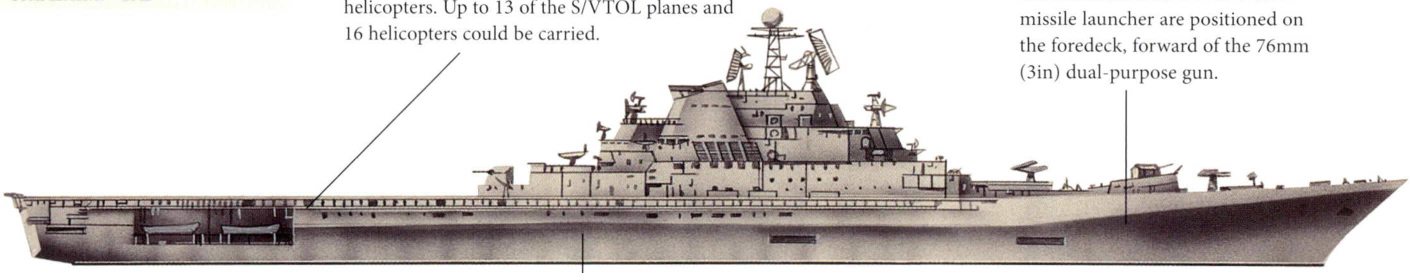

AIR WING
The air wing combined Yakovlev Yak-38 Forger A attack planes and Forger B trainers, with Ka-25 Hormone A and B helicopters and Ka-27 helicopters. Up to 13 of the S/VTOL planes and 16 helicopters could be carried.

ANTI-SUBMARINE MEASURES
Two RBU-6000 ASW rocket launchers and an SUW-N-1 ASW missile launcher are positioned on the foredeck, forward of the 76mm (3in) dual-purpose gun.

HEAT-RESISTANT TILES
Special tiles were fitted on the flight deck to absorb the heat of the Yak-38 Forger A's vertical-lift jet engines.

DEPLOYMENTS

Kiev and its companion ships were widely deployed; these were the first ships capable of giving air fighter cover to Soviet naval craft, with two assigned to the Northern and two to the Pacific fleet. *Kiev's* designation as a 'cruiser with aircraft' enabled it to pass through the Bosphorus (forbidden to carriers of any nation) despite protests, and conduct trials in the Mediterranean before making for its home port at Severomorsk.

BAKU
Baku was a *Kiev*-class aircraft carrier commissioned in December 1987. Here the ship is seen underway off the coast of southern Italy in 1988. *Kiev* and its companion ships were widely deployed, as these were the first ships capable of giving air fighter cover to Soviet naval craft.

USS *Nimitz* (1975)

TYPE • *Aircraft carrier* **COUNTRY** • *United States*

This was the lead ship of a class of 10 nuclear-powered supercarriers, considered the star of US naval strength, both in deterrence and attack. They were described as 'floating airfields' or 'movable pieces of sovereign territory'.

SPECIFICATIONS

Dimensions:	Length: 317m (1040ft); Beam: 40.8m (134ft); Draught: 11.3m (37ft)
Displacement:	110,250 tonnes (121,500 tons)
Propulsion:	2 Westinghouse A4W nuclear reactors; 4 steam turbines; 4 shafts
Speed:	31.5 knots (58.3km/h; 36.2mph)
Range:	Unlimited
Armament:	2 × Sea Sparrow launchers; 2 × RIM-116 missile launchers; 2 × Phalanx CIWS; 2 × .50-cal machine gun mountings
Aircraft:	90
Complement:	5680

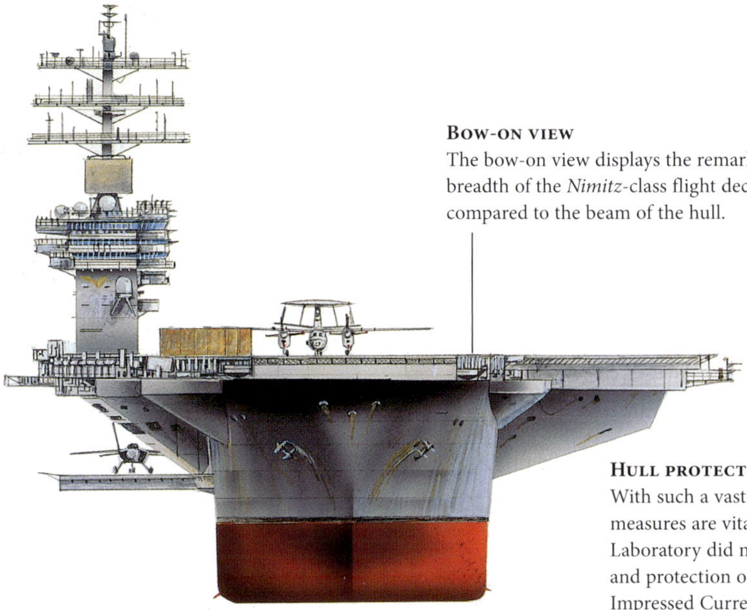

Bow-on view
The bow-on view displays the remarkable breadth of the *Nimitz*-class flight deck compared to the beam of the hull.

Hull protection
With such a vast hull, anti-corrosion measures are vital. The US Navy Research Laboratory did much work on the design and protection of CVN hulls, using Impressed Current Cathode Protection rather than zinc sacrificial anodes, to keep the problem to a minimum.

Armament
Installed armament is two Sea RAM missile launchers and two Mk 29 Sea Sparrow surface-to-air missile launchers.

Flight deck
Overall deck length is 332.8m (1092ft), and width is 76.7m (251ft 10in). It is covered with an anti-slip coating.

IRAQ INVASION

In March 2003, *Nimitz* supported the US–British invasion of Iraq in Operation *Iraqi Freedom* as flagship of a Carrier Battle Group. *Nimitz* launched more than 6500 sorties against targets in Iraq during this six-month mission, in which it launched the F/A-18F Super Hornet and E-2C Hawkeye 2000 aircraft on their first combat flights.

AIR WING

USS *Nimitz* (CVN 68) and its embarked Carrier Air Wing (CVW) 11 transit into San Diego prior to mooring at Naval Air Station North Island. *Nimitz*'s air wing comprises up to 90 planes. In 1991 it carried 24 Hornets, 24 Tomcats, six Vikings, four Prowlers, four Hawkeyes and 15 Intruders.

USS *Tarawa* (1976)

TYPE • *Assault ship* **COUNTRY** • *United States*

SPECIFICATIONS

Dimensions:	Length: 250m (820ft); Beam: 32m (106ft); Draught: 7.9m (26ft)
Displacement:	40,673 tonnes (44,845 tons) full load
Propulsion:	2 boilers; 2 geared steam turbines; 2 shafts; 57,419kW (77,000shp)
Speed:	24 knots (44.4km/h; 27.6mph)
Range:	10,000nmi (19,000km; 12,000 miles) at 20 knots
Armament:	2 × 8-cell Mk 25 Sea Sparrow BPDMS launchers; 1 × Mk 49 RAM missile system; 2 × Vulcan Phalanx CIWS mountings; 6 × 25mm automatic cannon
Aircraft:	43
Complement:	960

Ships in this class were designed to transport, land and support US Marine troops on any coastline worldwide, serving as the core of an Amphibious Readiness Group. *Tarawa* (LHA-1) was the first of these general-purpose amphibious assault ships.

Right: The amphibious assault ship USS Tarawa *(LHA 1) transits the Indian Ocean while deployed in support of maritime sercurity operations, 22 December 2007.*

Ballast pumps
Ballast pumps could fill and empty the tanks with 12,192 tonnes (13,228 tons) of sea water when lowering and raising the dock.

Design
Tarawa was constructed on carrier lines, with a full-length flight deck fitted with two deck-edge aircraft lifts, and a long island superstructure set to starboard.

Air wing
Tarawa carried up to eight AV-8B Harrier II V/STOL planes in addition to helicopters, on missions to the Persian Gulf and other world trouble spots.

TARAWA CLASS

Although nine ships were envisaged, only five were actually built. Laid down on 15 November 1972 at Ingalls Shipbuilding Corp., Pascagoula, Mississippi, *Tarawa* was launched on 1 December 1973 and commissioned at the shipyard on 29 May 1976. In terms of function it was identical to the British HMS *Fearless*; in terms of size it was a substantially bigger ship that was capable of landing a larger and more fully equipped force in brigade-size operations.

SETTING SAIL
On 16 July 2005, sailors and Marines aboard the USS *Tarawa* man the rails as the ship leaves San Diego, beginning a scheduled deployment in support of the Global War on Terrorism (GTOW).

Type 22 *Broadsword* class (1976)

TYPE • *Frigate* **COUNTRY** • *United Kingdom*

SPECIFICATIONS

Dimensions:	Length: 131.2m (430ft); Beam: 14.8m (48ft); Draught: 6.1m (20ft)
Displacement:	4500 tonnes (4960 tons) full load
Propulsion:	COGOG; 2 Rolls-Royce Olympus TM3B gas turbines, 40,000kW (54,000shp); 2 Rolls-Royce Tyne RM1C cruise gas turbines, 7200kW (9700shp); 2 shafts
Speed:	30 knots (56km/h; 34.8mph)
Range:	4500nmi (8890km; 5524 miles) at 18 knots
Armament:	2 × Sea Wolf SAM launchers; 2 × quad Harpoon missile launchers; 2 × triple 324mm (12.75in) ASW torpedo tubes; 1 × 114mm (4.5in) Mk 8 gun; 2 × 20mm (0.78in) GAM-B01 guns; 1 × Goalkeeper CIWS
Complement:	222

Broadsword **was the first general-purpose frigate designed to follow the** ***Leander*** **class. It was planned to build 26 units armed with missiles only, the main anti-submarine weapon being the** ***Lynx*** **helicopter.**

Left: A starboard side view of HMS Broadsword *(F88) tied up in port. In November 1994 the ship and three others in the class were sold to the Brazilian Navy; it entered Brazilian service on 30 June 1995 as the* Greenhalgh.

Exocet launchers
The four Exocet missile launchers on the foredeck were replaced in the Batch 3 ships by the classic 114mm (4.5in) Mk 8 gun.

Seawolf
At the rear of the ship are 2 × 6 GWS25 Seawolf SAM launchers, which provide the ships with a point-defence system against air attacks.

FALKLANDS SERVICE
Two Type 22 frigates served in the Falklands War in 1982: HMS *Brilliant* and HMS *Broadsword*. During their time in theatre, both ships came under heavy Argentine attack. *Brilliant* was slightly damaged by aerial cannon fire, while *Broadsword* was hit by a bomb that passed through the ship, disabling a Lynx helicopter on the way, but thankfully exiting the vessel before exploding. Between them the two vessels shot down five Argentine aircraft (*Broadsword* shared one with HMS *Antelope*).

Hull
The Type 22 design went through three evolutions, denoted by Batch numbers. In each evolution, the hull length was increased, with Batch 3 at 148.1m (486ft).

HMS *BROADSWORD* (F88)
In total, some 14 Type 22 frigates were produced for Royal Navy service between 1979 and 1990. They were initially intended to be anti-submarine vessels, but over the course of their development and deployment they performed many other roles, from providing air-defence screens to counterpiracy operations.

Kortenaer (1976)

TYPE • *Frigate* **COUNTRY** • *Netherlands*

SPECIFICATIONS

Dimensions:	Length: 130m (426ft 6in); Beam: 14.4m (47ft 3in); Draught: 4.4m (14ft 5in)
Displacement:	3538 tonnes (3900 tons) standard
Propulsion:	2 Rolls-Royce Tyne RM1C gas turbines, 3700kW (4900shp) each; or 2 Rolls-Royce Olympus TM3B gas turbines, 19,200kW (25,700shp) each; 2 shafts
Speed:	30 knots (56km/h; 35mph)
Range:	4700nmi (8704km; 5410 miles) at 16 knots
Armament:	1 × Goalkeeper CIWS; 1 × 76mm (3in) gun; 2 × twin Mk 46 ASW torpedo tubes; 2 × RGM-84 Harpoon anti-ship missile launchers; 1 × 8-cell Sea Sparrow SAM launcher
Complement:	176–196

During the Cold War, several European nations, particularly the Netherlands and the Federal Republic of Germany, collaborated on frigate design and production. Together, they built vessels like the frigate *Kortenaer* to meet shared operational requirements.

Left: An aerial photograph of Kortenaer *(F 807) underway. The* Kortenaer*-class frigates were highly successful vessels, and a powerful asset in NATO's anti-submarine warfare arsenal. Their turbines gave them the level of manoeuvrability and acceleration that are vital when hunting fast nuclear-powered submarines.*

Aircraft
As designed, the *Kortenaer*-class ships could carry two Westland Sea Lynx ASW helicopters, but from 1986 the landing pad was replaced by a Standard SAM missile launcher.

Missiles
The *Kortenaer* class originally carried two quadruple GMLS Mk 141 Harpoon surface-to-surface missile launchers with eight missiles, and one octuple launcher for the GMLS Mk 29 Sea Sparrow surface-to-air missile. The later *Kortenaers* were fitted with the Standard SAM system.

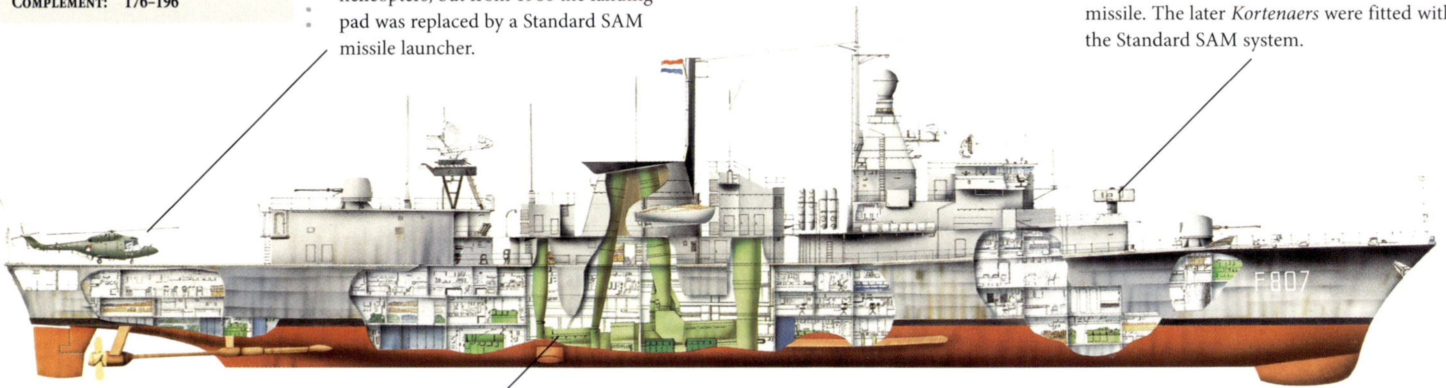

Hull
The hull followed the lines of contemporary French designs, with a continuous upper and main deck, a single rudder of the semi-balanced type, and a clipper bow with negative sheer.

INTERNATIONAL SALES

Ten of the *Kortenaer*-class vessels were built by de Schelde in Flushing and two by Wilten-Fijenoord in Schiedam. Of the 12, 10 served with the Royal Netherlands Navy. Two were sold to Greece while still under construction and replaced by two *Jacob van Heemskerck*-class frigates, which are basically an anti-aircraft variant of the *Kortenaer* class. In Greek service, the frigates were renamed the *Elli* class. Eight similar ships were built by Germany during the same period as the *Bremen* class. By 2003 all ships had been transferred to the navies of Greece and the UAE.

***ABRAHAM CRINJNSSEN* (F 816)**
This *Kortenaer*-class frigate is the *Abraham Crijnssen*, named after a 17th-century admiral. The original ship of that name became famous for its escape from Surabaya in World War II disguised as a tropical island.

USS *Los Angeles* (1976)

TYPE • *Submarine* **COUNTRY** • *United States*

SPECIFICATIONS

Dimensions:	Length: 110.3m (362ft); Beam: 10m (32ft 10in); Draught: 9.9m (32ft 6in)
Displacement:	6180 tonnes (6815 tons) surfaced; 7038 tonnes (7760 tons) submerged
Propulsion:	1 GE S6G pressurized water nuclear reactor; steam turbines delivering 26,000kW (35,000hp); auxiliary Magnetek motor; single shaft
Speed:	20 knots (37km/h; 23mph) surfaced; 33 knots (61.1km/h; 37.9mph) submerged
Range:	Unlimited
Armament:	4 × 533mm (21in) torpedo tubes; Tomahawk land-attack cruise missiles; Harpoon anti-ship missiles
Complement:	133

USS *Los Angeles* (SSN-688) was the lead ship of the *Los Angeles* class. Designated as fast attack submarines and escorts for carrier battle groups, they have proven to be highly versatile, undertaking a variety of combat, escort and intelligence-gathering missions over the years.

Above: In the 1990s, construction cost of a Los Angeles*-class submarine was $900,000,000. Extension of its service life by 12 years is estimated to cost $200,000,000.*

Hull construction
The hull is formed from HY-80 high-tensile steel 76mm (3in) thick, with a test depth of 290m (950ft).

Air and water processing
Atmosphere control equipment replenishes oxygen used by the crew, and removes carbon dioxide and other atmosphere contaminants. Two distilling plants convert salt water to fresh for drinking, washing and use in the propulsion plant.

Sail
In the forward top portion of the sail is the bridge. When the submarine is on the surface, the Officer of the Deck shifts his watch from the control room to here.

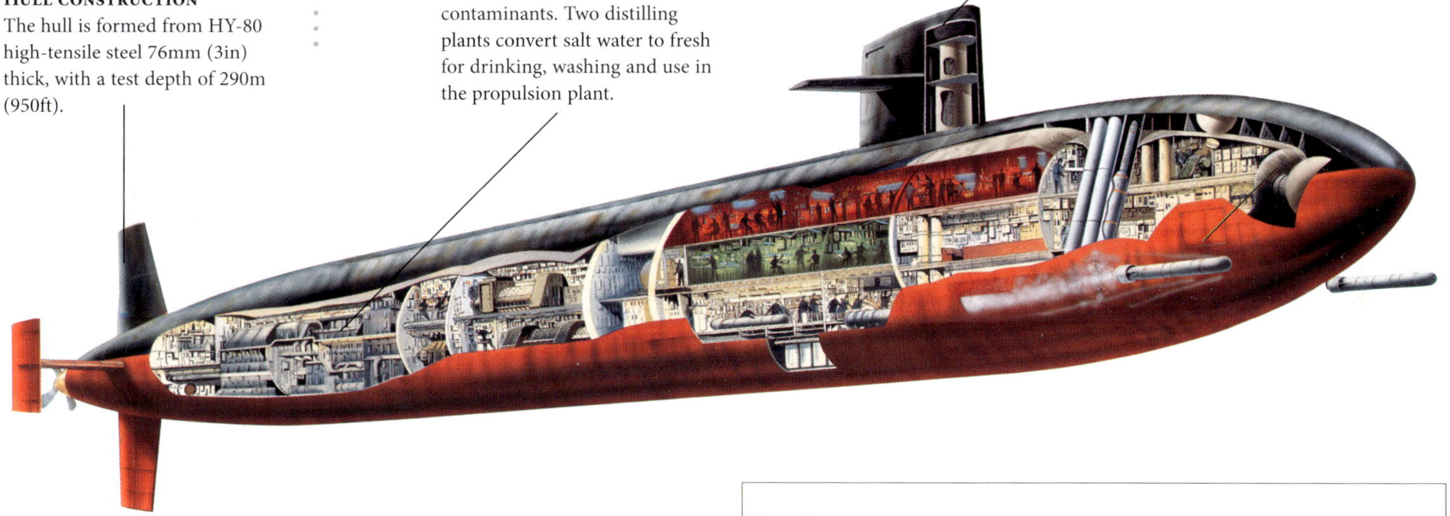

CLASS DEVELOPMENT

Of the 62 *Los Angeles* class, SSNs 688–718 were built to the original design. From SSN-719, commissioned in July 1985, 12 vertical launch tubes for Tomahawk cruise missiles were fitted. The last 23, SSN 751 to 773, which were commissioned between 1988 and 1996, are configured for under-ice operations, with forward diving planes moved from the sail to the bow, and the sail strengthened for ice breaking.

PACIFIC DEPLOYMENT
USS *Los Angeles* spreads its wake as it heads out to sea with the coast of Oahu, Hawaii, in the background. The nose cone is made of composite materials that allow external sounds to penetrate to the sonar sphere.

Kirov (1977)

TYPE • *Missile cruiser* **COUNTRY** • *Soviet Union*

SPECIFICATIONS

DIMENSIONS:	Length: 252m (827ft); Beam: 28.5m (94ft); Draught: 9.1m (30ft)
DISPLACEMENT:	25,401 tonnes (28,000 tons) full load
PROPULSION:	CONAS; nuclear-powered plant with steam-turbine boost; 2 shafts; 104,400kW (140,000shp)
SPEED:	32 knots (59.3km/h; 36.8mph)
RANGE:	1000nmi (1852km; 1151 miles) at 30 knots; essentially unlimited at 20 knots
ARMAMENT:	20 × SS-N-19 Shipwreck anti-ship cruise missiles; 14 × SS-N-14 anti-submarine cruise missiles; 10 × 533mm (21in) torpedo tubes; rocket launchers and naval guns
ARMOUR:	76mm (3in) plating around reactor compartment
COMPLEMENT:	727

Launched on 26 December 1977 and commissioned on 30 December 1980, *Kirov* was originally designed as a large anti-submarine warship to search for and engage enemy ballistic missile submarines.

Above: One of the prominent features of the Kirov-class warships was the sharply raked bow, a feature of all Soviet warships designed in the latter years of the Cold War.

SENSORS
Kirov carried a formidable array of sensors, including air/surface search radar, fire-control radar, and hull-mounted sonar. The Voskhod MR800 Top Pair search radar was mounted on the foremast.

AVIATION
The cruiser was equipped to carry three Kamov Ka-25 Hormone or Ka-27 Helix ASW helicopters. On board were 18 aircrew, two crews being assigned to each helicopter.

SURPRISES
In the last three decades of the Cold War, Soviet naval architects produced many surprises. In December 1977, the Soviet navy – which by then had expanded its operations worldwide – launched the nuclear-powered *Kirov*, the largest warship built by any nation since World War II apart from aircraft carriers. The vessel was more akin, in terms of appearance and firepower, to the obsolete battlecruiser category. Three more vessels of the Kirov class were completed: the *Frunze*, *Kalinin* and *Yuri Andropov*.

UNDERWAY
After the introduction of the Granit anti-ship missile system, the role of the Kirov role was expanded to engage large surface targets and to provide air and anti-submarine protection to naval forces.

Sovremenny (1977)

TYPE • *Destroyer* COUNTRY • *Soviet Union*

SPECIFICATIONS

DIMENSIONS:	Length: 156m (512ft); Beam: 17.3m (56ft 10in); Draught: 6.5m (21ft 4in)
DISPLACEMENT:	7203 tonnes (7940 tons)
PROPULSION:	2 steam turbines; 2 shafts; 74,600kW (100,000hp)
SPEED:	32 knots (59.3km/h; 36.8mph)
RANGE:	3406nmi (6307km; 3920 miles) at 18 knots
ARMAMENT:	2 × quad SS-N-22 Sunburn (Moskit) SSM launchers; 2 × single SA-N-7 Gadfly SAM launchers; 4 × 130mm (5.1in) guns; 4 × 30mm (1.18in) AK-630 Gatling CIWS; 4 × 533mm (21in) torpedo tubes; 2 × RBU-1000 ASW rocket launchers
COMPLEMENT:	350

The name *Sovremenny* translates from Russian as 'modern'. The designation was ideally suited to this formidable class of warship, its weapon systems upgraded to counter the American *Spruance*-class multi-role destroyers.

Left: A starboard quarter view of the Russian Sovremenny-*class guided missile destroyer* Bezuprechny *underway.*

COMBAT SYSTEMS
The ship's combat systems can use target-designation data from the ship's active and passive sensors, from other ships in the fleet, from surveillance aircraft, or via a communications link from the ship's helicopter.

HELICOPTER
Sovremenny-class ships are fitted with a helicopter pad for one Kamov Ka-25 ASW helicopter, which can operate up to 200km (125 miles) from the parent vessel.

ANTI-SUBMARINE WEAPON
The destroyer has two double 533mm (21in) torpedo tubes and two six-barrel RBU-1000 anti-submarine rocket launchers, with 48 rockets. The rocket is armed with a 55kg (122lb) warhead and has a range of 1000m (3280ft).

FIRST OUTINGS

The *Sovremenny*-class destroyers made their first appearance in strength in March 1984, during a large-scale exercise in the north Norwegian Sea, with about half the Soviet Northern Fleet's major surface combatants and a high proportion of the submarine order of battle participating. Two aggressor groups, simulating NATO forces, carried out an incursion into northern waters, one group deploying from the Northern Fleet and one from the Baltic. Defending forces included vessels of the *Sovremenny* class.

***OTCHAYANNY* 1986**
The *Otchayanny* was a *Sovremenny*-class vessel launched in 1980 and commissioned in 1982. The class were part of a drive by the Soviet navy to deploy a true blue-water fleet capable of sustaining combat operations anywhere in the world with powerful Surface Action Groups.

Project 705 *Lira* (*Alfa* class) (1977)

TYPE • *Submarine* **COUNTRY** • *Soviet Union*

In 1971, the Soviet Union completed the world's first titanium-hulled submarine, which also became the fastest. Revolutionary in almost every way, including a novel compact nuclear reactor, it was the precursor to a bold new class of attack submarines.

SPECIFICATIONS

Dimensions:	Length: 81m (265ft 9in); Beam: 9.5m (31ft); Draught: 8m (26ft 3in)
Displacement:	2845 tonnes (3137 tons) surfaced; 3739 tonnes (4121 tons) submerged
Propulsion:	1 BM-40A liquid-metal-cooled nuclear reactor; steam turbines; single screw
Speed:	20 knots (37km/h; 23mph) surfaced; 42 knots+ (77.7km/h+; 48.3mph+) submerged
Range:	Unlimited
Armament:	6 × 533mm (21in) torpedo tubes; up to 18 torpedoes or 21 × RPK-2 Vyuga missiles or up to 36 mines
Complement:	31

Above: Since the prime aim of Project 705 was to achieve high speed, other considerations were adapted to make this possible. A smaller, more streamlined hull was designed with a profile of unique form, the sail blending smoothly into the hull.

Armament
In addition to torpedoes, the submarine could carry the RPK-2 Vyuga (Blizzard) ASW missile. This was a ballistic missile capable of carrying a 200kt nuclear warhead, similar to the US Navy's SUBROC. Its range was 45km (28 miles).

Titanium hull
This material gave a 30 per cent lower mass, a 25 per cent lesser displacement and a 10 per cent increase in speed compared to steel plate.

Automated procedures
Only two central compartments, for control and living space, were manned. In the other four compartments procedures were automated. The original proposals envisaged a crew of only 18.

DESIGN

The design team for Project 705 at the Malakhit Design Bureau (TsKB-143) was led by Mikhail G. Rusanov. It was a double-hull design, with the inner hull able to withstand the enormous pressures that would be faced by extreme operating depths, and a lighter outer hull optimized for speed, manoeuvrability and acoustic stealth. Numbered K-162 (K-222 from 1978), codenamed *Papa* by NATO, it was the prototype of Project 705, the *Lira*-('Lyre'-) class submarine, commenced in 1974 and known to NATO as *Alfa*.

PORT VIEW
A port view of a Soviet *Alfa*-class nuclear-powered fleet ballistic missile submarine (SSN) underway. By 1985, the Soviet Navy had at least six operational *Lira*-class boats.

Oliver Hazard Perry class (1977)

TYPE • *Frigate* **COUNTRY** • *United States*

SPECIFICATIONS

Dimensions:	Length: 136m (445ft); 138m (453ft) long-hull variants; Beam: 14m (45ft); Draught: 6.7m (22ft)
Displacement:	4200 tonnes (4630 tons) full load
Propulsion:	2 General Electric LM2500-30 gas turbines, 31,000kW (41,000shp); 1 shaft; 2 retractable electric azimuth thrusters, 260kW (350hp) each
Speed:	30 knots (56km/h; 35mph)
Range:	4500nmi (8300km; 5200 miles) at 20 knots
Armament:	1 × Mk 13 single-arm missile launcher; Mk 38 Mod 2 machine gun systems fitted after Mk 13 removal; 2 × triple Mk 32 ASW torpedo tubes for Mk 46 or Mk 50 torpedoes (up to 24 torpedoes); 1 × OTO Melara 76mm (3in) naval gun; 1 × 20mm (0.79in) Phalanx CIWS
Complement:	176

These general-purpose frigates were frequently assigned to the outer edges of Carrier Strike Groups, offering anti-submarine and anti-aircraft capabilities. The US Navy took 51 of these ships and they were also built for or transferred to many international navies.

Left: The USS Kauffman *(FFG-59) was commissioned on 28 February 1987 and served until 2015. It made numerous operational deployments, including to the Persian Gulf during the 1990–91 Gulf War.*

Helicopters
The short hull vessels carried two SH-2F Seasprite (LAMPS I) helicopters (retired in 1993), while the long hull ships had two SH-60 Seahawk (LAMPS III) helicopters.

Missile launcher
The Mark 13 guided missile launching system (GMLS) is capable of firing RIM-66 Standard, RGM-84 Harpoon, and RIM-24 Tartar missiles, fed from a 40-missile magazine.

Hull length
The *Oliver Hazard Perry* frigates were produced in both short-hull and long-hull variants. Both types of vessel were capable of carrying two helicopters.

CARRIER STRIKE GROUP

The US Carrier Strike Group of the 1990s was typically centred upon a *Nimitz*-class carrier, operating with an assigned nuclear submarine of the *Los Angeles* class. The surface escort included two guided missile cruisers of the *Ticonderoga* class, with two destroyers of the *Arleigh Burke* or *Spruance* classes, armed primarily with SAMs, and two frigates of the *Oliver Hazard Perry* class, armed primarily for anti-submarine warfare, flanking the group.

USS *SIMPSON* FFG-56
The USS *Simpson* was the last of the *Oliver Hazard Perry*-class frigates to be retired from service. It was decommissioned on 29 September 2015 after a long and eventful career.

Ivan Rogov (1978)

TYPE • *Assault ship* **COUNTRY** • *Soviet Union*

SPECIFICATIONS

DIMENSIONS:	Length: 158m (518ft 3in); Beam: 24m (78ft 9in); Draught: 8.2m (26ft 11in)
DISPLACEMENT:	11,176 tonnes (12,318 tons) standard; 13,208 tonnes (14,559 tons) full load
PROPULSION:	2 gas turbines; 2 shafts; 37,175kW (50,000shp)
SPEED:	23 knots (42.5km/h; 26.5mph)
RANGE:	12,500nmi (23,125km; 14,375 miles) at 14 knots
ARMAMENT:	1 × twin 76mm (3in)/59-cal gun; 4 × 30mm AK-630 CIWS; 1 × SA-N-4 missile launcher; 1 × 40-tube 122mm (3in) rocket launcher
AIRCRAFT:	4 helicopters
COMPLEMENT:	750

The Soviet Navy's version of the amphibious assault ship in the 1970s was the four-strong *Ivan Rogov*-class, capable of landing a naval infantry battalion and its equipment. The ship's design made it adaptable either to direct beach assaults with land vehicles or to stand-off operations using helicopters and missiles.

Above: The capacities of the Ivan Rogov*-class substantially improved the Soviet Union's capacity to engage in amphibious warfare.*

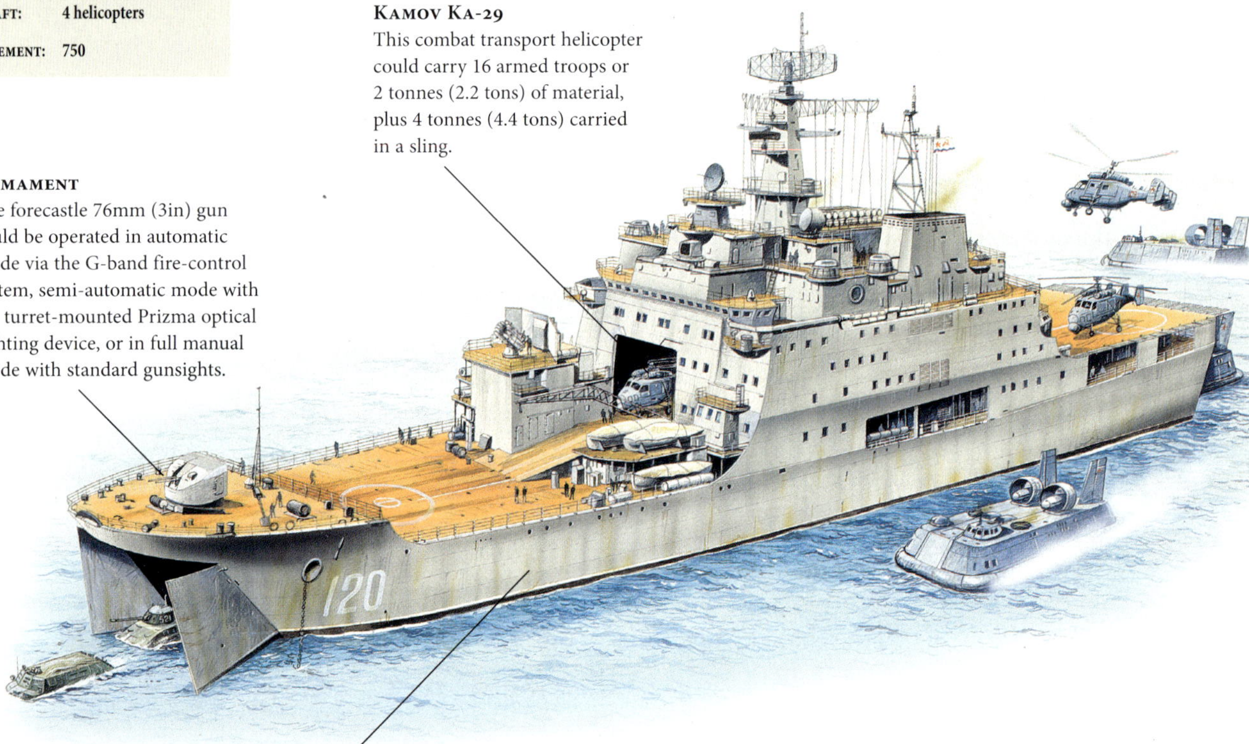

KAMOV KA-29
This combat transport helicopter could carry 16 armed troops or 2 tonnes (2.2 tons) of material, plus 4 tonnes (4.4 tons) carried in a sling.

ARMAMENT
The forecastle 76mm (3in) gun could be operated in automatic mode via the G-band fire-control system, semi-automatic mode with the turret-mounted Prizma optical sighting device, or in full manual mode with standard gunsights.

DESIGN
A well deck for landing craft and an aircraft hangar were built in. Bow doors were fitted, with an internal ramp leading to the main vehicle parking deck, which could take 25 tanks or a mix of tanks, trucks and support vehicles.

FIRST IN CLASS

Named for a former Bolshevik and political commissar of the Navy, Ivan Rogov was the first ship of Project 1174, launched at the Yantar Shipyard in Kaliningrad in 1976 and commissioned in 1978 as the leader of a class of three. Three times the size of the preceding *Alligator* class landing ships, in size it was closer to the British *Fearless* than the USS *Tarawa*.

FOREDECK

The foredeck can be used to hold around 22 trucks or light armoured vehicles, or can be cleared as an additional landing pad.

HMS *Invincible* (1980)

TYPE • *Aircraft carrier* **COUNTRY** • *United Kingdom*

SPECIFICATIONS

Dimensions:	Length: 210m (689ft); Beam: 36m (118ft); Draught: 8.5m (29ft)
Displacement:	16,256 tonnes (17,924 tons) standard; 22,352 tonnes (24,643 tons) full load
Propulsion:	4 Rolls-Royce Olympus gas turbines; 4 shafts; 84,500kW (112,000shp)
Speed:	28 knots (52km/h; 32mph)
Range:	7000nmi (13,000km; 8050 miles) at 18 knots
Armament:	2 × GAM-B01 20mm (0.79in) guns; 3 × Goalkeeper CIWS
Aircraft:	22
Complement:	1051

The development of the VTOL (Vertical Take-Off and Landing) jet fighter transformed HMS *Invincible* (RO5) from a helicopter carrier into a more potent and capable warship, underlined by its central role in the Falklands War in 1982.

Left: An aerial port quarter view of the Invincible *underway during the NATO exercise* Display Determination *in 1991.*

Refits
After post-1990 refits, the all-round capabilities of the *Invincible* class encompassed communication, control and command from ship to ship, ship to air and ship to shore.

Foredeck
Invincible was launched as a flat-top, but was modified with a 12 degree swept-up foredeck in order to fly off the new Sea Harrier jet.

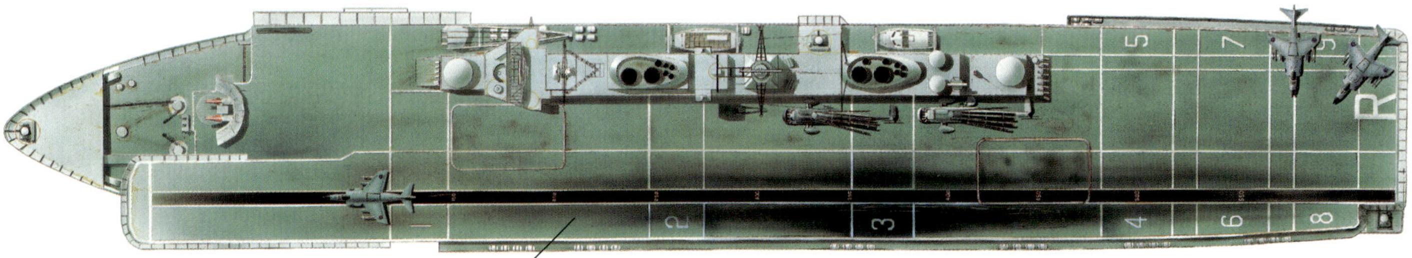

Aerial view
An aerial view of *Invincible*. The twin Sea Dart surface-to-air missile (SAM) launcher, later removed, can be seen alongside the ramp.

AIR GROUP
The air group carried varied according to mission. Up to 24 aircraft could be carried after modification, including nine Harrier GR7 or GR9 planes. Sea King ASaC Mk7 early warning reconnaissance, Merlin HM.1 ASW and Sea King and Chinook general-purpose helicopters were carried. *Invincible* was also understood to have storage facilities for nuclear weapons.

US VISIT
HMS *Invincible* arrives at the Naval Station in Mayport, Jacksonville, Florida on 20 May 2004 for a five-day port visit before continuing on to Norfolk, Virginia and New York City. *Invincible* was the sixth ship of the Royal Navy to bear the name and the first of the *Invincible* class of anti-submarine warfare carriers.

Yūshio (1980)

TYPE • *Submarine* **COUNTRY** • *Japan*

SPECIFICATIONS

DIMENSIONS:	Length: 76m (249ft 3in); Beam: 9.9m (32ft 6in); Draught: 7.5m (24ft 7in)
DISPLACEMENT:	2235 tonnes (2463 tons) surfaced; 2774 tonnes (3058 tons) submerged
PROPULSION:	Diesel-electric drive; single shaft; 2500kW (3400shp) surfaced; 5400kW (7200shp) submerged
SPEED:	12 knots (22.2km/h; 13.8mph) surfaced; 20 knots (37km/h; 23mph) submerged
RANGE:	9500nmi (17,603km; 10,940 miles) at 10 knots surfaced
ARMAMENT:	6 × 533mm (21in) torpedo tubes
COMPLEMENT:	75

Japan's 10-strong *Yūshio* class of attack submarines developed from the preceding *Uzushio* class. They are larger vessels and offer an enhanced diving capacity, with a test diving depth of 275m (902ft).

Left: From 1984, new boats were fitted to carry the Sub-Harpoon anti-ship missile, and all apart from Yūshio *were retrofitted for this weapon.*

EQUIPMENT
Electronic sensor equipment included ZQQ-5 bow sonar (modification of the US BQS-4) and the ZQR towed array.

TORPEDOES
The torpedoes carried on the *Yūshio*-class submarines were Type 89 active-passive with a maximum range of 50km (31 miles).

HULL PROFILE
The *Yūshio* class has what is known as a 'teardrop hull' configuration, meaning the hull is specifically designed for enhanced submerged performance.

DOUBLE HULL

Commissioned between 1980 and 1989, the *Yūshio* class are of double-hull construction and follow the US practice of midship-mounted, outward-angled torpedo tubes so as not to impede the operation of the bow sonar array. From 1984, new boats were fitted to carry the Sub-Harpoon anti-ship missile, and all but *Yūshio* were retrofitted for this weapon. *Yūshio* was used as a training boat from 1996, and the last of the class to be in service was withdrawn in 2008.

BREACHING
A *Yūshio*-class submarine makes a dramatic breach, a lively experience for the crew inside. *Yūshio* served as a training boat from 1996, and the last of the class was withdrawn from service in 2008.

USS *Ohio* (1981)

TYPE • *Submarine* **COUNTRY** • *United States*

SPECIFICATIONS

Dimensions:	Length: 170.7m (560ft); Beam: 12.8m (42ft); Draught: 11.1m (36ft 5in)
Displacement:	16,764 tonnes (18,479 tons) surfaced; 18,750 tonnes (20,668 tons) submerged
Propulsion:	1 S8G pressurized water nuclear reactor; 2 steam turbines delivering 45,000kW (60,000hp); single screw
Speed:	24 knots (44.4km/h; 27.6mph) surfaced; 28 knots (52km/h; 32.2mph) submerged
Range:	Unlimited
Armament:	4 × 533mm (21in) torpedo tubes; SSBN configuration: 24 × Trident C4 SLBMs; SSGN configuration: up to 152 Tomahawk cruise missiles
Complement:	155

The *Ohio* class will remain the foundation of America's strategic nuclear deterrent until the 2030s. The largest submarines ever built for the US Navy, all 18 underwent substantial modification and updating programmes in the early 21st century.

Above: Cutaway profile of an Ohio-class SSBN, with red and black paint scheme.

Trident
From the ninth boat, USS *Tennessee*, the class has carried the Trident II D5 SLBM, capable of carrying up to 12 MIRVs (multiple independent re-entry vehicles), and older boats have been retrofitted.

Armament
The Mk 48 ADCAP (Added Capability) torpedo has a 292kg (650lb) warhead and can be launched at a depth of 915m (3000ft) with a range of 32km (20 miles) and a speed of 50 knots (92.5km/h; 57.5mph).

Auxiliary power
A 325hp (242kW) Magnatek auxiliary prop motor is fitted, and auxiliary power also comes from a Fairbanks-Morse 38 8-1/8 two-stroke opposed-piston diesel engine.

NUCLEAR TIPPED

USS *Ohio*, the class leader, was laid down as SSBN-726 at the General Dynamics Electric Boat Yard, Groton, Connecticut, on 10 April 1976 and launched on 7 April 1979. The largest submarine yet built for the US Navy, it was commissioned on 11 November 1981 as SSBN-726 and equipped with 24 Trident C-4 ballistic missiles with multiple independently targeted warheads, set in parallel rows of 12. The class is powered by the S8G pressurized water reactor, providing steam to two geared turbines.

REPAIRS
The guided-missile submarine USS *Ohio* (SSGN 726) transits through the Puget Sound after departing Puget Sound Naval Shipyard after nearly a year spent in the yards.

Tripartite (1981)

TYPE • *Minesweeper* **COUNTRY** • *Netherlands*

SPECIFICATIONS

DIMENSIONS:	Length: 51.5m (169ft); Beam: 8.96m (29ft 5in); Draught: 3.6m (12ft)
DISPLACEMENT:	605 tonnes (667 tons) full load
PROPULSION:	1 Werkspoor RUB 215 V12 diesel; 18,638kW (1900shp); 1 shaft
SPEED:	15 knots (27.8km/h; 17.3mph)
RANGE:	3000nmi (5556km; 3452 miles) at 12 knots
ARMAMENT:	1 × 20mm (0.79in) modèle F2 gun; 2 × 12.7mm (0.5in) machine guns; 2 × 7.62mm (0.3in) machine guns
COMPLEMENT:	36

The *Tripartite* minehunter vessels were designed for NATO service. France, Belgium and the Netherlands each built its own hulls, which were fitted out in Belgium with French electronics and Dutch machinery.

Left: Alkmaar *(M850) was another of the* Tripartite *class; 'Alkmaar' was the class designation in the Netherlands.*

WINCH
A large hydraulic winch is mounted near the stern of the ship on the centreline, its function being to lower and recover the submersible.

BOW PROPELLER
The *Tripartite* vessels are fitted with a bow propeller to enhance their manoeuvrability when operating in confined areas, such as harbours and estuaries.

HULL
To reduce the magnetic and acoustic signatures of the vessels, their hulls were constructed in a sandwich of wood and foam, the outer skin being covered by a thin film of glass-reinforced plastic.

CONFIGURATION

Minehunting vessels must be heavily protected. Their hulls are designed with materials that reduce their acoustic and magnetic signatures, two common forms of trigger for mines, and they are often soundproofed by having their machinery mounted on shock absorbers. They are fitted with an auxiliary propulsion system in the bow, enabling the engine thrust to be transmitted in any direction and so enhance manoeuvrability.

***CASSIOPÉE* (M642)**
In France, the Tripartites were known as the *Éridan* class. Here we see one of the French vessels, *Cassiopée* (M642), which had been commissioned on 5 May 1984.

Natya (1981)

TYPE • *Minesweeper* **COUNTRY** • *Soviet Union*

SPECIFICATIONS

DIMENSIONS:	Length: 61m (200ft 2in); Beam: 10.2m (33ft 6in); Draught: 3.6m (11ft 10in)
DISPLACEMENT:	873 tonnes (963 tons) full load
PROPULSION:	2 M-503 diesel engines; 3700kW (5000hp); 2 shafts
SPEED:	17 knots (31km/h; 20mph)
RANGE:	1500nmi (2800km; 1700 miles) at 12 knots
ARMAMENT:	2 × AK-230 twin 30mm (1.18in) autocannons; 2 × 2M-3M twin 25mm (1in) autocannons; 2 × 5-barrel RBU-1200 ASW rocket launchers; up to 7 AMD-1000 naval mines or 32 depth charges; minesweeping gear (BKT, AT-3, TEM-4)
COMPLEMENT:	68

The *Natya* class were the near Soviet equivalent of the *Tripartite* minesweeper. Produced in significant numbers, they offered contact, acoustic and magnetic minesweeping capabilities. They were used heavily during the Cold War era, albeit mostly in coastal patrol duties.

Above: Some 45 Natya*-class vessels were constructed between 1970 and 1982. Many of these transferred to foreign navies.*

SONAR
The ship's sonar system was the MG-69/79 High Frequency, which was hull mounted. The ship also carried the Don 2 I-band air/surface search radar.

MAIN GUNS
The twin gun mounts were capable of firing 30mm (1.18in) shells at a cyclical rate of fire of 500rpm to a range of 5km (3 miles).

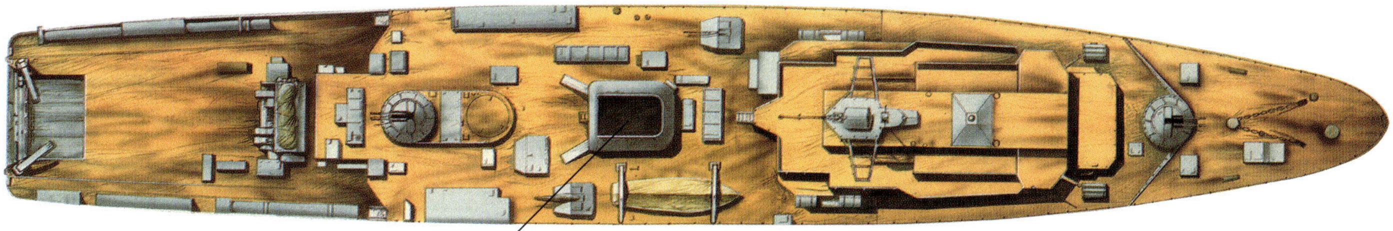

PROPULSION
The ship was driven to a maximum speed of 17 knots (31km/h; 20mph) by two M-503 diesel engines generating 3700kW (5000hp).

DESIGN

Known to the Russians as the Project 266M *Avkvamarin*, the design for the *Natya* class evolved from the *Yurka*-class minesweeper. However, they were fitted with improved minehunting equipment including more advanced sonar and closed-circuit TV. A stern ramp made recovering sweeps easier. The hull was built of low-magnetic steel to reduce the risk from magnetic mines.

RESUPPLY
A *Natya*-class minesweeper (in the foreground) is resupplied by a *Don*-class submarine tender in 1983. Several of the ageing minesweepers remain in service with the Russian Pacific and Northern Fleets.

Ticonderoga class (1981)

TYPE • *Missile cruiser* **COUNTRY** • *United States*

SPECIFICATIONS

DIMENSIONS:	Length: 173m (567ft); Beam: 16.8m (55ft); Draught: 10.2m (33ft 6in)
DISPLACEMENT:	approx. 9800 tonnes (10,802 tons) full load
PROPULSION:	4 General Electric LM2500 gas turbines; 60,000kW (80,000shp); 4 shafts
SPEED:	32.5 knots (60km/h; 37.4mph)
RANGE:	6000nmi (11,000km; 6835 miles) at 20 knots
ARMAMENT:	2 × 61-cell Mk 41 VLS (122 missiles total for SAM, SSM, ASW, and cruise missiles); 8 × Harpoon anti-ship missiles; 2 × 127mm (5in)/54 Mk 45 Mod 2 guns; 2 × 25mm (1in) Mk 38 guns; 2 × Phalanx CIWS; 2 × Mk 32 ASW torpedo tube mounts
COMPLEMENT:	approx. 400

The *Ticonderoga* class of guided-missile cruisers forms a critical element of today's US Navy. A total of 27 vessels were built, and at the time of writing, seven remain in active service – testament to the durability of the design.

Above, left to right: USS Chancellorsville *(CG-62) and USS* Anzio *(CG-6), both* Ticonderoga*-class cruisers. This aerial view shows the sleek lines of the primary design; the hull design is largely based on that of the earlier* Spruance *class.*

HELICOPTER DECK
The *Ticonderoga*-class vessels can operate two Sikorsky SH-60B or MH-60R Seahawk LAMPS III helicopters.

VERTICAL LAUNCHING SYSTEM
The Mk 41 VLS can fire missile types including cruise missiles, SAMs and anti-ship weapons, allowing multiple missiles to be kept in a state of readiness.

USS *TICONDEROGA*
Ticonderoga was commissioned on 22 January 1983 and served in the US Navy until it was decommissioned on 30 September 2004.

AEGIS
The *Ticonderoga* class incorporates the Aegis Ballistic Missile Defense System, which enables the vessels to track, engage and destroy incoming ballistic missiles and even, potentially, satellites in a low-Earth orbit. To test that capability, on 20 February 2008, *Lake Erie* fired an SM-3 missile at a dead US satellite as it was just about to re-enter Earth's atmosphere. The DoD announced that the missile had indeed achieved a direct hit on the target.

USS *NORMANDY* (CG-60)
USS *Normandy* was commissioned on 9 December 1989 and just a year later deployed to the Persian Gulf in support of Operations *Desert Shield* and *Desert Storm*.

Project 941 *Akula* class (*Typhoon* class) (1981)

TYPE • *Submarine* **COUNTRY** • *Soviet Union*

The largest submarines ever constructed, the six *Typhoon*-class vessels featured multiple pressurized hulls and carried Russia's ballistic missile deterrent until the 2000s. Notably, they did not have to submerge or even go to sea to launch their long-range missiles.

SPECIFICATIONS

Dimensions:	Length: 170m (557ft 9in); Beam: 24m (78ft 8in); Draught: 12.5m (41ft)
Displacement:	25,400 tonnes (28,000 tons) surfaced; 48,000 tonnes (52,944 tons) submerged
Propulsion:	2 OK-750 pressurized water nuclear reactors; steam turbines delivering 74,000kW (99,200hp); twin screws
Speed:	22 knots (40.7km/h; 25.3mph) surfaced; 27 knots (50km/h; 31mph) submerged
Range:	Unlimited
Armament:	20 × RSM-56 SLBMs; 4 × 650mm (25.4in) and 2 × 533mm (21in) torpedo tubes
Complement:	160

Above: The outer casing over the Akula*'s sail and pressure hulls is believed to be capable of breaking through ice up to 3m (13ft) thick.*

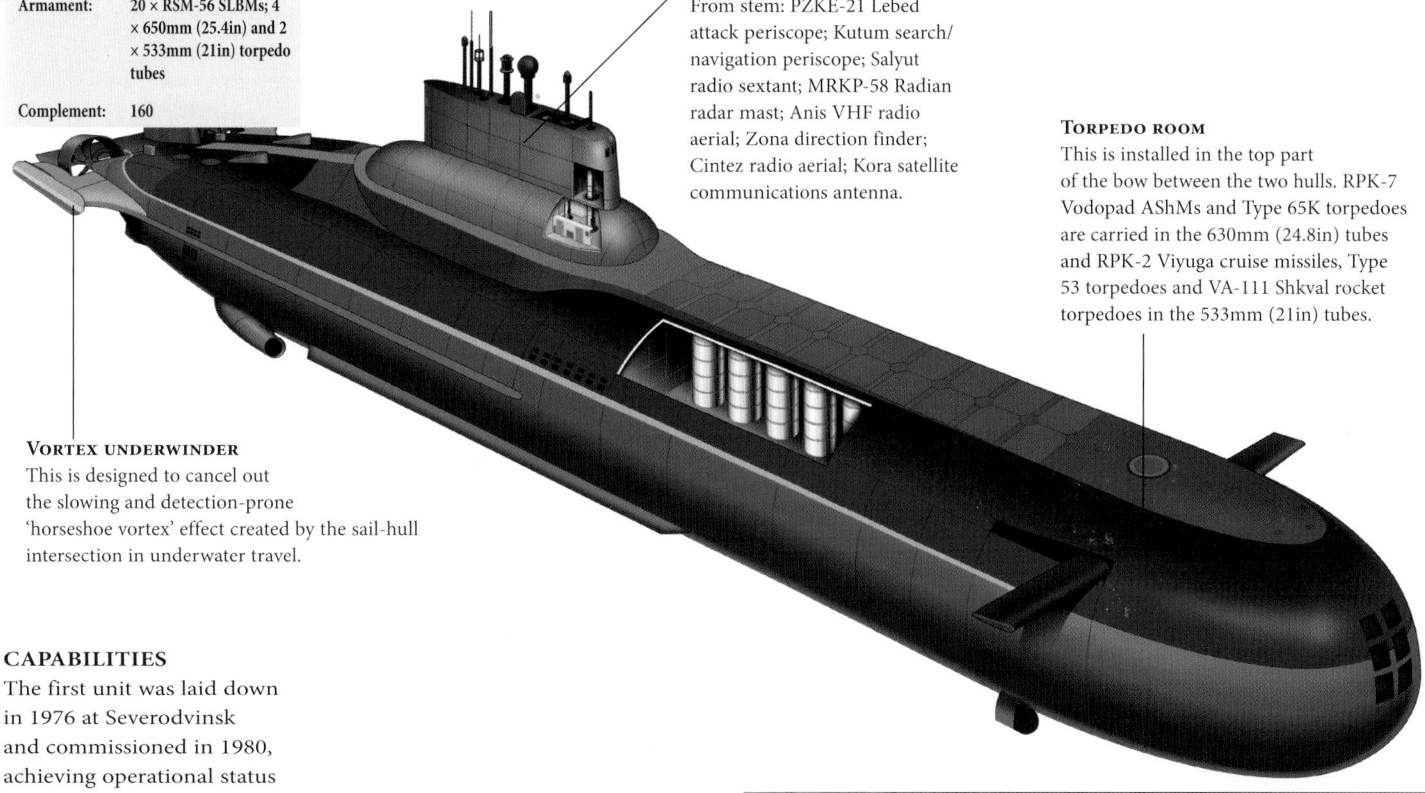

Masts and antennae
From stem: PZKE-21 Lebed attack periscope; Kutum search/navigation periscope; Salyut radio sextant; MRKP-58 Radian radar mast; Anis VHF radio aerial; Zona direction finder; Cintez radio aerial; Kora satellite communications antenna.

Torpedo room
This is installed in the top part of the bow between the two hulls. RPK-7 Vodopad AShMs and Type 65K torpedoes are carried in the 630mm (24.8in) tubes and RPK-2 Viyuga cruise missiles, Type 53 torpedoes and VA-111 Shkval rocket torpedoes in the 533mm (21in) tubes.

Vortex underwinder
This is designed to cancel out the slowing and detection-prone 'horseshoe vortex' effect created by the sail-hull intersection in underwater travel.

CAPABILITIES

The first unit was laid down in 1976 at Severodvinsk and commissioned in 1980, achieving operational status in 1981. Six more were constructed between 1981 and 1989. This was the first SSBN to have launch tubes installed forward of the sail. Its 20 RSM-52 intercontinental three-stage solid propellant ballistic missiles were each capable of holding nuclear warheads with an explosive force of 100 kilotons. It is also fitted with four 650mm (24.4in), two 533mm (21in) torpedo tubes and eight *Igla* (Needle) anti-aircraft missile systems.

TYPHOON ENDURANCE

The *Typhoons* could spend at least 120 consecutive days submerged. The class were built specifically for operations with the Soviet Northern Fleet in the Arctic ice pack. The reinforced sail, advanced stern fin with horizontal hydroplane fitted aft of the screws and retractable bow hydroplanes allow the submarine to break through areas of thin ice within the Arctic ice shelf.

Slava class (1982)

Ten of the Project 1164 *Atlant*, or *Slava* class guided missile cruisers, were intended for Soviet/Russian service. Only three made it to commission, and one was sunk during the Ukraine war in 2022.

SPECIFICATIONS

Dimensions:	Length: 186.4m (611ft 7in); Beam: 20.8m (68ft 3in); Draught: 8.4m (27ft 7in)
Displacement:	11,490 tonnes (12,670 tons) full load
Propulsion:	COGAG; 2 M70 cruise gas turbines; 4 M90 boost gas turbines; 97,000kW (130,000shp); 2 shafts
Speed:	32 knots (59km/h; 37mph)
Range:	6800nmi (12,600km; 7800 miles) at 18 knots
Armament:	16 × P-1000 Vulcan (SS-N-12 Sandbox) anti-ship missiles; 64 × S-300F Fort (SA-N-6 Grumble) SAM; 40 × Osa-M (SA-N-4 Gecko) SAM; 1 × twin AK-130 130mm (5in) guns; 6 × AK-630 CIWS; 2 × 12-barrel RBU-6000 ASW rocket launchers; 10 × 533mm (21in) torpedo tubes
Complement:	485

Right: Photographed in more peaceful times, the Marshal Ustinov *ties up in Naval Station Mayport, Jacksonville, Florida. on a four-day goodwill visit.*

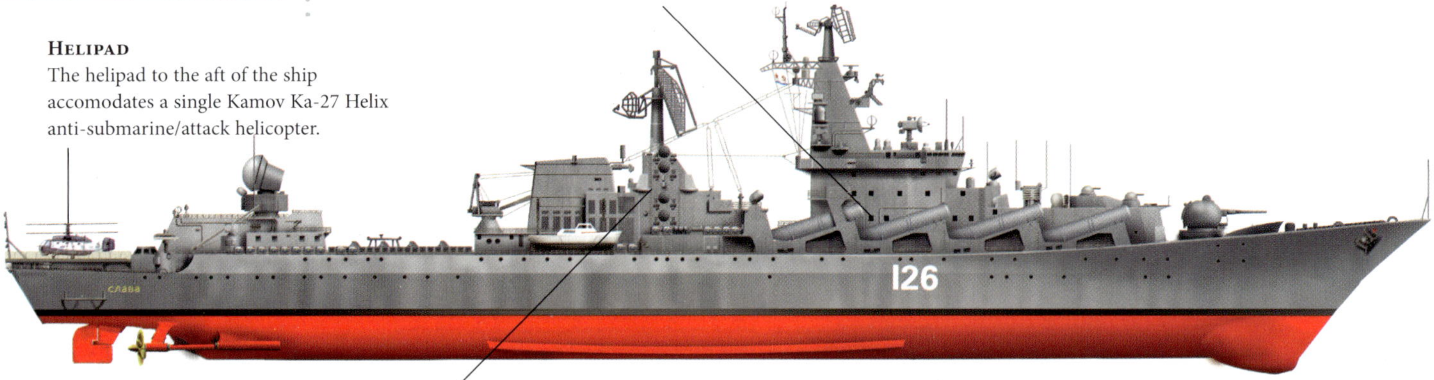

Sandbox
The P-500 *Bazalt* (SS-N-12 *Sandbox*) launchers send supersonic cruise missiles out to a range of 550km (342 miles) with a conventional or nuclear payload.

Helipad
The helipad to the aft of the ship accomodates a single Kamov Ka-27 Helix anti-submarine/attack helicopter.

Slava/Moskva
The modernized *Slava* class has an extensive sensor suite, including a MR-800 Voshkod/Top Pair, 3D air search radar, MR-700 Fregat/Top Steer 3D air/surface search radar, MG-332 Tigan-2T/Bull Horn hull mounted sonar and Mare Tail variable depth sonar.

UKRAINE SINKING

The *Slava*-class warship *Moskva* was the flagship of the Russian Navy's Black Sea Fleet, and was active in supporting the Russian invasion of Ukraine from February 2022. On 13 April, the ship was hit by two Ukrainian R-360 Neptune anti-ship missiles while east of Snake Island in the Black Sea. The missiles caused extensive damage and onboard fires, and the ship sank the next day while under tow. The ship became the largest Russian vessel sunk in combat since World War II.

MARSHAL USTINOV
The *Marshal Ustinov* was the second of the *Slava*-class cruisers to enter service, in September 1986. It was initially named *Admiral Lobov*, but was renamed on its commission to the Northern Fleet.

USS *Carl Vinson* (1982)

TYPE • *Aircraft carrier* **COUNTRY** • *United States*

SPECIFICATIONS	
Dimensions:	Length: 317m (1040ft); Beam: 40.8m (134ft); Draught: 11.5m (37ft 10in)
Displacement:	113,500 tonnes (125,110 tons)
Propulsion:	2 A4W nuclear reactors; 4 steam turbines; 4 shafts; 193,700kW (260,000shp)
Speed:	30 knots (56km/h; 35mph)
Range:	Unlimited
Armament:	2 × RIM-116 RAM launchers; 2 × Sea Sparrow Mk 57 Mod 3 SAM launchers; 3 × Phalanx 20mm (0.79in) CIWS
Aircraft:	90
Complement:	5680

Carl Vinson (CVN-70) is the third of the 10 _Nimitz_-class carriers. Its keel was laid at Newport News on 11 October 1975, and it was launched on 15 March 1980. It was the first US Navy ship to be named for a living person – the Georgia congressman who led the campaign for naval expansion in 1940.

Above: An elevated bow view of Carl Vinson. *On 2 May 2011 the body of the al-Qaeda leader Osama Bin Laden, killed by a SEAL team raid in Pakistan, was buried at sea at an undisclosed location from the ship's deck.*

Elevators
The deck edge elevators are vulnerable to occasional exceptional waves. Seven men were swept overboard from No.1 elevator, positioned at hangar deck level 8m (25ft) above the waterline, on 16 August 1986. All were rescued.

Markings
From overhead, the vast flight deck of *Carl Vinson* is marked out to show landing lines, catapults, lifts and aircraft parking areas.

RCOH
From 2005 to 2009 *Carl Vinson* underwent the scheduled refuelling and complex overhaul (RCOH), during which the nuclear reactors are refuelled and every aspect of the ship's systems is checked or upgraded. This included reconstruction of the two upper levels of the island, installation of a new antenna mast and radar tower; upgraded communications and navigation systems; refurbished radar; new ship defence missiles; new oxygen and nitrogen generation equipment; and enhancements to the aircraft landing and recovery systems.

Design
Standard for all 10 *Nimitz*-class carriers are four elevators and four catapults, and the two A4W reactors powering four steam turbines.

ENDURANCE
A *Nimitz*-class carrier can remain at sea for up to six months at a time, during which many replenishments are conducted to sustain crew living conditions and air operations.

Rubis class (1983)

TYPE • *Submarine* COUNTRY • *France*

SPECIFICATIONS

DIMENSIONS:	Length: 70.1m (236ft 6in); Beam: 7.6m (24ft 11in); Draught: 6.4m (21ft)
DISPLACEMENT:	2385 tonnes (2630 tons) surfaced; 2713 tonnes (2991 tons) submerged
PROPULSION:	1 CAS-48 pressurized water nuclear reactor; 2 propulsion turbo-generators, 3150kW each; 1 electric motor, 6300kW (8448shp); 1 auxiliary SEMT Pielstick 8PA4V185SM diesel generator, 450kW; 1 shaft
SPEED:	25 knots (46.3km/h; 28.8mph) submerged
RANGE:	Indefinite
ARMAMENT:	4 × 533mm (21in) torpedo tubes for F-17 Mod 2 torpedoes or Exocet SM39 missiles
COMPLEMENT:	70

The *Rubis* class dates back to 1983, but only the final two, *Améthyste* and *Perle*, commissioned in 1992 and 1993, remain in service. The *Rubis* class were designed for anti-submarine warfare in the ocean depths but also for convoy protection and littoral surveillance.

Left: Améthyste *(S605) represents a major upgrade of the* Rubis *class. Many of the improvements were subsequently retrofitted into older boats.*

SONAR SUITE
The sonar suite includes the Thales Underwater Systems (formerly Thomson Marconi Sonar) DMUX 20 multifunctional passive sonar operating at low frequency. The submarine's VLF passive towed array system is the DSUV 62C.

HULL
Hull length is 75m (246ft). The hulls are of 80 HLES high-elasticity steel, which allows for operation down to 300m (980ft).

REACTOR
The CAS 48 pressure water reactor provides 48MW, fed via two turbo generators to a 6299kW (8448hp) electric motor.

EXOCET
Four 533mm (21in) tubes are fitted, for torpedoes and Exocet missiles. The Exocet SM39 anti-ship missile is manufactured by MBDA (formerly EADS Aerospatiale). Target range and bearing data is downloaded from the submarine's SAT into the Exocet's computer. The missile approaches the target area in sea-skimming mode using inertial navigation (Sigma 40XP) and then active radar homing, at speeds over Mach 0.9 with a range of 50km (31 miles).

***SAPHIR* (S602)**
Saphir was laid down on 1 September 1979, launched 1 September 1981 and commissioned on 6 July 1984. While the submarine never saw combat, it was reported in 2015 that it had 'sunk' the USS aircraft carrier *Theodore Roosevelt* and its escort in a training exercise off Florida.

Project 945A *Kondor* class (1984)

TYPE • *Submarine* **COUNTRY** • *Soviet Union*

SPECIFICATIONS

Dimensions:	Length: 110m (360ft 11in); Beam: 14.2m (46ft 7in); Draught: 8.8m (28ft 10in)
Displacement:	7600 tonnes (8377 tons) surfaced; 9100 tonnes (10,033 tons) submerged
Propulsion:	1 pressurized-water nuclear reactor; 190MW; 1 shaft; 2 emergency electric motors, 747kW (1002hp) each
Speed:	10 knots (18.5km/h; 11.5mph) surfaced; 32 knots (59.3km/h; 36.8mph) submerged
Range:	Unlimited (limited only by provisions)
Armament:	6 × 533mm (21in) torpedo tubes; SS-N-21 Sampson SLCM; SS-N-15 Starfish ASW missile; SS-N-16 Stallion ASW missile; minelaying capability for up to 42 mines
Complement:	72

The Project 945A *Kondor* class is known by NATO as the *Sierra II* class. Consisting of two ships in the class – *Nizhniy Novgorod* and *Pskov*, both are nuclear-powered vessels with a mission endurance measured in months rather than days.

Left: Further additions to the Kondor *class were cancelled in favour of the new Project 971* Akula *class.*

PROJECT 945A *KONDOR*
A distinguishing feature of the Project 945A is its large sail, which is 5m (16ft 5in) longer than the previous *Sierra I* class.

ARMAMENT
Armament includes P-100 Oniks (SSN-22) anti-ship cruise missiles and Type 40 torpedoes, fired from four 533mm (21in) and four 650mm (25.6in) tubes.

HULL
The hull is made of titanium alloy, providing deep diving capability, to 700m (2300ft) with an outer maximum of 800m (2625ft) and the ability to avoid magnetic anomaly detection.

PROJECT 945
For 10 years, Project 945 had been working on an attack submarine that could outfight any other, resulting in a single SSN, commissioned in September 1984, with a second in 1987, both classified by NATO as Sierra I. With some different features, two more Project 945A *Kondor* (*Sierra II*) were completed in 1992 and 1993, after the fall of the Soviet Union.

***SIERRA* CLASS**
The first *Sierra I*, named first *Karp*, then *Tula*, has been withdrawn. Its companion, *Krab*, is on the active list, as are the two *Sierra II*, *Nizhniy Novgorod* and *Pskov*.

Giuseppe Garibaldi (1985)

TYPE • *Helicopter carrier* **COUNTRY** • *Italy*

SPECIFICATIONS

Dimensions:	Length: 180.2m (591ft); Beam: 33.4m (110ft); Draught: 8.2m (27ft)
Displacement:	10,100 tonnes (11,134 tons); 13,850 tonnes (15,272 tons) full load
Propulsion:	4 GE-Avio gas turbines; 2 shafts; 60,400kW (81,000shp)
Speed:	30 knots (56km/h; 35mph)
Range:	7000nmi (13,000km; 8100 miles) at 20 knots
Armament:	4 × AOSM Otomat Mk 2 SSM (removed 2003); 2 × eight-cell launchers for SARH Aspide missiles; 3 × OTO Melara Twin 40L70 DARDO CIWS; 2 × Albatross surface-to-air missile launchers
Aircraft:	18
Complement:	830

Italy's first through-deck carrier, *Giuseppe Garibaldi* (C-551), was initially denied the right to carry fixed-wing aircraft, although that position was later reversed. Laid down on 26 March 1981, it was launched on 11 June 1983 and commissioned on 30 September 1985.

Left: The flight control station, communications, sensory and defensive (Albatross SAM launcher) systems are all visible here, as well as the telescopic-armed hydraulic crane.

Hangar deck
The hangar deck, 105m (320ft) long, 15m (45ft 9in) wide, and 6m (18ft 4in) high, can be rapidly partitioned into three sections in the event of a fire.

Radar
Long-range SPS-768, surface scanning SPS-702, 3-D search SPS-52C, combined-watch SPS-774 and navigational SPN-749 radar systems were carried on the ship in 2009, plus fire-control SPG-74 for the 40mm (1.57in) guns and SPG-75 for the Albatross launchers.

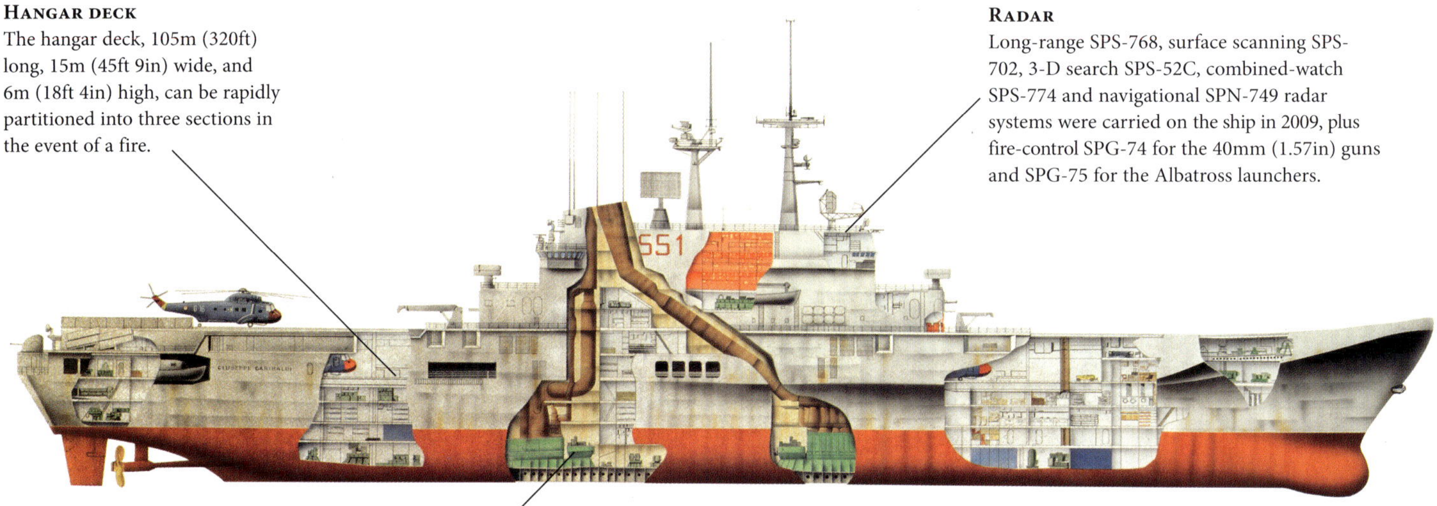

Powerplant
The ship is powered by four GE COGAG gas turbines, built under licence by Fiat, and driving two propeller shafts. Six Grandi Motori Triesti diesel generators are also installed.

BACKGROUND

A law of 1937 precluded the Italian Navy from operating fixed-wing aircraft, with the intention of keeping all air operations under control of the Air Force, as with the Luftwaffe in Germany. In the postwar years the Marina Militare focused its attention on helicopters, carried at first on frigates. Although in the 1960s successful test flights of British Sea Harrier vertical take-off jets were made on the cruiser *Andrea Doria*, *Giuseppe Garibaldi* was first planned as a helicopter carrier. However, final plans incorporated a bow ramp.

SEA KING
An SH-3D/H Sea King helicopter parked on the flight deck of *Giuseppe Garibaldi*. Anti-submarine operations were seen as the ship's main function, primarily to be carried out by helicopters, but two ILAS triple tube torpedo launchers were mounted, capable of firing 324mm (12.75in) MU-90 torpedoes.

Project 971 *Shchuka-B* class (1986)

TYPE • *Submarine* **COUNTRY** • *Soviet Union*

SPECIFICATIONS

Dimensions:	Length: 110.3m (362ft); Beam: 13.6m (45ft); Draught: 9.8m (32ft 2in)
Displacement:	8298 tonnes (9150 tons) surfaced; 10,668 tonnes (11,763 tons) submerged
Propulsion:	1 OK-650M pressurized water nuclear reactor; OK-9VM steam turbine; 2 electric motors; 2 diesel auxiliaries; single fixed-pitch propeller
Speed:	33 knots (61km/h; 38mph) submerged
Range:	Indefinite
Armament:	4 × 650mm (25.6in) and 4 × 533mm (21in) torpedo tubes for torpedoes and missiles
Complement:	73

The Russian Navy is working on life-extension programmes for boats that might have been marked for withdrawal during the 2020s. The Project 971 *Shchuka-B* class (*Akula* is its NATO reporting name) is one of those.

Right: Shchuka-B *class are double hulled, with hydrodynamic styling including the sail and a distinctive rudder pod that houses the towed sonar array.*

Wake diffuser
The Schshuka-B appears to have a wake diffuser positioned aft of the seven-bladed propeller, in order to minimize traceable noise.

Sail
The highly streamlined sail structure, unsuitable for installing modular equipment, has not been followed in subsequent Russian designs. The stern pod for the towed sonar array matches it in height.

Interior
The submarine has six compartments plus bow and stem parts, and 17 main ballast tanks. To help buoyancy, the first compartment is divided into two parts by a horizontal deck resistant to 10MPa (1450psi).

COMBAT ROLE

The Project 971 submarine was originally planned for searching, detection and shadowing of an adversary's ballistic missile nuclear submarines and aircraft carriers, and destroying them if hostilities became warfare. An additional aim was the ability to strike on land at ranges up to 3000km (1864mi) with 533mm calibre *Granat* subsonic cruise missiles similar to US Tomahawks.

AERIAL PHOTO
A starboard quarter view of a Soviet *Akula*-class nuclear-powered attack submarine underway. In 2012 the Russian Federation leased a nuclear-powered submarine for the first time, delivering the *Shchuka-B* class *Nerpa* (K-152) to India on a 10-year contract, shedding new light on the capabilities of the class.

Duke Type 23 (1987)

TYPE • *Destroyer* **COUNTRY** • *United Kingdom*

SPECIFICATIONS

Dimensions:	Length: 133m (434ft 4in); Beam: 16.1m (52ft 10in); Draught: 7.3m (24ft)
Displacement:	3810 tonnes (4200 tons)
Propulsion:	4 Paxman Valenta 12CM diesel generators, 1510kW (2025shp) each; 2 GEC electric motors, 2980kW (4000shp); 2 Rolls-Royce Spey SM1A gas turbines, 23,190kW (31,100shp); 2 shafts
Speed:	28 knots (51.9km/h; 32.3mph)
Range:	9000nmi (16,668km; 10,355 miles) at 15 knots
Armament:	2 × quad Harpoon missile launchers; 1 × Sea Wolf VLS; 2 × twin 324mm (12.75in) ASW torpedo tubes; 1 × 114mm (4.5in) Mk 8 gun; 2 × DS30B 30mm (1.18in) guns; 2 × CIWS
Complement:	181

The Type 23 light anti-submarine frigate was conceived in the late 1970s to replace both the *Leander* class of the 1950s and the 1960s-vintage Type 21. Designed as the backbone of the Royal Navy's surface anti-submarine force, it represented a major step forward in British capability.

Right: Sailors line the deck of Type 23 frigate HMS Sutherland *as the ship returns to Plymouth following a long deployment.*

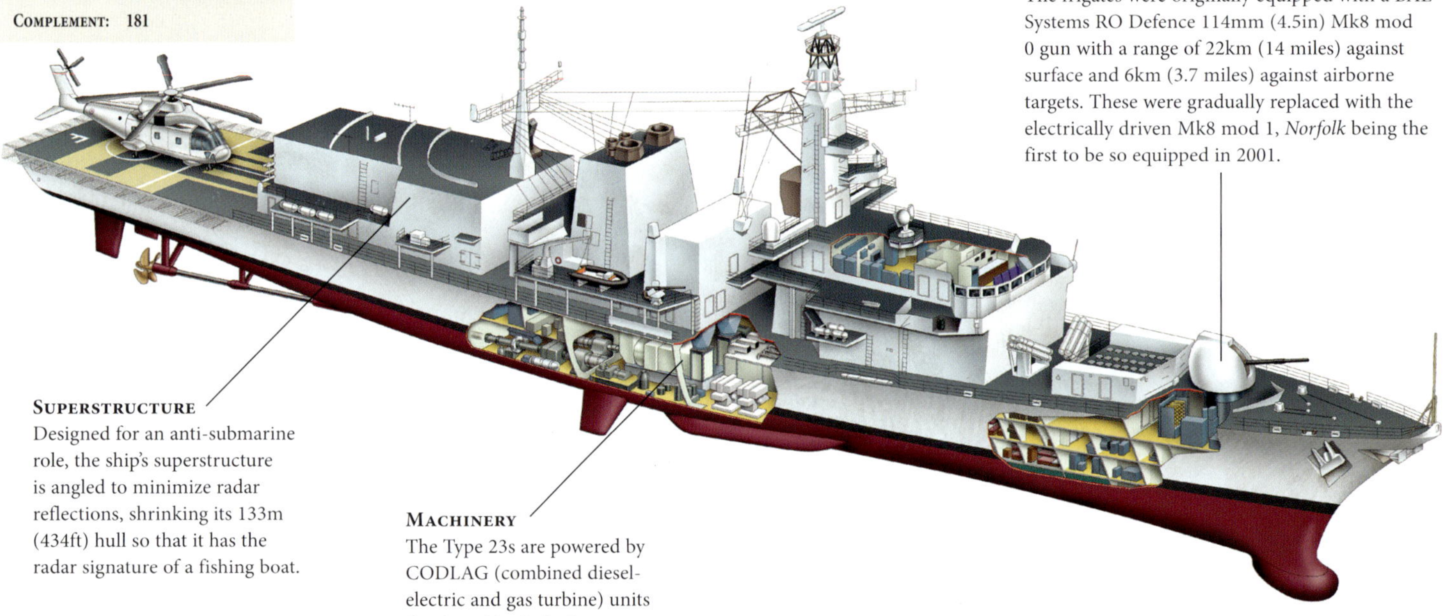

Guns
The frigates were originally equipped with a BAE Systems RO Defence 114mm (4.5in) Mk8 mod 0 gun with a range of 22km (14 miles) against surface and 6km (3.7 miles) against airborne targets. These were gradually replaced with the electrically driven Mk8 mod 1, *Norfolk* being the first to be so equipped in 2001.

Superstructure
Designed for an anti-submarine role, the ship's superstructure is angled to minimize radar reflections, shrinking its 133m (434ft) hull so that it has the radar signature of a fishing boat.

Machinery
The Type 23s are powered by CODLAG (combined diesel-electric and gas turbine) units developed by Rolls-Royce, Paxman and GEC. These provide quiet power during long-range sonar detection and tracking operations.

FRIGATES
The frigate-type warship was conceived in the 18th century, when its roles were defined as reconnaissance and commerce raiding. Today, the name is applied to a wide variety of vessels, ranging from expensive and highly specialized anti-submarine warfare ships like the Type 23 to cheaper ships like the US Navy's *Knox* class, designed to escort convoys and amphibious warfare task groups.

HMS *RICHMOND*
The Type 23 *Duke* class were originally intended to detect Soviet submarines in the North Atlantic, without mounting any defensive armament. The lessons learned in the Falklands War of 1982 reversed that design plan, and the Type 23s became heavily armed.

Västergötland (1987)

TYPE • *Submarine* **COUNTRY** • *Sweden*

SPECIFICATIONS	
DIMENSIONS:	Length: 48.5m (159ft 1in); Beam: 6.1m (20ft); Draught: 5.6m (18ft 4in)
DISPLACEMENT:	1070 tonnes (1180 tons) surfaced; 1150 tonnes (1267 tons) submerged
PROPULSION:	2 Hedemora V12A/15 diesel engines; 1 Jeumont-Schneider electric motor; 1 screw
SPEED:	11 knots (20km/h; 13mph) surfaced; 20 knots (37km/h; 23mph) submerged
RANGE:	Not known
ARMAMENT:	6 × 533mm (21in) torpedo tubes; 3 × 400mm (16in) torpedo tubes; minelaying capability
COMPLEMENT:	24

The *Västergötland* class were a four-ship class of diesel-electric submarines commissioned in the late 1980s and early 1990s. True multi-role submarines, they performed roles as varied as anti-ship and anti-submarine attacks, minelaying, surveillance and Special Forces deployment.

Left: The Swedish diesel-powered attack submarine HMS Gotland *transits through San Diego Harbour in October 2025. The* Gotland *class were constructed in the 1990s, and featured a Stirling engine air-independent propulsion (AIP) system to extend endurance.*

TORPEDOES
With six 533mm (21in) and three 400mm (16in) torpedo tubes, the submarines can fire six heavy torpedoes and six wire-guided light torpedoes simultaneously at different targets.

HULL
The *Archer* (previously *Västergötland*) class is single-hulled, divided into two separate pressure-tight compartments in its original form but with the AIP added section since 2008/10. No statement has been made about their future but they are likely to remain active into the 2030s.

VÄSTERGÖTLAND
The original form of *Västergötland*, first commissioned in 1987 and relaunched in 2009, in extended and modified form as RSS *Archer* in the Singapore Navy.

ATTACK CAPABILITY

The *Västergötland* class had a formidable warfighting capability. The ships were armed with six 533mm (21in) torpedo tubes and three 400mm (16in) tubes. In a remarkable display of fighting versatility, the were able to fire up to 12 torpedoes – six heavy and six light wire-guided – at the same time against different targets, something that no other submarine could perform.

***VÄSTERGÖTLAND* CLASS**
The *Västergötland* class of diesel-electric submarines was introduced in 1987 by the Swedish Navy. It has gone on to influence several subsequent classes of submarine.

Zeeleeuw (*Walrus* class) (1987)

TYPE • *Submarine* **COUNTRY** • *Netherlands*

SPECIFICATIONS

Dimensions:	Length: 67.7m (222ft 2in); Beam: 8.4m (27ft 7in); Draught: 6.6m (21ft 8in)
Displacement:	2450 tonnes (2700 tons) surfaced; 2800 tonnes (3086 tons) submerged
Propulsion:	3 diesel generators; 1 electric motor; 1 screw
Speed:	21 knots (39km/h; 24mph) submerged
Range:	10,000nmi (18,520km; 11,508 miles) at 9 knots
Armament:	4 x 533mm (21in) torpedo tubes for Mk 48 torpedoes or UGM-84 Harpoon SSM; minelaying capability for up to 40 mines
Complement:	Approx. 50

The Netherlands currently has three active *Walrus*-class attack boats, commissioned between 1990 and 1994. Two will be retained until the 2030s, when replacements are due to be delivered.

Left: The Walrus *boats are notably silent and although designed as attack boats, their prime function has been surveillance and intelligence-gathering.*

Zeeleeuw
The second boat in the class, *Zeeleeuw* was commissioned in 1990. *Walrus*-class boats are used by the international Submarine Command Course (SMCC) for the selection of officers who will go on to command non-nuclear powered submarines.

Periscope
A modernization programme from 2013 began with *Zeeleeuw* and involved replacement of the navigation periscope with an L3-KEO optronic one, and of the Gipsy combat system with an up-to-date Guardion modular counter-drone one.

Hulls
The *Walrus*-class hulls are fabricated from French MAREL high-tensile steel to enable diving to 300m (985ft). This was a late design change that required extensive changes to cooling and fuel systems but gave greatly enhanced performance.

DESIGN

Walrus was an innovative Dutch design, built at Rotterdamsche Droogdok, partially double hulled in teardrop shape, with a pressure hull of HY-100 steel and an X-form afterplane configuration with four independently controlled rudder planes, two of which are sufficient to maintain direction. Three main compartments hold respectively the weapons, the crew accommodation and control centre, and engine room. The level of automation was high for the time and the engine room is normally unmanned.

ENDURANCE

In the sail are mounted two communication masts, a radar mast, ESM mast and two periscopes. They can cruise below snorkelling level for up to six days at slow speed. Stores allow for 60-day patrols and bunks are provided for all crew.

Principe de Asturias (1988)

TYPE • *Aircraft carrier* COUNTRY • *Spain*

SPECIFICATIONS

DIMENSIONS:	Length: 196m (643ft); Beam: 24.3m (80ft); Draught: 9.4m (31ft)
DISPLACEMENT:	16,166 tonnes (17,813 tons); 16,967 tonnes (18,695 tons) full load
PROPULSION:	COGAG gas turbines; 1 shaft; 34,568kW (46,400shp)
SPEED:	26 knots (48km/h; 30mph)
RANGE:	6500nmi (12,000km; 7500 miles) at 20 knots
ARMAMENT:	4 × FABA Meroka Mod 2B CIWS systems with 12-barrel Oerlikon L120 20mm (0.79in) guns
AIRCRAFT:	29
COMPLEMENT:	830

Commissioned on 30 May 1988, with a name going back in naval history to the Spanish Armada in 1588, this short take-off and vertical landing (STOVL) carrier was the Spanish Navy's flagship from 1988 to 2013.

Left: Principe de Asturias *was planned not only to add power to the Spanish Navy,but also as a model compact carrier that might be sold to other navies.*

DECK SPACE
Restricted space on the flat deck kept the crew constantly moving aircraft about and delayed take-offs. This is one reason why the ship normally carried less than its potential aircraft wing.

MAST
The mast, a complex pylon structure heavily arrayed with communication and sensor equipment, was mounted directly on top of the navigation bridge.

FLIGHT DECK
The flight deck was axial with take-off and landing on the port side, and two lifts, one forward on the island, with a retractable crane, the other at the stern. The ship had no catapult, but a 12 degree bow ramp helped get the planes into the air.

SERVICE
The carrier was based at Rota Naval Station on the southern coast of Spain. It was not extensively used, but took part in numerous NATO exercises in the Atlantic and the Mediterranean, including one in 1996 when, together with the British *Invincible* and Italian *Giuseppe Garibaldi*, it formed a light carrier squadron in mock combat with American and French units. The ship was decommissioned in February 2013.

ON EXERCISE
Príncipe de Asturias (R-11) steams through the Atlantic Ocean while participating in *Majestic Eagle 2004*, a multinational exercise conducted off the coast of Morocco.

Wasp class (1989)

TYPE • *Assault ship* **COUNTRY** • *United States*

The eight *Wasp*-class amphibious assault ships (LHD) that came into commission between 1989 and 2009 were a development of the preceding *Tarawa* class. They are designed to support ship-to-shore movements through significant onboard air power.

SPECIFICATIONS

DIMENSIONS:	Length: 257m (843ft); Beam: 31.8m (104ft); Draught: 8.1m (27ft)
DISPLACEMENT:	41,150 tonnes (45,360 tons) full load
PROPULSION:	LHD 1–7: 2 boilers; 2 geared steam turbines; 52,000kW (70,000shp); 2 shafts
SPEED:	22 knots (41km/h; 25mph)
RANGE:	9500nmi (17,600km; 10,940 miles) at 18 knots
ARMAMENT:	2 × Mk 29 launchers for RIM-7 Sea Sparrow SAM; 2 × Mk 49 launchers for RIM-116 RAM; 2–3 × Mk 15 Phalanx CIWS; 3–4 × Mk 38 Mod 2 25mm machine gun systems
AIRCRAFT:	Varies according to mission
COMPLEMENT:	1070; embarked landing force 1687 Marines (+187 surge)

Right: USS Wasp, *the leader of its class, is here seen underway during a surge deployment to the Middle East carrying the Ospreys to its first combat deployment in 2007.*

SEAHAWK
The helicopter on the flight deck is a Sikorsky SH-60 Seahawk, a versatile multi-mission helicopter. Up to six can be carried onboard.

WELL DECK
In the well deck, the *Wasp* class can transport three Landing Craft Air Cushion, 12 Landing Craft Mechanized, or 40 Amphibious Assault Vehicles (AAVs), with another 21 AAVs on the vehicle deck.

USS *MAKIN ISLAND*
The last of the class, USS *Makin Island* (LHD-8) has a different powerplant configuration to the other vessels: two General Electric LM2500 gas turbines, 6 Fairbanks Morse diesel generators and two electric motors.

AIRCRAFT

The air complement carried by the *Wasp* class varies depending on the mission parameters. A typical complement would be six AV-8B Harrier II attack aircraft (currently being phased out) or six F-35B Lightning II strike-fighters, four AH-1W/Z Super Cobra/Viper attack helicopters, 12 MV-22B Osprey assault support tiltrotors, four CH-53E Super Stallion heavy-lift helicopters and 3–4 UH-1Y Venom utility helicopters.

MARINE OSPREYS
US Marine Corps MV-22 Ospreys, assigned to Marine Medium Tiltrotor Squadron (VMM) 263, Marine Aircraft Group 29, prepare for flight from the deck of the USS *Wasp* (LHD-1).

Tupi (1989)

TYPE • *Submarine* **COUNTRY** • *Brazil*

SPECIFICATIONS

Dimensions:	Length: 61.2m (200ft 9in); Beam: 6.25m (20ft 6in); Draught: 5.5m (18ft 1in)
Displacement:	1170 tonnes (1290 tons) surfaced; 1460 tonnes (1610 tons) submerged
Propulsion:	4 diesel-electric generators; 4500kW (6100shp); 1 screw
Speed:	11 knots (20km/h; 13mph) surfaced; 22 knots (41km/h; 25mph) submerged
Range:	11,000nmi (20,000km; 13,000 miles) at 10 knots surfaced; 8000nmi (15,000km; 9200 miles) at 10 knots snorkelling; 400nmi (740km; 460 miles) at 4 knots submerged
Armament:	8 × 533mm (21in) torpedo tubes; 14 torpedoes; provision for UGM-84 Harpoon SSM
Complement:	30

The German-designed Type 209 submarine was one of the most numerous diesel-electric submarines of the Cold War era. Brazil's *Tupi*-class patrol submarine was inaugurated in 1989, and *Tupi* is a standard Type 209/1400.

Left: Brazil's Tupi*-class patrol submarine was inaugurated in 1989. Designed and built by Howaldtswerke- Deutsche Werft (HDW) in Kiel, Germany, a variant on the German U-209 class, it was followed by three others built at the Arsenal de Marinha at Rio in 1994–99.*

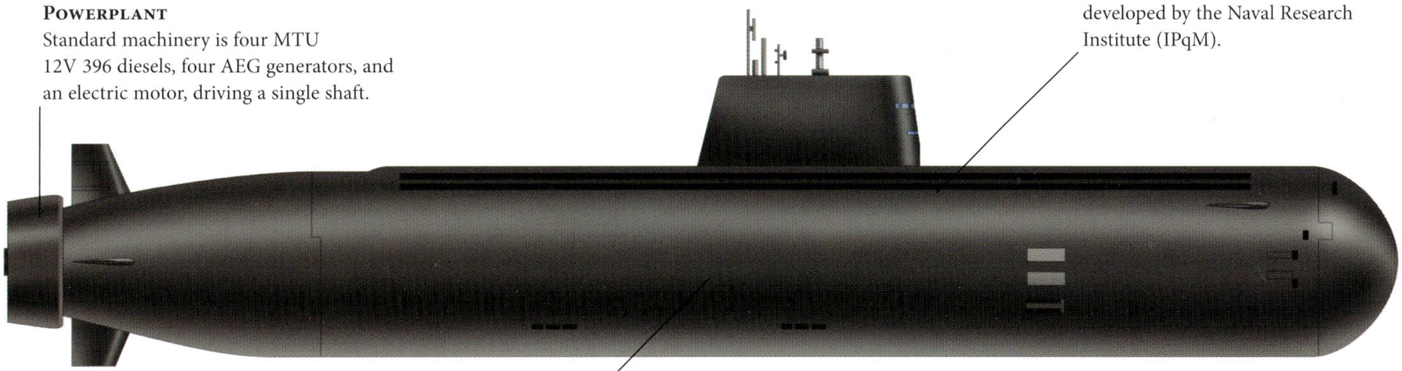

Armament
The *Tupi* boats were armed with British Tigerfish wire-guided torpedoes and Brazilian anti-submarine torpedoes developed by the Naval Research Institute (IPqM).

Powerplant
Standard machinery is four MTU 12V 396 diesels, four AEG generators, and an electric motor, driving a single shaft.

Battery compartment
To maintain balance, the two battery compartments are placed at keel level, forward and aft of the sail. Batteries account for 25 per cent of displacement.

***TUPI* CLASS**
One of the original four Type 209/1400 (commissioned 1989–2006) remains active. All except the first were built under licence in Brazil. *Tupi* is a standard Type 209/1400; *Tikuna* (commissioned 2005) was slightly longer, with greater surface displacement. It had ISUS 83-13 weapons control, improved silencing, more powerful MTU 12V 3906 diesels, a Siemens 3.4MW (4559hp) electric propulsion motor, and wire-guided torpedoes. Brazil later received new integrated combat systems from Lockheed Martin to enable use of the Mark 48 torpedo.

***TUPI* AT SEA**
The Brazilian Navy's Type 209/1400 *Tupi*, commissioned 1989, is the class leader of four. It was built in Germany, the others in Brazil. A modernization plan for the class, including open-architecture combat systems, was begun in 2008.

Ula class (1989)

TYPE • *Submarine* COUNTRY • *Norway*

SPECIFICATIONS

DIMENSIONS:	Length: 59m (193ft 8in); Beam: 5.4m (17ft 9in); Draught: 4.6m (15ft 1in)
DISPLACEMENT:	1056 tonnes (1040 tons) surfaced; 1168 tonnes (1150 tons) submerged
PROPULSION:	2 MTU 16V 396 diesel engines, 970kW (1300hp) each; 1 Siemens Permasyn electric motor, 2200kW (3000hp); 1 shaft
SPEED:	11 knots (20km/h; 13mph) surfaced; 23 knots (43km/h; 26mph) submerged
RANGE:	5000nmi (9300km; 4800 miles) at 8 knots
ARMAMENT:	8 x 533mm (21in) torpedo tubes with 14 AE DM2A3 torpedoes
COMPLEMENT:	21

Six diesel-electric submarines form the Norwegian-built *Ula* class, commissioned between 1989 and 1991. Although essentially a German design from Thyssen Nordseewerke, it is not considered part of the Type 209 stream (in Germany it is known as Type 210).

Right: The torpedo department of the submarine Ula *(S 300).*

UTSIRA
The *Ula*-class S-301 *Utsira*. The pair of stern-mounted casings on the class are unusual but no description of their function – perhaps to do with navigation in confined spaces – seems to be available.

ASSEMBLY
The hull sections were built in Norway and assembled by Thyssen at Emden in Germany. German MTU diesels and a Siemens Permasyn electric motor are fitted.

BOW TUBES
The *Ula* class is fitted with eight 533mm (21in) bow torpedo tubes for launching Atlas Elektronik DM2A3 torpedoes. The DM2A3 can carry a 260kg (570lb) warhead up to a maximum range of 28,000m (91,864ft).

COMBAT SYSTEM
The *Ula* class' MSI-90U combat/control system, currently being upgraded as part of a life-extension programme, is designed by the Norwegian Kongsberg company. It is based on a high-capacity database and a distributed computer system. The system uses a local area network or point-to-point connections to communicate with the on-board sensors, weapons and navigation system. The multi-function operator console provides tactical information based on the data available from sensors and other sources.

NATO EXERCISE
The Norwegian *Ula*-class submarine *Utstein* (KNM-302) negotiates rough seas while participating in the NATO Exercise *Odin-One* (2003), designed to test Allied and NATO submarine-hunting capabilities.

Giuliano Prini (*Nazario Sauro* class) (1989)

TYPE • *Submarine* **COUNTRY** • *Italy*

SPECIFICATIONS

Dimensions:	Length: 64.4m (211ft 3in); Beam: 6.83m (22ft 5in); Draught: 5.6m (18ft 4in)
Displacement:	1476 tonnes (1627 tons) surfaced; 1662 tonnes (1833 tons) submerged
Propulsion:	3 diesel engines, 2700kW (3620hp); 1 electric motor, 2686kW (3599hp); 1 shaft
Speed:	12 knots (22km/h; 14mph) surfaced; 19 knots (35km/h; 22mph) submerged
Range:	2500nm (4600km; 2900mi) at 12kt
Armament:	6 × 533mm (21in) torpedo tubes for 12 Black Shark Mod 3 torpedoes
Complement:	51

The final four of the eight *Nazario Sauro* class of attack submarines remained in service in 2025. Commissioned between 1988 and 1995, they are updated versions of a 1970s design built at the Fincantieri Yard in Monfalcone – the last wholly Italian submarine design.

Right: Nazario Sauro *(S-518) is no longer in service, having been decommissioned on 30 April 2002, after a career lasting more than two decades.*

MACHINERY
The machinery is diesel-electric, with three GMT (Grandi Motori Trieste) 2686kW (2003hp) diesels on all four boats, Magneti Marelli electric motors on S522 and 523, and ABB motors on S524 and 525, driving a single shaft.

GIULIANO PRINI
Commissioned in 1989, *Giuliano Prini* No. 523, substantial modifications were made between 1999–2004. It is one of the two Batch III boats, slightly longer and wider than the first two of the class.

DIVING PLANES
An unusual feature is three sets of diving planes: bow, sail, and stern. Two of the class are scheduled for an operational life-extension programme.

ONBOARD SYSTEM
An update of electronics was made between 1999 and 2005. The submarines carry the SMA MM/BPS 704-V2 radar and STN Atlas Elektronik-ISUS 90-20 series sonar. Kollmorgen periscopes and a Hagenuk Marinekommunikation IRSC communication system are fitted. Armament is six 533mm (21in) torpedo tubes, with 12 reloads of Whitehead A-184 Mod 3 torpedoes, or 24 naval mines.

***LEONARDO DA VINCI* (S 520)**
In the era of AIP and the preoccupation with stealth features, the *Sauros* are virtually museum pieces with a top speed of 19 knots (35kmh, 22mph). Nevertheless, they were excellent boats for their era and retain a value in training and exercises, operating alongside their more sophisticated successors.

21

Modern Era

Since the early 1990s, naval warfare has become as much about networked digitization as it is about raw firepower and manoeuvrability. Today, the ships of a professional navy function as interlinked nodes within a tactical digital framework, with unprecedented reaction times and communication speeds increasingly driven by artificial intelligence (AI). Yet the future of naval warfare is already looking uncertain. The development of long-range hypersonic anti-ship missiles threatens even the mighty carrier battle groups, while unmanned surface vessels (USVs) – demonstrated effectively in the war in Ukraine – show the potential for warships operating entirely without onboard crews. Meanwhile, carrier flight decks and surface-ship helipads are deploying unmanned aerial vehicles (UAVs). The evolving interface between human operators and AI-driven systems represents the next stage in the development of naval warfare.

USS *SEAWOLF*
The nuclear-powered attack submarine USS *Seawolf* (SSN-21), conducts sea trials off the coast of Connecticut in preparation for its commissioning in July 1997. Such submarines remain the ultimate form of national power projection.

HMS *Triumph* (*Trafalgar* class) (1991)

TYPE • *Submarine* **COUNTRY** • *United Kingdom*

SPECIFICATIONS

DIMENSIONS:	Length: 85.4m (280ft); Beam: 9.8m (32ft); Draught: 9.5m (31ft)
DISPLACEMENT:	4500–4800 tonnes (4960–5291 tons) surfaced; 5200–5300 tonnes (5732–5842 tons) submerged
PROPULSION:	1 Rolls-Royce PWR1 nuclear reactor; 2 GEC steam turbines; 2 WH Allen turbo-generators; 1 shaft
SPEED:	30 knots (56km/h; 35mph) submerged
RANGE:	Indefinite
ARMAMENT:	5 × 533mm (21in) torpedo tubes; Spearfish heavyweight torpedoes and Tomahawk Block IV cruise missiles
COMPLEMENT:	130

The last of the Royal Navy's seven-strong *Trafalgar*-class SSNs, HMS *Triumph* was decommissioned in 2025. The class incorporated a number of features that have been maintained or developed in the larger *Astute* successor class.

Right: The Royal Navy nuclear attack submarine HMS Triumph *is here moored in Den Helder, the Netherlands, on 26 October 2010. At this time it was conducting sea trials following a major six-year refit.*

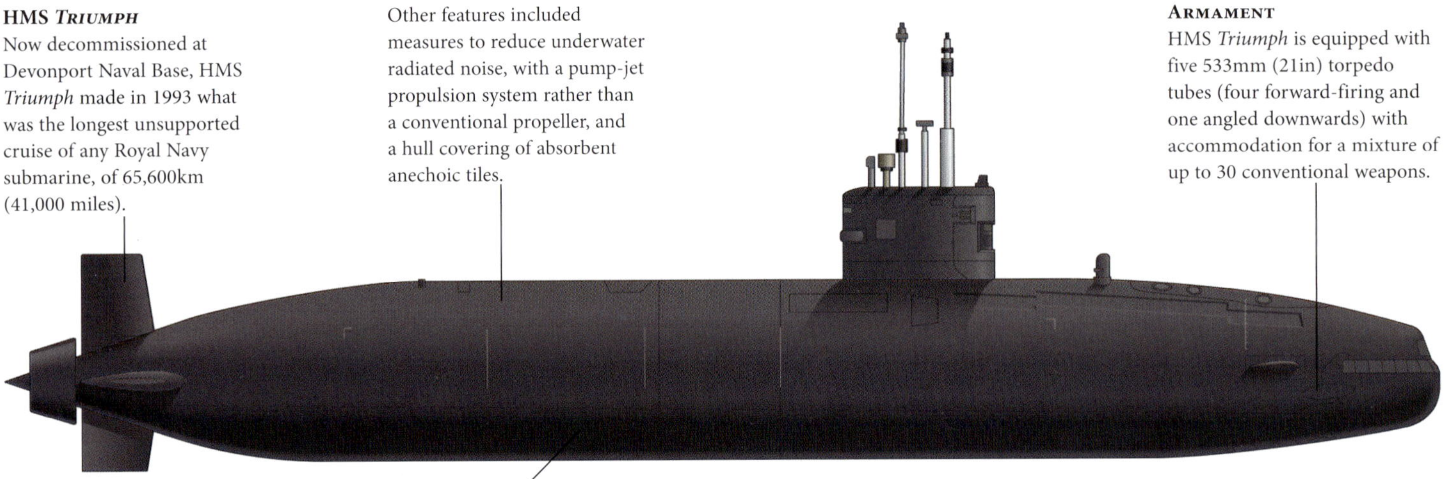

HMS *TRIUMPH*
Now decommissioned at Devonport Naval Base, HMS *Triumph* made in 1993 what was the longest unsupported cruise of any Royal Navy submarine, of 65,600km (41,000 miles).

HULL
Other features included measures to reduce underwater radiated noise, with a pump-jet propulsion system rather than a conventional propeller, and a hull covering of absorbent anechoic tiles.

ARMAMENT
HMS *Triumph* is equipped with five 533mm (21in) torpedo tubes (four forward-firing and one angled downwards) with accommodation for a mixture of up to 30 conventional weapons.

POWERPLANT
The design includes a first-generation Rolls-Royce PWR1 reactor, two GEC steam turbines, and two W.H. Allen turbo generators (3198kW; 4291hp).

ADVANCED WEAPONS

In the 1990s a 'first night' land attack capability with Tomahawk Land Attack Missiles (TLAM) was added to the *Trafalgar* class. These are now the improved Block IV model. The main underwater weapon is the Spearfish torpedo. This is a dual-purpose weapon able to hit both submarines and surface targets. It is in the same class as the US Navy's Mk48 ADCAP. The latest version, the Spearfish Mod-1, has improved homing, warhead and tactical systems. It is guided by a high bandwidth fibre-optic data link.

RETURN TO BASE
A submariner fixes HMS *Triumph*'s nameplate as the *Trafalgar*-class submarine returns to Plymouth following its successful participation in Operation *Ellamy* off Libya in 2011, supporting the enforcement of a no-fly zone over the country.

USS *Arleigh Burke* (1991)

TYPE • *Destroyer* **COUNTRY** • *United States*

SPECIFICATIONS (Flight I)

Dimensions:	Length: 154m (505ft); Beam: 20m (66ft); Draught: 9.3m (30ft 6in)
Displacement:	8315 tonnes (9167 tons)
Propulsion:	4 General Electric LM2500 gas turbines, each 19,570kW (26,250bhp), 2 shafts
Speed:	30 knots (56km/h; 35mph)
Range:	4400nmi(8100km; 5030 miles) at 20 knots
Armament:	Mk 41 Vertical Launch System (various missile fits); 2 × Mk 141 Harpoon missile launchers; naval guns and CIWS (various fits); 2 × or 6 × torpedo tubes (configuration-dependent)
Complement:	c. 300

This class, designed to replace the *Adams*- and *Coontz*-class destroyers, entered service in the early 1960s. *Arleigh Burke* was commissioned in 1991, to provide effective anti-aircraft cover, for which the SPY 1 D version of the Aegis system was fitted.

Right: The Arleigh Burke*-class guided missile destroyer USS* Shoup *(DDG 86) and* Nimitz*-class aircraft carrier USS* Abraham Lincoln *(CVN 72) perform a strait transit exercise in the Pacific Ocean, 26 January 2008. The strike group was participating in joint task force exercises off the coast of southern California.*

Phalanx
The Phalanx CIWS is a radar-guided multibarrel 20mm (0.79in) cannon, used for engaging fast incoming threats with a stream of shells, fired at between 3000 and 4500rpm.

Design
The destroyers use smooth-angled surfaces as much as possible to reduce the radar signature to enemy vessels and aircraft.

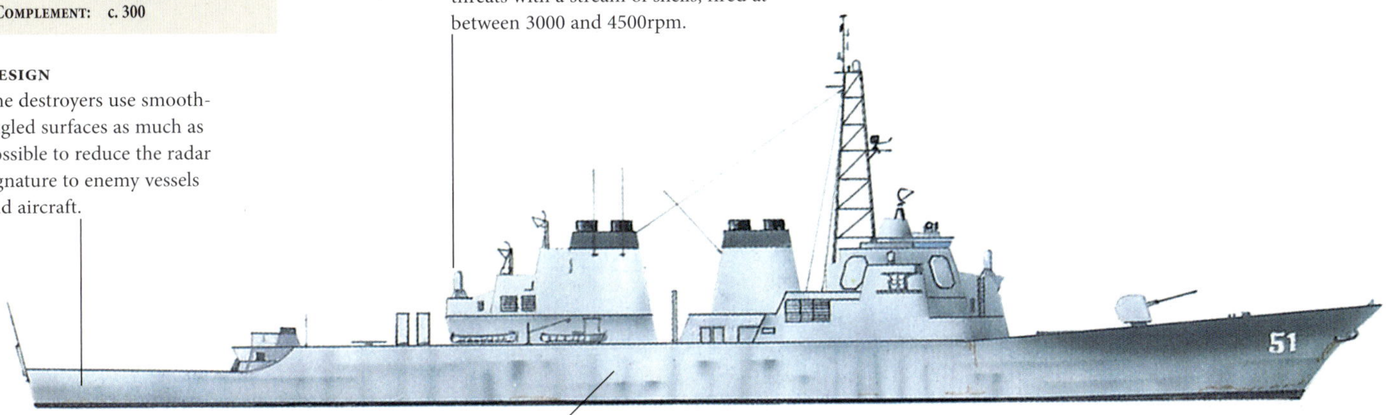

Flights
The *Arleigh Burke*-class destroyers have been built in several evolutions known as 'Flights', the most modern being Flight III.

FLEET MAINSTAY
Named after a distinguished US Navy admiral, the *Arleigh Burke* class of guided missile destroyers are the foundation of the US Navy's combat surface fleet, with some 74 vessels completed by 2025 and more planned for the future (leading to a planned total of 99 vessels). Each ship can deploy more than 100 missiles of various types, ranging from Tomahawk cruise missiles through to RIM-161 ballistic defence missiles and Harpoon SSMs.

USS *JAMES E. WILLIAMS* (DDG 95)
The *Arleigh Burke*-class (Flight IIA) destroyer USS *James E. Williams* is pictured here underway in the Gulf of Mexico during the builder's sea trials in 2004. The ship was commissioned in December 2024.

Vanguard class (1994)

TYPE • *Submarine* COUNTRY • *United Kingdom*

The four *Vanguard* class of nuclear-powered submarines are the largest submarines in the Royal Navy. Commissioned between 1993 and 1999, they were built by Vickers (now BAE Systems) at Barrow-in-Furness.

SPECIFICATIONS	
DIMENSIONS:	Length: 149.9m (491ft 10in); Beam: 12.8m (42ft); Draught: 12m (39ft 4in)
DISPLACEMENT:	c.14,150 tonnes (15,600 tons) surfaced; 16,154 tonnes (17,810 tons) submerged
PROPULSION:	1 Rolls-Royce PWR2 nuclear reactor; 2 GEC steam turbines developing 27,500shp (20.5MW); 1 shaft with pump-jet propulsor
SPEED:	25 knots (46.3km/h; 28.8mph) submerged
RANGE:	Indefinite
ARMAMENT:	16 × Trident II (D5) SLBMs; 4 × 533mm (21in) torpedo tubes

Right: HMS Vanguard *(seen here) was involved in a glancing mid-Atlantic collision with the French submarine* Le Triomphant *in 2009 when both sustained minor damage.*

SONAR
The submarines carry the Thales Underwater Systems Type 2054 composite sonar, a multi-mode, multifrequency system, which incorporates the 2046, 2043 and 2082 sonars.

HMS VENGEANCE
Commissioned in 1999, HMS *Vengeance* and its crew completed one of the longest sustained patrols recorded, with 210 days at sea, in 2023–24. *Vengeance* carries up to 16 Trident II D5 missiles.

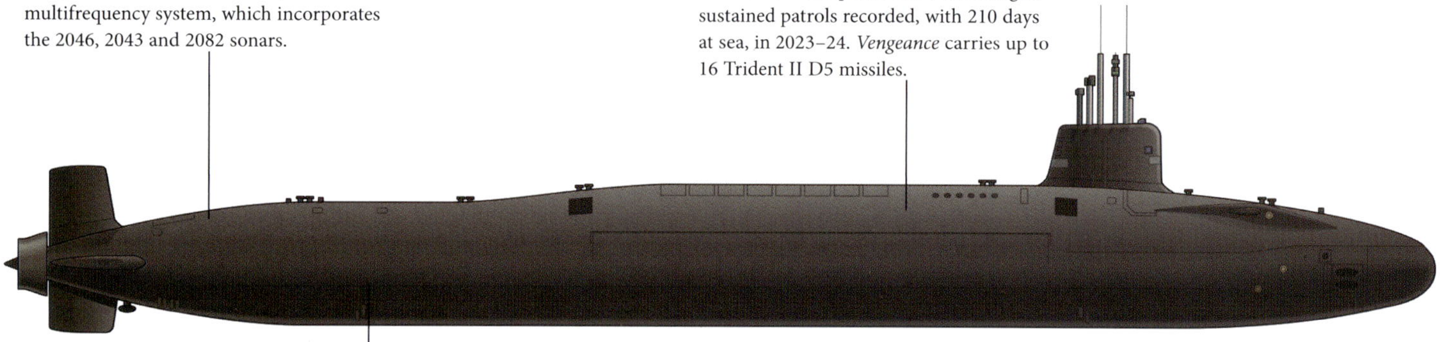

AUXILIARY POWER
Auxiliary power is provided by a pair of 6MW (7884hp) steam-turbine generators supplied by W.H. Allen (later Allen Power & Rolls-Royce), and two 905kW (1214hp) Paxman Valenta diesel generators for provision of backup power supply.

WEAPON LOAD

The *Vanguard* class are equipped with 16 ballistic missile tubes. However, as of the 2010 Strategic Defence and Security Review, the Royal Navy loads only eight of the missile tubes with the Trident II D5 submarine-launched ballistic missiles, each armed with up to eight nuclear warheads. In addition, the submarines are fitted with four 533mm (21in) torpedo tubes and Tigerfish Mk 24 Mod 2 and Spearfish heavyweight wire-guided torpedoes.

HMS *VANGUARD*
The *Vanguards*, all now exceeding their original planned 25-year life, and increasingly subject to maintenance and availability issues, are to be kept in service until the successor *Dreadnought* class becomes operational in the 2030s.

Kursk (Project 949A) (1994)

TYPE • *Submarine* **COUNTRY** • *Russian Federation*

A total of 11 Project 949A *Antei*-class submarines were built at the Sevmash yard in Severodvinsk between 1985 and 1999. Five or six are believed to remain active and the ill-fated *Kursk* belonged to this class-

SPECIFICATIONS

Dimensions:	Length: 154m (505ft 2in); Beam: 18.2m (60ft); Draught: 9m (29ft 6in)
Displacement:	13,600 tonnes (14,991 tons) surfaced; 18,300 tonnes (20,173 tons) submerged
Propulsion:	2 OK-650B pressurized-water nuclear reactors, approx. 190MW (254,790hp) thermal each; steam turbines developing approx. 74.6MW (100,000shp) total; 2 screws
Speed:	32 knots (59km/h; 37mph) submerged
Range:	Indefinite
Armament:	2 × 650mm (25.6in) torpedo tubes; 4 × 533mm (21in) torpedo tubes; 24 × P-700 3M45 Granit missiles; 24 torpedoes
Complement:	118

Left: Voronezh *(K-119) was one of the* Oscar II-*class SSGNs, serving from 1989 until it was decommissioned in 2020.*

Power
Power was supplied by two VM-5 190MW (254,790hp) pressurized water reactors, via two steam turbines generating 67,113kW (90,000hp) and turning two seven-bladed propellers.

Missile compartment
The modernized 949A boats will have the existing launchers replaced by universal launchers, able to fire Kalibr, Oniks or Zircon missiles, and carrying a larger weapons stock. The hypersonic Zircon, 9m (29ft) long and 60cm (23in) wide, is a multi-purpose missile designed for submarine-launched anti-ship and land attacks.

Kursk
The long sail of Project 949A holds an emergency escape exit. On *Kursk* this had been intentionally disabled in the course of a previous exercise.

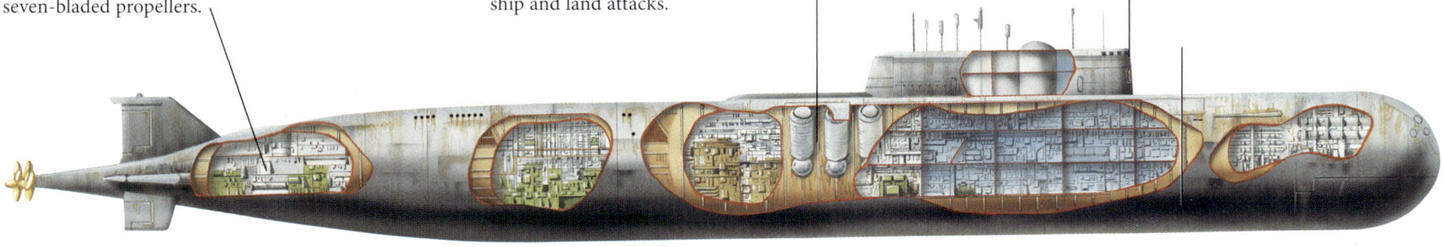

DESIGN
The NATO reporting name for this class is *Oscar II*. These boats are very large, displacing up to 14,700 tonnes (16,204 tons) surfaced. They are double hulled, with a hydrodynamic outer hull that helps to achieve an underwater speed of 30 knots (56km/h; 35mph). The inner hull is divided into 10 independent compartments, with five deck levels. The sail is reinforced for ice-breaking, with all masts retractable, and there is both an enclosed and open bridge.

KURSK
The Russian nuclear submarine, *Kursk*, at the vessel's home base of Vidyayevo, in Murmansk Oblast, Russia, March 1995. The *Kursk* later sank when a torpedo exploded on board during naval exercises in the Barents Sea off northwestern Russia on 12 August 2000. All 118 personnel on board were killed.

Admiral Kuznetsov (1995)

TYPE • *Aircraft carrier* **COUNTRY** • *Russian Federation*

SPECIFICATIONS

Dimensions:	Length: 305m (1001ft); Beam: 72m (236ft); Draught: 11m (36ft)
Displacement:	53,848 tonnes (59,360 tons) standard; 59,537 tonnes (65,660 tons) full load
Propulsion:	8 boilers; 4 steam turbines; 4 shafts
Speed:	29 knots (54km/h; 33mph)
Range:	8000nmi (14,800km; 9200 miles)
Armament:	12 × P-700 Granit (SS-N-19) anti-ship missile launchers; 24 × vertical Klinok (SA-N-9) SAM launchers; Kashtan combined gun-missile CIWS; 6 × AK-630 30mm (1.18in) AA guns
Aircraft:	26 fixed wing, 24 helicopters
Complement:	Approx. 2100

The *Admiral Kuznetsov* is designated as an 'aviation cruiser' and was commissioned for active service in 1995. The design and construction of the warship, however, are from the pre-1990 Soviet period.

Right: The most recently recorded air group for Admiral Kuznetsov *was 18 Su-33 (two are seen here conducting take-off) and six MiG-29K jet fighters, with four Ka-31 and two Ka-27 helicopters.*

Fuel
The fuel used is mazut, heavier and more viscous than diesel, with a higher sulphur content, and prone to generating thick black smoke.

Propulsion
Eight KGV-4 turbo-pressure boilers feed steam to four GTZA-674 turbines, (150,000kW/200,000hp) driving four shafts with fixed-pitch five-blade propellers.

Flight deck
Total flight deck area is 14,700 sq m (158,230 sq ft), with an angled deck and arrestor wires fitted on the landing area. There are two lifts on the starboard side.

FIRE CONTROL
The ship's armament reflects the 'cruiser' function: cAK-630 AA guns; CADS-N-1 Kashtan CIWS; P-700 anti-ship missiles; 3K95 Kinzhal surface to air VLS missiles; and RBU-12000 UDAV-1 ASW rocket launchers. *Admiral Kuznetsov* is equipped with D/E-band air-and-surface target-acquisition radar, F-band surface search and G/H-band flight control radars, as well as I band navigation radar and four K-band fire-control radars for the Kashtan CIWS.

EN ROUTE NORTH
A port beam view of the *Admiral Kuznetsov* underway south of Italy, en route to duty with the Soviet Northern Fleet. *Kuznetsov* (Project 1143 *Orel*, meaning 'Eagle') was designed by the Nevsky Bureau in Leningrad and built in modular 1400-tonne sections.

La Fayette class (1996)

TYPE • *Frigate* **COUNTRY** • *France*

SPECIFICATIONS

Dimensions:	Length: 125m (410ft 1in); Beam: 15.4m (50ft 6in); Draught: 4.1m (13ft 5in)
Displacement:	3200 tonnes (3527 tons) standard; 3800 tonnes (4189 tons) full load
Propulsion:	4 SEMT Pielstick 12PA6V280 STC2 diesel engines, 16,000kW (21,000hp), 2 shafts
Speed:	25 knots (46km/h; 29mph)
Range:	4000nmi (7400km; 4600 miles) at 15 knots
Armament:	1 × 100mm (3.9in) TR automatic gun; 2 × 20mm (0.79in) Modèle F2 guns; 8 × Exocet MM40 Block II anti-ship missiles; 1 × Crotale CN2 CIWS; 2 × 6-cell Sadral launchers with Mistral Mk 3 missiles
Complement:	164+

First commissioned in 1996, the French *La Fayette* class of frigates currently has 20 vessels in service (including export versions). The smooth and angular profile of the ship is part of its 'stealth' configuration, designed to reduce the ship's radar signature.

Left: The frigate Guépratte *(F 714) was the fifth vessel of the* La Fayette *class to serve in the French Navy. The class have also been exported to Saudi Arabia, Singapore and Republic of China (Taiwan).*

Noise reduction
The class is fitted with a Prairie-Masker noise reduction system, which masks the detectable acoustic signature of the propellers, reducing it to little more than the sound of rain water hitting the surface.

Forward gun
The Modèle 100 TR 100mm (3.9in) gun is a variant of the standard Modèle 68 weapon but set in a stealth turret. It can fire up to 60rpm cyclical rate.

Construction
The class was designed for rapid production, the inner parts of the ship constructed from 11 factory-completed prefabricated modules, assembled at the shipyard. One frigate can be built in less than two years.

STEALTH SHIPS

The *La Fayette* class was designed from the ground up with stealth characteristics in mind. The near avoidance of vertical lines and the lack of protruding fittings gives radar little to 'hook' itself onto. The superstructure is also formed from radar-absorbent synthetic materials, while the engines are configured to give a low thermal output. It is thought that the radar signature of these frigates is equivalent to that of a small fishing boat.

***LA FAYETTE* (F710)**
The lead ship of the class, *La Fayette*, was commissioned on 22 March 1996. Given its age, it received a major life extension upgrade in October 2021, receiving improved sensors and armament.

Collins class (1996)

TYPE • *Submarine* **COUNTRY** • *Australia*

SPECIFICATIONS

Dimensions:	Length: 77.8m (255ft 3in); Beam: 7.8m (25ft 7in); Draught: 7.0m (23ft 0in)
Displacement:	3100 tonnes (3418 tons) surfaced; 3470 tonnes (3827 tons) submerged
Propulsion:	3 Garden Island-Hedemora V18B/15UB diesel engines driving generators; 1 Jeumont-Schneider electric motor delivering approx. 5400kW (7200hp); single shaft with skew-back propeller
Speed:	Submerged 20kt (37km/h; 23mph)
Range:	11,500nmi (20,372km; 13,236 miles)
Armament:	6 × 533mm (21in) launch tubes; Mk 48 Mod 4/6/7 heavyweight torpedoes; UGM-84C Sub-Harpoon Block 1B missiles; up to 44 naval mines
Complement:	48

The Royal Australian Navy (RAN) has a submarine force of six Collins-class diesel-electric boats, which entered service between 1996 and 2003. While all are classed as active, not all are believed to be currently operational.

Right: HMAS Collins *(S-73) is the lead ship of the class. The RAN refers to them as 'guided missile submarines', which emphasizes their strategic role.*

Forward armament
Six forward 533mm (21in) launch tubes are fitted, and a mix of 22 Mk 48 torpedoes and USGM 84C Sub-Harpoon missiles, or 44 naval mines, can be carried.

X-rudders
The sail form and the X-rudders show the Swedish antecedents of the *Collins* class. An A$4–5 billion life extension programme for all six is set to start in 2026.

Weapons control
The class is fitted with a range of high-level electronic equipment, with AN-BYG 1 Link 11 weapons control to combine intelligence from sensors, compute input, and launch and direct weapons.

LIFE EXTENSION

After some troubled early years, the *Collins* boats have emerged as highly effective patrol/attack submarines. It is intended to give the entire class a life-of-type extension into the 2040s, while Australia's next generation of submarines is being developed, in conjunction with the USA and UK. This programme will include the replacement of power and propulsion systems, fitting of optronic masts and upgrading of weapons systems, perhaps including Raytheon Tomahawk missiles launched through the torpedo tubes.

RIMPAC EXERCISE
The crew of HMAS *Rankin* (SSG 78) enters Pearl Harbor on 17 July 2018 for a brief personnel stop during the *Rim of the Pacific* (RIMPAC) exercise. Twenty-five nations, 46 ships, five submarines, about 200 aircraft and 25,000 personnel participated in the exercise.

Improved Kilo class (Project 636.3) (1997)

TYPE • *Submarine* **COUNTRY** • *Russian Federation*

SPECIFICATIONS

Dimensions:	Length: 73.8m (242ft 1in); Beam: 9.9m (32ft 6in); Draught: 6.2m (20ft 4in)
Displacement:	2388 tonnes (2634 tons) surfaced; 3170 tonnes (3494 tons) submerged
Propulsion:	2 diesel engines driving generators; 1 main electric propulsion motor rated at approx. 4100kW (5500hp); single shaft with fixed-pitch seven-blade low-noise propeller
Speed:	Submerged 20 knots (37km/h; 23mph)
Range:	7500nmi (13,890km; 8630 miles)
Armament:	6 × 533mm (21in) torpedo/missile tubes; 3M-14 Kalibr land-attack cruise missiles and 3M-54 anti-ship missiles
Complement:	52

The original *Kilo* class of attack submarine was designed by the Rubin Bureau in Soviet Russia in the mid-1970s. The basic design, Project 877, has been updated several times, including the *Improved Kilo* Project 636.3.

Above: ORP Orzeł *(291) is a Polish Navy 'Project 877E' (*Kilo*-class) submarine.*

Sonar suite
The class has an up-to-date MGK-400EM Rubikon sonar suite with sail-mounted MG- 519M Arfa-M and MGK-400V.1 active sonars and forward-mounted MGK-400V.1 passive and intercept sonars.

Torpedoes
Six torpedo tubes enable the submarine to fire a fan of high-speed torpedoes, with greater chance of striking a moving target. Two can launch wire-guided torpedoes.

Project 636.3
Early *Kilos* had three diesel generators and three electric motors. Project 636.3 has two diesels and a single electric motor.

VARIANTS
Currently active *Kilo* boats go back to the mid-1990s when the Project 636 *Varshavyanka* type appeared, coded by NATO as *Improved Kilo*. Further variants have appeared since 2019 with the Project 636.3 PPK class (*Improved Kilo II*), and new *Kilos* are still being added, built by the Admiralty Yard in St Petersburg. The Russian Navy currently deploys 10 *Kilo* Project 877 (original) and 12 *Improved Kilo* Project 636.3, with six *Improved Kilo* under construction.

PETROPAVLOVSK-KAMCHATSKY
Three sailors of the Russian Navy stand on the *Kilo*-class submarine *Petropavlovsk-Kamchatsky* as they take part in the Navy Day parade, celebrating the 325th anniversary of the Russian Navy, in St Petersburg on 25 July 2021.

HTMS *Chakri Naruebet* (1997)

TYPE • *Aircraft carrier* **COUNTRY** • *Thailand*

Built in Spain as a smaller version of *Principe de Asturias*, the ship was intended for tactical defence and is now described as an Offshore Patrol Helicopter Carrier. *Chakri Naruebet* is the only carrier among the navies of the Southeast Asian region.

SPECIFICATIONS

Dimensions:	Length: 182.5m (599ft 3in); Beam: 22.5m (73ft 10in); Draught: 6.12m (20ft 1in)
Displacement:	11,670 tonnes (12,864 tons) full load
Propulsion:	2 gas turbines plus 2 diesel motors, 2 shafts
Speed:	25.5 knots (47.2km/h; 29.3mph)
Range:	10,000nmi (19,000km; 12,000mi) at 12 knots
Armament:	3 × sextuple Sadral launchers for Mistral missiles; 4 × 20mm (0.79in) Mk20 Rh-202 autocannon
Aircraft:	6 × Sikorsky SH-60 Seahawk helicopters; 2 (+4) × MH-60S Knighthawk; up to 14 additional helicopters when required
Complement:	455 plus 146 aircrew

Right: In 2012 the Royal Thai Navy (RTN) began an upgrade of the ship's command and control system, with Saab 9LV Mk4 command and control, and supply data-link equipment.

Capacity
This is the smallest working carrier to have hosted a fixed-wing air group, with a flight deck 174.6m (572.8ft) long. As a flagship, it has generous accommodation.

Sensors
Sensors include a Raytheon AN/SPS-52C 3-D medium-range air-search radar, Kelvin Hughes navigation and helicopter-control radar, Kelvin Hughes I-band navigation radar, MX 1105 Transit/GPS Omega satellite navigation system and an URN 25 Tacan system.

MBDA LAUNCHERS
The carrier has three MBDA (formerly Matra BAe Dynamics) Sadral six-cell launchers for the Mistral short-range anti-aircraft missile, able to intercept incoming sea-skimming missiles. The Sadral launcher is a stabilized turret equipped with a camera, which can be fitted with an infrared channel for target acquisition. Equipped with an infrared homing head supplied by SAGEM, the Mistral has a 3kg (6.6lb) high-explosive warhead loaded with tungsten balls. Its range is 4km (2.5m).

Hangars
Hangar space allows for 10 medium-size helicopters, though only five or six are carried. A centreline and deck-edge lifts are fitted.

UNDERWAY, 2001
The carrier's flight deck is elevated to a 12 degree ski jump at the bow to allow for short take-off aircraft. Initially with an air group of nine AV-8S Harriers (acquired from Spain), it has been a helicopter-only carrier since 2006.

Le Triomphant (1997)

TYPE • *Submarine* **COUNTRY** • *France*

SPECIFICATIONS	
Dimensions:	Length: 138m (453ft); Beam: 17m (55ft 9in); Draught: 12.5m (41ft)
Displacement:	12,842 tonnes (14,156 tons) surfaced; 14,335 tonnes (15,805 tons) submerged
Propulsion:	K15 pressurized-water nuclear reactor, 150,000MWth; turbo-reductor pump-to-shaft system delivering 30,519kW (40,927hp); 2 SEMT Pielstick 8PA4V200 SM emergency diesel generators, 700kW (940hp) each
Speed:	25 knots (46.3km/h; 28.8mph) submerged
Range:	Indefinite
Armament:	16 × M51-2 SLBMs; 4 × 533mm (21in) torpedo tubes
Complement:	112

France's four SSBNs form the *Le Triomphant* class, commissioned between 1997 and 2010. They are based at Ile Longue (Brest) in Brittany. The class has the distinction of being the most expensive submarines yet built.

Left: Triomphant *underway – power comes from a Type K15 pressure water reactor (150MW), made by Areva-TA, and burning low-enriched uranium (LEU) fuel, which involves six-yearly refuelling.*

Sail
The sail, 8.5m (27ft) high, is oblong in shape, with diving planes mounted in the forward part, and an emergency escape hatch and rescue chamber are located in the after part.

Noise reduction
The two turbo generators and the electric motor (30,500kW/40,900hp) are raft-mounted to minimize noise.

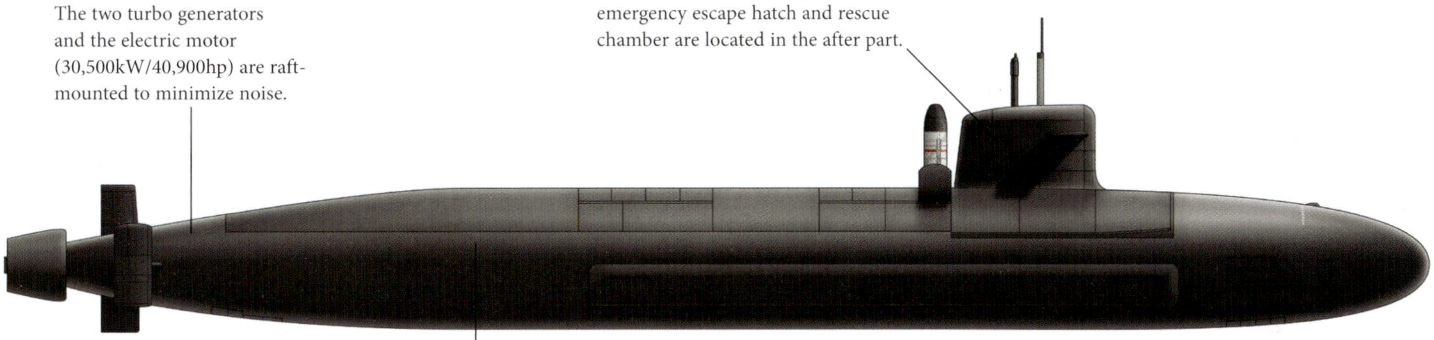

Construction and storage
For the non-pressurized bow, stern and upper deck hull sections, composite fibreglass/polymer material is used. The food stocks carried enable submerged patrols of up to 70 days duration

MODERNIZATION

Modernization of the class in 2012–18 saw *Le Triomphant* equipped with UMS-3000 sonar suite and SCOBS CCS, with armament of 16 M51-2 ballistic missiles to replace the previous M45s. These submarine-launched ballistic missiles (SLBMs) carry 6–10 independently targetable re-entry vehicles fitted with TNO (Tête Nucléaire Océanique) nuclear warheads. An advanced M51-3 version of the missile is under development.

LE TRIOMPHANT

Le Triomphant's keel was laid in 1986 at the DCN yard in Cherbourg, and the boat was commissioned in 1997. The pressure hull is of 100 HLES steel formed to withstand pressure exceeding 100kg per mm2 (1,422,33psi). Its French designation is SNLE-NG (*sousmarin lanceur d'engins nucléaires-nouvelle génération*).

USS *Seawolf* (1997)

TYPE • *Submarine* **COUNTRY** • *United States*

SPECIFICATIONS

Dimensions:	Length: 107m (351ft); Beam: 12m (39ft 4in); Draught: 10.9m (35ft 9in)
Displacement:	8738 tonnes (9634 tons) surfaced; 9285 tonnes (10,235 tons) submerged
Propulsion:	1 S6W pressurized water reactor (≈220MWth); 2 steam turbines delivering 42,500kW (57,000shp); secondary submerged propulsion motor; 1 shaft with pump-jet propulsor
Speed:	Submerged 33 knots (61km/h; 38mph)
Range:	Indefinite
Armament:	8 × 673mm (26.5in) launch tubes sleeved for 533mm (21in) weapons; up to 50 weapons including Tomahawk land-attack missiles, Harpoon anti-ship missiles, and Mk 48 torpedoes
Complement:	140

Two *Seawolf*-class boats were commissioned in 1997 and 1998, with a third in 2005 to a significantly different design. Intended as a replacement for the *Los Angeles* class, *Seawolf* was then discontinued in favour of the smaller, less expensive *Virginia* class.

Above: Commissioned in July 1997, USS Seawolf *(SSN-21) has seen numerous deployments north of the Arctic Circle where its ice-detection and ice-breaking abilities have been particularly valuable.*

Periscope
Seawolf's Kollmorgen Type 18 periscope is a sophisticated electro-optical instrument with multiple functional modes and a colour TV camera to assist intercept capability.

Sensors
The class has extensive equipment to permit shallow-water operations and make use of the advanced ARCI Modified AN/BSY-2 combat system, which includes a larger spherical sonar array, a wide aperture array (WAA) and a new towed array sonar.

Bow planes
Retractable bow planes are fitted and the sail is built for ice penetration (with active ice and mine avoidance sonar) and hydrostatic efficiency.

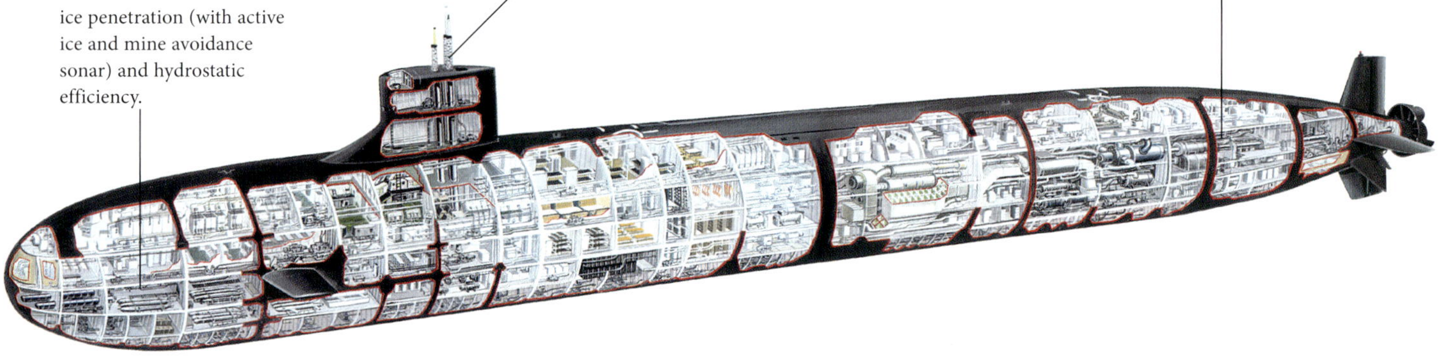

DESIGN
All three *Seawolf* class were built at the Electric Boat yard, Groton, Connecticut. The hulls are entirely constructed from HY-100 steel, which is stronger than the HY-80 steel employed in previous classes, in order to withstand water pressure at depths of 600m (1970ft) or more. Its lower length-to-beam ratio gives an improved speed and reduces noise, and the modular internal design was intended to make it easier to install upgraded equipment and weaponry.

SEA TRIALS
Then the nation's newest and most advanced attack submarine, *Seawolf* (SSN-21) puts to sea in the Narragansett Bay operating area for its first at-sea trial operations on 3 July 1996. Sea trials included various tests of the propulsion systems and the first underway submergence of the submarine.

HMS *Ocean* (1998)

TYPE • *Assault ship* **COUNTRY** • *United Kingdom*

SPECIFICATIONS

Dimensions:	Length: 203.4m (667ft); Beam: 35m (115ft); Draught: 6.5m (21ft)
Displacement:	21,500 tonnes (23,705 tons)
Propulsion:	2 Crossley Pielstick 16PC2.6 V200 diesels, 2 shafts, 17,557kW (23,904bhp)
Speed:	18 knots (33.3km/h; 20.7mph)
Range:	7,000nmi(13,000km; 8,000mi)
Armament:	3 × Vulcan Phalanx Mk 15 CIWS; 8 × Oerlikon/ BMARC GAM-B03 20mm guns in 4 twin mountings
Aircraft:	18
Complement:	365

Planned as a support vessel for amphibious landings, *Ocean* (L12) was unusually launched with its machinery fitted, and went under its own power from its Clyde launch site to Barrow-in-Furness for fitting out as a warship. The ship was sold to Brazil in 2017.

Above: High blank sides and a reduction of extraneous detail mark modern warship design in an era when 'stealth' is paramount and protection against radiation must be incorporated.

Flight deck
The flight deck, 170 x 33m (557 x 108ft) has six landing and six parking spots. The ship has full facilities for 12 EH101 Merlin and six Lynx helicopters and can also land and refuel Chinooks.

Hull form
The hull form is essentially that of a merchant vessel, with 'bolt on' features to give it some of the essential capabilities of a naval ship. These include such aspects as strengthening and fireproofing interior walls and coverings to withstand both heavy seas and battle conditions.

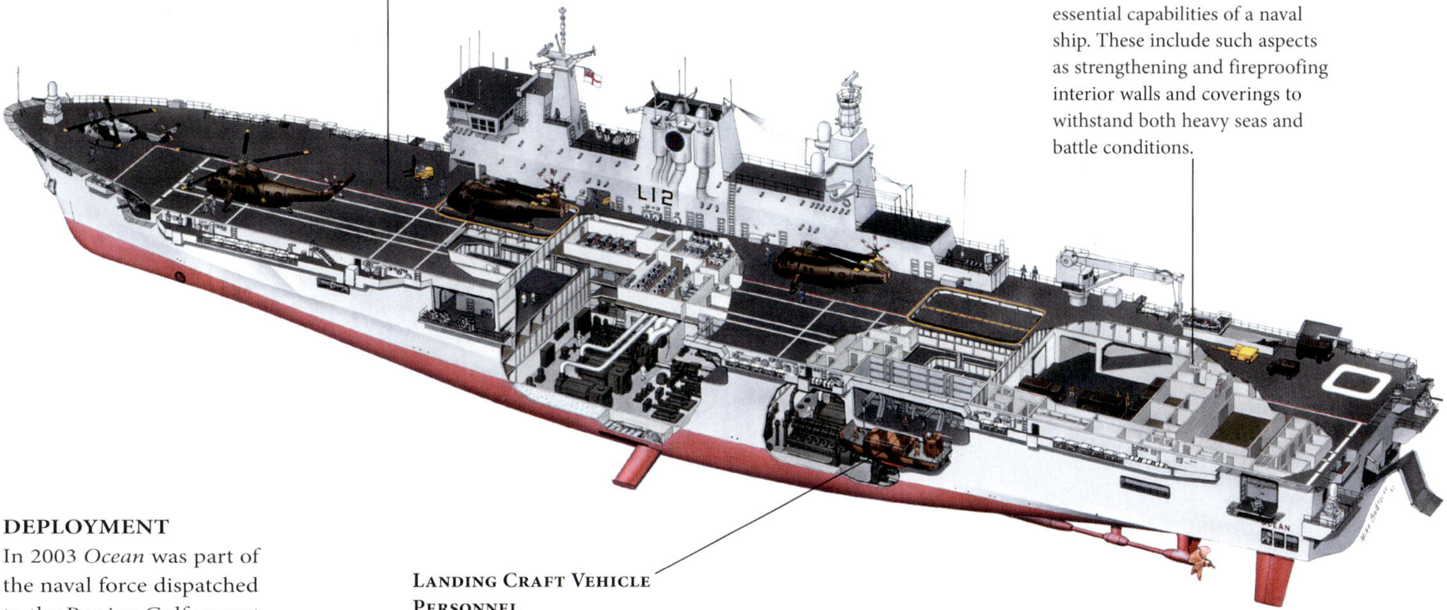

Landing Craft Vehicle Personnel
The LCVPs are mounted on box-section davit arms 6m (18ft 4in) long, each weighing 4 tonnes (4.4 US tons) with twin pivot points, operating on a winch and reeved rope system that must be kept permanently ready for use.

DEPLOYMENT
In 2003 *Ocean* was part of the naval force dispatched to the Persian Gulf as part of the British involvement in the Iraq War along with the carrier *Ark Royal*, three destroyers, a nuclear submarine and support ships. During the Iraq War, problems in the hydraulic and salt-water cooling systems were found with *Ocean*'s Mk5A landing craft. They were replaced with Mk 5B by March 2004. Also in 2004, between March and May, *Ocean* was the first ship to test the Apache AH Mk 1 helicopter at sea. A major refit was undertaken at Devonport in 2007.

CARIBBEAN EXERCISE
HMS *Ocean*'s embarked helicopters carry out training while heading for the Caribbean in 2017. *Ocean* was one of the ships to be tasked to support the government's Humanitarian Aid and Disaster Relief (HADR), providing assistance in the aftermath of Hurricanes Irma, Jose and Maria in the Caribbean.

Dolphin class (1999)

TYPE • *Submarine* COUNTRY • *Israel*

Israel's Type 800 or *Dolphin*-class submarine is based on the German Type 209, but modified to such an extent that it is really a class in its own right. Further development has produced the larger, AIP-fitted *Dolphin II* type.

SPECIFICATIONS (*Dolphin II*)

DIMENSIONS:	Length: 68.6m (225ft); Beam: 6.8m (22ft 3in); Draught: 6.2m (20ft 4in)
DISPLACEMENT:	2083 tonnes (2296 tons) surfaced; 2438 tonnes (2687 tons) submerged
PROPULSION:	Diesel-electric; 3 MTU V16 396 SE 84 diesels delivering 3120kW (4180hp); 3 Siemens alternators 750kW (1006hp); 1 Siemens main motor delivering 2850kW (3822hp); HDW/Siemens AIP system delivering 300kW (402hp); 1 screw
SPEED:	20 knots (37km/h; 23mph) submerged; snorkelling 11 knots (20.3km/h; 12.6mph)
RANGE:	8000nmi (14,816km; 9206 miles) at 8 knots
ARMAMENT:	4 × 650mm (25.5in) and 6 × 533mm (21in) launch tubes; torpedoes, Harpoon and Triton missiles, mines
COMPLEMENT:	35

Left: Based on a German design, constructed in Germany, but with substantial hi-tech instrumentation installed in Israel, the Dolphin-*class vessels operate as multi-function submarines, making regular patrols and engaging in special missions.*

BRIDGE
An observation/surface navigation bridge can be opened out at the forward brim of the sail, holding up to four personnel. A divers' lockout is positioned below it.

FIXED PLANES
Smaller horizontal fixed planes are placed each side of the screw on the *Dolphin IIs*, to allow for trailing wires or sonar arrays.

BEAM AND DRAUGHT
The profile shows *Dolphin I. Dolphin II*, although 11.6m (38ft) longer, is of the same beam and draught, with a similar outline.

CONSTRUCTION
The first of six boats, *Dolphin* was commissioned in 1999 and two more were in service by 2000. These are *Dolphin I*. A further three, of which two are delivered with a third under construction, are equipped with AIP and known as *Dolphin II*. With considerable input from the Israeli Navy, the boats were designed by Ingenieurkontor Lübeck Prof Gabler Nachf GmbH (IKL), and all are built by the Howaldtswerke-Deutsche Werft shipyard in Kiel, Germany.

INS *TANIN* (*DOLPHIN II*)
The basic design owes much to the German Type 209 submarine, and the second trio have benefitted from new automation systems developed for HdW's Type 212, but their increased size and numerous specialized features, some installed in Israel, make the *Dolphins* effectively a class on their own.

Type 051B *Luhai* class (1999)

TYPE • *Destroyer* **COUNTRY** • *China*

SPECIFICATIONS

Dimensions:	Length: 154m (505ft 3in); Beam: 16m (52ft 6in); Draught: 6m (19ft 8in)
Displacement:	6096 tonnes (6719 tons)
Propulsion:	Combined diesel or gas; 2 Ukrainian gas turbines, 35.7MW (47,900hp); 2 MTU 1163 TB 83 diesel engines, 6.5MW (8720hp); 2 shafts
Speed:	29 knots (54km/h; 33mph)
Range:	4500nmi(8300km; 5200 miles) at 14 knots
Armament	(post-2016): 16 × YJ-12 anti-ship missiles; 32-cell vertical launching system for HHQ-16 surface-to-air missiles; 2 × Type 1130 CIWS
Complement:	250

There is currently only one Chinese Type 051B *Luhai* class of guided missile destroyers – the *Shenzhen*, commissioned in December 1998. It was part of a broad experimentation in naval warship types by the People's Liberation Army Navy (PLAN) in the 1990s and 2000s.

Left: The PLAN guided-missile destroyer Shenzhen *(DDG 167) gets underway from Apra Harbor, Guam, following a three-day port call in October 2003.*

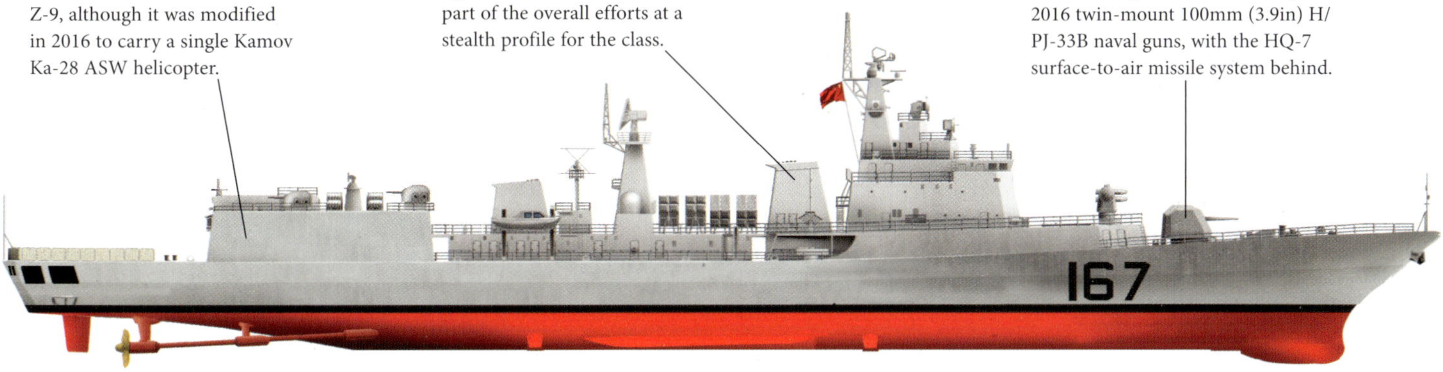

Hangar
The rear hangar (with flight deck above) can house two Harbin Z-9, although it was modified in 2016 to carry a single Kamov Ka-28 ASW helicopter.

Funnels
Shenzhen has two funnels, each of them fitted with infrared signature reduction devices, part of the overall efforts at a stealth profile for the class.

Main gun
Here the ship is shown with its pre-2016 twin-mount 100mm (3.9in) H/PJ-33B naval guns, with the HQ-7 surface-to-air missile system behind.

UPGRADE
Shenzhen received a major upgrade package in 2015–16, to equip it for future naval operations. Its four Type 76A 37mm (1.5in) AA guns were changed to two H/PJ-11 11-barreled 30mm (1.18in) CIWS and its SAM coverage was dramatically improved with a 32-cell vertical launch system (VLS) for extended range HHQ-16 missiles. The ship also received an extensive new sensor suite, including a Type 382 Radar on the forward mast and a Type 364 targeting radar on the aft mast.

TOKYO, 2007
Shenzhen sails through Tokyo Harbour in 2007, its crew marshalled on the decks. It was part of the PLAN's first goodwill visit to Japan. Note the octuple SAM system under the bridge.

Victoria (ex-*Upholder*) class (2000)

TYPE • *Submarine* **COUNTRY** • *Canada*

SPECIFICATIONS

DIMENSIONS:	Length: 70.26m (230ft 6in); Beam: 7.2m (23ft 7in); Draught: 7.6m (24ft 11in)
DISPLACEMENT:	2455 tonnes (2707 tons) surfaced
PROPULSION:	2 Paxman Valenta diesel generators developing 1.517MW (2035hp) each; 1 GEC electric motor; 1 shaft
SPEED:	12 knots (22km/h; 14mph) surfaced; 20 knots (37km/h; 23mph) submerged
RANGE:	8000nmi (15,000km; 9200 miles) at 8 knots
ARMAMENT:	8 × 533mm (21in) torpedo tubes; 18 Mk 48 Mod 7 heavyweight torpedoes
COMPLEMENT:	59

Canada has four diesel-electric submarines in service. Known as the *Victoria* class, they were originally the British Navy's last conventional submarine type, the *Upholder* class, with a surface displacement of 2438 tonnes (2687 tons) and six 533mm (21in) torpedo tubes.

Left: A casing party finish securing HMCS Windsor *in Barrow-in-Furness, Cumbria on 6 October 2001, ready for the voyage to its new home in the port of Halifax, Nova Scotia. Built for the Royal Navy as HMS* Unicorn *(*Upholder *class),* Windsor *is the second of four* Victoria*-class submarines in Canadian service.*

BATHYTHERMOGRAPH
Included in the full array of sensors is a submarine-launched expendable bathythermograph, used to determine the temperature and pressure of sea layers.

DIESEL ENGINES
The class is equipped with two Paxman Valenta 1600 RPS SZ diesel engines, each driving a 1.4-MW (1900hp) GEC electric alternator.

LIFE EXTENSION
A long-term life-extension programme is in progress to keep the class in service into the 2030s. Every aspect from hull condition through to propulsion, battery type, weapons control, combat management, communications and sensors will be upgraded.

CANADIZATION

First commissioned between 1990 and 1993, the British *Upholders* joined the Canadian fleet under new names between 2000 and 2015. Subsequent 'Canadianization' included exchanging British sensor and weapons systems for Canadian ones, many cannibalized from the previous *Oberon* class. It also included fitting USN Mk 48 heavyweight torpedoes and fire control system, together with new communications and periscope equipment fits. A Canadian-designed towed sonar array was also included in the programme.

HMCS *VICTORIA*
The Royal Canadian Navy submarine HMCS *Victoria* (SSK 876) arrives at Naval Base Kitsap-Bangor for a port call and routine maintenance. Crew conditions have improved considerably since the original 1980s design.

USS *Virginia* (2000)

TYPE • *Submarine* **COUNTRY** • *United States*

SPECIFICATIONS

Dimensions:	Length: 115m (377ft); Beam: 10m (33ft); Draught: undisclosed
Displacement:	7900 tonnes (8708 tons) submerged
Propulsion:	1 S9G pressurized-water nuclear reactor, 210MW (28,0000hp); steam turbines driving a single shaft with pump-jet propulsor
Speed:	25 knots (46.3km/h; 28.8mph) submerged
Range:	Indefinite
Armament:	4 × 533mm (21in) torpedo tubes for Mk 48 ADCAP torpedoes; vertical launch tubes for Tomahawk land-attack missiles
Complement:	134

The *Virginia* class is the US Navy's successor to the *Los Angeles* SSNs. By the end of 2025, 24 of the submarines were in service, with further vessels either under construction, awaiting delivery or undergoing trials.

Right: The Virginia*-class attack submarine USS* California *(SSN-781) underway during sea trials.*

Bow sonar
The Virginia Large Aperture Bow (LAB) sonar array is installed in Block III boats, with a TR353A active array containing 188 transducers, a DT-574B passive array with 1822 DT-547-B hydrophones, and a high-frequency CHIN array.

Rudders
Forward of the propulsor are the rudders, the towed array mounting, aft hydroplane and a sensor fin.

Drive system
A *Virginia*-class boat, USS *Montana* (SSN-794), has been experimentally fitted with magneto-hydrodynamic drive (MHD), also called 'caterpillar drive'. Superconducting magnets are used to accelerate the passage of water through the propulsion system. No moving mechanical parts are involved, so the submarine's movement is 'supersilent'.

BLOCKS

From USS *Virginia* (SSN-774, commissioned in 2004) onwards, the class is grouped in five 'blocks', with essentially the same specification and features, up to SSN-801 *Utah* (the bow change between Blocks II and III being an exception). Block V *Virginias* will be larger – displacing 10,400 tonnes (11,464 tons) compared to Block I's 7900 tonnes (8708 tons) – and with an advanced range of capacities including Virginia Payload Module (VPM) from SSN-803.

USS *VIRGINIA* (SSN-774)
The USS *Virginia* (SSN 774) returns to the General Dynamics Electric Boat shipyard in July 2004 following the successful completion of its first voyage in open seas, known as the 'Alpha' sea trials.

Visby (2000)

TYPE • *Corvette* **COUNTRY** • *Sweden*

SPECIFICATIONS

DIMENSIONS:	Length: 72.6m (238ft); Beam: 10.4m (34ft); Draught: 2.5m (8.2ft)
DISPLACEMENT:	590 tonnes (650 tons)
PROPULSION:	CODAG; 2 Honeywell TF50A gas turbines, total 16MW (21,446shp); 2 MTU Friedrichshafen 16V 2000 N90 diesel engines, total 2.6MW (2385shp); 2 shafts
SPEED:	40 knots (74km/h; 46mph)
RANGE:	2500nmi (4600km; 2900mi) at 15 knots
ARMAMENT:	1 × 57mm (2.24in) Mk3 gun; 8 × RBS15 Mk2 anti-ship missiles; mines and depth charges
COMPLEMENT:	43

The *Visby*-class is one of the latest class of corvettes to be deployed with the Royal Swedish Navy after the *Göteborg* and *Stockholm* classes. Becoming operational early in 2010, the vessel has many stealth features.

Right: The Swedish Visby*-class corvette HSwMS* Harnosand *(K33) transits the Baltic Sea during an exercise with the US* Wasp*-class amphibious assault ship USS* Kearsarge *(LHD 3) on 30 August 2022.*

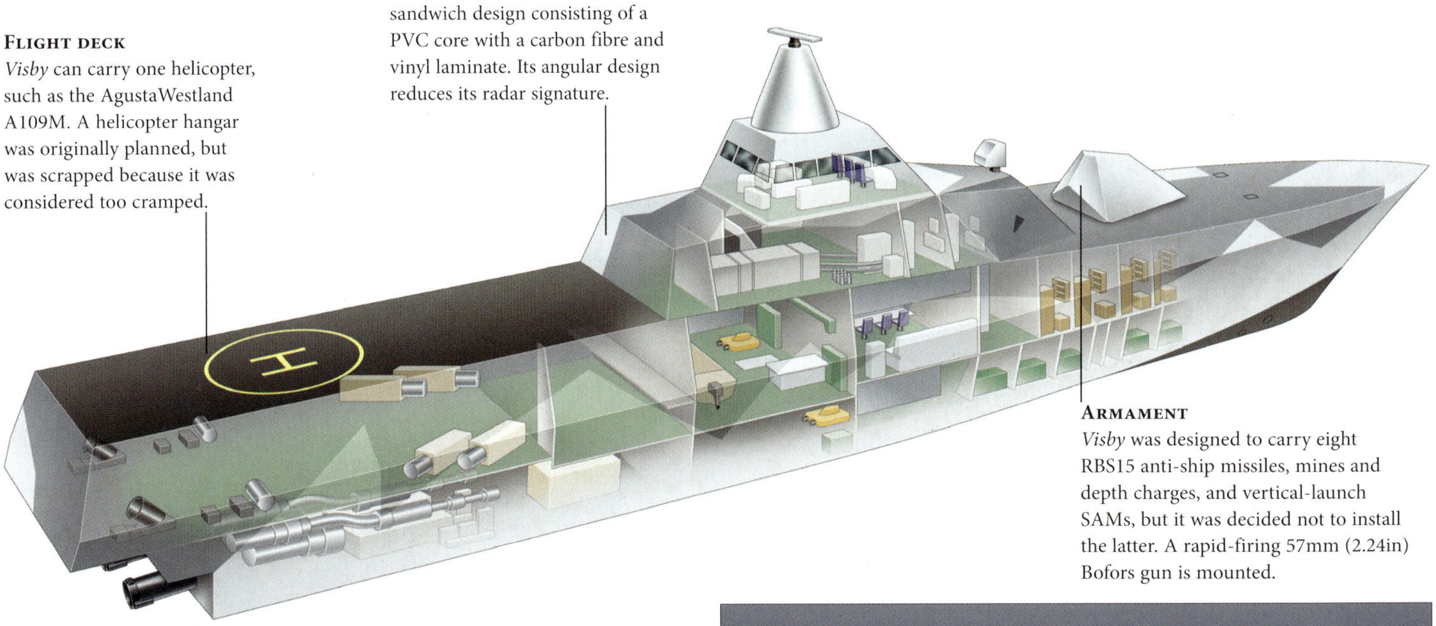

DELAYS

Visby was the lead vessel, launched in 2000. However, the entire progamme was immediately beset by delays, caused not least by a reduction of funding at the end of the Cold War. It was not until 2009 that *Visby* and a second vessel, *Helsingborg* (K32), were delivered to the Royal Swedish Navy. Even then, they lacked some of the planned weapon systems.

HSWMS *HELSINGBORG* (K32)

The Royal Swedish Navy *Visby*-class corvette HSwMS Helsingborg (K32) conducts manoeuvres during the Northern Coasts multinational exercise in September 2025.

Type 039 *Song* class (2001)

TYPE • *Submarine* **COUNTRY** • *China*

SPECIFICATIONS

Dimensions:	Length: 74.9m (245ft 8in); Beam: 8.4m (27ft 6in); Draught: 5.3m (17ft 4in)
Displacement:	1700 tonnes (1874 tons) surfaced; 2286 tonnes (2520 tons) submerged
Propulsion:	4 MTU 16V 396 SE84 diesel engines developing 18,148kW (24,360hp); single screw
Speed:	15 knots (27.7km/h; 17.2mph) surfaced; 22 knots (40.7km/h; 25.3mph) submerged
Range:	Undisclosed
Armament:	6 × 533mm (21in) torpedo tubes; 18 torpedoes or missiles or 24–36 mines
Complement:	60

Given the reporting name *Song* by NATO, the Type 039 – an ocean-going multirole SSK – is the first submarine to have been designed, built and fitted out in China. The *Song* class shows similarities with the Russian *Kilo*-class diesel-electric submarine.

Above: A Type 039G Song *class underway. The Type 039 is the first Chinese submarine built to a teardrop hull design. The body is water-drop shaped and the hull is covered with rubber tiles to absorb active sonar sound waves.*

Cruise missiles
The 039 class include YJ-82 cruise missiles for underwater launching. The YJ-82 is a highly effective surface skimmer with a 120km (75-mile) range, and a warhead of 165kg (363lb).

Sonar
To enhance passive search capabilities, a low frequency sonar of French Thomson-CSF TSM-2255 design is mounted on the flanks. Its maximum range is in excess of 30km (18.6 miles) and it can simultaneously track four targets.

Skew propeller
The propeller is designed to minimize pressure fluctuations on the blade to reduce cavitation (the formation of bubbles and consequent noise).

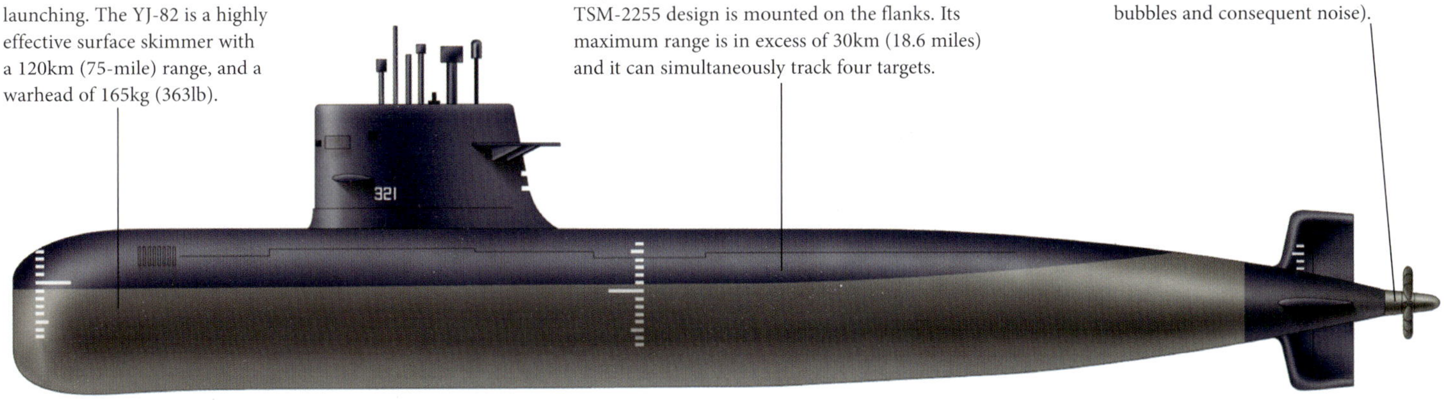

TYPE 039G
The first Type 039G, No. 321, was launched in November 1999 and commissioned in April 2001. This boat is characterized most obviously by a sail without the stepped-down forward section of No. 320. The new design reduced the acoustic signature and improved the underwater performance of the submarine. Estimated diving depth is 300m (980ft). A series of 12 submarines were built at the two shipyards.

TYPE 039G
A multirole combat and command system provides all the data needed for control of the boat and for the firing of torpedoes and missiles. This is likely to be an updated derivative of the combat/command system used in the *Ming*-class submarines.

HMS *Astute* (2001)

TYPE • *Submarine* COUNTRY • *United Kingdom*

SPECIFICATIONS	
Dimensions:	Length: 97m (318ft); Beam: 11.3m (37ft); Draught: 10m (33ft)
Displacement:	7000 tonnes (7716 tons) surfaced; 7400 tonnes (8160 tons) submerged
Propulsion:	1 Rolls-Royce PWR2 Core H nuclear reactor; two steam turbines; single shaft; pump-jet propulsor
Speed:	29 knots (53.7km/h; 33.3mph) submerged
Range:	Unlimited
Armament:	6 × 533mm (21in) torpedo tubes for Spearfish heavyweight torpedoes and Tomahawk Block IV TLAM; up to 38 weapons
Complement:	135

Britain's latest fast-attack SSN class – named after HMS *Astute* – is a class of seven boats designed to replace the *Trafalgar*-class fleet submarines. The service life of the new class is intended to extend into the mid-century.

Right: Measuring nearly 100m (328ft) from bow to stern, Astute *is longer than 10 London buses. When fully loaded, it displaces a volume of sea water equivalent to 65 blue whales.*

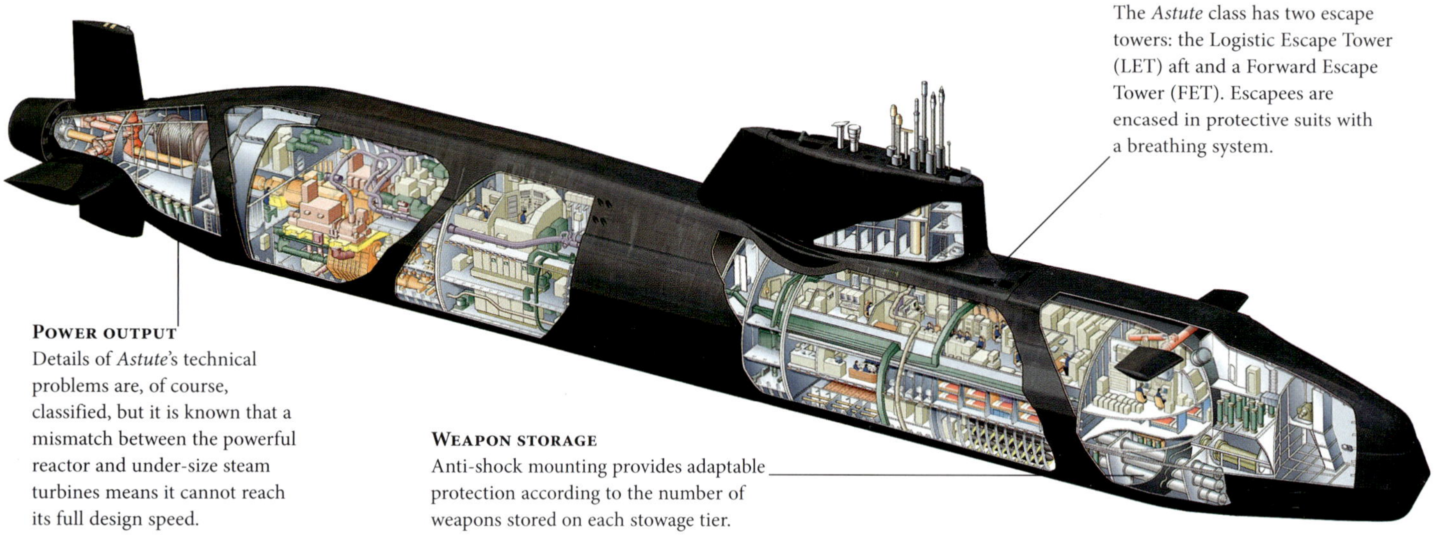

Escape towers
The *Astute* class has two escape towers: the Logistic Escape Tower (LET) aft and a Forward Escape Tower (FET). Escapees are encased in protective suits with a breathing system.

Power output
Details of *Astute*'s technical problems are, of course, classified, but it is known that a mismatch between the powerful reactor and under-size steam turbines means it cannot reach its full design speed.

Weapon storage
Anti-shock mounting provides adaptable protection according to the number of weapons stored on each stowage tier.

FIRST DEPLOYMENT
Astute's first deployment on active service was from February to September 2014. Its cruise took it to the Mediterranean Sea where it joined units of the US Mediterranean Fleet in exercises using the Tomahawk Land Attack Missile (TLAM) for potential anti-terrorist operations, then proceeded to the Gulf of Oman to conduct ASW exercises. This cruise was said to confirm the boat as a fully operational front-line unit.

DESIGN SECRETS
Few details of *Astute*'s design and structure have been published. The steel hull is said to be 31.78cm (12in) thick in places, and is covered with a total of 39,000 anechoic tiles.

Charles de Gaulle (2001)

TYPE • *Aircraft carrier* **COUNTRY** • *France*

For a long time the *Charles de Gaulle* was the only non-US nuclear-powered carrier. Designed in the 1980s and laid down at Brest in 1987, the carrier's ever-mounting costs, along with some technical problems, delayed its commissioning until 2001, seven years after its launch.

SPECIFICATIONS

Dimensions:	Length: 261.5m (858ft); Beam: 64.3m (211ft 3in); Draught: 9.4m (30ft 11in)
Displacement:	37,085 tonnes (40,880 tons); 42,500 tonnes (46,848 tons) full load
Propulsion:	2 K15 nuclear reactors, 2 Alstom steam turbines, 2 shafts, 61,000kW (83,000shp)
Speed:	27 knots (50km/h; 31mph)
Range:	Unlimited
Armament:	2 × eight-cell Sylver launchers for MBDA Aster 15 SAMs; 2 × six-cell Sadral launchers for Mistral missiles; 8 × GIAT 20F2 20mm (0.79in) cannon
Aircraft:	40

Right: Charles de Gaulle *(R91) transits the Suez Canal, its deck crowded with aircraft. The ship can operate a wing of up to 40 aircraft: Rafale M, Super Etendard and three E-2C Hawkeye airborne early warning aircraft. The ship also supports Dauphin PEDRO, AS 565 Panther or NH 90 helicopters.*

Island
With no need to incorporate a funnel to vent the engines, the island is placed well forward and is notably high, containing the bulk of the ship's detection and communication systems. The tip of the mast is 75m (228ft 6in) above the waterline.

Refuelling
Charles de Gaulle runs for 7.5 years between refuellings and can carry oil to supply three escorts for 10 days as well as 3400 tonnes (3346 tons) of aviation fuel.

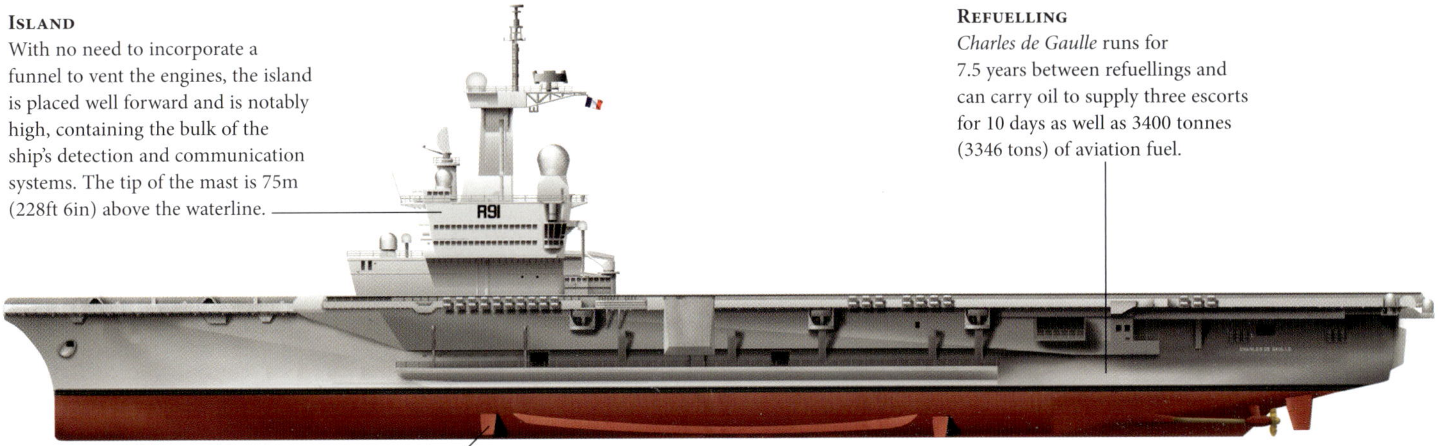

REFIT
De Gaulle's home port is Toulon in the south of France. The ship has been involved in numerous missions to world troublespots, including its first deployment in November 2001 to support the US-led Operation *Enduring Freedom* in Afghanistan. A 15-month refit began in September 2007, which included refuelling of the nuclear reactors and the fitting of new propellers. Defects in the original propellers had forced the ship to use the spare propellers of the smaller *Foch* and *Clemenceau* carriers.

Fins
The SATRAP (Système Automatique de Tranquillisation et de Pilotage) system, designed to launch planes in high winds and rough seas, combines lateral stabilization and two sets of retractable fins with the movement of 12 22-tonne rolling weights mounted on rails below the flight deck. The motion of these weights compensates for shifts in balance during take-off.

FLIGHT OPERATIONS
A French Rafale fighter jet lands on the flight deck of the *Charles de Gaulle* (R 91) as the ship conducts flight operations in the Mediterranean in November 2022.

USS *Ronald Reagan* (2003)

TYPE • *Aircraft carrier* **COUNTRY** • *United States*

SPECIFICATIONS

Dimensions:	Length: 332.8m (1092ft); Beam: 40.8m (134ft); Draught: 11.3m (37ft)
Displacement:	113,600 tonnes (125,200 tons)
Propulsion:	2 A4W nuclear reactors, 4 steam turbines, 4 shafts, 194,000kW (260,000shp)
Speed:	30 knots (56km/h; 35mph)
Range:	Unlimited
Armament:	2 × eight-cell Mk 29 launchers for RIM-162 Evolved Sea Sparrow SAMs; 2 × 21-cell RIM-116 RAM launchers; CIWS
Aircraft:	90
Complement:	5680

Ninth ship of the *Nimitz* class, *Ronald Reagan* (CVN-76) was commissioned on 12 July 2003. Flagship of the world's largest navy, this huge carrier displays a number of alterations and additions to the earlier ships of the *Nimitz*-class.

Right: The Precommissioning Unit (PCU) Ronald Reagan *(CVN-76) is taken out to sea for the first time during builder's sea trials off the coast of Virginia, 5 May 2003.*

Bow
A new form of bulbous bow was fitted to this ship, adding 122 tonnes (134 tons) to bow weight compared to its class-fellows, improving forward buoyancy and smoothing and stabilizing the vessel's passage through the water.

Flight deck
The island, mounted as a single pre-constructed unit, had many modifications, including a new design of mainmast and rearranged antennae. *Reagan*'s island has one deck less than the preceding *Nimitzes*, but greater internal height.

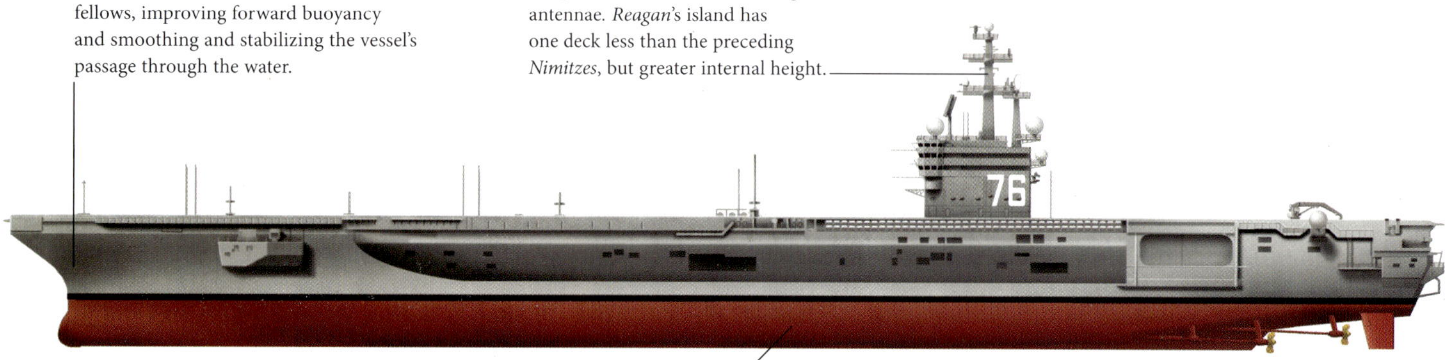

Powerplant
The ship is powered by the well-tried A4W reactors built by Westinghouse and with HEU 93.5 per cent fuel, rated for 25 years without refuelling. Combined power of the two reactors is 104MW (140,000hp) for propulsion.

DEPLOYMENTS

Ronald Reagan made its first deployments in 2006, performing maritime security operations in the Persian Gulf region and later taking part in Exercise *Valiant Shield*, which tested naval proficiency in detecting, tracking and engaging suspected threats at sea and on land, as well as in the air and cyberspace. On 28 August 2008 it launched its first combat sorties in Operation *Enduring Freedom*, supporting ground forces in southern Afghanistan.

PEARL HARBOR, 2009
Sailors aboard the USS *Ronald Reagan* man the rails as the ship pulls into Pearl Harbor, Hawaii, 13 October 2009. The convention of naming supercarriers after US presidents began with *John F. Kennedy* (CV-67) in 1968.

Type 212A (2005)

TYPE • *Submarine* **COUNTRY** • *Germany*

SPECIFICATIONS

Dimensions:	Length: 56m (183ft 9in); Beam: 7m (23ft); Draught: 6m (19ft 8in)
Displacement:	1450 tonnes (1598 tons) surfaced; 1830 tonnes (2018 tons) submerged
Propulsion:	1 MTU 16V 396 turbocharged diesel generator set, 2150kW (2881hp); 1 Siemens Permasyn Type FR6439 electric motor, 3000kW (4023hp); HDW/ Siemens AIP system, 300kW (402hp); single shaft
Speed:	12 knots (22.2km/h; 13.8mph) surfaced; 20 knots (37km/h; 23mph) submerged
Range:	8000nmi (14,816km; 9206 miles) at 8 knots
Armament:	6 × 533mm (21in) torpedo tubes for up to 24 Atlas Elektronik DM2A4 torpedoes or Siluri Whitehead A184 Mod 3
Complement:	27–30

In commission with the German and Italian navies, the Type 212 is a multirole submarine with highly sophisticated electronics. It was the first submarine to use AIP propulsion in regular service, making it capable of long-distance submerged passage to an area of operation.

Left: The Salvatore Todaro *(S526) is a Type 212A that came into service with the Italian Navy in March 2006. This forward view shows the slim profile of the class.*

Hull design
As with most other modern designs the hull has a prismatic cross-section and smoothly faired transitions to the sail.

Hydrogen fuel
In the interest of safety the bulk of the hydrogen fuel is carried outside of the inner hull and the reservoirs can be jettisoned outwards away from the boat.

SONAR
The Type 212A carries leading-edge technology. The first batch are equipped with an integrated DBQS sonar system that has a cylindrical array for passive medium-frequency detection, a sail-mounted TAS-3 low-frequency towed sonar array, FAS-3 flank array sonar for low/medium-frequency detection, passive ranging sonar and a hostile sonar intercept system.

GERMAN NAVY
U-31 (S181) is a Type 212A submarine, one of six serving with the German Navy. The original brief for the Type 212 came in the early 1990s from the Bundesmarine (German Navy) specifying a high-performance diesel-electric submarine that would also have AIP drive. Later, Italian participation in the project resulted in a change of spec and its designation as 212A.

Scorpène class (2005)

TYPE • *Submarine* **COUNTRY** • *Chile*

Originally a French–Spanish project and now solely French, the *Scorpène* ('Scorpionfish') class of diesel-electric attack submarines has been sold to several navies and remains a strong contender in competition for further orders.

SPECIFICATIONS

Dimensions:	Length: 70m (230ft); Beam: 6.2m (20ft 4in); Draught: 5.8m (19ft)
Displacement:	1870 tonnes (2061 tons) surfaced; 2032 tonnes (2241 tons) submerged
Propulsion:	2 diesel engines developing 1250kW (1675hp); electric motor developing 2900kW (3886hp); air-independent propulsion module; single screw
Speed:	12 knots (22.2km/h; 13.8mph) surfaced; 20 knots (37km/h; 23mph) submerged
Range:	6500nmi (12,038km; 7480 miles) at 8 knots
Armament:	6 x 533mm (21in) torpedo tubes for up to 18 torpedoes or SM39 Exocet anti-ship missiles or up to 30 mines
Complement:	31

Above: The Scorpène *hull can carry 18 heavyweight torpedoes or missiles, or 30 mines. It can fire the latest wire-guided torpedo types and can be adapted as an anti-ship, anti-submarine or dual-purpose attack boat.*

Armament
Black Shark torpedoes from Finmeccanica subsidiary Whitehead Alenia Sistemi Subacquei (WASS) are carried by the Chilean and Malaysian *Scorpènes*.

Rudder
The lower rudder is shorter to allow for 'bottoming' on the sea floor. DCNS also offers *Scorpène* with an X-rudder configuration.

Profile
The profile shows the AM-2000 form of *Scorpène*. Incorporation of AIP propulsion lengthens the hull by 8.3m (27ft 2in) compared with the CM-2000.

LOW NOISE
The forms of hull, sail and appendages, including the 'albacore' bow shape and optimized propeller, have been designed to produce minimum hydrodynamic noise. Internal decks have suspended mounts, and all mechanical and electrical items of equipment are mounted on elastic supports, which are in turn mounted on uncoupled blocks and suspended platforms, both minimizing noise and vibration and also providing better shock protection to the machinery.

US AND CHILEAN EXERCISE
A US helicopter hovers over a Chilean *Scorpène*-class submarine. Two *Scorpène* submarines, *O'Higgins* and *Carrera*, have been in service with the Chilean Navy since 2006.

Type 093 *Shang* class (2006)

TYPE • *Submarine* **COUNTRY** • *China*

SPECIFICATIONS

Dimensions:	Length: 110m (361ft 11in); Beam: 11m (36ft 1in); Draught: 10m (32ft 10in)
Displacement:	6090 tonnes (6714 tons) surfaced; 7110 tonnes (7839 tons) submerged
Propulsion:	2 pressurized-water nuclear reactors rated at approximately 70–75MW each; steam turbines driving a single shaft (exact turbine power output not publicly disclosed)
Speed:	35 knots (64.8km/h; 40mph) submerged
Range:	Indefinite
Armament:	6 × 533mm (21in) torpedo tubes for heavyweight torpedoes and submarine-launched cruise missiles; up to 36 mines
Complement:	100

The Type 093 SSN/SSGN is coded by NATO as the *Shang* class. The first was laid down in 1994 and the second shortly after. It is reported that Russian technicians contributed to the development of the Type 093.

Left: A Chinese Type 093B nuclear-powered attack submarine surfaces in October 2024. Chinese submarine technology apparently matches that of Russia and the USA and may exceed them in some respects.

Sail
The Type 093 is double-hulled. The 093A has a more streamlined sail, 2.5m (8.2ft) longer and set partly over a bulge believed to contain the winding gear of the towed sonar array.

Hull
The pressure hulls appear to be formed of HY-980 116ksi steel, with a yield strength of 800MPa (116,030psi), with high-strength welding applied via automation both for joining and monitoring the finish.

Reactors
The reactors on Types 093 and 094 are China's second-generation compact PWR types as on other nuclear submarines. They are believed to use LEU fuel and to have an integrated core and cooling system within the pressure vessel and a passive safety system.

DEVELOPMENT

A modified version of the Type 093, the Type 093A, formed four boats commissioned between 2015 and 2019, and four Type 093B, the first Chinese submarines with pump-jet propulsion, are also in service or under test. As with all Chinese nuclear submarines, they were built at the Bohai Yard in the city of Huludao, in Liaoning Province. The Russian Rubin Central Design Bureau is believed to have made substantial contributions to the design, including hull shape, noise reduction and combat and countermeasures systems.

FIGHTING SHIP
The Type 093s are designed both for oceanic and land attack depending on the weapon load. Six launch tubes of 533mm (21in) and 650mm (25.6in) diameter are fitted. The basic torpedo/mine load is 20 torpedoes or 36 mines.

Mistral (2006)

TYPE • *Assault ship* **COUNTRY** • *France*

SPECIFICATIONS

Dimensions:	Length: 199m (653ft); Beam: 32m (105ft); Draught: 6.2m (20ft 6in)
Displacement:	16,500 tonnes (18,188 tons); 21,300 tonnes (23,479 tons) full load
Propulsion:	2 Alstom Mermaid thruster pods; 14,000kW (18,774hp)
Speed:	18.8 knots (35km/h; 21.6mph)
Range:	5800nmi (10,000km; 6670mi) at 18 knots
Armament:	2 × Simbad sea-air missile launchers; 1 × twin 30mm (1.18in) Breda-Mauser cannon; 4 × 12.7mm (0.5in) Browning M2-HB guns
Aircraft:	16 (large) or 32 (light) helicopters
Complement:	310

Mistral **is lead ship of a class of three multi-purpose amphibious assault vessels (*bâtiment d'intervention polyvalent*) designed in the 1990s. This vessel was the first all-electric large warship, and the first to be driven by thruster pods.**

Right: Mistral's design is distinctive with its faired bow form, slab sides and squared stern. Operational needs required a relatively shallow draught.

Aviation wing
The aviation wing can be formed of 16 heavy or 30 light helicopters, or a mix. All types of French military helicopters can be carried.

Propulsion
Propulsion is from two Alstom Mermaid azimuth thruster pods, driven by electricity generated from five Wärtsilä 16V32 diesel alternators, the three main ones generating 4850kW (6504hp), with an auxiliary of 3000kW (4025hp) and an emergency type of 800kW (1072hp).

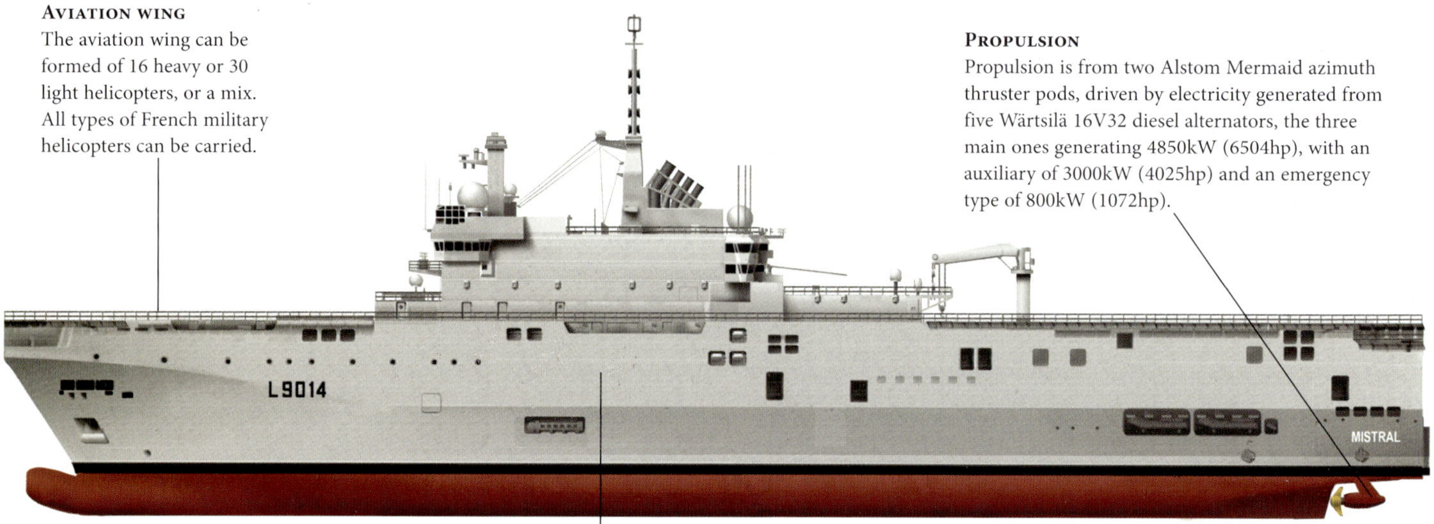

Command and control
A control and command centre is under the island on the second deck below the flight deck, with 150 work stations. Information from the sensors is centralized in the SENIT 9 tactical information system.

DESIGN

The French Navy has three ships of the *Mistral*-class. Work on the design began in 1997. The aim was for a ship capable of supporting an air wing and of carrying and landing military vehicles and armed troops sufficient for landing and supporting four combat companies and 280 vehicles within a 100km (62-mile) deep sector for up to 10 days. The design was planned to be modular in order to keep costs down and make it easy to update electronic equipment.

PURPOSE

The hull was built in two sections, the forward part at St Nazaire and the stern block at Brest, where the final assembly was completed. Like other ships of this type it provides waterborne, air and ground support for localized military operations.

HMS *Daring* (Type 45) (2006)

TYPE • *Destroyer* **COUNTRY** • *United Kingdom*

SPECIFICATIONS

Dimensions:	Length: 152.4m (500ft); Beam: 21.2m (69ft 6in); Draught: 7.4m (24ft 3in)
Displacement:	8092 tonnes (8919 tons) deep load
Propulsion:	2 Rolls-Royce WR-21 gas turbines, 21.5MW (28,819shp); 2 Converteam electric motors, 20MW (26,808shp); 2 shafts
Speed:	29 knots (53.7km/h; 33.3mph)
Range:	7000nmi (12,964km; 8050 miles) at economical speed
Armament:	1 × PAAMS air-defence system with SYLVER VLS for Aster 15 and Aster 30 SAMs; 1 × 115mm (4.5in) Mk 8 Mod 1 gun; 2 × 30mm (1.18in) guns; 2 × Phalanx CIWS
Complement:	190

The Type 45 destroyer is a state-of-the-art air defence destroyer programme, designed to replace the Type 42 destroyers in service with the British Royal Navy. The Type 45s are also known as the *Daring* class after the lead ship of the same name.

Radar
Daring's SAMPSON radar is said to be capable of tracking an object the size of a tennis ball travelling at three times the speed of sound.

Stealth features
The Type 45 incorporates signature reduction features, including the elimination of right angles and reduced equipment on deck. The infrared signature is reduced by cooling devices on the funnels.

Air defence
The Principal Anti-Air Missile System (PAAMS), a joint British, French and Italian design, is able to control and coordinate several missiles in the air at once.

Accommodation
The Type 45 is able to embark up to 60 troops (in addition to its own complement) and their equipment, and support them with a modern medical facility that can provide surgical capability.

SAM SYSTEM
Daring's main armament is the Principal Anti-Air Missile System, a surface-to-air missile system developed under a tri-national programme by France, Italy and the UK. This advanced weapons system will defend the Type 45, its consorts, and other task force vessels against highly manoeuvrable hostile incoming aircraft approaching at subsonic and supersonic speed, individually or in salvoes.

HMS *DARING*
The *Daring*-class destroyers are significantly larger than the Type 42s that they replaced, displacing about 7350 tonnes (8101 tons) compared to 5200 tonnes (5732 tons) of the Type 42. Their range is also about one-third greater, thanks to their advanced propulsion system.

Type 039C *Yuan* class (2006)

TYPE • *Submarine* **COUNTRY** • *China*

SPECIFICATIONS

DIMENSIONS:	Length: 77.2m (253ft 4in); Beam: 8.4m (27ft 7in); Draught: 6.7m (22ft)
DISPLACEMENT:	2768 tonnes (3051 tons) surfaced; 3657 tonnes (4031 tons) submerged
PROPULSION:	Diesel-electric propulsion with air-independent propulsion system; electric propulsion motor driving a single shaft;
SPEED:	20 knots (37km/h; 23mph) surfaced and submerged
RANGE:	6000nmi (11,112km; 6900 miles)
ARMAMENT:	6 x 533mm (21in) torpedo tubes for torpedoes, anti-ship cruise missiles and mines
COMPLEMENT:	65

The Type 039C *Yuan* is the most advanced expression of the Type 039 class. The strategic aim behind the Type 039 has a dual focus: first, to prevent any attempts to blockade Chinese ports, and second, to enable China to impose a blockade on Taiwan if necessary.

Above: A docked Type 039C. The profile of the vessel reflects the stealth prerequisite of most modern submarines.

MAST
Type 039C is China's first submarine fitted with photonic masts rather than conventional periscopes. These retract within the sail and convey information electronically to the control room in the pressure hull.

SONAR
Unusually for a 'conventional' submarine, Type 039C has a towed sonar array deployed from the upper rudder fin.

HULL
The 039s have a double-hull construction, indicated by the lateral flooding apertures ('limber holes').

FEATURES

The most notable feature of the Type 039C is a recontoured sail (the 'chine' form similar to that of the upcoming Swedish A-26 *Blekinge* class) with permanently extended diving planes, intended to help maximize the stealth features that form a key aspect of the entire design. A towed sonar array is fitted, deployable through the upper rudder structure.

YUAN CLASS SSKN

A further variant of the class, noted in 2024, appears to be a guided missile (SSG) boat with vertical launch tubes: a feature formerly associated with nuclear submarines, but one that other navies may also be developing on AIP boats.

Horizon class (2007)

TYPE • *Frigate* **COUNTRY** • *Italy/France*

SPECIFICATIONS

Dimensions:	Length: 152.87m (501ft 7in); Beam: 20.3m (67ft); Draught: 5.4m (18ft)
Displacement:	7050 tonnes (7770 tons) full load
Propulsion:	CODOG; 2 GE/Avio LM2500Plus gas turbines 20,500kW (27,491hp) each; 2 SEMT Pielstick 12PA6 STC diesels 4320kW (5793hp) each; 4 Isotta Fraschini V1716T2ME diesel generators 1600kW (2146hp) each; 2 shafts
Speed:	29 knots (54km/h; 33mph)+
Range:	6100nmi(11,000km) at 18 knots
Armament:	1 × SYLVER A50 VLS; 2 × 76mm (3in) guns; 8 × MBDA MM40 Exocet Block 3 SSM; 1 × SADRAL CIWS for Mistral SAM; 3 × Nexter Narwhal 20 Naval Weapon Stations; 2 × WASS/EuroTorp B515/1 torpedo tubes
Complement:	255

The four *Horizon*-class frigates (a further four are planned) are the product of the Common New Generation Frigate (CNGF) programme, a joint naval development programme between France, Italy and Britain, although the latter withdrew in 1999.

Right: The French Classe Horizon *frigates are known in Italy as the* Classe Orizzonte. *The first of the vessels came into commission in 2007.*

VLS
The SYLVER (*SYstème de Lancement VERtical*) is a vertical launching system (VLS), with 48 cells for Astor 15 or Astor 30 surface-to-air missiles.

Hangar
The class has a hangar for one AW101 or NH90 Caïman, with a flight deck aft. The helicopters perform lift and ASW duties.

Chevalier Paul
Sensors include the Selex ES EMPAR phased array G-band multi-purpose radar, BAE/Thales S1850M 3D long-range radar and Selex ES RTN-30X fire direction radar.

JOINT EFFORT
The Common New Generation Frigate (CNGF) was intended for Britain, France and Italy to replace key frigates and destroyers with a state-of-the-art air defence vessel. Development quickly became problematic, however, with each country having different commercial, organizational and technical requirements, even down to fundamentals such as the size of the ship. Britain withdrew in 1999 to focus on the Type 45 destroyer, and in 2000 France and Italy signed an agreement to produce four ships together.

FS *CHEVALIER PAUL* (D 621)
There are currently four vessels of the class in service. The French Navy operates the *Forbin* (D620) and *Chevalier Paul* (D621) while the Italian Navy has the *Andrea Doria* (D553) and the *Caio Duilio* (D554).

Type 094 *Jin* class (2007)

The Type 094 (NATO *Jin* class) is China's latest SSBN challenger. China's first SSBN design was Type 092 (*Xia* class), a single boat commissioned in 1983 that required many modifications and underwent several refits.

SPECIFICATIONS (Type 094A)

Dimensions:	Length: 137m (449ft 6in); Beam: 11.8m (38ft 8in)
Displacement:	8128 tonnes (8962 tons) surfaced; 11,176 tonnes (12,320 tons) submerged
Propulsion:	Pressurized-water nuclear reactor; steam turbines; single shaft
Speed:	20 knots (37km/h; 23mph) submerged
Range:	Indefinite
Armament:	12 × JL-2 submarine-launched ballistic missile tubes; 6 × 533mm (21in) torpedo tubes
Complement:	142

Right: A Type 094 Jin-class nuclear submarine, flying the flags of the Chinese People's Liberation Army Navy (PLAN), participates in a naval parade to commemorate the 70th anniversary of the founding of China's PLAN in the sea near Qingdao in eastern China's Shandong province, 23 April 2019.

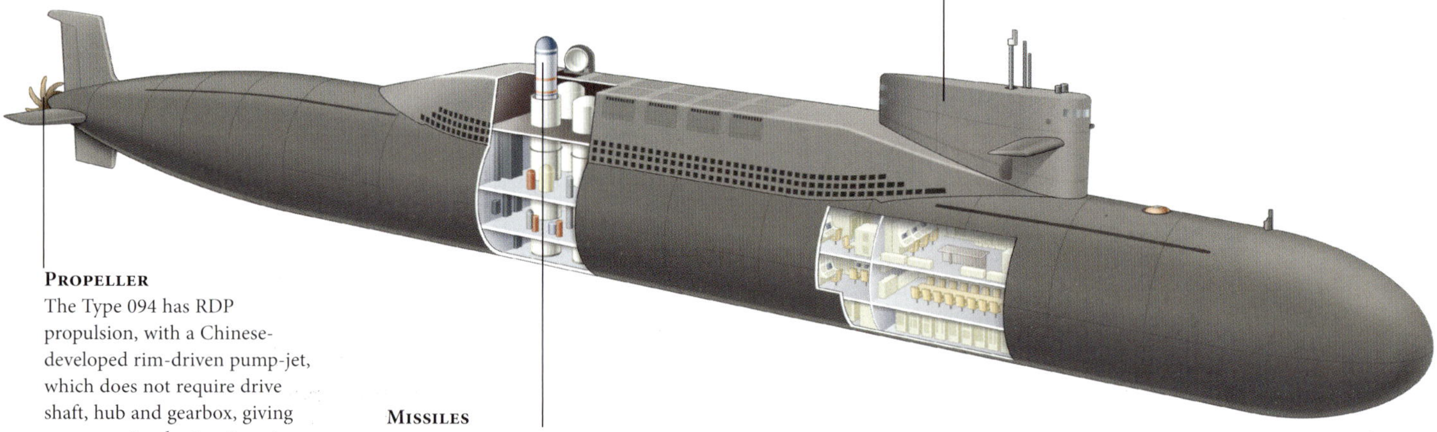

Sail
The Type 094A is said to have an improved sail, with features tested on a Type 093. These are likely to consist of sonar side panels and perhaps an anti-radar system such as STCM (space-time coding metasurface).

Propeller
The Type 094 has RDP propulsion, with a Chinese-developed rim-driven pump-jet, which does not require drive shaft, hub and gearbox, giving consequent reduction in noise and detectability.

Missiles
The missile compartment houses Ju Lang JL-3 missiles, solid-fuel powered using ammonium perchlorate, aluminium powder and hydroxyl-terminated polybutadiene (HTPB) binder, and with a guidance system linked to China's BeiDou navigation satellite system.

TYPE 094A

Four Type 094 boats were commissioned between 2007 and 2015. They did not rate well on the stealth scale compared to Western or Russian vessels, and an improved variant, 094A, was commissioned from 2020, with a second in 2021. It is China's most silent SSBN, fitted with rubber-synthetic hull tiles and improved internal damping systems, including automated 'floating damper' technology pioneered from the 1960s by the US Navy.

TYPE 094 SSBNN

The Type 094 is a second-generation SSBN, the development of which began in 1994 in the design offices of Wuhan Second Ship Design Institute. After the first boat's completion in 2004, it was subjected to tests and trials for several years.

Type 214 *Dolphin II* (2007)

TYPE • *Submarine* **COUNTRY** • *Germany/Israel/Greece*

The Israeli Navy operates six diesel-electric attack submarines, of which three are AIP-enabled. The first three, known as the *Dolphin I* class, were commissioned in 1999–2000; the others, commissioned between 2014 and 2024 and with AIP, are known as *Dolphin II*.

SPECIFICATIONS

Dimensions:	Length: 68.6m (225ft 1in); Beam: 6.8m (22ft 4in); Draught: 6.2m (20ft 4in)
Displacement:	2083 tonnes (2296 tons) surfaced; 2438 tonnes (2688 tons) submerged
Propulsion:	3 diesel generators; electric propulsion motor driving a single shaft; HDW fuel-cell air-independent propulsion system
Speed:	25 knots (46.3km/h; 28.8mph) submerged
Range:	Undisclosed
Armament:	4 × 650mm (25.6in) and 4 × 533mm (21in) torpedo tubes for heavyweight torpedoes and submarine-launched cruise missiles
Complement:	35 plus up to 10 special forces personnel

Above: Papanikolis *was completed in 2008 but not actually commissioned until 2010. In the Greek Type 214s, four tubes have expulsion systems for Harpoon missiles; the other four effect swim-out launches.*

Class features
Notionally of Type 214, the *Dolphin II* attack submarines have enough special modifications to merit being considered a separate class, especially the final member, *Drakon*, which is equipped with VLS launchers for cruise or ballistic missiles.

Dolphin II
On the *Dolphin II* subs, three additional stabilizers are fitted around the X-form rudders.

Dry shelters
Particular attention has been paid to servicing Special Forces activity. A dry shelter can be mounted aft of the sail. A lockout divers' chamber is located in the forward base of the sail. Fifteen supernumerary personnel can be accommodated.

MULTI-FUNCTION

Within a relatively limited geographical range (eastern Mediterranean and Red Sea), these are multi-function submarines. Notable design features are the prismatic hull cross-section and smoothly faired transitions from the hull to the sail, improving the boat's stealth characteristics. The hull and internal features are constructed of non-magnetic materials, significantly reducing the chances of detection by magnetometers or setting off magnetic naval mines. An external special forces hangar can be mounted aft of the sail.

***PAPANIKOLIS* (U214)**
The *Papanikolis* conducts trials off the coast of Norway. All the Type 214s are built by Thyssen-Howaldtswerke in Germany, and combine features of the Types 209 and 212, modified to Israeli specifications.

Cavour (2008)

TYPE • *Aircraft carrier* **COUNTRY** • *Italy*

SPECIFICATIONS

DIMENSIONS:	Length: 244m (800ft 6in); Beam: 39m (128ft); Draught: 8.7m (28ft 6in)
DISPLACEMENT:	27,100 tonnes (29,872 tons); 30,000 tonnes (33,069 tons) full load
PROPULSION:	COGAG; 4 gas turbines; 2 shafts
SPEED:	28 knots (52km/h; 32mph)
RANGE:	7000nmi (13,000km; 8057 miles)
ARMAMENT:	2 × SYLVER A-43 16-cell VLS for Aster-15 SAM; 2 × Oto Melara 76/62mm (3in) guns; 3 × Oto Melara/Oerlikon KBA 25/80mm (0.98/3.1in) guns
AIRCRAFT:	8 Lockheed Martin F-35B Lightning II or 12 helicopters (EH101 Merlin, NH90)
COMPLEMENT:	654 plus additional troops

Built by Fincantieri at Sestri Levante in Italy, the carrier *Cavour* was laid down in 2001, launched in 2004 and commissioned in July 2008. It is a multi-function vessel designed to deliver and support amphibious operations.

Above: The Cavour, *as part of Standing NATO Maritime Group 2, visits Civitavecchia, Italy after participating in an exercise in 2022.*

DEFENCE
For defence there are two SYLVER 8-cell VLS launchers for the Eurosam system, with Aster missiles, two Oto Melara 76mm (3in) super-rapid guns and three 25mm (1in) AA guns.

SENSORS
A wide range of integrated combat and sensor systems includes 2D surveillance radar, 3D multifunction and long radars, two navigation radars and mine avoidance sonar.

ANTI-TORPEDO SYSTEM
The French SLAT anti-torpedo system, which forces an attacking torpedo to follow a course that keeps it away from the ship until its endurance expires, is installed,

COGAG

Cavour is powered by a COGAG (combined gas) system, with four Fiat Avio (General Electric) LM2500 gas turbines each of 22,000kW (29,500hp), driving two shafts through two gear units each of 44,742kW (60,000shp). Six Wärtsilä CW 12V200 generators rated at 2200kW (2950hp) each provide power for stabilizers, bow and stern thrusters, desalinators, ramps, lifts, and two rudders.

ATLANTIC TRANSIT
The Italian aircraft carrier ITS *Cavour* (CVH 550) transits the Atlantic Ocean in March 2021, alongside (outside the photograph) the *Ford*-class aircraft carrier USS *Gerald R. Ford* (CVN 78), marking the first time a *Ford*-class carrier and an Italian carrier had operated together at sea.

Type 054 (2008)

SPECIFICATIONS (Type 054)

DIMENSIONS:	Length: 134m (439ft 7in); Beam: 16m (52ft 6in); Draught: c.5m (16ft 5in)
DISPLACEMENT:	4118 tonnes (4540 tons) full load
PROPULSION:	CODAD; 4 SEMT Pielstick 16PA6 STC diesel engines, 5700kW (7600hp) each; 2 shafts
SPEED:	27 knots (50km/h; 31mph)
RANGE:	8025nmi (14,900km; 9258 miles)
ARMAMENT:	2 × quadruple YJ-83 anti-ship missile launchers; 1 × 8-cell HQ-7 short-range SAM system; 4 × 6-barrel 30mm (1.18in) AK-630 CIWS; 2 × Type 87 6-tube ASW rocket launchers; 1 × 100mm (3.94in) gun
COMPLEMENT:	c. 165

This 'stealth' frigate class carries formidable missile armament and a full range of radar systems, with many features developed from French originals – the Chinese based their design heavily on the French *La Fayette*-class frigates.

Above: The PLAN Type 054A Yueyang *(FF 575) was commissioned on 3 May 2013 and has since participated in national and international naval exercises.*

HUANGSHAN (570)
The *Huangshan* is a Type 05A frigate commissioned on 13 May 2008. Its primary armament is a 32-cell vertical launch system for SAM and anti-submarine missiles.

TYPE 054A+
The ships from frigate 577 have been designated Type 054A+, which added variable depth and towed array sonar to improve anti-submarine capabilities.

HULL FORM
The hull form of the Type 054A has the same general stealth profile as the Type 054, with a mixture of sensors derived from French and Russian sources, as well as indigenous designs.

DEVELOPING CLASS

The Type 054 class consists of two ships – *Ma'anshan* (525) and *Wenzhou* (526) – built between 1999 and 2005 as anti-ship and air defence frigates with an overall stealth design. The class served as the foundation to two improved classes. By far the most prolific is the Type 054A, of which 40 are in service and more planned for delivery. This sub-class featured enhanced air-defence capability. The 054 family has been expanded by two longer and wider Type 054B, commissioned in 2025.

TYPE 054A *DAQING* (FF 576)
The original Type 054 is known by the NATO codename *Jiangkai I*, whereas the Type 054A (the *Daqing* pictured) and Type 054B are the *Jiangkai II* and *Jiangkai III* respectively.

America class (2008)

TYPE • *Assault ship* **COUNTRY** • *United States*

SPECIFICATIONS

Dimensions:	Length: 257.3m (844ft); Beam: 31.8m (104ft); Draught: 8.1m (27ft)
Displacement:	46,432 tonnes (51,225 tons) full load
Propulsion:	2 General Electric LM2500+ gas turbines, 52,000kW (70,000shp); 2 diesel-electric 3700kW (5000hp) auxiliary engines 2 shafts
Speed:	22 knots (41km/h; 25mph)
Range:	9500nmi(17,600km; 10,936mi) at 18 knots
Armament:	2 × RIM-116 RAM launchers; 2 × RIM-162 ESSM launchers; 2–3 × Phalanx CIWS; 3–4 × Mk 38 Mod 2 25mm guns
Aircraft:	Up to 20 × F-35B Lightning II plus rotary craft (mission dependent)
Complement:	1059, plus up to 1687 embarked Marines

At the time of writing (2026), two ships of the *America* class are in service and three more are being completed or are under construction. Defined as landing helicopter assault (LHA) types, they can act as compact fixed-wing and rotary-wing carriers.

Right: USS Tripoli *(LHA-7) – the second of the class – was built by Ingalls Shipbuilding (Huntington Ingalls Industries) in Pascagoula, Mississippi, and was commissioned into service on 15 July 2020.*

Mixed aviation
The *America* class does not have a fixed complement of aircraft. The types and number of aircraft vary according to the operational brief.

Design
The core influence on the design of the *America* class is the USS *Makin Island*, an improved ship of the *Wasp*-class amphibious assault ships.

CODLOG
The class uses a combined diesel-electric or gas (CODLOG) system. The ships can use gas-turbines for delivering high speeds and diesel-electric engines as alternative, steadier power.

AMPHIBIOUS ASSAULT
Amphibious assault ships are an increasingly important category with modern navies. Easily confused with aircraft carriers, they are rather multirole platforms for amphibious missions, providing the craft and embarkation point for assault troops and landing craft, while aviation assets deliver logistical and combat support across a beachhead and deep inland.

USS *AMERICA* (LHA-6)
This aerial photograph of the USS *America* shows the variety of aircraft types that can be operated, including the F-35B Lightning II, MV-22B Osprey, CH-53E Super Stallion and MH-60S Knighthawk.

Aquitaine (2008)

TYPE • *Frigate* COUNTRY • *France*

SPECIFICATIONS	
Dimensions:	Length: 142m (466ft); Beam: 20m (66ft); Draught: 5m (16ft)
Displacement:	6000 tonnes (6614 tons) full load
Propulsion:	CODLOG; 1 gas turbine, 32,000kW (42,900hp); 2 electric motors, 2500kW (3353hp) each; 4 diesel generators, 2200kW each; 2 shafts
Speed:	27 knots (50km/h; 31mph) maximum
Range:	6000nmi (11,000km; 6800 miles) at 15 knots
Armament:	VLS systems for SAM and naval cruise missiles; 1 × 76mm (3in) OTO Melara Super Rapid gun; 8 × MBDA MM40 Exocet Block 3 SSM; 2 × Nexter Narwhal 20mm(0.79in) naval weapon stations; 2 × twin 324mm (12.8in) torpedo tubes for MU90 torpedoes
Complement:	145

A product of French-Italian co-operation, the FREMM multi-purpose frigate can be adapted to anti-air, anti-ship and anti-submarine operations. More than 20 of the vessels are active today, with many more planned or under construction.

Right: The FREMM vessels are designed around a noticeably 'stealth' profile, with a minimal radar cross-section.

VLS
The *Aquitaine* class is fitted with a 16-cell SYLVER VLS for Aster 15 air-defence missiles (using A43 or A50 launchers depending on the ship), and a 16-cell SYLVER A70 VLS for MdCN land-attack cruise missiles.

Main gun
Although principally relying on missile armament, the Leonardo OTO Melara 76mm (3in)/62-calibre gun is common to the FREMM vessels.

Aquitaine (D650)
Aquitaine is the lead ship of the French class, commissioned in November 2012. In Italy the FREMM vessels are known as the *Bergamini* class.

FREMM
The *Aquitaine* class is the label applied to French Navy frigates of the Franco-Italian Frégate Européenne Multi-Mission (FREMM) programme. FREMM was initiated during the 2000s to provide replacements for ageing frigate types within the French and Italian navies and for export markets. FREMM ships have been configured for different roles, particularly land-attack, anti-submarine and air defence capabilities.

NAVAL SUCCESS
The FREMM project has largely been a success, notwithstanding contractual and budget issues. The ships have been exported to Egypt, Greece, Indonesia and Morocco. The United States has begun developing the *Constellation* class based on the Italian FREMM vessels.

Sōryū (2009)

SPECIFICATIONS

Dimensions:	Length: 84m (275ft 7in); Beam: 9.1m (29ft 10in); Draught: 8.5m (27ft 11in)
Displacement:	2900 tonnes (3197 tons) surfaced; 4200 tonnes (4630 tons) submerged
Propulsion:	2 diesel engines; 4 air-independent propulsion engines; electric propulsion motor driving a single screw
Speed:	20 knots (37km/h; 23mph) submerged
Range:	6100nmi (11,297km; 7020 miles) at 6.5 knots
Armament:	6 × 533mm (21in) torpedo tubes for heavyweight torpedoes, anti-ship missiles and mines

Japan's Maritime Self-Defence Force's (JMSDF) most numerous class – the *Sōryū* – is formed of 12 boats commissioned between 2009 and 2021. They are multi-role boats, equipped and armed for ASW, surface warfare and land attack, as well as for surveillance.

Left: The JS Sōryū *on exercise in the Pacific. The last two in the* Sōryū *class (2020–21) were the first submarines to be equipped with lithium-ion battery sets to feed the AIP motors. Earlier boats have lead-acid batteries.*

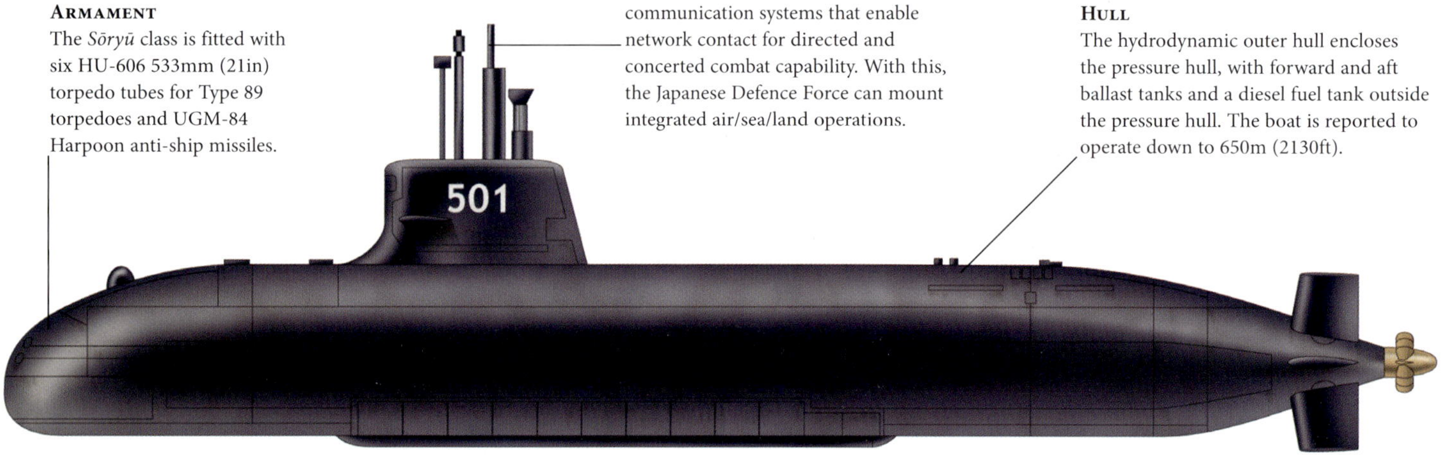

Communications
The class is fitted with communication systems that enable network contact for directed and concerted combat capability. With this, the Japanese Defence Force can mount integrated air/sea/land operations.

Armament
The *Sōryū* class is fitted with six HU-606 533mm (21in) torpedo tubes for Type 89 torpedoes and UGM-84 Harpoon anti-ship missiles.

Hull
The hydrodynamic outer hull encloses the pressure hull, with forward and aft ballast tanks and a diesel fuel tank outside the pressure hull. The boat is reported to operate down to 650m (2130ft).

HULL FORM
The hull form of the *Sōryū* class is made of high-tensile NS110 steel, covered with an anechoic coating to reduce the reflection of acoustic waves. Its hydrodynamic design is based on that of the *Oyashio* class. It has a larger displacement than any other submarine class in JMSDF's service. The submarine features computer-aided X-form control planes, stealth capabilities, and enhanced safety measures, including snorkel equipment.

JS *SŌRYŪ* (SS-501)
At Polaris Point, Guam, the Japan Maritime Self-Defence Force submarine JS *Sōryū* (SS-501) is pulled away from the submarine tender USS *Frank Cable* (AS 40) after a touch-and-go exercise on 10 June 2018.

Hyuga class (2009)

TYPE • *Helicopter carrier* **COUNTRY** • *Japan*

SPECIFICATIONS

Dimensions:	Length: 197m (646ft); Beam: 33m (108ft); Draught: 7m (23ft)
Displacement:	14,173 tonnes (15,620 tons) standard; 19,304 tonnes (21,282 tons) full load
Propulsion:	COGAG; 4 GE LM2500-30 gas turbines; 2 shafts
Speed:	30 knots (56km/h; 35mph)
Range:	Undisclosed
Armament:	1 × Mk 41 16-cell VLS for RIM-162 ESSM and RUM-139 VL-ASROC; 2 × Mk 15 Phalanx CIWS; 2 × Mk 32 324mm (12.75in) torpedo tubes
Aircraft:	Up to 18
Complement:	360

Designated as a 'helicopter destroyer', *Hyuga* is in all but name a light aircraft carrier. *Hyuga* and companion ship *Ise* are the largest Japanese warships since the giants of World War II. *Hyuga* serves as the Japan Maritime Self-Defense Force (JMSDF) flagship.

Left: Hyuga *at sea in the Pacific in 2023. The ship's primary concern is detection and destruction of hostile submarines, and the hangar deck can service 15 to 18 helicopters.*

Island
The island is topped by a six-level antenna tower, and two louvred exhaust uptakes are incorporated into the superstructure, with a gap between.

Control station
An aviation control station is placed aft with views fore and aft along the flight deck, which has four landing spots marked.

Propulsion
The power system is COGAG-type, combined gas turbine, with four GE LM2500-30 turbines of 44,742kW (100,000hp) driving two shafts with variable-pitch propellers.

THROUGH-DECK
Built by Ishikawajima Harima Industries Marine United at Yokohama and commissioned in 2009, *Hyuga* has a through-deck design to maximize potential aircraft space, enabling it to launch and retrieve its helicopter complement. It has no bow ramp or catapult provision, but there is speculation that it could be fitted for VTOL aircraft such as the F-35B Lightning.

PACIFIC EXERCISE
JS *Hyuga* (DDH 181) participates in manoeuvring exercises as part of Rim of the Pacific exercise in 2016. Twenty-six nations, more than 40 ships and submarines, more than 200 aircraft, and 25,000 personnel were involved, in and around the Hawaiian Islands and southern California.

Juan Carlos I (2010)

TYPE • *Aircraft carrier* **COUNTRY** • *Spain*

SPECIFICATIONS

Dimensions:	Length: 230.8m (757ft 3in); Beam: 32m (105ft); Draught: 6.9m (23ft)
Displacement:	26,000 tonnes (25,584 tons) standard; 32,000 tonnes (31,488 tons) full load
Propulsion:	1 GE gas turbine; 2 MAN diesel engines; 2 shafts; 2 thruster pods
Speed:	21 knots (39km/h; 24mph)
Range:	9000nm (17,000km; 10,000mi) at 15kt
Armament:	4 × Sentinel 2.0 RWS; 2 × BPDMS (FBNW); 1 VLS (FBNE); 4 × 12.7mm (0.5in) machine guns
Aircraft:	30 helicopters, or 10–12 AV-8B Harrier II or F-35 Lightning II plus 10–12 helicopters
Complement:	433 plus troops

This multirole warship reflects its era in two ways: it is designed both for an uncertain and unpredictable international environment, requiring a versatile ship, and to be affordable in terms of construction and outfitting.

Right: ESPS Juan Carlos I *in the Adriatic Sea in February 2023. The ship's LHD/STOVL design has been successfully adapted for use by other navies.*

Aviation
The ship can carry 30 helicopters, or 10/12 McDonnell Douglas AV-8B Harrier II or Lockheed Martin F-35 Lightning II and 10/12 helicopters, using the light vehicles bay as an additional aircraft storage zone.

CODLAG propulsion
CODLAG (diesel-electric and gas) propulsion is employed, with a single GE LM2500 19.75MW (26,484hp) gas turbine and two Navantia-MAN 32/40 16V 7.86MW (10,540hp) diesel generators raising power for two Siemens 11MW (14,751hp) azimuth thruster pods.

Capacity
The ship has a dock of 69.3m (227.3ft) length and 16.8m (55ft) width. The flooding deck can carry four LCM-1E or two LCAC landing craft. *Juan Carlos I* is able to transport 144 containers. It has full medical facilities including two operating theatres.

LHD DESIGN
This LHD design has four decks: a dock and garage for heavy cargo, a habitability deck, a hangar and light cargo garage. It is a STOVL type carrier with a through flight deck of 202m (663ft), and a port-side 12 degree ski-jump. The flight deck has eight landing spots. The island is blank-sided with forward navigation and air control decks, a centrally placed radar and communications mast, an integrated exhaust system, and a third mast set aft. A crane is mounted to starboard forward of the island.

DOCKED IN PIRAEUS
Juan Carlos I is pictured here docked at the port of Piraeus, Greece in 2024. The ship's great size is evident – able to carry equipment and support elements for 1200 troops.

Lada (Project 677) (2010)

TYPE • *Submarine* **COUNTRY** • *Russian Federation*

SPECIFICATIONS

Dimensions:	Length: 66.8m (219ft 2in); Beam: 7.1m (23ft 3in); Draught: 6.6m (21ft 8in)
Displacement:	1793 tonnes (1977 tons) surfaced; 2692 tonnes (2968 tons) submerged
Propulsion:	Diesel-electric propulsion; battery-powered when submerged; electric propulsion motor driving a single seven-blade screw
Speed:	21 knots (39km/h; 24mph) submerged
Range:	7500nmi (13,890km; 8630 miles) at 3 knots
Armament:	6 × 533mm (21in) torpedo tubes for heavyweight torpedoes and Kalibr cruise missiles
Complement:	35

Russia is supplementing and ultimately replacing the *Kilo* class with the new fourth-generation *Lada* class. It is designed for multi-function: detection, attack, self-defence, reconnaissance, and surveillance, in different sea environments.

Left: The Lada-*class diesel-electric submarine* Kronstadt *takes part in a naval display during Russian Navy day in St Petersburg, Russia, on 28 July 2019.*

Propulsion
Power comes from two diesel generators each producing 1250kW (1675hp), feeding an all-mode electric motor of 3020–3730kW (4050–5000hp).

Armament
The class is armed with Kalibr cruise missiles, carrying warheads of up to 400kg (880lb), with a range of 300km (187 miles).

Hull
A single-hull design is used, made of new AB-2 strength steel, with a Molniya hull anechoic covering. Hydroplanes are fitted on the forward part of the sail, which is fitted mid-hull on a long flat platform-like casing, which most likely holds ASW and other devices.

FEATURES

Stealth and detection features show the latest in Russian technology. The *Lada* class is highly automated, with the Lira sonar system (the first Russian bow sonar with a conformal antenna) plus a towed array, the Litiy automated combat management system, a Parus-98 optronic mast, and the Distantsiya integrated communications system. Given the level of automation, the crew numbers only 35 personnel.

***LADA* CLASS, ST PETERSBURG**
The *Lada* class was a stop-and-go process from 1997 to 2024, when a satisfactory version went into commission (B-586 *Kronstadt*, seen here). With only 75 per cent of the displacement of the *Kilo* class, it is more effective as a multirole boat.

San Antonio class (2011)

TYPE • *Assault ship* **COUNTRY** • *United States*

The *San Antonio* class are designated as either a Landing Platform Dock (LPD) or an amphibious transport dock. By 2026, the US Navy and the US Marine Corp were operating 12 of the vessels, with five more Flight II ships planned for service.

SPECIFICATIONS

Dimensions:	Length: 208m (684ft); Beam: 32m (105ft); Draught: 7.0m (23ft) full load
Displacement:	25,300 tonnes (27,900 tons) full load
Propulsion:	4 Colt-Pielstick turbocharged marine diesel engines, (41,600shp) 2 shafts,
Speed:	22 knots (41km/h; 25mph)+
Range:	7800nmi (14,445km; 8975 miles) at 15 knots
Armament:	2 × 30mm Bushmaster II close-in guns; 2 × RIM-116 RAM launchers; multiple twin 12.7mm M2 Browning machine gun mountings
Complement:	361 crew; 699 embarked troops

Right: The San Antonio-*class amphibious transport dock ship USS* New York *(LPD-21) arrives at Marine Hydraulic Industries (MHI) shipyard for maintenance in September 2025.*

Landing craft
The *San Antonio* class can hold and deploy two Landing Craft Air Cushion (LCAC) hovercrafts or one Landing Craft Utility (LCU) vessel and 14 Amphibious Assault Vehicles (AAVs).

USS *New York* (LPD-21)
Steel salvaged from the World Trade Centre was incorporated in the construction of this fifth ship in the *San Antonio* class.

Propulsion
Power comes from four sequentially turbocharged marine Colt-Pielstick diesel engines driving two shafts, achieving speeds beyond 22 knots (25mph; 41km/h).

LPD VESSELS
The US Landing Platform Dock (LPD) is another vessel type designed to support amphibious or littoral operations. LPDs are sizeable ships, capable of transporting hundreds of troops in addition to its crew, plus the landing craft to make ship-to-shore landings. The *San Antonio* class is also well-armed with SAM and SSM missile systems, increasing survivability in expeditionary warfare.

USS *PORTLAND* (LPD-27)
USS *Portland* is a *San Antonio*-class vessel commissioned in 2018. That same year, *Portland* was test fitted with the Technology Maturation Laser Weapon System Demonstrator (LWSD) MK 2 MOD 0, destroying a UAV with the laser weapon during trials in 2020.

Liaoning (2012)

TYPE • *Aircraft carrier* **COUNTRY** • *China*

SPECIFICATIONS

Dimensions:	Length: 304.5m (1000ft); Beam: 73.15m (240ft); Draught: 10.97m (36ft)
Displacement:	53,050 tonnes (52,215 tons) standard; 60,900 tonnes (59,938 tons) full load
Propulsion:	8 Harbin boilers; 4 Erzhong steam turbines; 4 shafts
Speed:	32 knots (59km/h; 37mph)
Range:	3850nmi (7130km; 4430mi) at full speed
Armament:	3 × Type 1130 CIWS; 3 × HQ-10 SAM missile systems
Aircraft:	36
Complement:	2586

Type 001, the PLAN Navy's first carrier, was originally planned in the 1980s as *Varyag*, companion ship to the Soviet Navy's *Admiral Kuznetsov*. In the aftermath of the Soviet Union's collapse in 1991, the hull was sold by Ukraine to China. From 2002 it became a carrier.

Left: China's first aircraft carrier, the impressive Liaoning, *arrives in Hong Kong on 7 July 2017, for a five-day visit to celebrate the 20th anniversary of the territory's return to Chinese rule.*

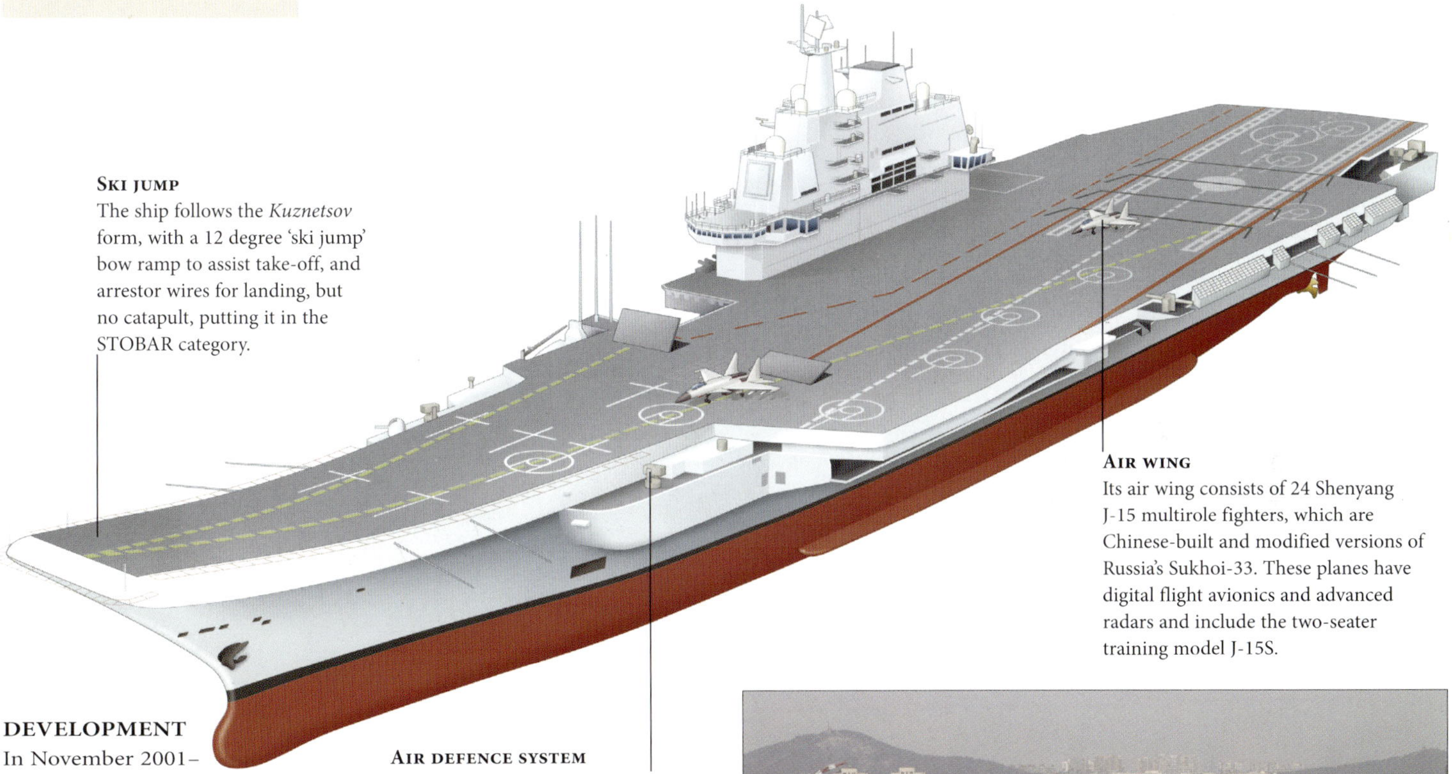

Ski jump
The ship follows the *Kuznetsov* form, with a 12 degree 'ski jump' bow ramp to assist take-off, and arrestor wires for landing, but no catapult, putting it in the STOBAR category.

Air wing
Its air wing consists of 24 Shenyang J-15 multirole fighters, which are Chinese-built and modified versions of Russia's Sukhoi-33. These planes have digital flight avionics and advanced radars and include the two-seater training model J-15S.

Air defence system
The FL 3000N CIWS is similar to the Raytheon RIM-116 Rolling Airframe Missile. Its fire-and-forget design locks on to the target immediately after launch and it travels at a speed of Mach 2.5.

DEVELOPMENT

In November 2001–February 2002 the ship was towed to the state-owned Dalian Shipyard in Liaoning Province. In June 2005 it was dry-docked and work began to transform it into an aircraft carrier. On 25 September 2012 it was commissioned and named after Liaoning Province. By then it had undergone sea trials, but much had still to be done, including installation of flight control, fire control, radar and weapons systems.

DALIAN, 2024
Liaoning sets out for sea trials at Dalian shipyard with the help of towboats after nearly one year of maintenance. Little is known outside the PLAN about key features of the ship, other than those that are apparent to external view.

Borei class (Project 955 and 955A) (2013)

TYPE • *Submarine* COUNTRY • *Russian Federation*

SPECIFICATIONS

Dimensions:	Length: 170m (557ft 8in); Beam: 13.5m (44ft 3in); Draught: 10m (32ft 10in)
Displacement:	14,720 tonnes (16,230 tons) surfaced; 24,000 tonnes (26,455 tons) submerged
Propulsion:	1 OK-650B pressurized-water nuclear reactor; steam turbine; pump-jet propulsor
Speed:	29 knots (53.7km/h; 33.4mph) submerged
Range:	Indefinite
Armament:	16 × RSM-56 Bulava submarine-launched ballistic missiles
Complement:	10

The successor to the Russian *Delta* class is the fourth-generation SSBN, Project 955 *Borei* class, named for the lead boat. The first boat was commissioned in 2013 and the class in 2026 numbered eight, with more under construction.

Left: The Project 955 submarine Knyaz Oleg *at Severomorsk, the main naval base for the Russian North Sea Fleet.*

Propulsion
The class is powered by a single OK-650 pressurized water reactor generating 190MW and providing steam to a GTZA OK-9VM single-shaft steam turbine plant, rated at 37,285kW (50,000shp).

Sonar
Russia claims that the *Borei*-class Irtysh-Amphora-B-055 sonar system is more than 1.5 times more effective than the US Navy's *Virginia* class, able to detect potential targets up to 100km (62 miles) further away.

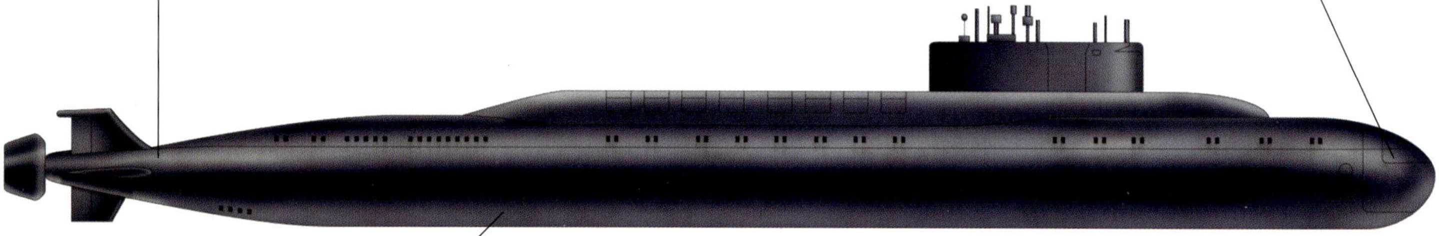

Hull
The pressure hull is formed from AB-2 steel, a high-yield alloy containing nickel, molybdenum and chromium. Its yield strength of 690MPa (6433 tons/sq ft) matches the USA's HY-100 rating.

DESIGN EVOLUTION
Work on the *Borei* class began in 1996. Design work is carried out by the Rubin Design Bureau in St Petersburg and construction is at Sevmash in Severodvinsk. The first boat was commissioned in 2013. While the first three boats were built using bow sections originally intended for *Akula* and *Oscar* SSGNs, the fourth and subsequent boats, designated Project 955A, are entirely new-build. These later boats show clear differences, with smoother, free-flowing hull lines and a more rounded sail.

***KNYAZ VLADIMIR* (K-549)**
The crew parade on the deck of the *Knyaz Vladimir* (K-549). As with the *Delta IV* boats, the *Borei* class has improved crew facilities, including a sauna holding up to four persons at a time.

Project 885 *Yasen* (*Granei* class) (2013)

TYPE • *Submarine* **COUNTRY** • *Russian Federation*

SPECIFICATIONS

DIMENSIONS:	Length: 119m (390ft 5in); Beam: 13.5m (44ft 3in); Draught: 8.4m (27ft 7in)
DISPLACEMENT:	8737 tonnes (9632 tons) surfaced; 14,021 tonnes (15,455 tons) submerged
PROPULSION:	1 KPM pressurized-water nuclear reactor; steam turbine; single shaft
SPEED:	35 knots (65km/h; 40mph) submerged
RANGE:	Indefinite
ARMAMENT:	8 vertical launch silos for submarine-launched cruise missiles; 8 × 650mm (25.6in) and 2 × 533mm (21in) torpedo tubes
COMPLEMENT:	85

Russia's newest SSGN is the formidable fourth-generation *Yasen* class, known as Project 885 *Yasen* and 885M *Yasen-M*. In effect this class is intended to fulfil the roles both of an SSGN and an attack SSN, replacing the remaining *Sierra II* and *Victor III* boats.

Left: A Yasen-class submarine slips out to open waters. The class is claimed to be virtually silent even at a submerged speed of 28 knots (52km/h; 32mph).

HATCHES
Hatches from which electronic countermeasures can be released are mounted forward of the submarine's sail.

MISSILE LAUNCHERS
The vertical launch tube arrangement is the UKSK (3P-14B), comprised of 8SM-346 modules, 10m (33ft) high and with a 2m (6.6ft) width that enables them to hold five Kalibr or four Oniks missiles.

PROFILE
The relative compactness of the 885M boats is partly explained by more extensive use of automatic equipment and AI systems. In addition, the monobloc reactor does not require separate steam generators – another space-saver.

POWER GENERATION

The monobloc KTP-6 reactor on the *Yasen-M*, built by OKBM Afrikantov, has a 25–30 year core life and will not have to be refuelled. Power generation is up to 200MW. A Kaluga Turbine Works steam turbine delivers 32,000kW (43,000hp). The new-generation machinery significantly reduces noise level: the reactor's primary cooling loop does not require continuous operation of the main circulation pumps, which are the key noise factor on a nuclear submarine.

HULL DESIGN

The hulls are formed of low-magnetic steel and are of partial double-hull construction, from the missile silos to the stern, with four decks in the single-hull midships section. Operational depth is down to a maximum of 658m (2160ft).

HMAS *Canberra* (2014)

TYPE • *Helicopter carrier* **COUNTRY** • *Australia*

SPECIFICATIONS

Dimensions:	Length: 230.8m (757ft 3in); Beam 32m (105ft); Draught: 7m (23ft)
Displacement:	27,534 tonnes (30,364 tons) standard; 27,940 tonnes (30,804 tons) full load
Propulsion:	1 GE LM2500 gas turbine; 2 MAN 16V32/40 diesel generators; 2 Siemens azimuth thrusters
Speed:	19 knots (35km/h; 22mph)
Range:	9000nmi (17,000km; 10,000mi) at 15 knots
Armament:	4 × Rafael Typhoon RWS; 3 × Mk 15 Phalanx CIWS; 6 × 12.7mm (0.5in) machine guns
Aircraft:	18
Complement:	358

Australia's largest warship is a state-of-the-art amphibious warfare vessel, capable of multiple roles, particularly enhancing the Australian Navy's off-shore capabilities. *Canberra* was modelled on Spain's *Juan Carlos I.*

Right: Showing its mighty proportions, HMAS Canberra *arrives at Joint Base Pearl Harbor-Hickam for the RIMPAC 20 exercise. The ship retains the bow ramp as on* Juan Carlos I*, though the ship is otherwise equipped for helicopters only.*

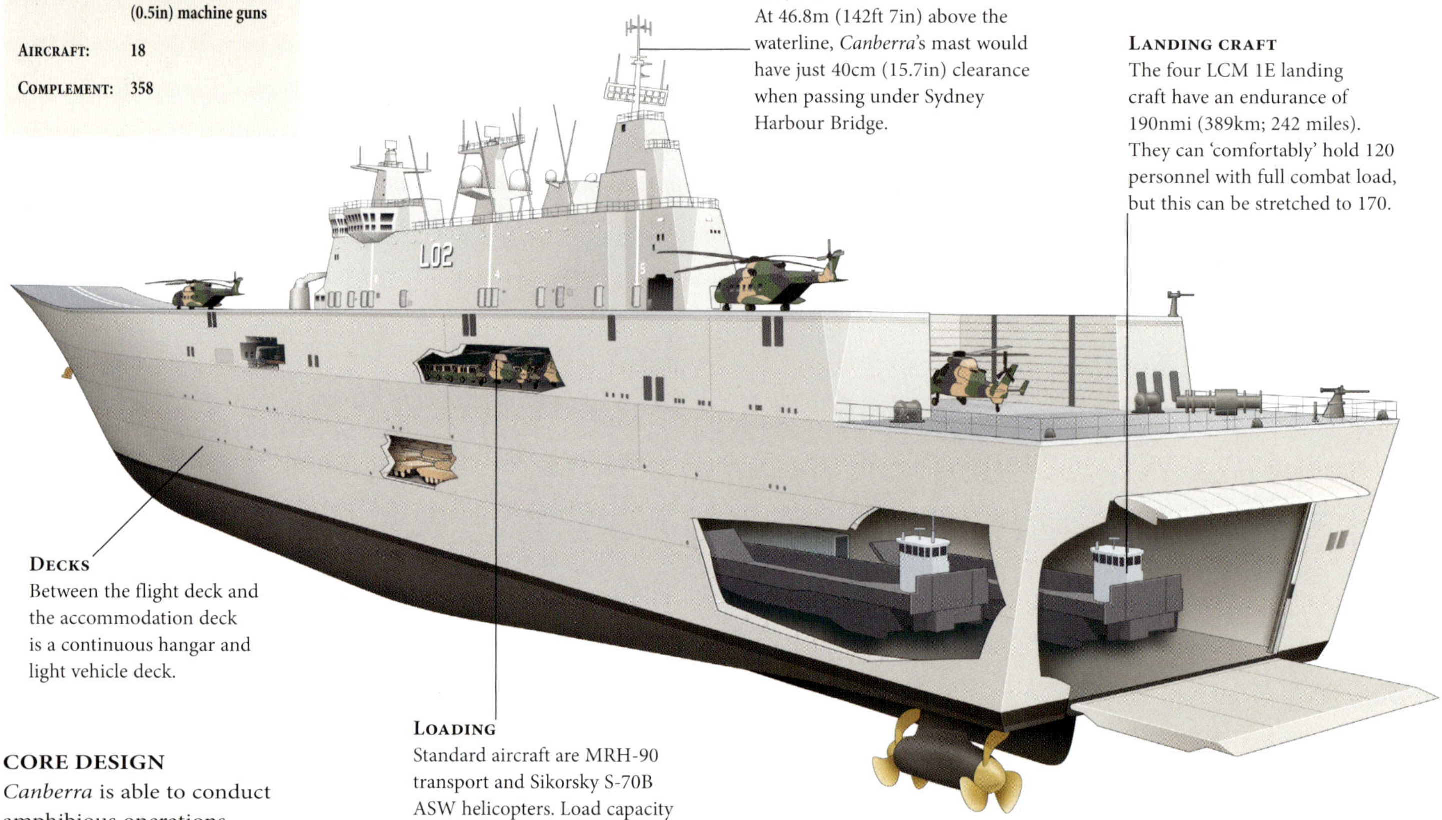

Mast height
At 46.8m (142ft 7in) above the waterline, *Canberra's* mast would have just 40cm (15.7in) clearance when passing under Sydney Harbour Bridge.

Landing craft
The four LCM 1E landing craft have an endurance of 190nmi (389km; 242 miles). They can 'comfortably' hold 120 personnel with full combat load, but this can be stretched to 170.

Decks
Between the flight deck and the accommodation deck is a continuous hangar and light vehicle deck.

Loading
Standard aircraft are MRH-90 transport and Sikorsky S-70B ASW helicopters. Load capacity is 18 helicopters, four LCM-1E landing craft, 110 military vehicles and 12 Abrams battle tanks.

CORE DESIGN

Canberra is able to conduct amphibious operations and land a force of more than 1,000 embarked personnel along with all their weapons, ammunition, vehicles and stores by LCM-1E landing craft, helicopters or a combination of both. Like *Juan Carlos I*, it has a conventional steel monohull design with the superstructure located on the starboard side of the flight deck. The appearance of both ships varies only in minor details.

LEAD SHIP

Canberra, although lead ship and first to be commissioned on 28 November 2014, carries pennant No. L02 while the other class member, *Adelaide*, is L01. The latter was commissioned on 4 December 2015.

Invincible class (Type 218SG) (2014)

TYPE • *Submarine* **COUNTRY** • *Singapore*

The new Sinaporean *Invincible* class of conventionally powered attack submarines currently consists of two Type 218SG, in service 2023–24, representing a switch by Singapore from Swedish to German models.

SPECIFICATIONS

Dimensions:	Length: 70m (229ft 8in); Beam: 6.3m (20ft 8in); Draught: 6.2m (20ft 4in)
Displacement:	2000 tonnes (2205 tons) surfaced; 2200 tonnes (2425 tons) submerged
Propulsion:	3 diesels; electric propulsion motor driving a single shaft
Speed:	20 knots (37km/h; 23mph) surfaced
Range:	Undisclosed
Endurance:	Up to 30 days
Armament:	8 × 533mm (21in) and 2 × 650mm (26in) forward-firing torpedo tubes
Complement:	28

Right: The Submarine Impaccable *lies in a ship dock shortly before the launching ceremony of two new 218SG class submarines for the Republic of Singapore Navy in December 2022 at Kiel, Germany. Classed as Type 218SG by the builders, the* Invincibles *are homed at Changi Naval Base. Crews may include both male and female sailors.*

RUDDERS
The class has been customized in various ways, including an X-form rudder system. Sensors have been optimized for operation in relatively shallow and congested sea lanes.

AIRLOCK CHAMBER
A horizontal multi-person airlock chamber allows deployment both of special forces and of submarine-launched cruise missiles (SLCMs).

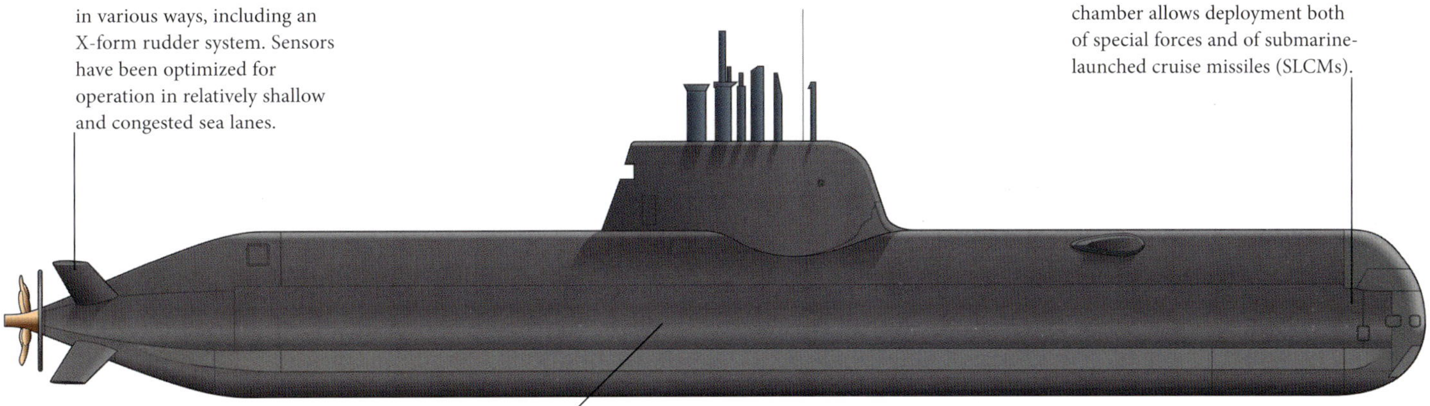

AI EQUIPMENT
Sophisticated automation and AI equipment installed under the direction of Singapore's Defence Science and Technological Agency allows for three eight-hour watches rather than the traditional two 12-hour stints.

FEATURES

Features of Types 212 and 214 are combined to fit local requirements in the *Invincible* class. All have fuel cell AIP drive in addition to the standard MTU diesels, with lithium-ion batteries, giving a submerged endurance of up to 42 days. The design is intended to operate in the shallow, warm waters around Singapore, with X-shaped rudders for greater manoeuvrability. Included in the bow section is a horizontally mounted multi-purpose release chamber for swimming out robotic vehicles or combat teams.

OUT OF WATER
This *Invincible*-class submarine is seen out of water in Kiel, Germany, in 2007. Two are in service and a further two are on order, for delivery by 2030, replacing the two *Challenger*-class boats withdrawn at the end of 2024.

Izumo (2015)

TYPE • *Multirole cruiser* **COUNTRY** • *Japan*

SPECIFICATIONS

Dimensions:	Length: 248m (814ft); Beam: 38m (124ft 8in); Draught: 7.5m (24ft 7in)
Displacement:	19,800 tonnes (21,829 tons) standard; 27,432 tonnes (30,250 tons) full load
Propulsion:	COGAG system; 4 GEC/IHI LM2500IEC gas turbines; 2 shafts
Speed:	30 knots (55.5km/h; 34.5mph)
Range:	Undisclosed
Armament:	2 × Phalanx CIWS; 2 × SeaRAM missile launchers
Aircraft:	28
Complement:	4700

Classed as 'helicopter destroyers' to avoid contravention of Japan's constitution, the two ships of the *Izumo*-class are fast and powerful multirole warships. *Izumo* was laid down in 2012 and commissioned in 2015.

Right: A US resupply ship conducts a replenishment-at-sea with the Izumo*-class JS* Kaga *(DDH 184) while underway in the Pacific Ocean on 18 November 2024.* Kaga *had recently completed F-35 Lightning II aircraft developmental testing in the US 3rd Fleet area of operations.*

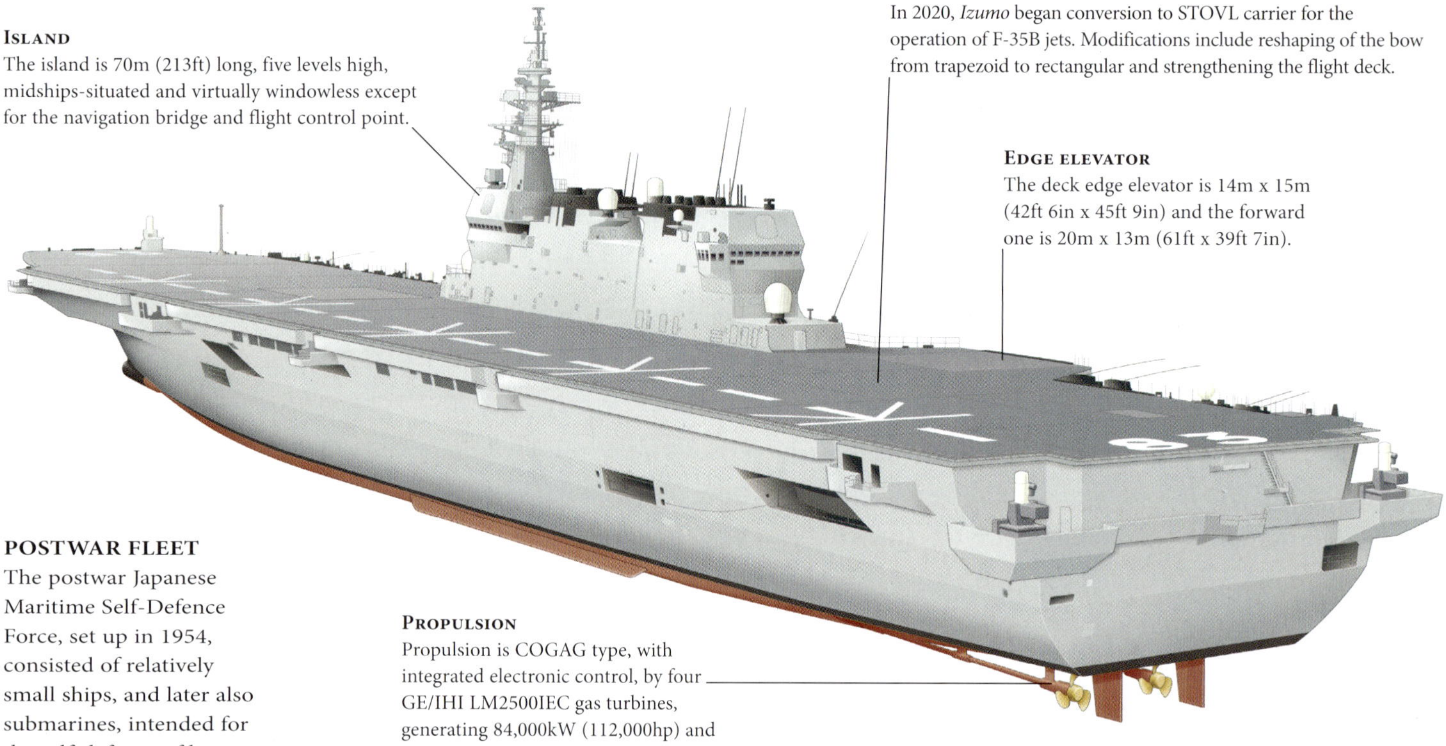

Island
The island is 70m (213ft) long, five levels high, midships-situated and virtually windowless except for the navigation bridge and flight control point.

Izumo
In 2020, *Izumo* began conversion to STOVL carrier for the operation of F-35B jets. Modifications include reshaping of the bow from trapezoid to rectangular and strengthening the flight deck.

Edge elevator
The deck edge elevator is 14m x 15m (42ft 6in x 45ft 9in) and the forward one is 20m x 13m (61ft x 39ft 7in).

Propulsion
Propulsion is COGAG type, with integrated electronic control, by four GE/IHI LM2500IEC gas turbines, generating 84,000kW (112,000hp) and driving two shafts with controllable-pitch propellers.

POSTWAR FLEET

The postwar Japanese Maritime Self-Defence Force, set up in 1954, consisted of relatively small ships, and later also submarines, intended for the self-defence of home waters. Later in the 20th century, growing strategic tensions prompted the Japanese government to take a wider view of the situation. This led to the construction of the large *Shirane*-class destroyers, built from 1977 to 1981. Due for withdrawal, their replacement began in 2009–11 with *Hyuga* and *Ise*, larger still, with a full-length flight deck and capacity for up to 18 helicopters.

***KAGA*, SAN DIEGO HARBOR**
JS *Kaga* (DDH-184) enters San Diego Harbor in 2024. Unlike US ships, the Japanese carriers are rarely seen publicly with aircraft parked on the flight deck.

Zumwalt class (2016)

TYPE • *Destroyer* **COUNTRY** • *United States*

SPECIFICATIONS

Dimensions:	Length: 186m (610ft); Beam: 24.6m (80ft 8in); Draught: 8.4m (28ft)
Displacement:	15,995 tonnes (17,631 tons) full load
Propulsion:	2 Rolls-Royce MT30 gas turbines, 35.4MW (47,500hp) each; 2 Rolls-Royce RR4500 turbine generators, 3.8MW (5100hp) each; Curtiss-Wright electric generators; 2 shafts;
Speed:	30.3 knots (56km/h; 35mph)
Range:	Undisclosed
Armament:	20 Mk 57 Peripheral VLS modules (80 cells) for Tomahawk land-attack cruise missiles, SM-6, ESSM and ASROC; 2 × 57mm Mk 110 guns; 155mm Advanced Gun System fitted (non-operational)
Complement:	175

The *Zumwalt* class consists of three advanced multi-mission missile destroyers for the United States Navy. The project has been periodically revised to meet changing requirements and to incorporate new technologies.

Right: The US Navy Zumwalt*-class USS* Michael Monsoor *(DDG 1001) is here seen underway during the Rim of the Pacific (RIMPAC) 2022 exercise.*

Aviation deck
The large aviation deck and hanger can accommodate one MH-60R helicopter and one MQ-8 Fire Scout VT-UAVs (Vertical Take-Off and Landing Tactical Unmanned Aerial Vehicle).

Stealth features
Thermal and acoustic emissions are kept to a minimum, and advanced low-probability-of-intercept radars are utilized, improving 'stealth' characteristics even when actively emitting radar pulses.

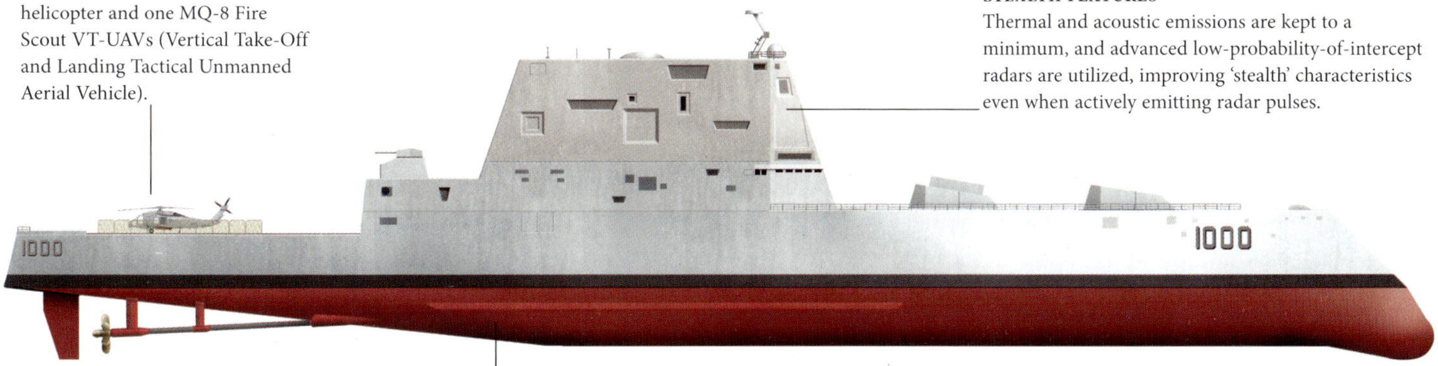

Protection
The Integrated Undersea Warfare System consists of two sonar arrays in one hull-mounted system, protecting the ship from mines, submarines and torpedoes.

WEAPONS PACKAGE
The *Zumwalt* class was designed with two 155mm (6.1in) guns capable of firing 12 rounds per minute at targets up to 100 nautical miles (185km; 115 miles) distant, although currently these weapons are non-operational. Missile armament includes Tactical Tomahawk cruise missiles in vertical launch tubes around the hull. This arrangement, rather than concentrating them in one area, reduces the effects of damage on the ship's ability to fight and the possibility of a single hit crippling the vessel's armament.

STEALTH HULL
The stealth features of the *Zumwalt* class are so sophisticated that despite being larger than an *Arleigh Burke* destroyer, they have the same radar signature as a small civilian vessel. However, the US Navy plans to build more of the less expensive *Arleigh Burke* class, which follows a far more traditional design.

Sea Hunter (2016)

TYPE • *Autonomous unmanned surface vehicle* **COUNTRY** • *United States*

SPECIFICATIONS	
Dimensions:	Length: 40m (132ft); Beam: 8.5m (27ft 11in); Draught: 1.6m (5ft 3in)
Displacement:	135 tonnes (298 tons) standard; 145 tonnes (320 tons) full load
Propulsion:	2 diesel engines driving twin screws; fully autonomous navigation and engineering control
Speed:	27 knots (50km/h; 31mph)
Range:	10,000nmi (18,520km; 11,507 miles)
Armament:	None (mission-payload capable but normally unarmed)
Complement:	None (uncrewed)

The Sea Hunter is an early expression of the emerging future of naval warfare. It is an entirely new class of unmanned sea surface vehicle, developed in partnership between the Office of Naval Research (ONR) and the Defense Advanced Research Projects Agency (DARPA).

Left: In 2019, Sea Hunter *completed an autonomous sail from San Diego to Hawaii and back – the first ship ever to do so autonomously.*

Propulsion
Powered by two diesel engines, the *Sea Hunter* can achieve a maximum speed of 27 knots (50km/h; 31mph), with a 30–90 day endurance without resupply.

Cost savings
The cost savings of a crewless ship are radical, with estimates of $15,000-$20,000 a day compared with $700,000 a day to operate a fully crewed destroyer.

Roles
The *Sea Hunters* are currently unarmed. At present they are used as sensor platforms for manned ships, providing an expanded operational awareness and contributing to anti-submarine defence.

AUTONOMOUS
Sea Hunter emerged from the Office of Naval Research's Medium Displacement Unmanned Surface Vehicle (MDUSV) project. The vessel is designed with a fully autonomous capability. Once programmed with its mission parameters, it uses onboard sensors and computerized navigation to steer its way safely to its destination, avoiding obstacles and threats along the way. The absence of crew offers radical savings in space and budget.

***SEA HUNTER*, PACIFIC, 2022**
This photograph shows *Sea Hunter*'s distinctive trimaran design. The advantage of the trimaran hull is that it gives the vessel excellent stability without the displacement, draught and payload penalties associated with a weighted keel.

INS *Arihant* (2016)

TYPE • *Submarine* **COUNTRY** • *India*

With the home-built SSBN *Arihant*, India has become the sixth country to construct and operate nuclear submarines, and an ongoing programme for further vessels, both missile carriers and attack boats, has been announced by the Indian government.

SPECIFICATIONS

DIMENSIONS:	Length: 110m (361ft); Beam: 11m (36ft); Draught: 9m (29ft 6in)
DISPLACEMENT:	6000 tonnes (6614 tons) surfaced; submerged undisclosed
PROPULSION:	1 pressurized-water nuclear reactor, 83MW (111,000hp); steam turbine; single shaft
SPEED:	15 knots (27.7km/h; 17.2mph) surfaced; 24 knots (44.4km/h; 27.6mph) submerged
RANGE:	Unlimited
ARMAMENT:	6 x 533mm (21in) torpedo tubes; up to 30 torpedoes; 12 x K-15 or 4 x K-4 submarine-launched ballistic missiles
COMPLEMENT:	96

Left: India's Advanced Technology Vessel (ATV) is intended to be the precursor of further SSBNs, whose design and equipment will be adapted from experience gained with Arihant.

PROPELLER
Arihant is driven via a seven-bladed fixed-pitch propeller with cruciform vortex dissipators. An emergency back-up drive is provided by two diesel engines of 370kW (496hp) giving a speed of 4 knots (7.4km/h; 4.6mph).

REACTOR COMPARTMENT
The reactor compartment is 42m (137ft 9in) long and 8m (26ft 3in) in diameter. The compartment holds the cooling-water tank, the reactor itself and the cooling system. There is a main control room and an auxiliary control room.

MISSILES
Arihant has successfully test-fired an unarmed Sagarika B-05 missile produced by India's Defence Research Development Organization (DRDO), with a 700km (440 mile) range and capable of carrying a nuclear warhead.

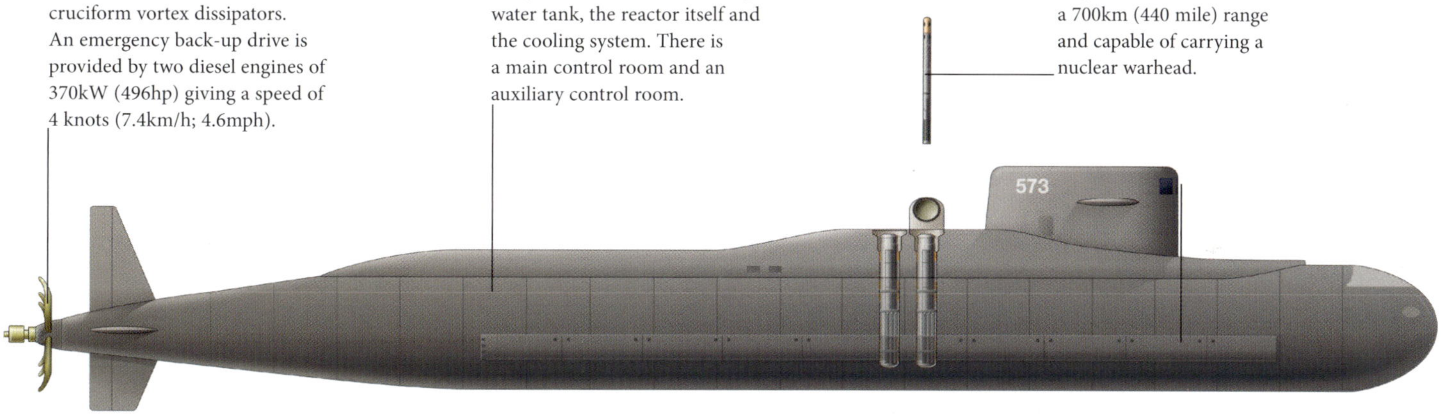

ATTACK CAPABILITIES
As a multirole platform, *Arihant* is designed for long-distance interdiction and surveillance. It can also facilitate Special Forces operations by covertly landing such forces ashore. With underwater ballistic missile launch capability, the boat carries 12 K-15 Sagarika submarine-launched ballistic missiles (SLBMs) with a 700–750km (434–465 mile) range or four K-4 missiles with an extended range of 3500km (2200 miles).

ARIHANT AT SEA
An Indian-developed anechoic covering said to be made of 100mm (4in) thick rubber-based tiles, each containing thousands of tiny voids, is applied to the outer hull. It is designed to absorb active sonar pulses and distort the return signal, while also dampening sounds emitted by the vessel itself.

USS *Lewis B. Puller* (2017)

TYPE • *Expeditionary mobile base vessel* **COUNTRY** • *United States*

SPECIFICATIONS

Dimensions:	Length: 233m (764ft); Beam: 50m (164ft); Draught: 7.8m (25ft 7in)
Displacement:	96,764 tonnes (106,664 tons) full load
Propulsion:	Diesel-electric integrated power system; 2 shafts
Speed:	15 knots (28km/h; 17mph)
Range:	9500nmi (17,600 km; 10,900 miles)
Armament:	12 × 12.7mm (.50 cal) machine-gun stations
Complement:	250

With the appearance of a floating factory, the USS *Lewis B. Puller* (ESB-3) is classified as an expeditionary mobile base, designed to act as a massive logistical and operational hub in far-flung corners of the globe:

Left: Sailors direct MH-60S Sea Hawk helicopters, attached to Helicopter Sea Combat Squadron (HSC) 26, aboard the USS Lewis B. Puller *(ESB-3) in April 2004, the ship then deployed to the US 5th Fleet area of operations to support maritime security and stability in the Middle East.*

Design
The industrial appearance of the USS *Lewis B. Puller* is no coincidence: the design is based on the hull of the civilian *Alaska*-class oil tanker.

Flight deck
The spacious flight deck can land four heavy-lift CH-53 transport helicopters at once; the deck also features a large aircraft hangar forward.

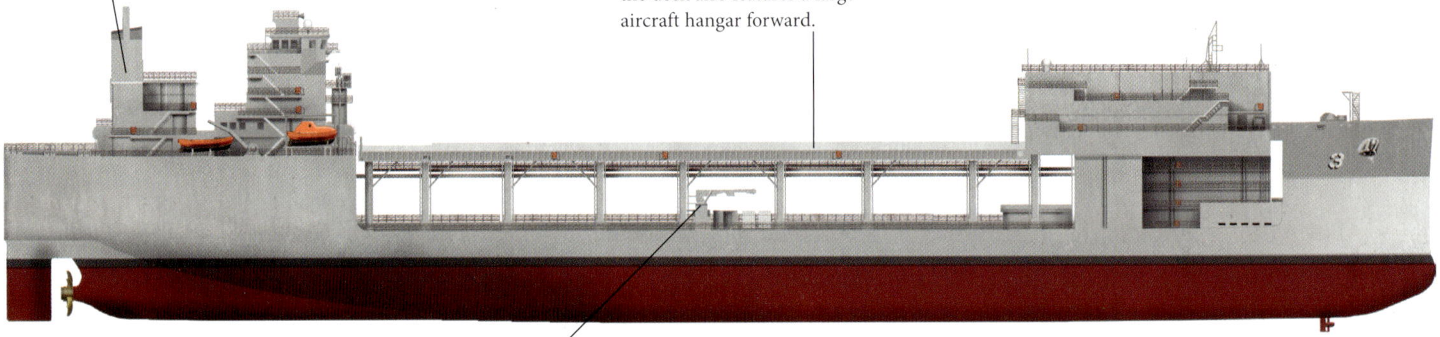

Mission deck
The mission deck is used to accommodate stores and small boats; the latter are lifted down onto the water by cranes.

SEA BASE
The Expeditionary Sea Base (ESB) is one of various classes of forward floating staging bases in use within the US Navy, with previous classifications including Mobile Landing Platform. The ship is not heavily armed – it has 12 machine-gun stations dotted around the deck – but it has a huge capacity for facilities and stores to support other naval forces operating in the region.

FIRST DEPLOYMENT
The *Lewis B. Puller* (T-ESB 3) departs Naval Station Norfolk to begin its first operational deployment on 10 July 2017. There are a total of six vessels in the *Lewis B. Puller* class.

Queen Elizabeth class (2017)

TYPE • *Aircraft carrier* **COUNTRY** • *United Kingdom*

SPECIFICATIONS

DIMENSIONS:	Length: 284m (931ft 9in); Beam: 73m (239ft 6in); Draught: 11m (36ft 1in)
DISPLACEMENT:	71,730 tonnes (79,080 tons) standard; 80,600 tonnes (88,850 tons) full load
PROPULSION:	Integrated Electric Propulsion (IEP); 2 gas turbines, 4 diesel generators, 4 electric propulsion motors; 2 shafts
SPEED:	27 knots (50km/h; 31mph)
RANGE:	10,000nmi (18,500km; 11,500 miles)
ARMAMENT:	3 × Phalanx CIWS; 4 × 30mm (1.18in) DS30M Mk2 guns; 6 × 12.7mm (0.5in) Browning heavy machine guns
AIRCRAFT:	40
COMPLEMENT:	679 plus aircrew

Six separate shipyards joined forces to build major sections of HMS *Queen Elizabeth*, and its companion ship *Prince of Wales*. Rosyth in Scotland was chosen as the yard for assembly because its dock was the only one big enough.

Below: HMS Prince of Wales *(R09) is the second* Queen Elizabeth-*class aircraft carrier.*

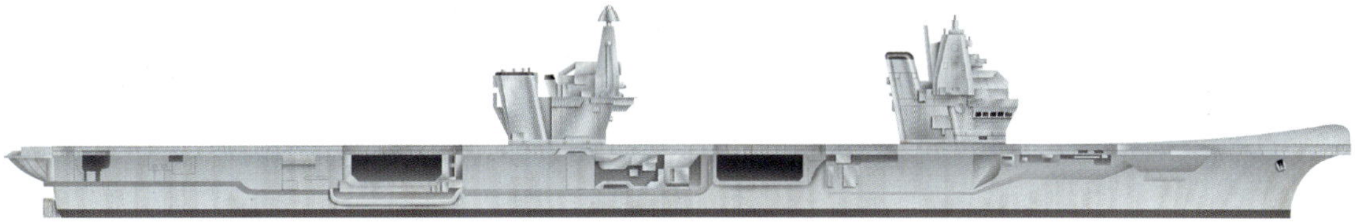

HMS *QUEEN ELIZABETH*
Other up-to-date equipment includes reverse osmosis equipment for 508 tonnes (560 tons) of fresh water daily, and a pyrolysis unit to compact and burn waste.

RUDDERS
The steering gear is supplied by Rolls-Royce as an integral system, including hydraulic power units, control and alarm systems. The twin rudders have 'twisted design' blades to reduce cavitation.

MOORING DECKS
The forward mooring deck with winches and cables is at the bow below the ramp. The stern has port and starboard mooring decks.

HULL COMPARTMENTS
Internally the hull has 3000 compartments on nine decks below the fiight deck. Some compromises had to be made, and a number of features such as armoured bulkheads and side armour plates were dropped from the original project design due to budget constraints. The design emphasis is on modularity to allow for changes in space requirements over the ship's life. A highly automated weapons handling system moves pallets of munitions from the magazines deep in the ship to a weapons preparation area and then up to the flight deck for fitting to the aircraft.

JET LANDING
An F-35B Lightning Jet recovers to HMS *Queen Elizabeth.* No catapult or arrestors are fitted, but the carrier is built to accommodate a future back-fit to CATOBAR (Catapult Assisted Take-Off But Arrested Recovery) format.

USS *Gerald R. Ford* (2017)

TYPE • *Aircraft carrier* **COUNTRY** • *United States*

SPECIFICATIONS

DIMENSIONS:	Length: 337m (1106ft); Beam: 41m (134ft 6in); Draught: 12m (39ft 4in)
DISPLACEMENT:	Approx 101,600 tonnes (111,995 tons)
PROPULSION:	2 A1B nuclear reactors, 4 steam turbines; 4 shafts
SPEED:	30 knots (56km/h; 35mph)
RANGE:	Indefinite
ARMAMENT:	2 × RIM-162 ESSM launchers; 2 × RIM-116 RAM launchers; 3 × Phalanx CIWS
AIRCRAFT:	75
COMPLEMENT:	4317

Commissioned in July 2017, *Gerald R. Ford* (CVN-78) is first of a four-strong class, with 23 new or upgraded systems compared to the *Nimitz*-class carriers. The new technologies are claimed to lead to a 30 per cent reduction in maintenance requirements.

Left: The Gerald R. Ford*-class, originally planned as three ships, is designed to supplement the* Nimitzes*, enabling greater operational flexibility and preparing for the retirement of the oldest ships.*

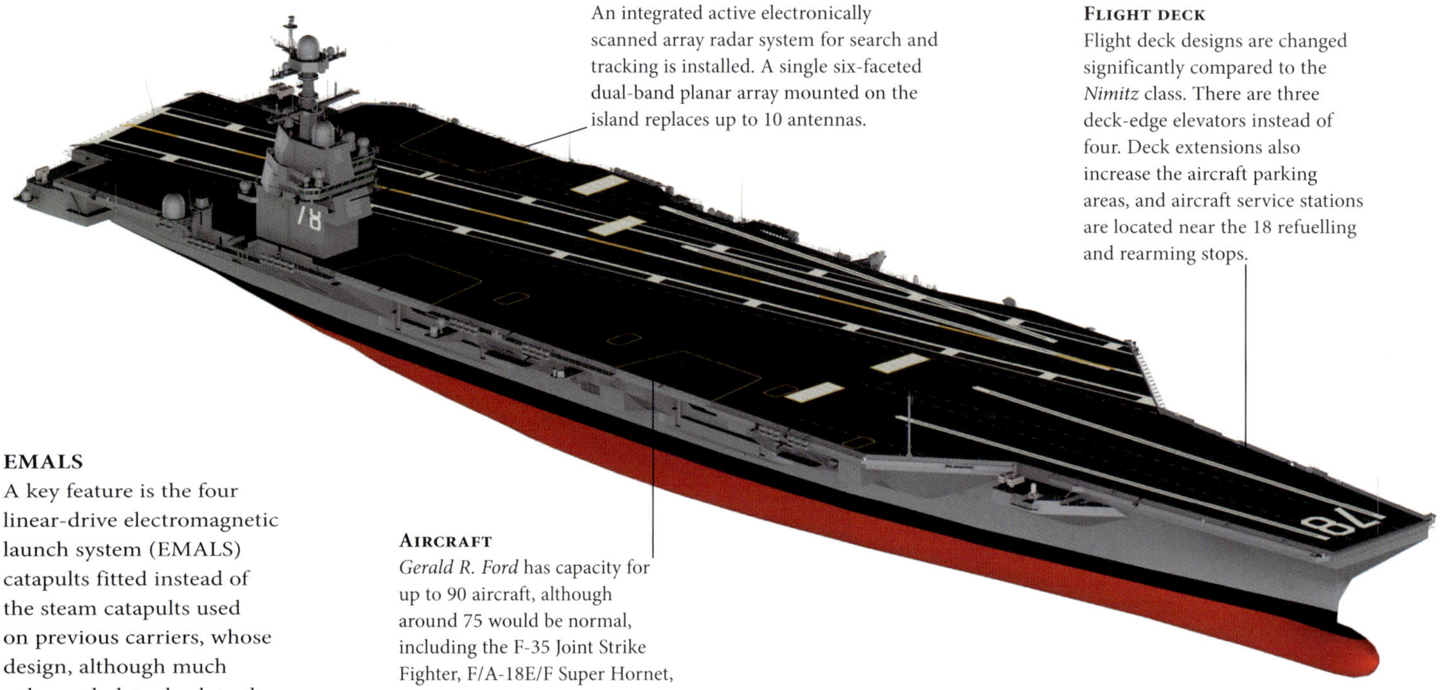

RADAR
An integrated active electronically scanned array radar system for search and tracking is installed. A single six-faceted dual-band planar array mounted on the island replaces up to 10 antennas.

FLIGHT DECK
Flight deck designs are changed significantly compared to the *Nimitz* class. There are three deck-edge elevators instead of four. Deck extensions also increase the aircraft parking areas, and aircraft service stations are located near the 18 refuelling and rearming stops.

EMALS
A key feature is the four linear-drive electromagnetic launch system (EMALS) catapults fitted instead of the steam catapults used on previous carriers, whose design, although much enhanced, dates back to the 1950s. While promising great improvements in reliability and efficiency, development has had to overcome many problems. The precision aircraft landing system (PALS) already in use on the *Nimitz* class has been upgraded. Advanced Arresting Gear incorporating water-twister components is installed. The aim is to achieve an increase of 33 per cent in the sortie generation rate of strike aircraft over the *Nimitz* class.

AIRCRAFT
Gerald R. Ford has capacity for up to 90 aircraft, although around 75 would be normal, including the F-35 Joint Strike Fighter, F/A-18E/F Super Hornet, E-2D Advanced Hawkeye, EA-18G Growler electronic attack aircraft, MH-60R/S helicopters, as well as unmanned air and combat vehicles.

HELICOPTER FLY-BY
Two Carrier Air Wing 8 MH-60S Sea Hawks, attached to Helicopter Sea Combat Squadron 9, fly in formation over what is the world's largest aircraft carrier, USS *Gerald R. Ford* (CVN-78), in 2025.

INS *Vikrant* (2018)

TYPE • *Aircraft carrier* **COUNTRY** • *India*

SPECIFICATIONS

DIMENSIONS:	Length: 259m (850ft 5in); Beam: 58m (190ft 3in); Draught: 8.4m (27ft 7in)
DISPLACEMENT:	38,100 tonnes (42,000 tons) standard; 40,640 tonnes (44,800 tons) full load
PROPULSION:	4 GE LM2500+ gas turbines; 2 shafts
SPEED:	28 knots (52km/h; 32mph)
RANGE:	7500nmi (13,875km; 8625mi) at 18 knots
ARMAMENT:	4 × 76mm (3in) OTO Melara dual-purpose guns; Barak-1 and Barak-8 SAM launchers; AK-630 CIWS
AIRCRAFT:	36
COMPLEMENT:	1560

***Vikrant* is a STOBAR-type carrier, built at India's Cochin Shipyard between 2009 and 2013, and commissioned in 2022. The ship has been designed with an advanced high degree of automation for machinery operation, navigation and survivability.**

Left: Vikrant *was laid down at Cochin Shipyard, Kochi, on 28 February 2009 and launched on 12 August 2013. Operationally it is a STOBAR carrier, fitted with a bow ramp to assist short take-off planes and with three arrestor wires to allow them to land.*

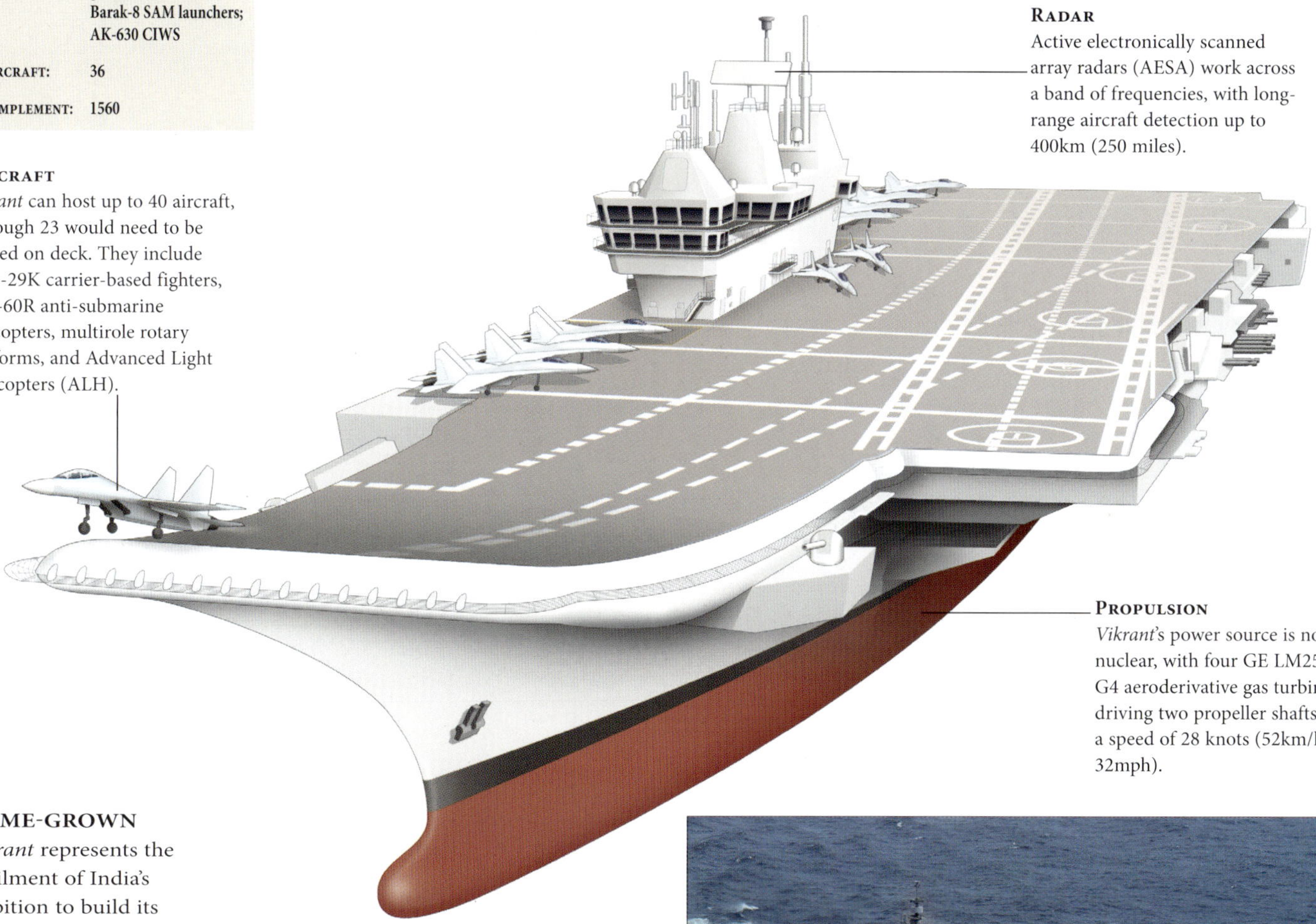

RADAR
Active electronically scanned array radars (AESA) work across a band of frequencies, with long-range aircraft detection up to 400km (250 miles).

AIRCRAFT
Vikrant can host up to 40 aircraft, although 23 would need to be parked on deck. They include MiG-29K carrier-based fighters, MH-60R anti-submarine helicopters, multirole rotary platforms, and Advanced Light Helicopters (ALH).

PROPULSION
Vikrant's power source is non-nuclear, with four GE LM2500+ G4 aeroderivative gas turbines driving two propeller shafts, with a speed of 28 knots (52km/h; 32mph).

HOME-GROWN
Vikrant represents the fulfilment of India's ambition to build its own carriers, with 76 per cent of its technological equipment home-produced, including the main switchboard, steering gear, air conditioning, gearboxes and pumps, and platform management systems. The remaining 24 per cent advanced technology, however, is vital to the effective operation of the carrier as a warship.

***VIKRANT* ON SEA TRIALS**
The island is sloping-sided as a stealth feature, with sensory and communications structures and masts at each end. There are 14 decks in all, including five in the superstructure.

Admiral Gorshkov class (Project 22350) (2018)

TYPE • *Frigate* **COUNTRY** • *Russian Federation*

The fifth vessel in the *Admiral Gorshkov* (Project 22350) class, *Admiral Amelko*, was launched in August 2025, expanding the series of advanced guided missile frigates. Ten are planned for the class, the remainder coming into service before 2030.,

SPECIFICATIONS

Dimensions:	Length: 135m (442ft 11in); Beam: 16m (52ft 6in); Draught: 4.5m (14ft 9in)
Displacement:	5400 tonnes (5,952 tons) full load
Propulsion:	CODAG; 2 10D49 cruise diesel engines 3900kW (5200shp); 2 M90FR gas turbines 20,500kW (27,500shp); 2 shafts
Speed:	29.5 knots (54.6km/h; 33.9mph)
Range:	4850nmi (8,980km; 5580mi) at 14 knots
Armament:	Multiple VLS systems for anti-ship, anti-submarine and surface-to-air missiles; 2 × Palash CIWS; 2 × 4 330mm (13in) torpedo tubes; 2 × 14.5mm (0.57in) MTPU pedestal machine guns
Complement:	210

Left: This photo taken on 21 April 2019, shows the Admiral Gorshkov *arriving in Qingdao in China's eastern Shandong Province to attend the 70th anniversary celebrations of the founding of the Chinese People's Liberation Army Navy (PLAN).*

Armament
The latest ships of the class will feature 32-cell 3S-14 UKSK launchers for 3M55 Onyx and 3M22 Zirkon anti-ship missiles and 3M14 cruise missiles.

Helicopter
The ship carries a single Kamov Ka-27 series dual-rotor helicopter, configured for anti-submarine duties.

Propulsion
The class features a CODAG propulsion system, combining two 10D49 cruise diesel engines and two M90FR boost gas turbines.

5P-42 FILIN
Some of the *Admiral Gorshkov* vessls have received the unusual 5P-42 Filin ('Eagle Owl') electro-optic countermeasure system. Developed by Roselectronics, it emits a powerful and strobing beam of light, which at night disrupts the eyesight and orientation of attackers. The system has a range of 500–700m (1640–2297ft) visually, but it can also be used to disrupt enemy night-vision devices.

ADMIRAL GORSHKOV
The lead ship of the class, *Admiral Gorshkov* was commissioned on 28 July 2018. It was the first ship in the Russian Navy to be armed with 3M22 Zircon hypersonic anti-ship cruise missiles.

F125 *Baden-Württemberg*-class (2019)

TYPE • *Frigate* **COUNTRY** • *Germany*

SPECIFICATIONS

Dimensions:	Length: 149.52m (490ft 7in); Beam: 18.8m (61ft 8in); Draught: 5.40m (17ft 9in)
Displacement:	7200 tonnes (7937 tons) standard
Propulsion:	CODLAG; gas turbine 20,000kW (26,820shp); 2 electric motors, 9400kW (12,610shp); 4 diesel generators, 11,600kW (15,550shp); 3 gearboxes; 2 shafts
Speed:	20 knots (37km/h; 23mph) diesel-electric; 26 knots (48km/h; 30mph) maximum
Range:	4000nmi (7408km; 4603 miles)
Armament:	1 × 127mm (5in) lightweight gun; 2 × 27mm (1in) remote autocannon; 7 × 12.7mm machine guns; 2 × RAM Block II launchers; 8 × RGM-84 Harpoon
Complement:	190 (standard crew 110)

The German F125 *Baden-Württemberg* frigates entered service starting in 2019, and four ships are currently operational. However, questions have been raised over what some perceive as limitations in the armament configuration of the class.

Left: Baden-Württemberg *participates in a missile-firing exercise in the North Sea on 13 October 2025 near Harstad, Norway. The exercise took place from 6 to 30 October and was the biggest live-fire missile exercise the German Navy had undertaken in 30 years.*

Baden-Württemberg
The lead ship of the class (F222) was laid down in November 2011, launched in December 2013 and commissioned in June 2019.

Deckhouses
A distinguishing visual feature of the class is its two pyramidal deckhouses, set fore and aft. This configuration enhances ship survivability in case of battle damage.

CODLAG
The combined diesel-electric a nd gas (CODLAG) system offers the high-speed performance of gas turbines combined with the reliability and economy of diesel power.

CHALLENGES

The F125 class frigates were ordered in 2007 to replace the F122 *Bremen*-class frigates and provide the German Navy with enhanced land-attack capabilities. The ships have encountered several physical and tactical issues, however. Most problematic was a 1.3 degree list to starboard in the lead vessel (*Baden-Württemberg*), which combined with excessive weight led to a two-year delay in its commissioning. The class in general also has no VLS armament system, limiting it combat potential.

BADEN-WÜRTTEMBERG
The German Navy frigate FGS *Baden-Württemberg* (F 222) is here seen underway in July 2024, in formation off the coast of Hawaii during Exercise Rim of the Pacific (RIMPAC) 2024.

Belgorod (2019)

TYPE • *Submarine* **COUNTRY** • *Russian Federation*

A unique variant of the Project 949A design is Project 9852 *Belgorod*, which differs sufficiently from the original boats to be seen as a class of its own. Russia's longest submarine at 184m (604ft), it is a powerful indicator of Russia's ambition to lead in deepsea operations.

SPECIFICATIONS

Dimensions:	Length: 184m (603ft 8in); Beam: 15m (49ft 3in)
Displacement:	Up to 17,000 tonnes (18,739 tons) surfaced; up to 30,000 tonnes (33,069 tons) submerged
Propulsion:	2 OK-650M.02 pressurized-water nuclear reactors; steam turbines; 2 shafts
Speed:	32 knots (59km/h; 37mph) submerged
Range:	Indefinite
Armament:	Up to 6 × Poseidon autonomous nuclear missile launch vehicles
Complement:	110

Left: Belgorod*'s operational movements since its commissioning in 2022 are unclear, although it has served in both the Northern and Pacific Fleets.*

Dive depth
Belgorod is capable of reaching depths beyond 520m (1700ft) and can achieve a submerged speed of 32 knots (59km/h; 37mph).

Torpedoes
Belgorod can also deploy six 2M39 Poseidon torpedoes, described as 'giant underwater drones with practically unlimited range and nuclear warheads'.

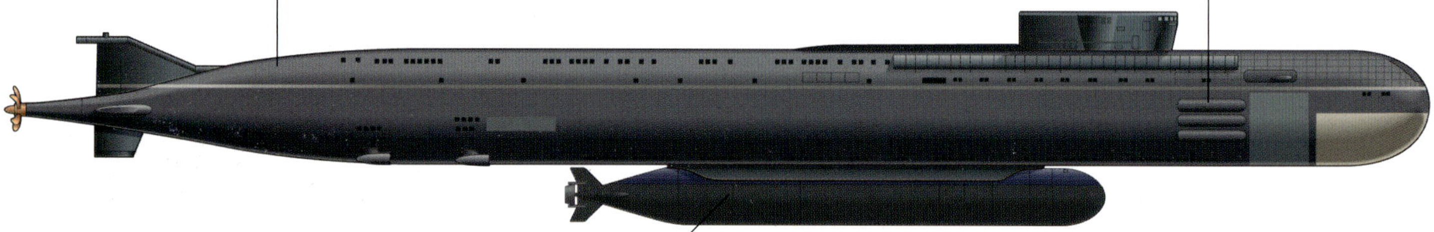

Payloads
Belgorod in profile, with an unmanned nuclear-powered 'deep-sea station' secured below the hull. Internal or hull-borne payloads could include a deep-sea rescue vehicle or a seabed nuclear power plant.

DEVELOPMENT

Belgorod's keel was laid in 2012 as a special purpose and research vessel, but the design was radically altered in 2017. The missile compartment was removed and the central section extended to form a longer and larger hold. With access through the double hull to the water, this space can be configured to hold crewed rescue and exploring vehicles, remotely controlled cable cutters, and other ROV machines, but also new strategic weapons.

LARGEST SUBMARINE
Belgorod is seen here at Severodvinsk, Arkhangelsk region, in July 2021, prior to its commissioning one year later. At the time of writing, *Belgorod* is the largest submarine in the world.

Suffren (*Barracuda* class) (2019)

TYPE • *Submarine* **COUNTRY** • *France*

SPECIFICATIONS

DIMENSIONS:	Length: 99.5m (326ft 5in); Beam: 8.8m (28ft 11in)
DISPLACEMENT:	4841 tonnes (5337 tons) surfaced; 5385 tonnes (5938 tons) submerged
PROPULSION:	1 pressurized-water nuclear reactor rated at approx. 150MW; 2 steam turbines; turbo-generators; electric propulsion motors driving a pump-jet propulsor
SPEED:	25 knots (46.3km/h; 28.8mph) submerged
RANGE:	Indefinite
ARMAMENT:	4 × 533mm (21in) launch tubes for heavyweight torpedoes and cruise missiles
COMPLEMENT:	63 plus up to 15 special forces personnel

***Suffren* is the first of a new-generation French SSN class that is intended to match the latest attack boats of other navies. Two are in commission: *Suffren* (2020) and *Duguay-Trouin* (2024), with six further planned boats.**

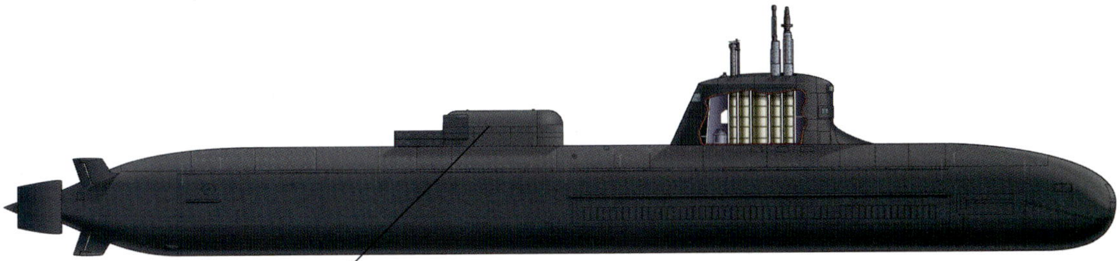

Above: Although nearly twice the size of the Rubis *class,* Suffren *and its companions operate with a smaller crew and, for the first time in French submarines, include women in their personnel.*

DECK HANGAR
The deck shelter, with a lock-out chamber, can accommodate drones and such vehicles as the French-designed PSM3G Special Forces Swimmer Delivery Vehicle with 12 swimmers. The special operations control room is situated directly below.

SONAR ARRAY
Descriptions of the Thales sonar equipment include provision of a thin line passive towed array, although no external fitting is seen.

BARRACUDA
While the optronic masts do not enter the pressure hull, a snorkel mast and diesel exhaust are required for the two emergency diesel generators.

FLANK ARRAY
A long flank array is part of the sonar equipment. This is applied after the 21 steel rings forming the hull have been welded together.

POWER
The aim is for this class to spend 250 days a year at sea between overhauls. The reactor is a more compact version of the TechnicAtome K15, with a power output of 150MW. Refuelling of the reactor is at 10-year intervals. Power production has been described as hybrid steam-electric. Steam produced from the nuclear process is used to power two turbo generator groups each of 10MW (13,000hp).

***SUFFREN*, TOULON**
Suffren enters Toulon on 6 July 2021, with crew members on the deck and sail. *Suffren* has deep-strike cruise missiles, advanced stealth and special forces capability. Designed for versatility and discretion, it is built to operate effectively across a wide range of missions.

Type 002 *Shandong* (2019)

TYPE • *Aircraft carrier* **COUNTRY** • *China*

SPECIFICATIONS

Dimensions:	Length: 304.5m (1000ft); Beam: 75m (246.1ft) (flight deck)
Displacement:	c. 67,000 tonnes (66,000 tons) full load
Propulsion:	8 boilers, 4 steam turbines, 4 shafts
Speed:	31 knots (57km/h; 36mph)
Range:	Undisclosed
Armament:	3 × Type 1130 CIWS; 3 × HQ-10 missile systems (18-cell)
Aircraft:	40+ fixed-wing and rotary-wing aircraft
Complement:	c. 2500

Built at China's Dalian shipyard, launched in 2016 and commissioned in 2019, the *Shandong* (Type 002) is a STOBAR system carrier, designed for short take-offs but arrested landings. It marks an intermediate stage between China's first and third carriers.

Left: Shandong *arrives in Hong Kong on 3 July 2025.* Shandong *was named* Kuznetsov 2 *by NATO, but in many ways it is a more advanced carrier, based on experience with* Liaoning.

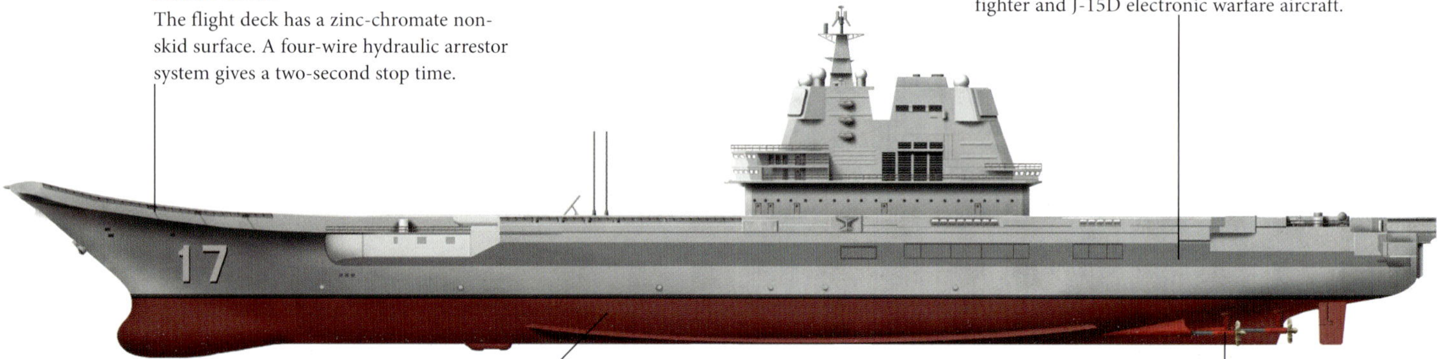

Flight deck
The flight deck has a zinc-chromate non-skid surface. A four-wire hydraulic arrestor system gives a two-second stop time.

Air wing
Air wing capacity is 36, including Shenyang J-15, Changhe Z-18 and Harbin z-9. *Shandong* is understood to be receiving J-15T fighter and J-15D electronic warfare aircraft.

Electrical power
Nine 1500kW (2011hp) turbogenerators and six 1500kW (2011hp) diesel generators supply shipboard power. An estimated six days steaming is possible before refuelling is required.

Propulsion
Propulsion is believed to be a modified version of *Liaoning*'s machinery, with eight oil-fired Harbin boilers and four Erzhong steam turbines each 37,285kW (50,000shp), driving four shafts.

FEATURES

A bow ramp is fitted to assist take-off, and the ship has an angled flight deck, but no catapults. A single nine-deck island mounted on a narrow podium just aft of midships carries a pyramidal mast, the funnel, and a range of antennas and domes. Two windowed command decks control navigation and air operations. The ramp, at 12 degrees, is designed to help heavy jets such as the J-15 into the air. Three lifts are fitted, two laterally fore and aft of the island.

***SHANDONG* AT SANYA**
This photograph, taken on 17 December 2019, shows the *Shandong* aircraft carrier at a naval port in Sanya, Hainan Province. The ski-jump take-off method limits the amount of weaponry and fuel that a plane like the J-15 can carry.

Type 055 *Renhai* class (2019)

TYPE • *Destroyer* **COUNTRY** • *China*

SPECIFICATIONS

Dimensions:	Length: 180m (590ft 7in); Beam: 20m (65ft 7in); Draught: 6.6m (21ft 8in)
Displacement:	12,000–13,000 tonnes (26,455–28,660 tons) full load
Propulsion:	COGAG; 4 QC-280 gas turbines, 112,000kW (150,000hp) total; 2 shafts, controllable-pitch propellers
Speed:	30 knots (56km/h; 35mph)
Range:	5000nmi (9300km; 5770 miles)
Armament:	112 × VLS cells for SAM, SSM and land-attack missiles; 1 × 130mm (5.1in) H/PJ-45 naval gun; 1 × H/PJ-11 CIWS; 1 × HHQ-10 point-defence missile system; 2 × triple 324mm (12.75in) torpedo tubes
Complement:	300+

The People's Liberation Army Navy (PLAN) classifies the Type 055 class as a guided-missile destroyer. However, the US Navy judges what it calls the *Renhai* class as cruisers, giving an indication of the size and power of these impressive combat vessels.

Left: The Type 055 increases its capability by embarking two medium helicopters, typically the Harbin Z-9 or Changhe Z-18.

SAM system
The Type 055 features a 24-cell HHQ-10 point-defence SAM launcher atop the aft helicopter hangar.

VLS
The 112-cell VLS system can take HHQ-9 SAMs, YJ-18 anti-ship and land-attack missiles, CJ-10 land-attack cruise missiles and CY-5 missile-launched anti-submarine torpedoes.

Performance
The four gas turbines in COGAG configuration deliver a top speed of 30 knots (56km/h; 35mph) and an estimated range of 5000nmi (9300km; 5770 miles) cruising at 18 knots (33km/h; 20.5mph).

VERTICAL LAUNCH

The Vertical Launch System (VLS) is now a common installation on international fighting vessels. Prior to the VLS, ships were fitted with a collection of individual missile types, each on its own dedicated launcher unit. The VLS, however, features multiple launch cells that can launch different missile systems from the same unit, giving warships greater armament flexibility and combat readiness.

***YAN'AN*, 2005**
At the time of writing (2026), there are eight Type 055 vessels in service, forming one of the most powerful components of the Chinese surface fleet. Pictured is the *Yan'an*, which was commissioned in March 2022.

Dokdo class (2020)

TYPE • *Aircraft carrier* **COUNTRY** • *South Korea*

SPECIFICATIONS	
DIMENSIONS:	Length: 199m (652ft 11in); Beam: 31m (101ft 8in); Draught: 7m (23ft)
DISPLACEMENT:	14,300 tonnes (15,764 tons) standard; 19,500 tonnes (21,496 tons) full load
PROPULSION:	CODAD; 4 SEMT Pielstick 16 PC2.5 STC diesels; 2 shafts
SPEED:	23 knots (43km/h; 26.4mph)
RANGE:	10,000nmi (18,500km; 11,510 miles) at 18 knots
ARMAMENT:	2 × Mk 15 Phalanx CIWS; 1 × 4-cell K-VLS for K-SAAM Sea Bow missiles
AIRCRAFT:	Up to 15 helicopters
COMPLEMENT:	330 (ship only)

The South Korean *Dokdo* class is a fusion between a landing platform helicopter (LPH) and an amphibious assault ship. There are two ships in the class: the *Dokdo* (LPH-6111), commissioned in 2007, and the *Marado*, commissioned years later in 2021.

Left: The Dokdo *(LPH-6111) assists in search and rescue operations in response to the the sinking of the ferry* Sewol *off the coast of the Republic of Korea on 20 April 2014.*

MARADO
Second of the two-ship *Dokdo* class, built by Hanjin Heavy Industries at Busan, *Marado* looks identical to *Dokdo,* but has a number of updated fittings.

WEAPONS SUITE
Marado's weapons suite consists of two 20mm (0.79in) Phalanx CIWS and a K-VLS at the rear of the superstructure for the South Korean-developed K-SAAM missile.

CAPACITY
Internally, *Marado* has a hangar for up to 15 helicopters, vehicle storage for six tanks and up to 200 vehicles, and a well deck, plus accommodation for a marine battalion of 720.

LAYOUT

Both ships of the class are through-deck, with a parking deck to port balancing a long, largely blank-sided island. The island has a squat tapering tower mast positioned above the navigation and air-control bridge. Two low funnels, integrated with the island structure, are separated by a taller sensor and communications tower, its angular form clearly shaped to reduce detectability. The hull is sharply flared below the foredeck, which is squared off above the bow.

***DOKDO* UNDERWAY**
Although essentially a helicopter carrier, the ship has a flight deck fitted with urethane pads and can take VTOL planes like the F-35 Lightning II. However, it is not fitted to service such aircraft.

Type 075 Landing Helicopter Dock (2021)

TYPE • *Helicopter carrier* **COUNTRY** • *China*

SPECIFICATIONS

DIMENSIONS:	Length: 232m (761ft 2in); Beam: 36.8m (120ft 9in); Draught: 8.5m (27ft 11in)
DISPLACEMENT:	35,000–40,000 tonnes (38,580–44,090 tons) full load
PROPULSION:	Diesel engines driving electric motors, 2 shafts
SPEED:	24 knots (44.4km/h; 27.6mph) estimated
RANGE:	Undisclosed
ARMAMENT:	2 × 30mm (1.18in) H/PJ-14 CIWS; 2 × HQ-10 short-range SAM systems
AIRCRAFT:	Up to 35 helicopters
COMPLEMENT:	Undisclosed

Western powers took note when, in April 2021, the People's Liberation Army Navy (PLAN) commissioned the first of three Type 075 class Landing Helicopter Dock (LHD) vessels, which significantly upgraded China's capabilities in amphibious warfare.

Right: The Type 075 LHD Hainan *is here docked at China Merchants Wharf in Hong Kong on 21 November 2024.* Hainan *(31) is the lead ship of the class.*

HELICOPTERS
The ship can deploy 28–35 helicopters of various types, particularly the Harbin Z-20, Changhe Z-18, Changhe/Harbin Z-8 and Changhe Z-10

CAPACITY
Upper estimates for the Type 075's vehicle capacity are: 10 main battle tanks, 20-35 Type 05 amphibious assault vehicles, 20 infantry fighting vehicles and 50 field mobile trucks.

FACILITIES
The Type 075 ships are massive. Onboard facilities include, laundry services, shops, postal services, barber shops, a conference space and a 100-bed hospital.

POWER PROJECTION

The Type 075 is designed specifically to deliver and support amphibious assault actions. Its full-length flight deck and extensive hangar deck means it can host up to 35 helicopters – crucial for ship-to-shore combat and logistical support – while its combined well deck and vehicle deck mean have the capacity to deploy a PLAN Marine Corps amphibious mechanized infantry company augmented by platoon-sized tank or artillery elements, with a further 900–1200 troops on board.

***HAINAN*, 2025**
This impressive aerial view of *Hainan* gives a sense of the operational capacity of the Type 075. The class rivals the capbilitiesof advanced US Navy LHDs.

KSS-III (*Dosan Ahn Changho*) (2021)

TYPE • *Submarine* **COUNTRY** • *South Korea*

The South Korean *Dosan Ahn Chango* class are also known as KSS-III attack submarines, with three Batch 1 boats successfully commissioned between 2018 and 2021 and the first Batch 2 boat under contract.

SPECIFICATIONS (Batch 1)

Dimensions:	Length: 83.5m (273ft 11in); Beam: 9.6m (31ft 6in); Draught: 7.62m (25ft)
Displacement:	3358 tonnes (3701 tons) surfaced; 3750 tonnes (4134 tons) submerged
Propulsion:	Diesel-electric propulsion with fuel-cell air-independent propulsion system; electric propulsion motor driving a single shaft
Speed:	20 knots (37km/h; 23mph) submerged
Range:	10,000nmi (18,520km; 11,508 miles)
Armament:	6 vertical launch system cells for submarine-launched ballistic missiles; 6 × 533mm (21in) torpedo tubes for LIG Nex1 K761 Tiger Shark torpedoes
Complement:	50

Above: A launch ceremony is held on 22 October 2025 for the Jang Yeong-sil, *the first of the three Batch-II class submarines, at shipbuilder Hanwha Ocean's dockyard in Geoje, South Korea.*

Missiles
A new design will accommodate the Hyunmoo 4-4, an upgraded version of the Korean-designed Hyunmoo 2B surface-to-surface ballistic missile, with a range of approximately 500km (311mi) and payload of approximately 1000kg (2200lb).

Power
The power system is essentially the same as for the KSS-I and KSS-II classes, with three MTU 16V396 SE84L diesel generators and four PH1 PEM fuel cells each of 150kW (201hp), combining oxygen (stored as LOX) and hydrogen, stored in metal-hydride cylinders.

Batch II variation
Batch 2 of the KS-III will retain their AIP-capable systems but with the electric motor drawing power direct from Samsung SDI lithium-ion fuel cells giving a much higher output than the lead-acid type carried by Batch 1 and earlier classes.

STRATEGIC FOCUS

The builders of the KSS-III are Daewoo SME and Hyundai Heavy Industries. Although hailed as a South Korean design, its links with the previous German-designed KSS-I and KSS-II classes are clear. This class is considerably larger than its predecessors, and holds 10 vertical launch tubes for ballistic or cruise missiles. The nine KSS-III boats represent the completion (scheduled for 2029) of South Korea's ambitious programme to develop a powerful modern submarine force.

LOCAL THREATS

The development of KSS-III class is linked to the introduction of SLBM-equipped submarines by North Korea. To monitor the entry and exit of such vessels from military ports and to track and attack them if necessary, new submarines are needed that can remain undetected and stable for long periods.

Thaon di Revel class (2021)

TYPE • *Offshore patrol vessel* **COUNTRY** • *Italy*

SPECIFICATIONS

DIMENSIONS:	Length: 143m (469ft 2in); Beam: 16.5m (54ft 2in); Draught: 5.1m (16ft 9in)
DISPLACEMENT:	5880 tonnes (6483 tons) full load
PROPULSION:	CODLAG; diesel engines and gas turbine driving electric motors; 2 shafts
SPEED:	32 knots (59km/h; 37mph)
RANGE:	5000nmi (9260km; 5750 miles) at 15 knots
ARMAMENT:	1 × 127mm (5in) gun; 1 × 76mm (3in) gun; VLS for SAMs and anti-ship missiles; 2 × 25mm (0.98in) remote weapon stations
COMPLEMENT:	171 core crew plus mission personnel

The ships of the *Thaon di Revel* class are classified as *Pattugliatore Polivalente d'Altura* (PPA), or Multipurpose Offshore Patrol Vessel. They have potent armament, however, and are replacing two ageing classes of light frigate and corvette in the Italian Navy.

Left: A forward view of the Giovanni delle Bande Nere *(P434), showing the OTO 127/64 LW (lightweight) gun assembly. The fully automatic weapon is capable of firing the Vulcano long-range guided ammunition, with range of 70–120km (43–75 miles).*

FLIGHT FACILITIES
The rear flight deck measures 25.5 × 16.5m (83ft 8in × 54 ft 2in). The hangar just behind it has capacity for one AgustaWestland AH-101 or two NHIndustries SH90 helicopters.

DOUBLE BOW
The striking double-bow design is intended to increase the ship's length-to-beam ratio to optimize its speed but also minimize drag and vertical motions, resulting in a drier deck in heavy seas.

PAOLO THAON DI REVEL
The *Paolo Thaon di Revel* (P430) is the lead ship of the class, launched in June 2019 and commisioned in March 2022.

CODLAG PROPULSION
The *Thaon di Revel* class employs a CODLAG (Combined Diesel-Electric and Gas) propulsion system. High-speed performance is delivered by a General Electric–Avio LM2500+G4 gas turbine, while normal operations use electric propulsion motors supplied by onboard diesel generator sets. Electrical power is generated by diesel generator sets, supplemented by auxiliary generators.

***RAIMONDO MONTECUCCOLI* (P432)**
The *Thaon di Revel*-class vessels are delivered in three different configurations: full, light+ and light, defined by different packages of equipment, sensors and armament.

Taigei class (2022)

TYPE • *Submarine* COUNTRY • *Japan*

SPECIFICATIONS

Dimensions:	Length: 84m (275ft 7in); Beam: 9.1m (29ft 10in); Draught: 10.4m (34ft 1in)
Displacement:	3000 tonnes (3307 tons) surfaced
Propulsion:	Diesel-electric propulsion; lithium-ion battery system for submerged operation; electric propulsion motor driving a single shaft
Speed:	20 knots (37km/h; 23mph) submerged
Range:	Undisclosed
Armament:	6 x 533mm (21in) torpedo tubes for heavyweight torpedoes, anti-ship missiles and mines
Complement:	70

The Japanese *Taigei* class comprises three boats commissioned between 2022 and 2024, with more to come. Intensive development work on virtually every aspect of structure and performance began in 2004 and resulted in the launch of *Taigei* in 2022.

Left: JS Taigei *(S-513) is the first Japanese submarine to include women in the crew. The first boat has been designated a test platform for the many novel features of the class.*

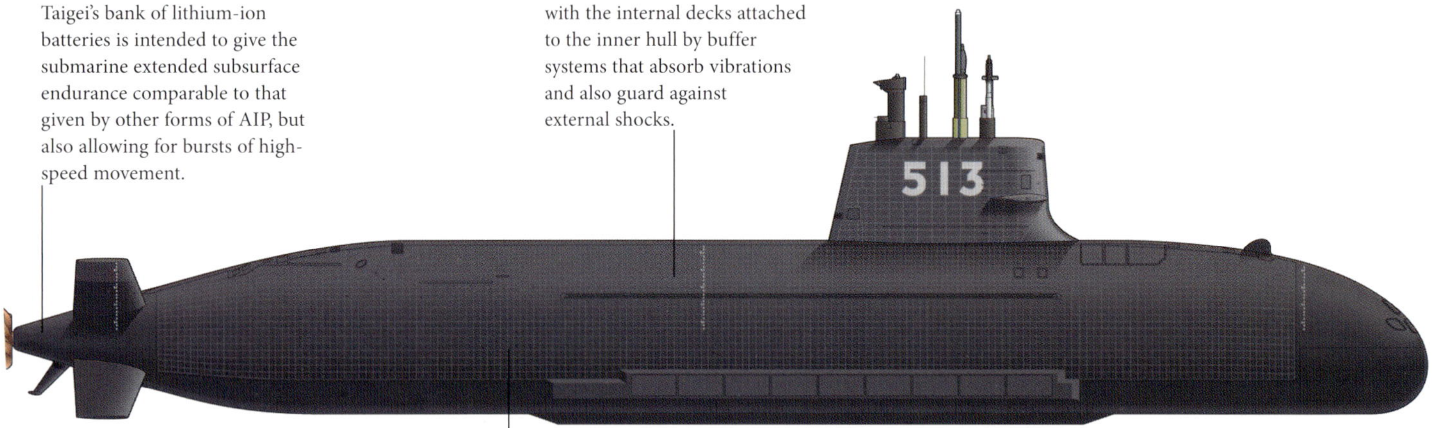

Propulsion
Taigei's bank of lithium-ion batteries is intended to give the submarine extended subsurface endurance comparable to that given by other forms of AIP, but also allowing for bursts of high-speed movement.

Floating decks
'Floating floors' are incorporated, with the internal decks attached to the inner hull by buffer systems that absorb vibrations and also guard against external shocks.

Displacement
Surface displacement is around 3000 tonnes (3307 tons), slightly heavier than the *Soryu* class.

ADVANCED TECH

Many other features are new in the class, including the snorkel design, which is said to enable rapid battery recharge; a high-powered, highly sensitive fibre-optic sonar set to counter the stealth abilities of other submarines; and a new combat management system. The *Taigei*-class submarines are equipped with six HU 606 533mm (21in) launch tubes, which can fire Mitsubishi Type 89 heavyweight homing torpedoes or new Type 18 heavy torpedoes for anti-submarine operations, as well as UGM-84 Harpoon anti-ship missiles.

UPGRADES
The next variant of the *Taigei* class is expected to include a vertical launch system (VLS) for SLBMs. The sixth vessel of the class was launched on 14 October 2025.

Mogami class (2022)

TYPE • *Frigate* COUNTRY • *Japan*

SPECIFICATIONS

DIMENSIONS:	Length: 142m (465ft 11in); Beam: 17m (55ft 9in)
DISPLACEMENT:	6200 tonnes (6834 tons) full load
PROPULSION:	CODAG; 1 Rolls-Royce MT30 gas turbine; 2 MAN Diesel V28/33DD STC diesel engines; total 52,198kW (70,000hp); 2 shafts
SPEED:	30 knots (55.6km/h; 34.5mph)+
RANGE:	10,000nmi (18,520km; 11,508 miles)
ARMAMENT:	1 × 127mm (5in) main gun; 1 × SeaRam SAM; Type 12 torpedoes; 16-cell Mk 41 VLS
COMPLEMENT:	90 (accommodation for up to 138)

Over the last 20 years, the Japan Maritime Self-Defense Force (JMSDF) has undergone a radical period of modernization. A notable outcome is the *Mogami* class frigate, a new generation of stealth warships equipped with cutting-edge air defence and anti-submarine warfare capabilities.

Left: This photograph of JS Yahagi *(FFM-5) shows the clean lines inherent in the stealth design. Note the unobstructed layout of the foredeck and how the forward gun also has a mount with stealth characteristics.*

AUTOMATION
The designers of the class have invested heavily in the digital automation of systems, which results in the ship having a human complement of just 90 personnel.

HELIDECK
For anti-submarine duties, the *Mogami* class operate a single Mitsubishi SH-60L helicopter from the rear helideck; the helicopter has its own hangar.

STEALTH DESIGN
The ship's design has been influenced by Japanese research into the experimental Mitsubishi X-2 Shinshin stealth fighter aircraft.

UPGRADED *MOGAMI*

By 2027, the Japanese Navy is expected to have 12 *Mogami*-class frigates. A the same time, plans are underway for a next-generation variant, known as the 'Upgraded *Mogami*' or 'New FFM', which Japan intends to build alongside similar ships for the Australian navy. It will have a longer, heavier and wider hull plus significant additions to the sensor suite and armament, the latter including Type 23 SAMs and Type 12 surface-to-ship missiles. Commissions are planned from 2028.

JS *KUMANO* (FFM-2)
The JS *Kumano* enters Kinuura Port, Honshu. The ship was commissioned on 22 March 2022 and deployed to Yokosuka Naval Base, a collection of ports and facilities in eastern Japan, shared with US warships.

Isaac Peral (2023)

TYPE • *Submarine* **COUNTRY** • *Spain*

SPECIFICATIONS

DIMENSIONS:	Length: 81.05m (265ft 11in); Beam: 11.68m (38ft 4in); Draught: 7.3m (23ft 11in)
DISPLACEMENT:	2965 tonnes (3269 tons) submerged
PROPULSION:	3 bio-diesel engines; electric propulsion motor developing 3500kW (3970hp) driving a single screw; fuel-cell air-independent propulsion system developing 300kW (402hp)
SPEED:	19 knots (35km/h; 22mph) submerged
RANGE:	Undisclosed
ARMAMENT:	6 × 533mm (21in) torpedo tubes for DM2A4 heavyweight torpedoes, UGM-84 Sub-Harpoon Block II anti-ship missiles and SAES seabed mines
COMPLEMENT:	32 plus up to 8 special forces personnel

The Spanish Navy currently has a single submarine, the *Isaac Peral,* of the new S-80 Plus class. Designed and built by Navantia at Cartagena, it is one of the largest non-nuclear boats in service in the world today.

Left: View of the Isaac Peral *(S-81) on 17 November 2023 at Navantia, Cartagena, Spain, during the completion of the snorkel manoeuvring sea trials.*

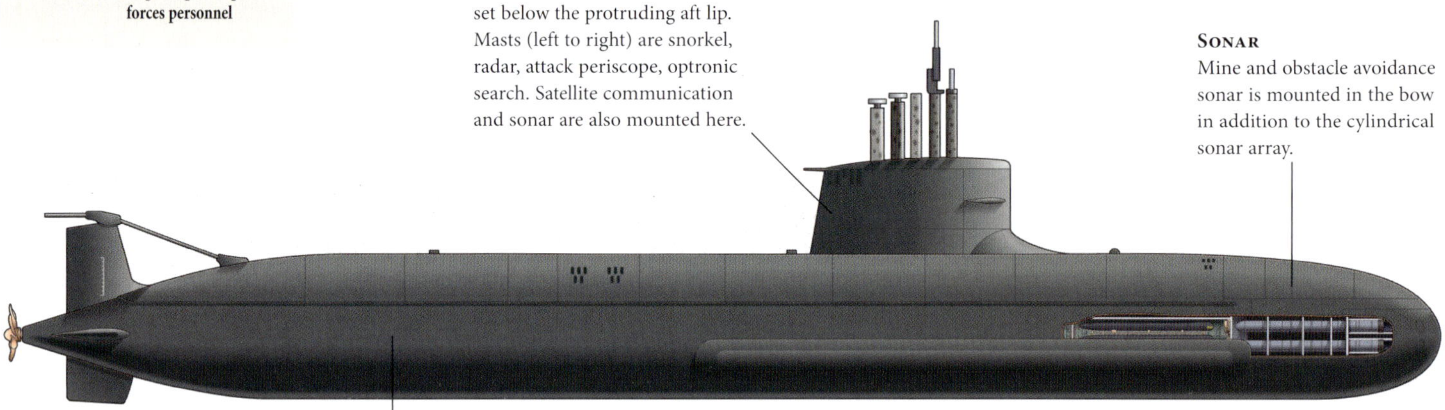

SAIL
The diesel exhaust diffuser is set below the protruding aft lip. Masts (left to right) are snorkel, radar, attack periscope, optronic search. Satellite communication and sonar are also mounted here.

SONAR
Mine and obstacle avoidance sonar is mounted in the bow in addition to the cylindrical sonar array.

ARMAMENT
Armament in the S-80 Plus class can include DM2A4 Sea Hake heavy torpedoes, UGM-84 Sub-Harpoon anti-ship missiles, and SAES mines. It was also planned to equip the subs with the UGM-109 Tomahawk land attack cruise missile (TLAM). While the TLAM order no longer applies, the submarine retains the capability to carry them in the future. This capability is unique among non-nuclear NATO submarines. The Spanish Navy is also considering adoption of the Naval Strike Missile (NSM) developed by Norwegian company Kongsberg Defence & Aerospace.

HULL
In 2013 the assistance of US-based Electric Boat Co. was called in to resolve a design error that resulted in lengthening the hull by 10m (33ft) to accommodate an additional pressure hull section.

ISAAC PERAL
Isaac Peral is the first submarine to be entirely built at the Navantia shipyard in Cartagena. Three more are scheduled for completion between 2026 and 2029.

TCG *Anadolu* (L-400) (2023)

TYPE • *Assault ship* **COUNTRY** • *Turkey*

SPECIFICATIONS

Dimensions:	Length: 232m (761ft 2in); Beam: 32m (105ft); Draught: 6.9m (22ft 8in)
Displacement:	25,054 tonnes (27,620 tons) as VSTOL; 27,875 tonnes (30,726 tons) full load (LHD)
Propulsion:	5 8000kW (11,000hp) Navantia MAN diesel generators; 2 11MW (15,000hp) Siemens SiPOD azimuthing propulsion units; 2 1500kW (2000hp) bow thrusters
Speed:	21 knots (39km/h; 24mph)
Range:	9000nmi (17,000km; 10,000 miles) at 16 knots
Armament:	1 × RAM launcher; 2 × Phalanx CIWS; 5 × 25mm (0.98in) Aselsan STOP RCWS; 4 × 12.7mm (0.5in) Aselsan STAMP RCWS
Aircraft:	30 helicopters; 30–50 UCAVs
Complement:	371 plus up to 800 aviation crew and troops

TCG *Anadolu* (L-400) is a multi-purpose amphibious assault ship, the largest vessel in the Turkish Navy. Designed and constructed in conjunction with Navantia, it has many similarities to the Spanish carrier *Juan Carlos I*.

Left: Amphibious assault ship TCG Anadolu *(L400) steams in the Mediterranean Sea, August 2023.* Anadolu *sailed in formation with the world's largest aircraft carrier USS* Gerald R. Ford *(CVN-78) and Turkish naval ships.*

UAVs
For aerial operations, in addition to manned aircraft, the ship is a platform for the TB-3 armed drone and the Kizilelma combat UAV, both Turkish made. Six landing spots are marked out on the flight deck.

Well deck
The floodable well deck holds four LCMs (Landing Craft Mechanized) or two LCVPs (Landing Craft Vehicle Personnel).

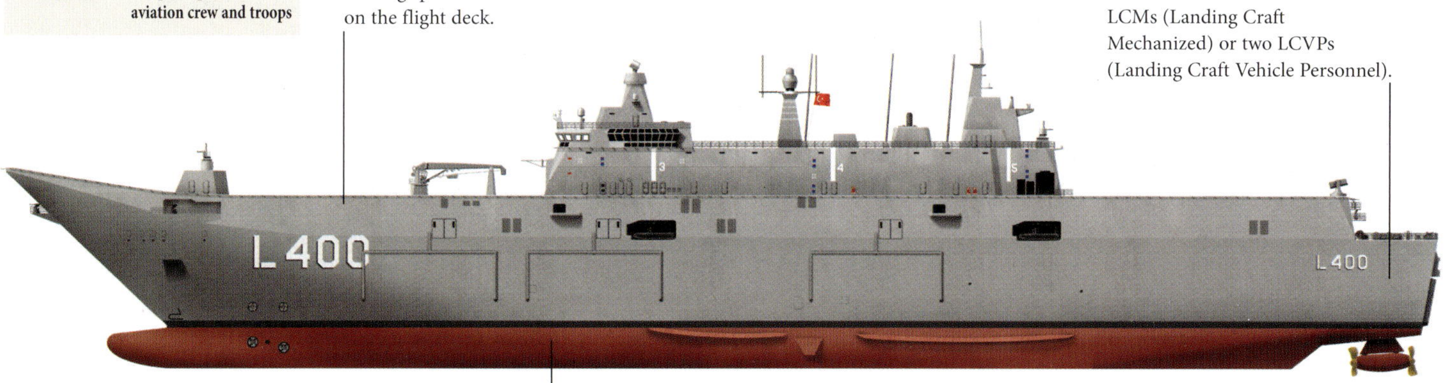

Powerplant
Anadolu is powered by five Navantia MAN diesel generators, each of 8MW (11,000hp) and two Siemens SIPOD electric motors (11MW/15,000hp), with two 1500kW (2000hp) bow thrusters.

AMPHIBIOUS ROLE
For amphibious sea-land operations the hull below hangar level serves both as garaging and dock space. A battalion unit of 1200 personnel, eight utility helicopters and three drones can be transported. A light vehicle deck of 1800 sq m (19,375 sq ft) and a heavy vehicle deck of 1400 sq m (15,000 sq ft) are fitted. The ship can deploy 13 tanks, 27 amphibious assault vehicles, six armoured personnel carriers, 33 various light and heavy vehicles, and 15 trailers.

***ANADOLU*, 2023**
The design enables the ship to operate either as an amphibious assault ship or as a VSTOL aircraft carrier. The hull resembles that of a car ferry – a vast straight-sided box with a sharply raked bow formation.

Type 003 *Fujian* (2024)

TYPE • *Aircraft carrier* **COUNTRY** • *China*

SPECIFICATIONS

Dimensions:	Length: 315m (1036ft 9in); Beam: 39m (128ft); Draught: undisclosed
Displacement:	c. 84,330 tonnes (c. 93,000 tons) full load
Propulsion:	Integrated power system (IPS); gas turbines and diesel engines driving electric motors; shafts undisclosed
Speed:	c. 30+ knots (55.5km/h; 34.5mph)
Range:	Undisclosed
Armament:	4 × CIWS; 4 × point-defence missile systems
Aircraft:	50+ aircraft
Complement:	c. 2500 (estimate)

Entering service in 2025, after 10 years of development, construction and sea trials, China's third carrier – the Type 003 *Fujian* – is the first carrier outside the United States to have electromagnetic catapults.

Left: A superb top view of the Fujian*'s flight deck. Its air wing, 50–60 strong, can be expected to include Shenyang J-15 fighter jets and KJ-600 early warning aircraft, plus the advanced stealth-capable Shenyang J-35 fighter.*

Flight deck
Scant official detail has been made public, but observers have noted three EMALS catapults, two lifts, four recovery lines and five helicopter/rotary winged landing spots.

Catapults
The electromagnetic catapults will enable the carrier to launch a plane every 45 seconds. The catapults deliver smoother acceleration, putting less stress on airframes and ship structure.

Island
Its small pyramid-shaped island, positioned to midships starboard, shows radar-deflecting features, and the lack of external aerials and antennas is striking.

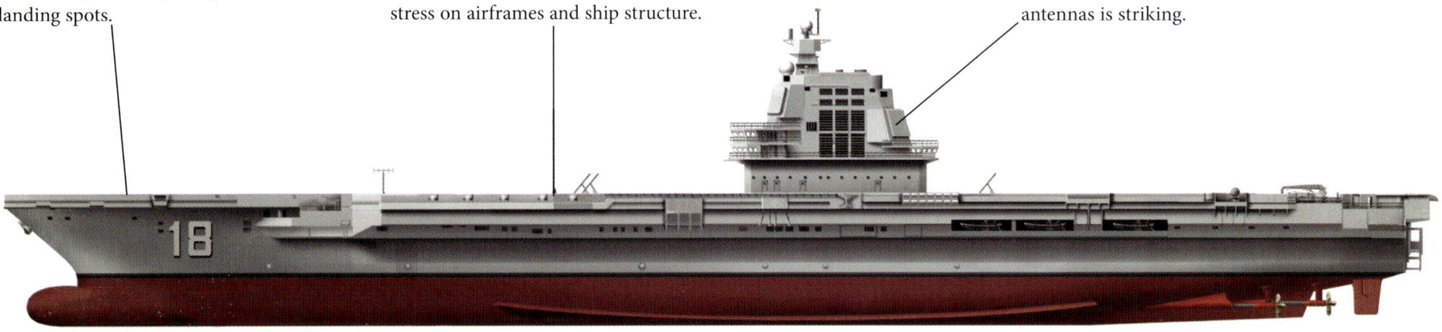

ELECTRIC PROPULSION
Propulsion in *Fujian* is by 'integrated electric propulsion', which suggests gas turbines and diesel generators powering electric motors, in an integrated power system (IPS) like the UK's Queen Elizabeth. To move an 84,330 tonne (83,000 ton) ship at 30 knots (56kmh/34.5mph) and to power the catapults and other power-hungry systems, maximum power generation well over 74,570kW (100,000hp) would be required.

SEA TRIAL
China's newest aircraft carrier, the *Fujian*, sets out on its maiden sea trial off Shanghai on 1 May 2024. Construction was modular, with component sections moved from assembly sheds to a vast dry dock.

Trieste (2024)

TYPE • *Helicopter carrier* **COUNTRY** • *Italy*

SPECIFICATIONS

DIMENSIONS:	Length: 245m (803ft 10in); Beam: 47m (154ft 3in); Draught: 7.2m (23ft 7in)
DISPLACEMENT:	38,000 tonnes (41,887 tons) full load
PROPULSION:	CODOGOL; 2 MT30 gas turbines; 2 MAN 20V32/44CR diesel engines; electric motors; 2 shafts
SPEED:	25 knots (46km/h; 29mph)
RANGE:	7000nmi (13,000km; 8100 miles)
ARMAMENT:	3 × OTO Melara 76/62mm (3in) AA guns; 3 × OTO Melara 25/80mm (3.2in) RCWS with Oerlikon KBA 25mm (0.98in) cannon
AIRCRAFT:	Usually 12 × helicopters
COMPLEMENT:	460 plus 604 marines

***Trieste* is a multi-function warship classed as LHD (Landing Helicopter Dock) and is the largest ship in the Italian Marina Militare. Built at the Fincantieri Castellamare yard, it was laid down in 2014 and commissioned in December 2024.**

Left: The double-island structure on Trieste *presents an arresting profile. The flight deck is set up for 5th generation STOVL aircraft, with a thermal metal spray treatment, acoustic protection and servicing areas.*

ISLANDS
The through deck has two separate island units to starboard; the ship's bridge and other control platforms are on the forward block.

WELL DECK
The most important feature of the *Trieste* is its floodable well deck, a 50m × 15m (160ft × 50ft) space designed to launch and recover amphibious vehicles, including Landing Craft Air Cushion (LCACs) and Landing Craft Mechanized (LCMs).

DECK 3
The main operational areas are concentrated on Deck 3, with Command Information Centre, Propulsion Operational Centre, air operations planning and traffic control centres, and F-35 operational secured facilities all in a large modular space. UAV operations are also controlled from here.

AMPHIBIOUS FORCE
Trieste can embark an amphibious force of 600 troops. Deck 7 accommodates both the garage and well deck, linked by a large ramp and two overhead rails each with two 5-tonne (5.5-ton) trolleys. The well deck has capacity of up to four new LC 23 LCMs (Landing Craft Mechanized) or a single US LCAC (Landing Craft Air Cushion). The LC23s offer roll-on, roll-off capabilities and can transport a maximum load equal to one Italian Army Ariete main battle tank or five LINCE light multirole vehicles.

TRIESTE (L-9890)
Like the British *Queen Elizabeth*, carrier, but smaller, *Trieste* has two islands, partly to avoid extensive ducting from the separate machinery compartments. Although classed as LHD, the flight deck can support and launch F-35B STOVL aircraft.

Constellation class (2025)

TYPE • *Frigate* **COUNTRY** • *United States*

SPECIFICATIONS

Dimensions:	Length: 151.18m (496ft); Beam: 19.81m (65ft); Draught: 7.3m (24ft)
Displacement:	7291 tonnes (7167 tons) full load
Propulsion:	CODLAG; 1 General Electric LM2500+G4 gas turbine, 32,000kW (43,000hp); 2 electric propulsion motors, 6800kW (9120hp) total; 4 Rolls-Royce MTU 20V 4000 M53B diesel generators, 12,000kW (16,100hp) total; 2 shafts
Speed:	26 knots (48km/h; 30mph)+
Range:	6000nmi (11,000km; 6900 miles) at 16 knots
Armament:	32 × Mk 41 VLS cells for SM-2, ESSM Block 2, Tomahawk and future missiles; 16 × over-the-horizon anti-ship missiles; 1 × Mk 49 RAM launcher (21 missiles); 1 × Mk 110 57mm (2.24in) gun
Complement:	c. 200

The *Constellation*-class frigate is a next-generation class of guided-missile frigates being developed for the US Navy. Its design is based on the European multi-purpose FREMM concept. The lead ship of the class, USS *Constellation*, is planned to enter service in 2029.

Left: An artist's impression of USS Constellation *(FFG-62) underway. The ship was laid down on 12 April 2024.*

Aegis Weapon System
The *Constellation* will be fitted with the proven Aegis Weapon System, a combat management system that enables the crew to track and engage multiple air and surface targets at one.

Radar
The Enterprise Air Surveillance Radar (AN/SPY-6(V)3) is an advanced active electronically scanned array radar system for detecting long-range air threats.

Anti-submarine systems
The Anti-Submarine Warfare Combat Suite includes a Combined Active Passive Towed Array Sonar and an Undersea Warfare Decision Support System.

MULTI-MISSION

The *Constellation*-class frigates are intended as true multi-role vessels. Tactical functions they might perform include air defence, anti-submarine warfare, anti-ship surface engagements, land attack and electromagnetic warfare. This broad operational remit reflects the philosophy of 'modularity' that is influential in the design of modern warships.

***CONSTELLATION* CLASS**
An artist's impression of a vessel from the FFG(X) programme, the US Navy's effort (announced in 2017) to produce the next generation of frigate design, which led to the *Constellation* class.

Project 17A *Nilgiri*-class (2025)

TYPE • *Frigate* COUNTRY • *India*

The Project 17A *Nilgiri*-class frigates are a new generation of Indian multi-mission stealth guided-missile frigates, the first of which came into commission in 2025. The class will ultimately comprise seven vessels, considerably strengthening India's surface fleet.

SPECIFICATIONS

Dimensions:	Length: 149m (488ft 10in); Beam: 17.8m (58ft 5in); Draught: 5.22m (17ft 2in);
Displacement:	6670 tonnes (7354 tons) full load
Propulsion:	CODAG; 2 HAL-GE LM2500+ gas turbines, 30,200kW (40,499h) each; 2 MAN Diesel 12V28/33D STC diesels, 6000kW (8046hp) each; 2 shafts
Speed:	32 knots (59km/h; 37mph)
Range:	5500nmi (10,200km; 6338 miles) at 16–18 knots
Armament:	32 × Barak 8 SAM in 4 × 8-cell VLS; 8 × BrahMos SSM in 1 × 8-cell VLS; 2 × triple 324mm (12.75in) torpedo tubes for TAL Shyena; 2 × RBU-6000 ASW rocket launchers; 1 × 76mm (3in) Strales gun; 2 × 30mm (1.18in) AK-630M CIWS; multiple 12.7mm (0.5in) remote-controlled guns
Complement:	226

Above: From front to back: INS Nilgiri *(F33),* Himgiri *(F34) and* Udaygiri *(F35), three ships of the* Nilgiri *class currently in service.*

Flight deck
The stern flight deck can accommodate two medium helicopters, such as HAL Dhruv or Sea King Mk. 42B.

Gun armament
While the armament is configured mainly towards missiles, the class retains a powerful OTO Melara 76mm (3in) Strales gun.

Modular
The ship is constructed using an integrated modular construction process, in which 'blocks' of the ship are manufactured separately and later assembled in place.

STEALTH SHIP

The *Nilgiri* class was designed by the Indian Warship Design Bureau to reflect best practice in stealth and low observability. The profile of the ship has a low radar cross-section, with many weapon systems mounted flush to the decks and reduced numbers of antennae. External surfaces have received radar-absorbent coatings, and systems are fitted to reduce both infrared (i.e. heat emissions) and acoustic signatures.

COMMISSIONING
The ship in the foreground is the INS *Nilgiri*, the lead ship of the class, pictured here at the time of its commissioning on 15 January 2025.

Glossary

AA Anti-aircraft, as in 'anti-aircraft artillery' (AAA); air-to-air, as in 'air-to-air missile' (AAM).

Armour Plates of iron, later steel and alloys, later still more exotic metals such as titanium, added to the hull and essential components of a ship to protect its integrity, its vital parts and its crew from battle damage.

ASM Anti-submarine missile; anti-submarine mortar (also air-to-surface missile, air-to-ship/anti-ship missile).

ASW Anti-submarine warfare.

Axial fire Gunfire ahead or astern, along the major axis of the vessel.

Ballast The weight added to a ship or boat to bring her to the desired level of floatation and to increase stability.

Battlecruiser The made-up designation for a hybrid warship armed like a battleship but sacrificing passive protection in the form of armour plate for speed.

Battleship Originally the biggest and most powerful ships of the fleet, mounting guns of usually 10in (254mm) or larger calibre (the biggest were those of the Japanese Yamato class, which were 18.1in (460mm)), and heavily armoured.

Beam The width of a ship's hull.

Boiler A device for heating water to boiling point, to turn it into steam so that it could be employed to power machinery.

Boom A spar used to extend the foot of a sail; also a floating barrier, usually across the entrance to a harbour.

Bore The diameter of a cylinder or gun barrel.

Bowsprit A spar protruding over the bows of a sailing vessel, to serve as an outboard anchorage for the tack of a flying jib.

Breech-loading (BL) Guns loading from the breech, via a removable segment.

Broadside The side of the ship; the simultaneous firing of the guns located there.

Bulges/Blisters Chambers added to the outside of a warship's hull to provide protection against torpedos; outer sections were generally water-filled, to absorbe splinters, inner, air-filled, to diffuse blast.

Bulkhead/Water-tight bulkhead A vertical partition employed to divide up a ship's internal space, both longitudinally and transversely. These partitions may be water-tight, in which case the openings in them to allow passage must be capable of being sealed, preferably by remote control.

Bunker/Bunkerage The part of a ship allocated to the storage of fuel; the fuel itself, and the quantity carried.

Calibre the diameter of the bore of a gun barrel; the number of times that diameter fits into the length of the barrel, expressed as 'L/(calibre)'; eg, a gun of 10in bore with a barrel 300in long would be described as '10in L/30', or just '10in/30'.

CAM ship Catapult-Armed Merchant ship A merchant ship equipped to fly off a fighter aircraft; a temporary expedient adopted during 1940.

Capital ship A term coined around 1910 to describe the most important naval assets, and group together battleships and battlecruisers (chiefly to give extra credibility to the latter); it was later extended to include monitors.

Carrier battle group A force designation coined during World War II; it was made up of one or more fleet aircraft carriers together with associated defensive elements – destroyers and cruisers – but often included battleships, which had by then largely been relegated to the shore bombardment role.

Carronade A short-barrelled, lightweight muzzle-loading gun, produced by the Carron Ironworks in Scotland from the 1770s.

Casemate A fixed armoured box within which a gun was mounted. It allowed the weapon to be elevated and trained, and usually protruded from the hull or superstructure to increase its arc of fire.

Catamaran A boat (and later ship) with two hulls joined by a continuous deck or decks.

Citadel/Central citadel A heavily-armoured redoubt within which the ship's main battery was housed.

Clipper An ultimately meaningless term used to describe any fast sailing ship, particularly one engaged in the grain, opium or tea trades, widely used in the mid-19th century.

Compound engine A multi-cylinder steam engine in which the steam is employed at least twice, at decreasing pressures; theoretically, all multi-cylinder engines are compound engines.

Corvette Originally a (French) sailing ship of war, too small to warrant a rate (and thus the equivalent of the British sloop); more recently, a warship smaller than a frigate or destroyer-escort.

Cruiser A warship, larger than a frigate or destroyer, much more heavily armed and often armoured to some degree, intended for independent action or to act as a scout for the battlefleet. Modern cruisers operate as defensive elements within carrier battle groups.

Cutter Originally a small, decked boat, lightly-armed, with a single mast and bowsprit, carrying a fore-and-aft (gaff) mainsail and a square topsail, with either two jibs or a jib and staysail, frequently used as an auxiliary to the fleet or on preventive duties.

CVA Attack aircraft carrier.

DDG Guided-missile destroyer.

DDH Helicopter destroyer.

Deadrise The angle to the vertical of the planking in the floor of a vessel's hull, and thus a measure of the 'sharpness' of her lines.

Deck The continuous horizontal platforms, the equivalent of floors in a building, which separate a ship.

Derrick A form of lifting gear comprising a single spar attached at its heel low down to a mast or kingpost and pivoted there, equipped with stays, a topping lift and guy pendants, so that its attitude and position may be positively controlled, and with a block attached at its head, through which a runner may be rove and led, if necessary, to a winch.

Destroyer Originally torpedo-boat destroyer; a small warship of little more than 200 tons displacement, itself equipped to launch torpedoes, but also armed with light guns.

Destroyer-escort A small warship, bigger than a corvette of the period, designed and constructed to guard convoys of merchant ships, especially on the Atlantic routes, during World War II.

Diesel A form of internal-combustion engine which ignites its fuel/air mixture by compression alone, invented by Rudolf Diesel in 1892.

Diesel-electric A form of propulsion in which the compression-ignition internal-combustion engine drives a generator, which in turn drives an electric motor which turns the propeller shaft.

Displacement A measure of the actual total weight of a vessel and all she contains obtained by calculating the volume of water she displaces.

Draught (also Draft) The measure of the depth of water required to float a ship, or how much she 'draws'.

Dreadnought The generic name given to a battleship modelled after HMS Dreadnought, the first with all-big-gun armament; it fell into disuse once all capital ships were of this form.

Driver An additional sail hoisted on the mizzen.

ECM Electronic Countermeasures; measures taken to decoy or confuse an enemy force's sensors.

EEZ Economic Exclusion Zone; a coastal strip, 200nm wide measured from the mean low water mark, within which a nation is held to have the exclusive right to exploit natural resources; it supplements, but does not replace, the territorial limit (originally three miles, the effective range of coastal artillery, later extended to 12 miles).

Fire control A (centralised) system of directing the firing of a ship's guns, based on observation of the fall of shot relative to the target, taking into account movements of both the target and the firing platform.

Flare The outward curve of the hull of a ship towards the bow.

Flotilla In the Royal Navy up to World War II, an organised unit of (usually eight) smaller warships – destroyers and submarines in particular, but also minesweepers and fast attack craft; cruisers and capital ships were grouped into squadrons, and squadrons and flotillas made up fleets – derived from the diminutive of the Spanish flota, fleet.

Flush decked Commonly, a ship with no forecastle or poop.

Forced draught A means of increasing the efficiency of a ship's boilers by forcing air at higher than normal pressure through the furnace element.

Fore-and-aft Sails which, when at rest, lie along the longitudinal axis of a vessel; a vessel rigged with such sails.

Forecastle Originally the superstructure erected at the bows of a ship to serve as a fighting platform, later the (raised) forward portion and the space beneath it, customarily used as crews' living quarters. Pronounced fo'c'sle.

Frames The ribs of a vessel, upon which the hull planking is secured, set at right-angles to the keel.

Freeboard The distance between the surface of the water and the upper deck of a ship or the gunwale of a boat.

Frigate Originally, fifth- or sixth rate ships carrying their guns on a single deck, employed as scouts, and the counterpart of the later cruiser.

Gas turbine A rotary internal-combustion engine in which a fuel/air mixture is burned and the rapidly-expanding gas thus produced used to drive turbine blades arranged upon a shaft; a gas turbine bears the same relationship to a steam turbine that a reciprocating internal-combustion engine does to a reciprocating steam engine.

Gundeck The name given to the main deck in sailing warships of the Royal Navy.

Head The bows of a ship (and as in '(down) by the head'), and by extension in the plural, because they were traditionally located there, the latrines; also the top of a four-sided sail.

Headsail Those sails – jibs and staysails in the main, but also spinnakers – which are set before the (fore)mast.

Heeling tanks Ballast tanks set on a ship's sides to allow her to be deliberately heeled over.

Horsepower A measure of the power produced by an engine; one horsepower = 550 foot/pounds per second ('the power required to raise 550 pounds through one foot in one second') as defined by James Watt.

Ironclad The contemporary name for wooden warships clad with iron, and by extension, to the first iron warships; it continued in use up until the arrival of the dreadnought.

Keel The main longitudinal timber of a ship or boat, effectively her spine and certainly her strongest member. In yachts, a downwards extension in the form of a wing (or wings) which balances the pressure of the wind upon the sails; also, the flat-bottom lighters used in ports in the northeast of England.

Ketch A two-masted sailing vessel; in modern terms, a yacht with a reduced mizzen stepped before the rudder post (see yawl).

Knot Internationally, the measure of a ship's speed – one nautical mile per hour.

Leeboard A primative form of drop-keel, hung over the side of a vessel from a forward pivot. Two were carried, one to port and one to starboard, though only the one on the lee side was employed.

Liner A ship carrying passengers to a fixed schedule, usually on trans-oceanic routes; the term became current from the mid-1800s. A cargo liner also operates on fixed schedules, with space for a limited number of passengers.

Magazine Secure storage for explosives.

Masts Spars, mounted vertically or close to it, normally stayed (guyed) at either side and fore and aft, employed to allow sails to be carried.

MGB/MTB Motor gun boat/motor torpedo boats.

Minesweeper A small ship, roughly the size of a trawler (many were, in fact, converted fishing boats originally) adapted and equipped to locate and neutralise submarine mines. Later supplemented by specialist minehunters.

Mizzen The third mast, counting from the bows; since most ships had three masts, it was also the aftermost, and invariably carried a fore-and-aft steadying sail.

Nautical mile Internationally, the measure of distance at sea which has become standardised at 6080ft (1852m).

Oerlikon A Swiss arms manufacturer whose 20mm cannon, widely acknowledged to be the best of its type, was adopted by both sides during World War II. Improved versions were still in production at the end of the 20th century.

Periscope An optical device allowing an observer to change his plane of vision. At sea they were commonly used to allow submerged submarines a view of the surface, and also in gun turrets.

Poop The short raised deck at the stern of a vessel, originally known as the aftercastle.

Propeller Properly speaking, the screw propeller; as essential to steam- and motor ships as their powerplant, the rotation of the screw propeller and the angle of its blades or vanes combine to generate thrust against the mass of water, which pushes the vessel through it.

Quarterdeck That part of the upper deck abaft the mainmast (or where the mainmast would logically be in a steam- or motor ship), traditionally the reserve of commissioned officers.

Quick-firing A designation applied to small- and medium-calibre guns to indicate that they used unitary ammunition (ie, with projectile and propellant cartridge combined).

Radar An acronym for Radio Direction and Range – a means of using electromagnetic radiation to locate an object.

Rudder A vertical board or fin hung on the centreline of the vessel at the stern post, originally (and still, in small boats) from simple hinges known as pintels, and connected either directly to the tiller or by ropes or chains to the steering wheel, which, when it is angled relative to the vessel's course, causes a change of direction.

SAM Surface-to-air missile.

Schooner-rigged A boat or ship with two or more masts of equal height (or with the foremast lower than the main and others), fore and aft rigged on all of them, with or without topsails.

Schnorkel/Snorkel A tube with a ball-valve at its upper extremity, which allows a submarine to take in air, and thus continue to operate its internal-combustion engines, while remaining below the surface.

Smooth-bore A gun with a smooth (ie, unrifled) barrel, used as naval ordnance until the second half of the 19th century.

Sonar An acronym for Sound Navigation and Ranging, a technique of using sound waves to detect objects underwater, and by extension, to the hardware employed; see also ASDIC.

Spanker Originally an additional sail hoisted on the mizzen mast to take advantage of a following wind, later taking the place of the mizzen course; see Driver.

Spar deck Strictly speaking, a temporary deck, but later used to describe the upper deck of a flush-decked ship.

Sponson A platform built outside the hull, at main- or upper deck level, usually to allow guns on the broadside to be sited so as to allow them to fire axially.

Squadron In the Royal Navy, originally an organised unit of (usually eight) major warships – cruisers and capital ships, but in the US Navy (and the practice became widespread), an organised unit of ships of any type, from minesweepers upwards, the term having taken over from flotilla.

Square-rigged A sailing vessel whose sails are set on yards, which when at rest are at right-angles to the longitudinal axis of the hull.

SSM Surface-to-surface missile.

SSN Nuclear-powered submarine.

Standing rigging That portion of a ship's rigging – stays and shrouds, for example – which is employed to steady her masts; see also Running rigging.

Steam turbine A rotary engine in which steam is used to drive turbine blades arranged upon a shaft.

Stem The foremost member of a ship's frame, fixed at its lower extremity to the keel.

Stern post The aftermost member of a ship's frame, fixed at its lower extremity to the keel.

Submarine A vessel capable of indefinite (or at least very prolonged) underwater operation.

Tiller A wooden or metal bar attached rigidly to the rudder and used to control its movement.

Tonnage The load carrying capacity of a merchant ship or the displacement of a warship.

Topmast The second section of a mast, stepped above the lower mast, carrying the (upper and lower) topmast yard(s).

Topsail In a square-rigged vessel, the square sails set immediately above the course, from the topmast yard(s) (bigger ships carried paired topsails, upper and lower, for ease of working); in a fore-and-aft rigged vessel, the topsails may be either square or themselves fore-and-aft.

Torpedo A self-propelled explosive device, with or without some form of guidance, running on or below the surface of the sea.

Triple-expansion A type of reciprocating compound engine, with a minimum of three cylinders of graduated sizes housing pistons connected to a common crankshaft; the steam, introduced at very high pressure into the smallest, is condensed, and passes to the second, slightly larger, cylinder, where it is condensed once more, and then passed to the largest cylinder(s).

Turbo-electric A form of propulsion in which the steam turbine drives a generator, which in turn drives an electric motor which turns the propeller shaft.

Turret Originally an armoured shell or covering for a gun, which rotated with the platform upon which the gun is mounted; later the armoured cover became an integral part of the rotating mounting, and itself supported the gun or guns.

Yawl A two-masted sailing yacht with a reduced mizzen stepped abaft the rudder post (see ketch).

Index

INDEX

INDEX

Picture Credits

Alamy: 12 (Prisma Archivo), 17 T (Ajax News & Feature Service), 18 (Timewatch Images), 19 (Chronicle), 22 (Walker Art Library), 24 (Jon Helgason), 27 (Heritage Image Partnership), 31 B (Chronicle), 32 T & B (Historic Illustrations), 35 (Ian Dagnall Computing), 103 T (Chronicle), 111 T (PA Images), 112 T (Sueddeutsche Zeitung Photo), 176 T (piemags), 182 B (colaimages), 184 T (The Picture Art Collection), 206 T (Chronicle), 206 B (dpa picture alliance), 215 B (piemags), 236 T (Associated Press), 237 B (Chronicle), 252 T (piemags), 258 B (Smith Archive), 262 T (Old Paper Studios), 287 (Pictorial Press), 301 B (Smith Archive), 306 T (Ajax News & Feature Service), 312 (Mirrorpix), 314 T & 319 B (PA Images), 334 T (Pictorial Press), 396 T (Associated Press), 408 T (Kremlin), 422 T & B (Sergey Yakovlev), 424 T (Newscom), 424 B (Imago), 427 B (CNSphoto), 428 T (Newscom), 428 B (The Canadian Press), 432 T (Associated Press), 432 B (Album)

Amber Books: 168 T & B, 216, 241 T & B, 242 B, 244 T, 249, 278 B

ASL Photos: 8, 9 T, 34, 42, 44 T, 47 T, 49, 50 T, 54, 56 T, 58 T, 61, 62 B, 65 T, 71, 73 T, 74 T, 75 B, 77 B, 78, 79 B, 81 T, 82, 86 T & B, 88 T & B, 90 B, 92 T, 93 B, 95 T, 96 B, 97, 98 T, 99 T, 100 T, 101 T, 102 T, 104 T & B, 105 B, 109, 113 T, 118 T, 120 T & B, 121 T & B, 124 B, 125 T right, 127 T, 130 T, 133 T, 134 T, 135 T, 137 T, 138 T & B, 139 T, 140, 141, 142 B, 144 T & B, 145 B, 146 B, 147 T, 148 T & B, 149 T, 150, 151 T & B, 152 B, 153 T, 154 T, 155 T, 156 T, 157 T, 158 T, 159 B, 164 T, 165 T & B, 170, 171 B, 172 T & B, 173 T, 175 T & B, 176 B, 177 T & B, 179 T & B, 180, 181 T & B, 182 T, 183 T & B, 185 T & B, 186, 187, 188 B, 191 T, 192, 194 B, 195 T & B, 196 T & B, 198-202 T, 203 T & B, 205 T & B, 207 T, 208-211, 212 B, 214 T & B, 215 T, 219 T & B, 221 B, 223 T & B, 228 T & B, 230 T & B, 231 T & B, 233, 234 T & B, 235 B, 237 T, 242 T, 245 T & B, 247 B, 251 T, 252 B, 253 T & B, 254 T & B, 257 T, 259 T & B, 260 T, 261, 262 B, 265 T & B, 266 T & B, 268 T & B, 272 T & B, 273 T, 274 B, 275 T, 277 T, 278 T, 279 T & B, 280 T, 281 B, 282 T & B, 288 T & B, 289 B, 290, 297 T & B, 298 B, 300 T & B, 301 T, 303 T, 304 B, 307 B, 308 B, 309 T & B, 311 B, 314 B, 317 B, 319 T, 320 B, 321 B, 322, 323 T, 325 T & B, 326 T & B, 327 B, 328-330, 331 B, 332 T & B, 333 B, 334 B, 338 T, 340, 341 B, 342 T, 343 T & B, 344, 346 T & B, 348 T & B, 349 T & B, 350 T & B, 352 T, 354 T & C, 359 T, 361 T, 364 T, 365 T, 370 T & B, 371 T, 372 T, 373 B, 374 T, 377 T & B, 380 T & B, 385 T & B, 389 T & B, 394 T & B, 396 B, 397, 409 T, 419 B, 421 B, 429 T, 430 T & B, 431 T & B, 435 B

BAE Systems: 386 T

British Library/Mechanical Curator Collection: 33

Canadian Forces Maritime Command: 382 T (Alan Rowlands)

Creative Commons Attribution 3.0 Unported License: 365 B (Jorge Guerra Moreno)

Creative Commons Attribution-Share Alike 2.0 Generic License: 163 B (Marinho do Brasil)

Creative Commons Attribution-Share Alike 4.0 International License: 321 T (ValtTeck), 411 B (HenSti), 435 T (Marina Militare Italiana)

Dreamstime: 13 (Vlad Ghiea), 17 B (Per Bjorkdahl), 23 T (Ajv123ajv), 50 B (Lei Xu), 359 B (Per Bjorkdahl), 413 B (zapperSiR)

Fincantieri Marine Group: 436 T

Getty Images: 14 (Heritage Images), 16 (Photo Josse/Leemage), 20 (Photo 12), 37 (Heritage Images), 52 B (Bettmann), 70 T (Illustrated London News), 108 (Popperfoto), 111 B (Arkivi), 117 T (1970s), 124 T (ullstein bild), 160 (Central Press), 173 B (Fox Photos), 174 (Corbis Historical), 232 (De Agostini), 303 B (Mirrorpix), 304 T (HUM Images), 305 (Pascal Parrot), 313 (H Armstrong Roberts/ClassicStock), 317 T (Mondadori Portfolio), 318 T (Mirrorpix), 318 B (Corbis), 368 T (Marcel Antonisse/AFP), 371 B (AFP), 375 (Alexey Nikolsky/AFP), 405 T & B (Bloomberg), 407 T (Kyodo News), 407 B (VCG), 408 B (Alexey Nikolsky /AFP), 411 T (Morris MacMatzen), 419 T (Anshuman Poyrekar/Hindustan Times), 420 T (CN-STR/AFP), 421 T (Sean Gallup), 423 (Alexis Rosenfeld), 425 T (China News Service), 425 B (VCG), 427 T (Peter Parks/AFP), 434 T (VCG), 434 B (Kyodo News)

Indian Navy - GODL: 437 T & B

Library of Congress: 80

Marinmuseum: 320 T

Ministerie van Defensie: 360 T & B (Henry Westendorp)

National Gallery of Art, Washington: 31 T

Naval History & Heritage Command: 26, 29, 30, 38, 41, 44 B, 45, 46 T & B, 47 B, 51 T & B, 52 B, 53, 55, 56 B, 58 B, 59 T & B, 62 T, 63 T & B, 64, 65 B, 66 T & B, 68 T & B, 69, 76, 81 B, 85 B, 89, 90 T, 91, 94 T & B, 95 B, 96 T, 99 B, 101 B, 102 B, 103 B, 105 T, 112 B, 116, 118 B, 119 T & B, 122 T & B, 123 T, 125 T left & B, 126, 127 B, 128 T & B, 130 B, 131 T & B, 132 T & B, 133 B, 134 B, 135 B, 136 T & B, 137 B, 139 B, 142 T, 143, 146 T, 149 B, 152 T, 154 B, 157 B, 159 T, 162, 163, 164 B, 167, 169 B, 171 T, 178 T & B, 184 B, 190 T & B, 191 B, 193 T & B, 202 B, 204 T, 213 T & B, 217 T & B, 218 T & B, 220 T & B, 221 T, 222, 224-227, 229 T & B, 235 T, 236 B, 238 B, 239 T & B, 240, 243, 244 B, 246 T & B, 247 T, 248 T & B, 250, 255, 256 B, 257 B, 258 T, 260 B, 263 B, 264, 267 T & B, 269 T & B, 270, 271 T, 273 B, 274 T, 275 B, 276, 281 T, 283-286, 289 T, 292 T & B, 295 B, 296 all, 298 T, 302 T & B, 307 T, 310 T & B, 316 T & B, 331 T, 333 T

Public Domain: 10, 15, 21, 25 T & B, 28, 36, 29, 40, 48, 57 T & B, 60 T & B, 67, 70 B, 72 T & B, 73 B, 74 B, 75 T, 77 T, 79 T, 83 T & B, 84 T & B, 85 T, 87, 92 B, 93 T, 98 B, 100 B, 106, 107 T & B, 110 T & B, 113 B, 117 B, 123 B, 129 T & B, 145 T, 147 B, 153 B, 155 B, 156 B, 158 B, 166 T & B, 169 T, 188 T, 189 T & B, 194 T, 197, 204 B, 207 B, 212 T, 238 T, 251 B, 256 T, 262 T, 277 B, 280 B, 287 T, 294 B, 299, 305 T, 308 T, 311 T, 324 T & B, 351, 391 T & B, 409 B, 415 T & B

UK MOD Crown Copyright: 293 T & B, 327 T (LA(Phot) Richard Harvey), 352 B (LPHOT Seeley), 358 T (LA(Phot) Vicki Benwell), 358 B (LPhot Bill Spurr), 368 B (PO(Phot) Angie Pearce), 379 (LPhot Kyle Heller), 386 B (J Massie), 393 (LA(Phot) Keith Morgan), 417 (LPhot Unaisi Luke)

Shutterstock: 294 T (Stanislaw1999), 363 T (A Paes), 373 T (Alexey Lesik), 404 B (George Stamatis)

U.S. Air Force: 410 B (1st Lt Joshua Thompson)

U.S. Army: 416 T (Pfc Kyron James)

U.S. Department of Defense: 337 T (OS2 John Bouvia), 337 B, 338 B (Don S Montgomery), 339, 340 T, 342 B (PH2 Beech), 345 T (PH2 R C Witham), 353 T & B (PH2 J P Protz), 355 T & B, 356 T & B (Don S Montgomery), 357 T & B, 368 (Jim Brennan/Electric Boat Corporation), 369 B (Ingalls Ship Building), 372 B (PH2 Paul A Vise), 374 B (Cmdr Cheryl Collins), 376 T & B (PH3 Alex C Witte), 378 (General Dynamics), 387 T, 406 T (Petty Officer 2nd Class Jesse Turner/Ricardo Navedo)

U.S. Marine Corps: 406 B (Cpl Avery Wayland), 426 T (Sgt Henry Antenor)

U.S. Navy: 9 B (Amn Danielle Sosa), 115, 271 B (PHAN J Alan Elliott), 295 T, 315 T & B (LCDR John Leenhouts), 323 B, 335 B (MCS3 John Philip Wagner), 336 T (MCSS Jon Husman), 336 B (PHOM3 Timothy F. Sosa), 345 B (PHOM2 Ch), 347 (MC1 Kenneth G Takada), 361 B (PH3 William Howell), 362 T & B (MC2 Zachary L Borden), 364 B, 369 T (MCS2 James R Evans), 382 B (LT Ed Early), 383 T (Chris Oxley), 383 B (General Dynamics Electric Boat Public Affairs), 384 T (PO2 Jesse Schwab), 384 B (MCSPO2 Jonathan P Nye), 387 B (MCS2 Danielle Baker), 388 T (PHOM2 James Thierry), 388 B (MCS3 Joshua Scott), 390 (MCS2 Chelsea D. Meiller), 392 T & B, 395 T (PO2 Nicholas Avis), 395 B (PO2 Jacob Mattingly), 398 T (Cmdr Fernando Estrella/NATO Allied Maritime Command), 398 B (PO2 Riley McDowell), 399 T (MC2 Tiarra Fulgham), 399 B (PO2 Alexander M Corona), 400 T (Derek Fountain), 400 B, 401 T (PO2 Malachi Lakey), 401 B (PO2 Terrin Hartman), 402 T & B (PO2 Randall Ramaswamy), 403 T (PO2 Larissa Dougherty), 403 B (Ryo Tanaka), 404 T (MCS3 Samuel Wagner), 410 T (MCS2 Jeff Troutman), 412 T (MCS2 Andrew Benvie), 412 B (CMCS Mark D. Faram), 413 T (MCS3 Aleksandr Freutel), 414 T (Moraima Johnston), 414 B (MCS3 Aleksandr Freutel), 416 B (Bill Mesta), 418 T (MCS2 Maxwell Orlosky), 418 B (MCS2 Jacob Mattingly), 420 B (M Nitz), 426 B (MCS3 Adam K. Thomas), 429 B (MCS2 Terrin Hartman), 433 T & B, 436 B (U.S. Navy graphic)

ARTWORK CREDITS:

All illustrations courtesy DeAgostini, Amber Books Ltd and Tony Gibbons, except the following:

David Bocquelet: 9 (top), 165, 169, 182, 196, 201, 204, 222 (both), 300, 306, 308, 311, 315, 319, 326, 330, 331, 333, 343, 350, 352, 362, 373, 376, 381, 387, 388, 392, 395, 398, 399, 400, 401, 403, 404, 406, 413, 414, 416, 417 (lower), 420, 421, 424, 425, 426, 427, 429, 431, 433, 434, 435, 436, 437